Protein Turnover

Protein Turnover

J.C. Waterlow

CABI is a trading name of CAB International

CABI Head Office
Nosworthy Way
Wallingford
Oxfordshire OX10 8DE
UK

Tel: +44 (0)1491 832111
Fax: +44 (0)1491 833508
E-mail: cabi@cabi.org
Web site: www.cabi.org

CABI North American Office
875 Massachusetts Avenue
7th Floor
Cambridge, MA 02139
USA

Tel: +1 617 395 4056
Fax: +1 617 354 6875
E-mail: cabi-nao@cabi.org

A catalogue record for this book is available from the British Library, London, UK.

A catalogue record for this book is available from the Library of Congress, Washington, DC.

ISBN-10: 0-85199-613-2
ISBN-13: 978-0-85199-613-4

Produced and typeset by Columns Design Ltd, Reading
Printed and bound in the UK by Biddles Ltd, Kings Lynn, UK

Contents

Foreword

When I first planned this book my idea was to produce an update of the book we published in 1978 on *Protein Turnover in Mammalian Tissues and the Whole Body* (Waterlow *et al.* 1978). It soon became clear that such a vast amount of work has been done in this field in the last 25 years that a new book was needed rather than a revision. But is there a need, since several books have already been produced, such as those of Wolfe (1984, 1992) and Welle (1999), together with numerous reviews and reports of conferences? None of these is entirely comprehensive, giving a conspectus of the whole field. There is, however, another and to me more compelling reason for embarking on this enterprise. Twenty-five years ago, with the increasing availability of stable isotopes and mass spectrometers, a huge new field was opening up for human studies. It extended also to experimental work on animals, since I have been told that it costs less to use stable isotopes than to provide all the facilities needed for working safely with radioisotopes. Good use has been made of these new developments, but I believe we are coming to the end of an era. Even a cursory look at the physiological and clinical journals shows that simple measurement of synthesis and breakdown rates is being overtaken by studies to unravel the molecular biology of these processes. The change of emphasis is part of scientific advance, and is to be welcomed, although many have expressed fears of excessive reductionism; but the pieces, after being taken apart, must be put together again to see how they work as a whole. Here kinetic studies may perhaps play a role. There may be an analogy with the contribution of metabolic control theory to our understanding of the rates of reaction through a sequence of enzymes. An interesting question that has not to my knowledge been tackled is whether the 'use' of an enzyme affects its rates of synthesis and breakdown.

This is looking forward, in the hope that protein kinetics at the molecular level may still have something to contribute. However, I have another aim in this book: to look back at the past and pay tribute to all who have contributed to our present knowledge, with studies that may be completely forgotten in the future. An example is the work on the turnover of plasma proteins labelled with radioactive iodine isotopes. This dominated two decades, from 1960 to 1980, and produced huge numbers of papers and reports on conferences. One of these, named Protein Turnover (Wolstenholme, 1970) was entirely devoted to plasma proteins, as if no others existed. Has all this work, and the mathematics that went with it, anything to offer us now? I believe that it has, though it would be hard to define exactly what.

It is possible that work on whole body protein turnover will meet the same fate as that on iodine-labelled plasma proteins, and disappear into a forgotten limbo. However, I hope that this will not happen, because if it is accepted that protein turnover is a biological process of great importance, equivalent to oxygen turnover, then we need to know more about it in different groups of people under different circumstances; we need to bring our knowledge to equal that of oxygen turnover or metabolic rate.

In citing references I have used the Harvard system because a name in the text not only refers to a particular paper but recalls a person or a group with whose work I am familiar. Some of these authors I know personally; others I do not, but I feel as if I did. The Harvard system has a human factor which the other systems lack. I apologize to authors whose relevant papers I have missed. Since readers may feel that too many references are cited, to them also I apologize: it is not easy to get the right balance.

This book is dedicated to Vernon R. Young, in recognition of his great contribution to the field, his stimulus and comradeship.

References

Waterlow, J.C., Millward, D.J. and Garlick, P.J. (1978) *Protein Turnover in Mammalian Tissues and in the Whole Body*. North-Holland, Amsterdam.

Welle, S. (1999) *Human Protein Metabolism*. Springer-Verlag, New York.

Wolfe, R.R. (1984) *Tracers in Metabolic Research. Radioisotope and Stable Isotope/Mass Spectrometry Methods*. Alan Liss, New York.

Wolfe, R.R. (1992) *Radioactive and Stable Isotopic Tracers in Biomedicine*. Wiley-Liss, New York.

Wolstenholme, G.E.W. and O'Connor, M. (1970) (eds.) *Protein Turnover*. CIBA Foundation Symposium no. 9. Elsevier, Amsterdam.

Acknowledgements

I acknowledge with gratitude the help and interest of Sarah Duggleby who compiled most of the data on the end-product method (Chapter 7) and of David Halliday in collating information for me from the British Library. I am deeply indebted also to Keith Slevin for the computer analysis of recycling in Chapter 6; and to the extraordinary endurance and efficiency of Mrs Constance Reed, who typed and re-typed numerous handwritten drafts; and to Dr Joan Stephen and my wife Angela for their encouragement and patience during the 3 years of writing this book.

1
Basic Principles

The concept that the standard components of the body are continually being replaced is not exactly new. Brown (1999: 17) tells us that, 'The idea of "dynamic permanence" was developed by Alcmaeon in the 6th century BC, according to which the structure of the body was continuously being broken down and being replaced by new structures and substances derived from food'. Nearly three millennia later the French physiologist Magendie wrote, 'It is extremely probable that all parts of the body of man experience an intestine movement which has the double effect of expelling the molecules that can or ought no longer to compose the organs, and replacing them by new molecules. This internal intimate motion, constitutes nutrition' (quoted by Munro, 1964: 7). It was not until 100 years later that the work of Schoenheimer and his colleagues put the concept on a scientific basis (Schoenheimer, 1942).

1.1 Definitions

1.1.1 Turnover

'Turnover' describes in a single word Schoenheimer's 'Dynamic State of Body Constituents' (Schoenheimer, 1942). It covers the renewal or replacement of a biological substance as well as the exchange of material between different compartments. In relation to protein, we use 'turnover' as a general term to describe both synthesis and breakdown. In the early days some authors equated turnover with protein breakdown, but this usage is now obsolete.

1.1.2 Compartment

A 'compartment' is a collection of material that is separable, anatomically or functionally, from other compartments. The term 'pool' refers to the contents of a compartment and implies that the contents are homogeneous. In studies of whole body protein turnover we refer to *the* pools of free and protein-bound amino acids, but this is a gross oversimplification of the real situation. In reality there are as many different protein-bound pools as there are different proteins, differing in their composition, structure and turnover rates. The free amino acid pools are separate in the intracellular and extracellular compartments and in the extracellular compartment they are separate in the plasma and extracellular space. The evidence for the reality of this separateness comes from tracer studies showing that a steady state of labelling at different levels can be observed in two compartments. There is much evidence also that the intracellular free amino acid pool is not homogeneous and is distributed between different sub-cellular compartments. It is entirely possible that within the cell there is no physical separation, but a gradient, with events occurring at different points along the gradient. Thus the defining of compartments and pools in the construction of models (see below) involves a high level of abstraction. Nevertheless, there is, of course, a real difference between pools of amino acids and pools of protein, and it is often convenient to distinguish between amino acids as the *precursor* and protein as the *product.* In the case of breakdown the reverse is of course the case: protein is the precursor and amino acids the product.

Another term that needs to be defined is *flux*, which refers to the rate of flow (amount/time) of material between any two compartments. Wolfe (1984) has objected to the word as being too vague. This is indeed true and the two compartments between which the flow is occurring need to be defined.

The exchanges between free amino acids and protein occur in both directions. They can therefore be looked at in two ways. The forward direction involves the disappearance or *disposal* (*D*) of amino acids into 'sinks' – protein synthesis and oxidation – from which the same amino acids do not return, at least within the duration of the measurement. This assumption is in practice largely justified: since the protein pool is many times the size of the free amino acid pool, the chance that a particular amino acid will be taken up into protein and come out again in a few hours is small and is usually neglected; this subject is discussed in more detail in Chapter 6. When a tracer is used it is disposed of along with the tracee, and the disposal rate is determined from the rate of disappearance of tracer. The reverse reaction involves the appearance (*A*) of amino acids in the free pool derived from protein breakdown, food or *de novo* synthesis. Since these amino acids are unlabelled, they dilute the tracer in the free pool, and the appearance rate is determined from the rate of dilution of the tracer. In the steady state *A* and *D* are the same – two sides of one coin. It is only when we are dealing with non-steady states that it becomes important to distinguish between them.

The term *enrichment* is used in this book both for specific radioactivity in the case of radioactive tracers and isotopic abundance for stable isotopic tracers.

1.2 Notation

Atkins (1969) published a table comparing the different systems of notation used by different authors. There is still no uniformity. In this book we use the following notation: capital letters signify tracee, lower case letters tracer.

M = amount of a substance in a given pool (units g or moles).
Subscripts, e.g. M_A, identify the pool.

Q = flux or rate of transfer (units amount/time). The italic capital designates a rate.
Subscripts identify the pools between which the exchange is occurring and its duration. Q_{BA} means flux to pool *B* from pool *A*.
Common variants of *Q* are V or F.

In accordance with much physiological practice, rates are sometimes designated by a superscript dot.

A = rate of appearance of tracee in a sampled pool.

D = rate of disposal of tracee from a sampled pool.
R_a and R_d are commonly used instead of *A* and *D*, but it is contrary to normal scientific practice to write R for rate. The relationship of *A* and *D* to rates of breakdown and synthesis are considered in Chapter 2.

S = rate of protein synthesis.

B = rate of protein breakdown.

Alternative terms with the same meaning are degradation and proteolysis; but since *D* refers to disposal it is best to use *B* for all these names.

O = rate of amino acid oxidation.

E = rate of nitrogen excretion.

I = rate of intake from food.

Lower case letters are used for tracer: e.g. m_A = amount of tracer in pool A.

ε = enrichment; either specific radioactivity or isotope abundance.
Subscript indicates what is enriched, e.g. ε_{leu}, but if it is obviously leucine, then one might write ε_p for the enrichment of leucine in plasma.

i = amount of tracer administered (moles).
Alternatively, it may be convenient to write d or *d* for tracer given by single dose or continuous infusion.

k = fractional rate coefficient:units fraction/time

k_{AB} = fraction of pool B transferred to A per unit time.

k_s, k_d = fractional rates of synthesis and breakdown of a pool of protein.
It would be more logical to use k_b rather than k_d for breakdown, but k_d has become imbedded in the literature.

$T\frac{1}{2}$ = half-life, = $\ell n\ 2/k$ = 0.693/k units: $time^{-1}$

λ_1, λ_2 ... = exponential rate constants; units time^{-1}.

X_1, X_2 = coefficients in exponential equations; units: amounts of activity or enrichment as per cent of tracer dose.

FSR, abbreviation for fractional synthesis rate, is widely used as the equivalent of k_s. The denominator of this fraction is often taken as 100 so that an FSR of 0.10 becomes 10% per day. This expression is unfortunately out of line with other quantities related to protein synthesis, such as RNA concentration, [RNA], usually expressed as mg RNA per gram of protein, or RNA activity (k_{RNA}) in units of g protein synthesized per g RNA. To be in line with these, a fractional synthesis rate of 0.10 should be expanded to be per thousand, i.e. 100 mg synthesized per g protein. We shall, however, retain the FSR expressed as a percentage because it is deeply embedded in the literature.

NOLD is an acronym for non-oxidative leucine disposal, used rather than 'synthesis' in studies with leucine, apparently to avoid confusion with *de novo* synthesis of leucine. This seems unnecessarily clumsy since, apart from the fact that there is no *de novo* synthesis of leucine, the synthesis *of* leucine *into* protein is a perfectly natural expression, obvious from the context.

1.3 Equivalence of Tracer and Tracee

It is a basic assumption that labelling a molecule does not alter its metabolism, so that the tracee behaves in exactly the same way as the tracer. This is not strictly correct: a small amount of biological fractionation has been found between, for example, deuterium and hydrogen or between ^{15}N and ^{14}N.

Similarly, there may be differences between the metabolism of a substance labelled with two different tracers. Bennet *et al.* (1993) found that fluxes obtained with [1-^{14}C] leucine were about 3–8% higher than those with [4.5 – ^{3}H] leucine. They concluded that the difference arose from discrimination *in vivo* rather than during the analytical procedures. Usually it will not matter, but it may become important when two tracers are used together to give a difference, as in measurements of splanchnic uptake (see Chapter 6, section 6).

1.4 The Kinetics of Protein Turnover

1.4.1 Random turnover

The word 'kinetics' in this context is not used with the same precision as in chemistry or enzymology. The complexity of the molecular processes of both synthesis and breakdown makes it difficult to see how the terminology of classical enzyme kinetics could have any real application. I do not think that in the field of protein turnover there are really many observations that appear to follow a particular reaction order. There are perhaps exceptions, such as plasma protein turnover (Chapter 15) and enzyme induction and decay, but they are few (see, for example, Schimke, 1970 and Waterlow *et al.*, 1978). On the contrary, I believe that it is no more than an assumption, for mathematical convenience, that the transfers of protein breakdown are considered to be first order reactions, occurring at a constant *fractional* rate, or k, which is usually referred to as the 'rate constant'. Glynn (1991) pointed out that this term is not appropriate: an analogy is with interest on money invested, which may be constant for a time, but may also change from time to time, and so he proposed instead the term 'rate coefficient'. A zero order process, by contrast, is one in which a constant *amount* of material is transferred, regardless of the size of the pool from which it comes. It might be better, to avoid unjustified assumptions, to refer to these two processes as 'constant amount' and 'constant fraction', rather than zero order and first order – but even this is not proved to be correct.

An essential feature of both processes is random selection of the molecules being metabolized. Randomness requires that all members of a molecular species in a pool be treated in the same way, whether unlabelled or labelled.

The behaviour of the tracer in a random constant fraction process is illustrated by the well-known analogy of a tank with constant and equal inflow and outflow of water, and hence constant volume, M, of water in the tank. If a bolus of some tracer, m, is added and instantaneously well mixed, the change with time of the amount of m in the tank is: dm/dt = V/M, where V is the rate of inflow or outflow.V/M = k; integrating gives $m_t/m_o = \exp(-kt)$, where m_o is the initial amount of tracer and m_t the amount remaining at time t. Since M is constant, the same holds for enrichment, m/M, as for amount of tracer. This relationship produces a straight line on a semi-log plot, sometimes referred to as 'exponential kinetics'.

Exponential kinetics can probably be regarded as proof of a random process, but the reverse does not apply. If the enrichment–time relationship is not exponential, the process may still be random. A situation in which input and output are not equal, so that M is changing, produces a curvilinear relationship, either concave or convex, according to whether the output is greater or less than the input (Shipley and Clark, 1972: 166), but the decay is still random.

In the tank analogy in a steady state a constant amount process also produces apparent exponential kinetics, since if M remains unchanged a constant amount is the same as a constant fraction. The two processes can only be differentiated in the non-steady state when the pool size M is changing.

A good example of a non-steady state is the flooding dose method of measuring protein synthesis, in which a large dose of tracee is given along with the tracer (see Chapter 14). The assumption of first order kinetics for synthesis has led some authors, e.g. Toffolo *et al.* (1993) and Chinkes *et al.* (1993), to propose that the increase in synthesis observed with the flood is the necessary consequence of the expansion of the precursor amino acid pool produced by the flood. This position is hardly tenable; there are many situations in which an increased amino acid supply stimulates protein synthesis, but we now recognize that the stimulus involves a complex signalling pathway, ending in an equally complex set of initiation factors. It is inconceivable that this regulatory chain should be describable by a simple (or, in the case of Toffolo *et al.*, not so simple) mathematical equation. On the other hand, when the amount of protein newly synthesized over a given time interval is determined experimentally, accurately or not, it is entirely acceptable to express this increment as a fraction of the existing protein mass – a fraction commonly denoted k_s: but the expression should not imply a constant fractional process. This convention is useful because it enables direct comparison between k_s and k_d, the fractional rate of degradation.

There are many observations suggesting that protein breakdown can be described with reasonable accuracy as a constant fractional process: an example is the early work on plasma albumin labelled with radioactive isotopes of iodine (see Chapter 15). An interesting relationship emerges that has been explored particularly by Schimke (1970) in relation to enzyme induction. Suppose that synthesis can be represented as a constant amount process and breakdown as a constant fractional process: M_o is the initial protein mass, and S_o and k_d the initial rates of synthesis and breakdown in a steady state, so that $S_o = k_d.M_o$. If S undergoes a finite change to S_t, then M will increase and a new steady state will be achieved at which $S_t = k_d.M_t$ and the amounts of synthesis and breakdown are equal. This will represent a change of steady state at the expense of mass M. Koch (1962) extended this idea to a non-steady state such as growth, in which both M and S are changing continuously. If after a bolus dose of tracer the protein mass moves from M_o to M_t but k_d remains unchanged, the exponential line describing the fall in *amount* of tracer *vs.* time will remain unchanged, but the process of synthesis dilutes the tracer, so the fall in enrichment will be steeper. Thus simultaneous measurements of amount and enrichment will allow determination of rates of both synthesis and breakdown. This principle has been applied to measuring the turnover rates of muscle protein in the growing rat (Millward, 1970).

In conclusion, k_s and k_d are useful ways of expressing experimental observations but no conclusion can be drawn from them about the underlying kinetics. It is wise to bear in mind Steele's (1971) dictum: 'It has become the custom to use reaction-order as a *simple description* of experimental observations.' Analysis of many of the models described in the next chapter goes well beyond this dictum.

1.4.2 Non-random turnover

Non-random implies selection. Synthesis of proteins is a non-random process *par excellence*, since amino acids are selected for synthesis by the genetic code. There are also interesting possibilities of non-random breakdown, of which the most important is life-cycle kinetics. The classical example is haemoglobin, which has a life cycle in an adult man of the order of 120 days, and is broken down when the red cell is destroyed. Another example is the epithelial cells of the gut mucosa which, over a period of about 4 days, migrate from the crypt to the tip of the villus and then fall off. The cells and their contained proteins are then broken down by the enzymes of the gastrointestinal tract. A particularly striking case, described by Hall *et al.* (1969), is the apoprotein of the visual pigment of the rods in the retina of the frog. If a pulse dose is given of a labelled amino acid a disc of labelled pigment appears at the base of the cell and gradually migrates to the apex, where it disappears (Fig. 1.1). The average life-span of the protein in this study was about 9 weeks. It is probable that life-span kinetics is commoner than has been thought, and occurs particularly in tissues with a high rate of cell turnover, such as the immune system and the epidermis.

It has also been suggested that breakdown might be best described by a power function which produces a linear relation between tracer concentration versus time on a log–log plot (Wise, 1978), but it is difficult to see the physiological meaning of such a relationship.

Another type of non-random breakdown would depend on the age of the molecules as well as their structure. Suppose that a protein molecule became susceptible to attack by degradative enzymes when it had been subjected to a certain number of stresses, which occurred at random. Perutz (personal communication) suggested that such stress might result from contraction and expansion of the molecule as its energy level changed. Garlick (in Waterlow *et al.*, 1978) calculated that when the average number of events needed to produce breakdown is large, with a relatively small coefficient of variation (cv) the resulting survivor curve (proportion of molecules not broken down at any time) resembles that of life-span kinetics. When the number of stresses needed is small, with a large coefficient of variation, the curve comes closer to the exponential (Fig. 1.2).

More work to distinguish between random and non-random kinetics of protein breakdown might well be rewarding, throwing light on the molecular dynamics of the process. However, there are difficulties; with fast turning over proteins labelling of a cohort of newly synthesized protein molecules is unlikely to be absolutely simultaneous. Moreover, if decay has to be studied over several half-lives, reutilization of tracer becomes a serious problem (see Chapter 6).

1.5 References

Atkins, G.L. (1969) *Multicompartment Models for Biological Systems*. Methuen, London.

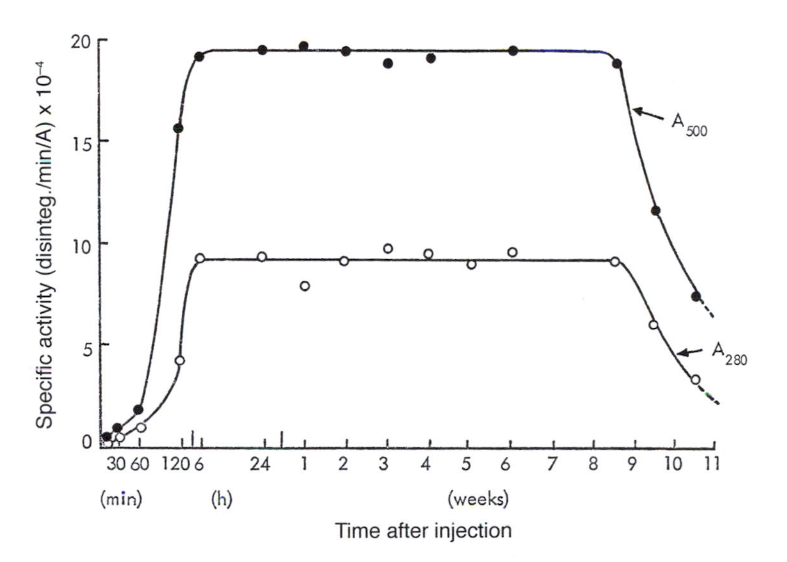

Fig. 1.1. Specific radioactivity of the purified visual pigment of frog retina as a function of time after injection of labelled amino acids. Top curve: dpm per unit absorbance at 500 nm. This represents the absorbance of the visual pigment. Lower curve: dpm per unit absorbance at 280 nm. This represents the absorbance of the apoprotein of the pigment. Reproduced from Hall *et al.* (1969), by courtesy of the *Journal of Molecular Biology*.

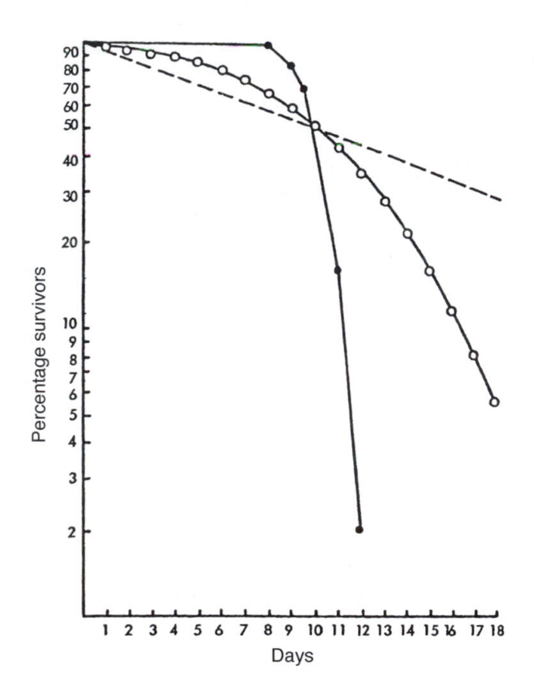

Fig. 1.2. Diagrammatic representation of different kinetic patterns of breakdown.
Abscissa: time; ordinate: per cent survivors.
– – – –, exponential breakdown; half-life 5 days.
•——•, 'multiple event' breakdown; mean life-span 10 ± 1 days (100 'events' required for breakdown).
o——o, 'multiple event' breakdown; mean life-span 10 ± 5 days (4 'events' required for breakdown).
Reproduced from Waterlow *et al.* (1978).

Bennet, W.M., Gan-Gaisano, M.C. and Haymond, M.W. (1993) Tritium and ^{14}C isotope effects using tracers of leucine and alpha-ketoisocaproate. *European Journal of Clinical Investigation* 23, 350–355.

Brown, G. (1999) *The Energy of Life*. Flamingo, London, p. 17.

Chinkes, D.L., Rosenblatt, J. and Wolfe, R.R. (1993) Assessment of the mathematical issues involved in measuring the fractional synthesis rate of protein using the flooding dose technique. *Clinical Science* 84, 177–183.

Garlick, P.J. (1978) Tracer decay by 'multiple event' kinetics. In: Waterlow, J.C., Garlick, P.J. and Millward, D.J. (eds) *Protein Turnover in Mammalian Tissues and the Whole Body*. North-Holland, Amsterdam, p. 215.

Glynn, J.M. (1991) The ambiguity of changes in the rate constants of fluxes. *Clinical Science* 80, 85–86.

Hall, M.O., Bok, D. and Bacharach, A.D.E. (1969) Biosynthesis and assembly of the rod outer segment membrane system. Formation and fate of visual pigment in the frog retina. *Journal of Molecular Biology* 45, 397–406.

Koch, A.L. (1962) The evaluation of the rates of biological processes from tracer kinetic data. 1. The influence of labile metabolic pools. *Journal of Theoretical Biology* 3, 283–303.

Millward, D.J. (1970) Protein turnover in skeletal muscle. I. The measurement of rates of synthesis and catabolism of skeletal muscle protein using [^{14}C] Na_2CO_3 to label protein. *Clinical Science* 39, 577–590.

Munro, H.N. (1964) Historical Introduction. In: Munro, H.N. and Allison, J.B. (eds) *Mammalian Protein Metabolism*, Academic Press, London, p. 7.

Schimke, R.T. (1970) Regulation of protein degradation in mammalian tissues. In: Munro, H.N. (ed.) *Mammalian Protein Metabolism* Vol. IV. Academic Press, New York, pp. 177–228.

Schoenheimer, R. (1942) *The Dynamic State of Body Constituents*. Harvard University Press, Cambridge, Massachusetts.

Shipley, R.A. and Clark, R.E. (1972) *Tracer Methods for In Vivo Kinetics*. Academic Press, New York.

Steele, R. (1971) *Tracer Probes in Steady State Systems*. C.C. Thomas, Springfield, Illinois.

Toffolo, G., Foster, D.M. and Cobelli, C. (1993) Estimation of protein fractional synthetic rate from tracer data. *American Journal of Physiology* 264, E128–135.

Waterlow, J.C., Millward, D.J. and Garlick, P.J. (1978) *Protein Turnover in Mammalian Tissues and in the Whole Body*. North-Holland, Amsterdam.

Wise, M.E. (1979) Fitting and interpreting dynamic tracer data. *Clinical Science* 56, 513–515.

Wolfe, R.R. (1984) *Tracers in Metabolic Research: Radioisotope and Stable Isotope/Mass Spectrometry Methods*. Alan R. Liss, NewYork.

2
Models and Their Analysis

2.1 Models

Metabolic models describe the dynamic aspects of metabolism, in contrast to the static descriptions of metabolic maps, which tell us of the pathways that exist but not of the traffic through them. In the words of Kacser and Burns (1973): 'These maps give information on the structure of the system: they tell us about transformations, syntheses and degradations and they tell us about the molecular anatomy. They tell us "what goes" but not "how much".' An anonymous editorial in the *Journal of the American Medical Association* (1960) said: 'A model, like a map, cannot show everything … the model-maker's problem is to distinguish between the superfluous and the essential.' The development of metabolic models was largely a consequence of the introduction of isotopes as tracers, without which dynamic measurements would not be possible. Schoenheimer makes no mention of models in his pioneer book (1942), but those who came after him soon realized that for quantitative analysis it was necessary to have a model as a simplified representation of a complex reality. The development and analysis of models have become so sophisticated that it requires a good knowledge of mathematics and statistics to understand them. More than 25 years ago Siebert (1978), in a paper with the title 'Good manners in good modelling', pointed out that 'The rise of the communication sciences has had much to do with stimulating the use of mathematical models (often as computer simulations)' and complained that 'Many models are implicated in forms that are difficult to comprehend by any but the modeller himself.' Here we shall confine ourselves to simple examples which have proved useful in the analysis of protein turnover.

There are two strands in the development of the models that are used in studies of protein turnover. The first is that the model should have some basis in the real physiological and anatomical properties of the system; the second is that it should be capable of mathematical analysis. The deductions from the analysis can then be compared with the observed data and the model adjusted to give the best fit. The difficulty is that although a good fit fortifies confidence in the validity of the model, there is still no way of being certain that the process of simplification, which is an essential part of model-building, may not have 'edited out' some important component. In the case of protein turnover there is no 'true' measurement of it that would act as a 'gold standard', in the way, for example, that analysis of cadavers is a gold standard for indirect measurements of body composition *in vivo*.

How can we tell that a model provides a 'true', if simplified, description of the kinetics that it is supposed to represent? Of course it increases confidence in the model if compartmental and stochastic approaches (see below) give the same answer, as was shown by Searle and Cavalieri (1972) for lactate kinetics. This does not, however, prove that the result obtained is 'correct'. The only way of testing for 'correctness' is to compare a result predicted from a model with one obtained independently without a model. The only test of this kind that we know of is an analysis by Matthews and Cobelli (1991) of a study by Rodriguez *et al.* (1986) of the effect on leucine kinetics of infusing trioctanoin. Measurement of the fraction of the infused tracer excreted in CO_2 showed that the octanoin increased the excretion nearly threefold. This measurement is a direct one, independent of any model. By comparison, a two-pool model of Nissen and Haymond (1981) of the kinetics of leucine and its transamination product showed no increase in labelled CO_2 output with the infusion of trioctanoin. The model was clearly inadequate.

A distinction is sometimes made between 'compartmental' and 'stochastic' models. 'Stochastic', according to the *Shorter Oxford Dictionary*, means 'pertaining to conjecture', from the Greek for aim or guess. According to the dictionary the word is rare and obsolete; the compilers could not have foreseen its future popularity! Stochastic implies a black-box approach, in which one is interested only in input and output, and not in what happens in between. This way of looking at it may have been useful in the early days, but is no longer appropriate. Both so-called compartmental and stochastic approaches require models, which may often be identical. The difference between them lies in the experimental method and the analysis. In the former, one or more tracers is given in a single dose, and the kinetic parameters determined from the curve(s) of enrichment with time in the sampled compartment(s). In the latter the tracers are given by continuous infusion and the parameters determined from the enrichment in the sampled compartments when an isotopic steady state has been achieved. The two approaches could be differentiated as isotopic non-steady and steady states, where 'steady' refers to the concentration of the tracer, not of the tracee. It is curious that the non-steady state was historically the first to be examined, although the steady state approach requires a less elaborate mathematical analysis. In what follows we shall retain the old terms because their usage is familiar.

Several assumptions are commonly made with both types of model. The first is that the pools are homogeneous. This assumption is necessary for analysis, but is incorrect. Even such a clear-cut entity as the extra-vascular part of the extracellular fluid is not homogeneous, part of it being bound to extra-cellular proteins (Holliday, 1999). The intracellular pool is even less homogeneous; the cell is a highly organized structure, not just a bag of enzymes – see Fig. 18.3 (Welch, 1986,1987), and there is much evidence which will frequently come up for the putative existence of sub-compartments or gradients within cells, between which mixing is not instantaneous or complete.

The second assumption is that transfers between compartments occur at constant fractional rates. This assumption is necessary for compartmental analysis, and was originally referred to as the 'rule' of the model (Waterlow *et al*., 1978). In the previous chapter it was argued that this 'rule' has no sound theoretical basis. It is anyway irrelevant for stochastic analysis, when a steady state of tracer has been achieved.

The third assumption is that the amount of metabolite in each pool remains constant throughout the period of observation, i.e. that there is a steady state of tracee. This assumption is convenient but not essential, and is probably accurate enough in many short-term studies.

Another usual assumption is that protein operates as a sink which is so large and turns over so slowly that once tracer has entered it, it does not return within the time of measurement, in spite of the continuing exchange of tracee with the precursor pool. This return of tracer is called 'recycling', and again it is not always justifiable to ignore it (see Chapter 6, section 6.7). In the description of models that follows we regard protein(s) as pool(s), just like any others, although some authors do not follow this convention.

2.2 Compartmental Analysis

Historically, compartmental analysis was the first technique to be applied to isotopic measurements of metabolic transfer rates. In the early days it was used particularly for studies on glucose metabolism, which is more complicated than that of amino acids and protein. The mathematics have been set out by Reiner (1953), Robertson (1957), Russell (1958), Zilversmit (1960), Steele (1971), Shipley and Clark (1972), Wolfe (1984), Cobelli and Toffolo (1984), and many others.

The simplest model is the tank described earlier: a single pool from which tracer given as a pulse dose disappears exponentially, i.e. linearly on a semi-log plot of concentration against time. A two-pool system gives a curve which is the sum of two exponentials; in general, the number of exponentials that can be extracted from the curve is equal to the number of separate compartments in the system. The general equation is therefore:

$$C = X_1.\exp(-\lambda_1 t) + X_2.\exp(-\lambda_2 t) + X_3.\exp(-\lambda_3 t) \ldots$$

where the units of C and X are activity or fractions of dose, such that the sum of the Xs = C, and the λs are exponential coefficients. In the days before computers the curves could be separated into their component semilog slopes by the process known as 'peeling' (Shipley and Clark, 1972: 24). Nowadays this is done by computer, but even so the experimental observations are seldom accurate enough for more than three slopes to be identified. For accuracy it is necessary that the slopes (exponential coefficients) should differ by a large factor, at least an order of magnitude. Myhill (1967) pointed out that in a two-compartment system with exponential coefficients differing by a factor of ten when the curve is defined by 11 points, a 5% random error in the measurements will produce an error of 44% in the value of the smallest exponential, which is generally considered to be the most important. If a further 20 measurements are made the error is still 32%. Atkins (1972) extended this analysis to show the enormous errors that may result in the derived values of the fractional rate coefficients, k, which describe the rates of exchange between the compartments. The k values can be derived from the slopes, λ, of the experimental curve by an algebra which becomes progressively more complicated as the number of exponentials increases (Shipley and Clark, 1972: Appendix I). Nowadays, of course, the solutions can be found by computer.

The total disposal, however, can be found quite simply from the area under the curve, calculated as:

$$D = \frac{d}{\Sigma X_i/\lambda_i}$$

Although both compartmental and stochastic analysis include reactions occurring in both directions between two pools, it is sometimes convenient to concentrate on one direction only, in which pool A is the precursor of the product in pool B. The concept of a precursor-product relationship is particularly useful in carbohydrate metabolism, where some reactions are irreversible and the product turns over rapidly, unlike the slowly turning over pool of protein. The treatment of the precursor-product relationship by Zilversmit (1960) leads to some rules of general application: (i) the activity curve of the product crosses that of the precursor at the point where the product curve is at its maximum; thereafter the two curves are parallel; (ii) the enrichments of all products derived from the same precursor are equal.

Two examples of compartmental analysis may be of interest to illustrate the early application of these principles to three-pool models. The first relates to studies of plasma albumin by Matthews (1957) (Fig. 2.1). The paper gives an example of curve-splitting or peeling as well as a detailed exposition of the mathematics. The specific activity curve suggested that the extravascular albumin pool could be divided into two compartments instead of one, as had previously been supposed. It is possible, as suggested by Holliday (1999), that the second compartment may be the extracellular water associated with connective tissues, where the water is partially bound to proteo-glycans. This is an example of the structure of a model being modified by the results.

Another instructive case is a study by Olesen *et al.* (1954) in which [^{15}N]-glycine was given in a single dose and the excretion of [^{15}N] measured in the urine over 2 weeks. Their model had three pools, an amino acid pool and two protein pools, one turning over fast and the other slowly. The slow pool was defined by the terminal part of the excretion curve. Examination of the results shows that it would be necessary to continue urine collection for 10 days before the curve deviated enough from that of a two-pool model for a clear distinction to be made between one and two protein pools. This illustrates the limitations of compartmental analysis. Other landmark studies of this period are those of Henriques *et al.* (1955), Wu *et al.* (1959) and Reilly and Green (1975).

In the 1980s, when computers arrived on the scene, compartmental models became more ambitious. If the information that could be obtained with a single tracer is in practice limited to exchanges between three pools, the next step was to use more than one tracer and more than one sampling site. Three examples of multicompartment models are summarized in Table 2.1 and Figs 2.2 to 2.4: they all include three additional pools concerned with CO_2 production and excretion. This may be treated as a separate process with its own kinetics and requiring its own tracer (see Chapter 8). It is the rest of the model that is interesting. The model of Umpleby *et al.* (1986) (Fig. 2.2) has three leucine

Table 2.1. Characteristics of three multi-compartment models.

	Umpleby	Irving	Cobelli
Number of pools:			
Total	7	11	13
Leucine	2	4	4
KIC	–	–	3
Bicarbonate	3	4	4
Protein	2[a]	2	2[b]
Intermediate	–	1	–
Tracer and route:	^{14}C-leucine IV ^{14}C-HCO_3 IV	^{13}C-leucine IV ^{13}C-HCO_3 IV ^{15}N-lysine, oral	^{14}C-leucine IV ^{3}H-KIC IV ^{13}C HCO_3 IV or ^{13}C-leucine IV ^{2}H KIC IV by constant infusion

From Umpleby *et al.* (1986); Irving *et al.* (1986); Cobelli *et al.* (1991).
All tracers given as IV bolus, except where indicated (Cobelli).
[a]The description of the model identified only one protein pool, but a second is implied and included here.
[b]The description of the model does not include any protein pools, but two are implied and are included here.

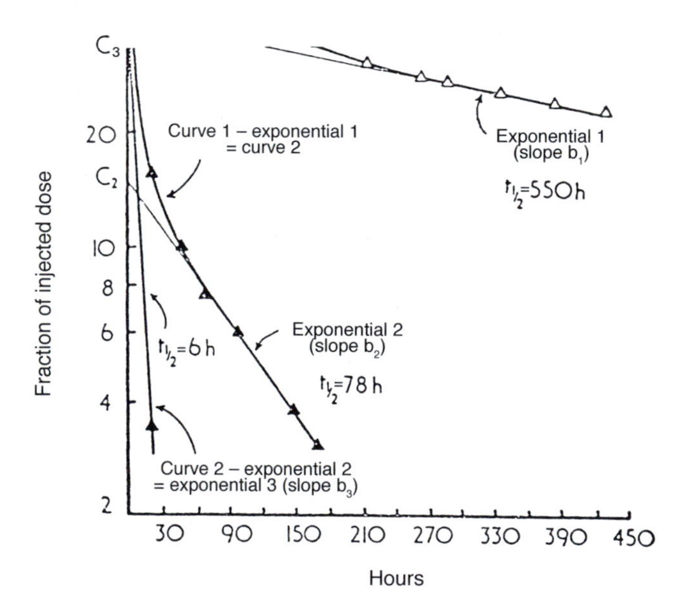

Fig. 2.1. Analysis by 'peeling' of plasma activity curve after a single injection of [^{131}I] albumin into a human subject. Reproduced from Matthews (1957), by courtesy of *Physics in Biology and Medicine.*

pools arranged in sequence; the unusual feature of it is that one of these pools is conceived as receiving the products of protein breakdown but is not the precursor pool for protein synthesis. The model of Irving *et al.* (1986) (Fig. 2.3) was designed to give separate information about the turnover of fast and slow proteins. It therefore had two precursor pools, one for visceral proteins, receiving an oral dose of

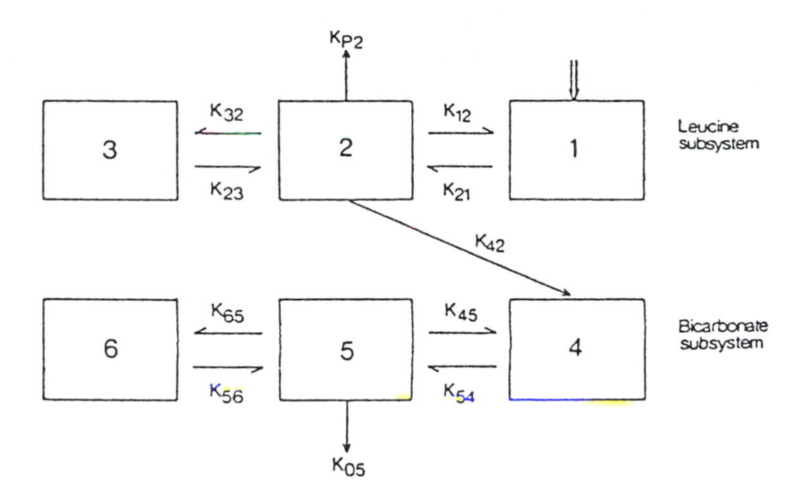

Fig. 2.2. Compartmental model of leucine and bicarbonate metabolism. The single arrows represent the direction of flux between compartments in or out of the system. The double arrow indicates the site of injection of tracer. Reproduced from Umpleby *et al.* (1986), by courtesy of *Diabetologia*.

tracer, and one for peripheral proteins, receiving tracer by the intravenous route. These two precursor pools were connected by a central pool, with flows in both directions. The most complex model is that of Cobelli *et al.* (1991) (Fig. 2.4). Like Irving's model it had four leucine pools, with three pools added on representing the metabolism of α-ketoisocaproate (KIC), the transamination product of leucine and the precursor for CO_2 production (see Chapter 4). Three of the leucine pools communicated with protein, one with rapid return of tracer, representing fast-turning over protein and two with no return of tracer. This model required the input of two tracers apart from that for CO_2, one of leucine and one of KIC. In some studies they were given in a single intravenous dose, in others by constant infusion.

A similar multi-compartmental model has been produced to describe the kinetics of VLDL-apolipoprotein β (Demant *et al.*, 1996).

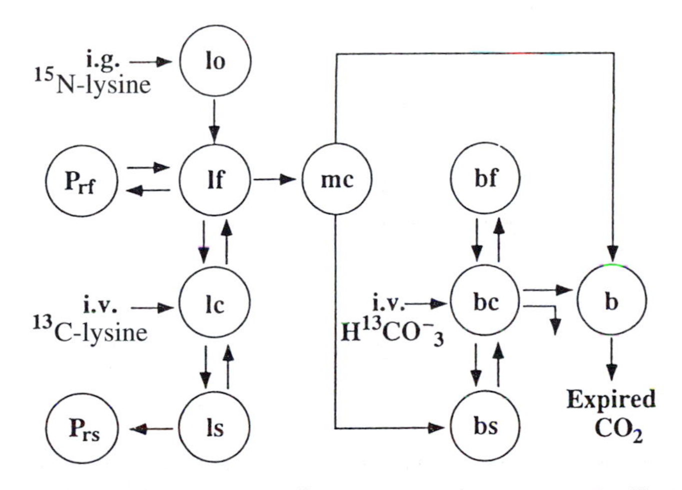

Fig. 2.3. Irving's model of lysine kinetics: L – [l-^{13}C] lysine was given intravenously, [^{15}N] lysine orally, and $NaH^{13}CO_3$ intravenously. Reproduced from Thomas *et al.* (1991), by courtesy of the *European Journal of Clinical Nutrition.*

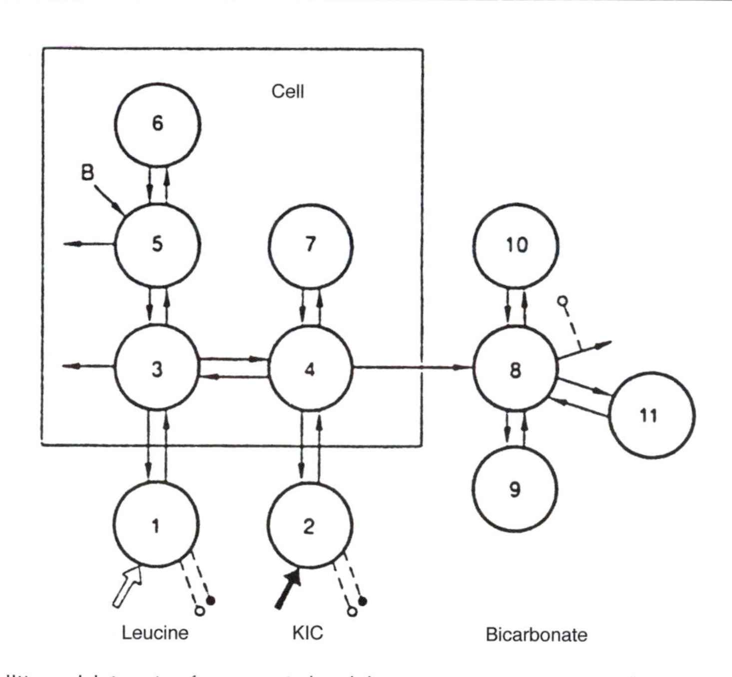

Fig. 2.4. Cobelli's model. Leucine from protein breakdown enters compartment 5; leucine incorporation into proteins takes place in compartments 3 and 5; oxidation occurs from compartment 4. Compartment 11 is a slowly turning over pool from which there is no return of tracer. Reproduced from Cobelli *et al.* (1991) by courtesy of the *American Journal of Physiology*.

In all these models physiological considerations governed the choice of what pools should be represented, but the arrangement of the pools was determined by computer analysis of the activity data to find the curve that fitted best. A practical disadvantage of this approach is that it is highly invasive. Cobelli's pulse dose experiments, for example, required 24 blood samples over 6 h, six of them in the first 5 minutes. It is difficult to put much reliance on the accuracy of the results from these early samples, which have an important influence on the shape of the whole curve. As a consequence the values of the derived parameters show very large inter-subject variations, sometimes as much as tenfold. There was variability of the same order, with coefficients of variation of 50% or more in the results with Irving's model.

Nevertheless, some useful information was obtained from these studies. That of Umpleby *et al.* (1986), designed to find the cause of raised plasma leucine concentrations in untreated diabetes, showed very clearly that it resulted from increased leucine production, presumably from protein breakdown, rather than from decreased utilization. Irving's model (Irving *et al.*, 1986) differentiated between fast and slowly turning over protein. An interesting relationship was found between the whole body flux and the net protein balance (synthesis–breakdown) in the fast and slow protein pools. As the flux became greater the net balance in the fast pool, presumably mainly the viscera, became more positive, whereas that in the slow pool, roughly equated with muscle, became more negative. A study based on Irving's model (Thomas *et al.*, 1991) was designed to show changes in protein metabolism during lactation. The main point that emerged was that synthesis of the slowly turning over proteins was decreased by nearly 40% during lactation. This might be a useful adaptation, favouring the production of milk protein.

To the best of our knowledge there has been no comparable study of a physiological problem with Cobelli's model. However, it was shown to be unnecessary to make separate measurements of bicarbonate kinetics, since the relevant information on oxidation could be obtained from the KIC data.

A model described recently by Fouillet *et al.* (2000) illustrates what can be achieved by modern computers and software (Fig. 2.5). They were interested in the distribution and fate of nitrogen after a meal of ^{15}N-labelled milk protein. Their model, firmly based on physiology, contained three subsections. The first, describing absorption, has three pools – gastric N content, intestinal N content and ileal effluent – and is sampled through a gastrointestinal tube. The second subsection, deamination, also has three pools – body urea, urinary urea and ammonia – and is sampled in the urine. The third subsection, retention, has five pools: a central free amino-acid pool, and free amino acid and protein pools for the splanchnic and peripheral areas. The sampling here is of plasma. The first subsection is connected with the second through the intestinal N pool, and the second with the third through the central free amino acid pool. As a preliminary stage the curves of ^{15}N enrichment for each subsection were analysed separately, and were then put together to get the best fit for all the parameters (rate coefficients), while ensuring that they matched in the connecting pools. Details of how the model was analysed and tested for uniqueness and validity are beyond the scope of this book. This example shows how extremely complex models can be analysed by modern methods; they are particularly effective in the isotopic non-steady state when tracer has been given as a bolus, and they give more information than can be obtained by stochastic methods. Against that, they are more invasive, because of the large number of samples needed, and can hardly be applied in routine studies.

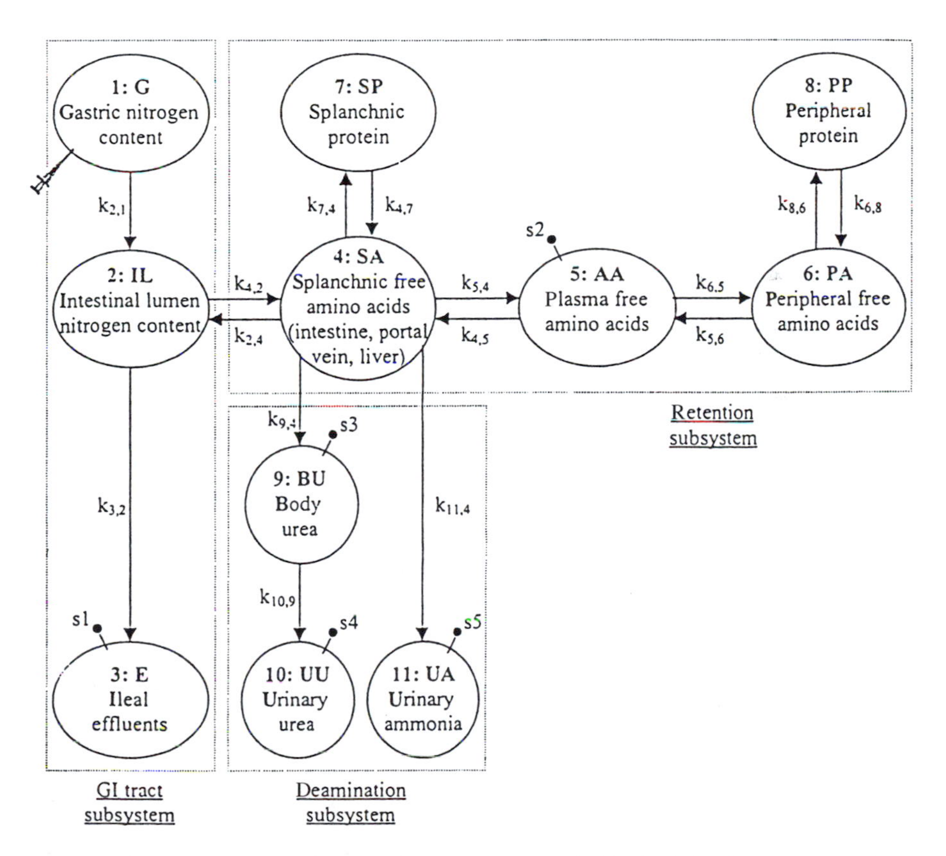

Fig. 2.5. Fouillet's model of nitrogen kinetics after a single meal of ^{15}N-labelled milk protein. Sampling is from three pools – gastrointestinal tract, urine and plasma. Reproduced from Fouillet *et al.* (2000), by courtesy of the *American Journal of Physiology.*

2.3. Stochastic Analysis

Good, although difficult, accounts of the stochastic approach have been given by Heath and Barton (1973) and Katz *et al.* (1974).

A useful practical distinction between compartmental and stochastic analysis is that the one depends on the slope of the enrichment curve, the other on the area under it. It is true that the compartmental approach can provide an estimate of the overall flux (see section 2.2). However, the main focus of compartmental analysis is to determine the rates of exchange between different pools, as is clear from the models illustrated above. This stochastic analysis cannot do, or only to a limited extent if there is more than one sampling site.

2.3.1. Determination of flux after a single dose of tracer in a two-pool model

This is the simplest of all models (Fig. 2.6). After a single dose of tracer the flux between the two pools over a given time interval can be determined from the area under the curve of enrichment in plasma or urine, without any need for an equation defining the way that enrichment changes with time. If the enrichment curve approached zero at time t, the total amount of tracer disposed of is:

$$\Sigma_{o\text{-}t}\,\varepsilon.\Delta t$$

and the disposal of tracee over the interval is given by:

$$Q_{o\text{-}t} = d/\Sigma_{o\text{-}t}\,\varepsilon.\Delta t$$

The area can be determined by cutting out and weighing or by dividing it into small segments of time. Heath and Barton (1973) give a general method for deciding on the number of samples that need to be taken and the intervals at which they should be spaced to provide an estimate of the total area to any given level of accuracy. The principle is that the total area should be divided into segments of equal area, so that samples are more widely spaced at later times. The method avoids the errors that inevitably occur in compartmental analysis in defining the slope of the terminal exponential. This approach is used in the end-product method of measuring whole body protein turnover (WBPT), in which excretion of tracer in urine is measured after a single dose (Chapter 7). It has also been used to good effect by Boirie *et al.* (1997) in studies of the response to a single meal containing biologically labelled protein.

2.3.2 Determination of flux by continuous administration of tracer

Stochastic analysis came into its own in the 1960s when tracer was given continuously rather than as a pulse dose (Gan and Jeffay, 1967, 1971; Waterlow and Stephen, 1967, 1968; Picou and Taylor-Roberts, 1969). When we first became interested in measuring whole body protein turnover in severely malnourished children we were deterred by the number of samples needed and the, to us, complex mathematics of compartmental analysis. It seemed that a simpler approach would be to infuse tracer at a constant rate until an isotopic steady state had been achieved. This state is referred to as a 'plateau', although it is really a pseudo-plateau. We did not realize that the method had been used for some years by those working in pharmacokinetics and endocrinology, e.g. for measuring the production rate of a hormone.

An isotopic steady state means that the rate of entry of tracer, d, into the sampled compartment, is equal to the rate at which it leaves, which is the product of the tracee flux, Q and its enrichment, ε. Thus:

$Q = d/\varepsilon$ where Q is the flux, d the rate of dosage and ε the activity of the tracer in plasma.

This simple relationship requires only a few measurements, which, at plateau, should fall on a straight line. If the tracer is a stable isotope which has to be given in amounts that are not negligible the dose has to be subtracted from the flux, and the equation becomes:

$$Q = d/\varepsilon - d = d\,(1/\varepsilon - 1)$$

This may be regarded as the first fundamental equation of stochastic analysis.

The so-called 'plateau' is really a pseudo-plateau, at which for practical purposes enrichment of free amino acid reaches a steady state within a few hours of infusion. Computer modelling shows that in the rat it would take about 6 years for a true steady state to be achieved in which all amino acids and proteins are equally labelled with the same activity as that of the infusate (Chapter 6). We can visualize the precursor pool 'filling up' with tracer in a few hours, to reach a pseudo-plateau, hereafter referred to simply as a plateau. Beyond that point its activity gradually increases as a result of recycling of tracer from labelled protein (see Chapter 6).

In our early experiments in which rats were infused with [U^{14}C]-lysine, the specific radioactivity of plasma free lysine rose to a steady state by a curve which could be approximately described by a single exponential equation:

$$\varepsilon_p/\varepsilon_{p\,max} = 1 - [\exp -k'_p t]$$

where ε_p is the activity in plasma, $\varepsilon_{p\,max}$ is the 'plateau' activity' and k'_p is a fractional coefficient for the net transfer of tracer out of the plasma pool. Clearly this transfer cannot be described by a single coefficient, since it is into many different protein pools, but a single value is an adequate approximation.[1] The value for k'_p of lysine in the rat was found to lie between 0.5 h^{-1} and 1 h^{-1} (Waterlow and Stephen, 1968).

It was originally supposed that the smaller the pool size of a free amino acid, the more rapid would be its turnover to achieve a given rate of synthesis. The equation shows that the larger k, the more quickly plateau is reached. This was one of the reasons for using tyrosine or leucine for measuring rates of protein turnover rather than lysine, because they have much smaller free pools. However, from such measurements as are available of the curve of enrichment to plateau with a constant infusion, there is little evidence of any constant difference between different amino acids (Chapter 3). This is presumably because the coefficient represents *net* transfer, not total outward transfer from the pool. Usually nowadays consideration of the curve to plateau is avoided by using a priming dose (see Chapter 6).

Once the flux has been determined from the plateau, it can be divided into its components. Using the notation of Chapter 1, section 1.2, in the steady state:

$$Q = A = D$$

$$A = B + I; D = S + E \text{ or } O$$

With amino acids of which substantial amounts are produced by *de novo* synthesis, an extra term, N, has to be added to the equation, so that:

$$A = B + I + N$$

In studies with ^{15}N, E is usually taken as total urinary nitrogen losses, the faeces and other sources of loss being ignored. In those with labelled carbon, oxidation is usually taken as the carbon in expired CO_2, that in urea and other possible routes of loss again being ignored.

This may be regarded as the second fundamental relation of stochastic analysis, first formulated by Picou and Taylor-Roberts in 1969.

2.3.3 The three-pool model and the precursor concept

Up to this point the two-pool model of Picou and Taylor-Roberts (1969) has been assumed, in which there is only a plasma pool and a protein pool.

However, the experimental work of the 1960s showed that with an infusion the plateau activities of free amino acids in the tissues were significantly lower than in plasma (Gan and Jeffay, 1967, 1971; Waterlow and Stephen, 1968) (Table 2.2). It was therefore necessary to modify the original two-pool model to include an intracellular free amino acid pool between the plasma and the tissues (Fig. 2.6). This pool provides the precursor for protein synthesis; attempts to define the precursor and its enrichment are described in Chapter 4. The calculation of flux from the equation: $Q = d/\varepsilon$, to be correct, must use the precursor enrichment, ε_i, rather than that in plasma, ε_p. The curve of the rise to plateau of ε_i is similar to that of ε_p, but a little slower.

The difference in enrichment at plateau between plasma and tissue pools (Table 2.2) arises because the amino acids produced by protein breakdown, which are not labelled, dilute the labelled amino acids entering the tissue from the plasma. Some of the amino acids from protein breakdown are taken up again by synthesis – a process called 'reutilization' of tracee, which has to be distinguished from 'recycling' of tracer (Chapter 6).

Another part of the amino acids from breakdown will enter the plasma, be carried to other tissues and contribute to synthesis of their proteins. We called this 'external' reutilization (Waterlow *et al.*, 1978) to emphasize that all parts

[1] In the original paper the referee described this as 'fudging'.

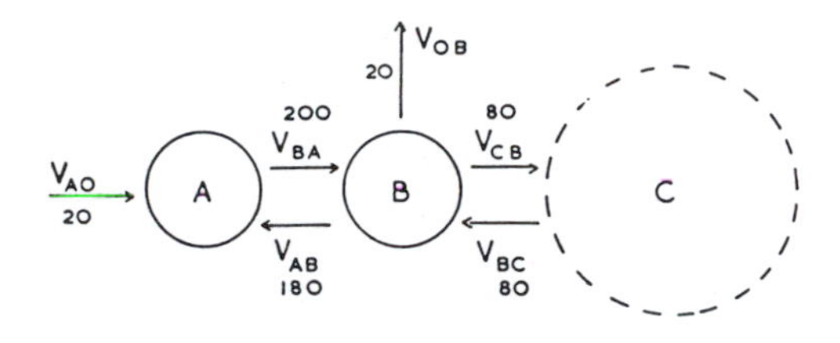

Fig. 2.6. Three-pool stochastic model: A, extracellular free amino acid pool; B, intracellular free pool; C, protein pool which acts as a sink, from which there is no return of tracer. Vs are amino-acid fluxes. V_{AO} = entry from food; V_{OB} = oxidation. The numbers are illustrative only.

of the body are constantly exchanging material. The shuttling of amino acids between synthesis and breakdown that goes on within the cell we call 'internal reutilization'.

The extent of this reutilization can be calculated from the steady-state plateau enrichments of tracer in plasma and tissue free amino acids. By applying mass balances to the model of Fig. 2.6 it can be seen that the extent of reutilization is the fraction of V_{CB} that is derived from V_{BC}. This fraction is:

$$V_{CB}/(V_{BA} + V_{BC}), \text{ or, since } V_{BC} = V_{CB},$$

$$V_{CB}/(V_{BA} + V_{CB})$$

If all three pools are in a steady state, it can easily be shown by mass balances of tracee and tracer that this fraction can be obtained from the enrichments in pools A and B:

$$V_{CB}/(V_{BA} + V_{CB}) = 1 - \varepsilon_B/\varepsilon_A$$

When applied to the whole body, if ε_A is taken as the enrichment in arterial plasma and ε_V as the enrichment in mixed venous plasma after it has drained the tissues, $\varepsilon_V/\varepsilon_A$ is usually about 0.75, indicating that about $^1/_4$ of the amino acids liberated by protein breakdown are re-used for synthesis. The extent to which amino acids are economized in this way in different situations has not been adequately investigated.

If, using different symbols, we write Q_A for V_{BA} in Fig. 2.6 (arterial outflow), Q_V for V_{AB} (venous return) and D for V_{CB}, then the tracer balance is:

$$Q_A.\varepsilon_A = Q_V.\varepsilon_V + D.\varepsilon_V$$

$Q_A.\varepsilon_{A} - Q_V.\varepsilon_V = D\varepsilon_V = d$ (assuming that ε_v is equal to ε of the intracellular precursor of D).

The power of this kind of stochastic analysis has been exploited in a number of ways. For example, in experiments on rats with a five-pool model (three amino acid pools, plasma, liver and muscle and two protein pools) it was possible to derive all the fluxes and rate coefficients from the plateau enrichments and estimates of pool sizes (Aub and Waterlow, 1970; Gan and Jeffay, 1971). More recently the method has been extensively used by Biolo, Tessari and others in studies on the human forearm, leg and visceral organs (Biolo *et al.*, 1995; Tessari *et al.*, 1995) (see Chapter 9). Biolo's model is reproduced in the next chapter (Fig. 3.1). From measurements of blood flow, arterial and venous tracee concentrations and plateau enrichments in artery, vein and muscle all the fluxes could be determined.

2.3.4 Predictive models

The models described so far are analytical in that they provide estimates of a variable that cannot be measured directly, such as whole body protein synthesis. Scientists working on animal production have developed a different kind of model

Table 2.2. Ratio of plateau specific activity of free lysine in liver and muscle to that in plasma in rats receiving intravenous infusions of [^{14}C] lysine.

	Liver	Muscle
Gan and Jeffay (1967)	0.40–0.45	0.65–0.75
Waterlow and Stephen (1968)		
Control	0.63	0.88
Starved 2 days	0.86	0.71

The effect of 2 days' starvation appears to be a reduction of protein breakdown in liver and an increase in muscle.

(e.g. France *et al.*, 1988; Maas *et al.*, 1997; Hanigan *et al.*, 2002, 2004). The outcome of interest is milk production; a model is then constructed and on the basis of a limited number of experimental measurements, including the output of milk protein, estimates are calculated of parameters that could not be measured directly. It is then possible to make a 'perturbation analysis' that shows which inputs, e.g. of individual amino acids, have the greatest effect on output.

These models have been called by France (personal communication) 'dynamic simulation' models. The number of parameters that they evaluate is remarkable.

2.3.5 Analysis in the non-steady state

The non-steady state means that the amounts of material, amino acids or protein, in the pools of the system are not constant. Changes in protein mass may be observed over times as short as a day or two in the growing rat, and in hours after a large meal (Garlick *et al.*, 1973). In man, however, changes in the mass of whole body protein are not likely to be detectable over the short period of measurement of protein turnover, although there may be significant changes in the mass of a single protein that turns over rapidly, such as plasma albumin or an enzyme.

Of greater interest in the present context are changes in the amino acid pool, for example as a result of a meal. Here stochastic analysis after a bolus dose of tracer, as in the studies of Boirie *et al.* (1997) on the effects of a single meal, is particularly useful. If the input of tracee changes, the enrichment curve will deviate from a straight exponential, becoming concave if the input rises and convex if it falls (Shipley and Clark, 1972: 166), but the relation $D = d/$(area under enrichment curve) still holds good.

With a constant infusion the rate of entry of tracer is fixed, but the size of the amino acid pool M and both input or appearance of tracee, A, and output, D, may change, either up or down. The equation for the changed rate over a short interval of time (Δt) given by Shipley and Clark (1972: Chapter 10, eqn 11), set out in our notation, is:

$$A_t = \frac{d \pm (M_{av} \cdot \Delta\varepsilon/\Delta_t)}{\varepsilon_{av}}$$

where A_t is the rate of entry or appearance of tracee at time t; d is the rate of tracer dosage; M_{av} is the mean mass of tracee in the pool over time interval Δ_t; $\Delta\varepsilon$ is the change in enrichment over that interval, and ε_{av} is the mean enrichment during the interval Δ_t.

If A is constant the disposal rate is given by:

$$D = A \pm \Delta M/\Delta_t$$

These equations reflect the fact that a change in input of tracee will alter the level of enrichment, as is obvious, so there will be no plateau; but the enrichment of the output must be the same as that of the pool from which it is derived, so that if A is constant a plateau will be maintained even if D is changing. For this reason Heath and Barton (1973) concluded that, at least for studies on glucose and ketone bodies, a single dose is preferable to a constant infusion, because a constant infusion gives no guarantee of a steady state.

The quantity M is the volume of distribution of amino acid, P, x its concentration, c. The original formulation of the non-steady state equation by Steele (1971) was applied to glucose turnover. Because glucose concentrations are very large, it was thought that mixing might be incomplete, and therefore a correction factor, p, was introduced into the estimate of distribution space. The basic estimate of P would be equal to total body water, i.e. about 0.7 l per kg body weight with a value for p of 1.0; Miles *et al.* (1983) with alanine, Boirie *et al.* (1996) with leucine and Kreider *et al.* (1997) with glutamine found that, varying p from 0.05 to 0.5 had no important effect on estimates of turnover rate.

It is interesting that in the last few years there has been a movement pioneered by the French school (Boirie *et al.*, 1996; Fouillet *et al.*, 2000) away from constant infusions to single dose of tracer, given with or as part of a meal – a very physiological approach. However, the analysis is still stochastic, not compartmental, and makes extensive use of corrections for non-steady state.

2.4 References

Atkins, G.L. (1972) Investigation of the effect of data error on the determination of physiological parameters by means of compartmental analysis. *Biochemical Journal* 127, 437–438.

Aub, M.R. and Waterlow, J.C. (1970) Analysis of a five-compartment system with continuous infusion and its application to the study of amino acid turnover. *Journal of Theoretical Biology* 26, 243–250.

Biolo, G., Fleming, R.Y.D., Maggi, S.P. and Wolfe, R.R. (1995) Transmembrane transport and intracellular kinetics of amino acids in human skeletal muscle. *American Journal of Physiology* 268, E75–84.

Boirie, Y., Gachon, P., Corny, S., Fauquant, J., Maubois, J.-L. and Beaufrère, B. (1996) Acute postprandial changes in leucine metabolism as assessed with an intrinsically labelled milk protein. *American Journal of Physiology* 271, E1083–1091.

Boirie, Y., Dangin, M., Gachon, P., Vasson, M.-P., Maubois, J.-L. and Beaufrère, B. (1997) Slow and fast dietary proteins differently modulate postprandial protein accretion. *Proceedings of the National Academy of Sciences, USA* 94, 14930–14935.

Cobelli, C. and Toffolo, G. (1984) Compartmental vs noncompartmental modeling for two accessible pools. *American Journal of Physiology* 247, R488–496.

Cobelli, G., Saccomani, M.P., Tessari, P., Biolo, G., Luzi, L. and Matthews, D.E. (1991) Compartmental model of leucine kinetics in humans. *American Journal of Physiology* 261, E539–550.

Demant, T., Packard, C.J., Demmelmair, H., Stewart, P., Bedynek, A., Bedford, D., Seidel, D. and Shepherd, J. (1996) Sensitive methods to study human apolipoprotein β metabolism using stable isotope-labeled amino acids. *American Journal of Physiology* 270, E1022–1036.

Fouillet, H., Gaudichon, C., Mariotti, F., Mahe, S., Lescoat, P., Hunceau, J.F. and Tomé, D. (2000) Compartmental modelling of post-prandial dietary nitrogen distribution in humans. *American Journal of Physiology* 279, E161–175.

France, J., Calvert, C.C., Baldwin, R.L. and Klasing, K.C. (1988) On the application of compartmental models to radioactive tracer kinetic studies of *in vivo* protein turnover in animals. *Journal of Theoretical Biology* 133, 447–471.

Gan, J.C. and Jeffay, H. (1967) Origins and metabolism of the intracellular amino acid pools in rat liver and muscle. *Biochimica et Biophysica Acta* 148, 448–459.

Gan, J.C. and Jeffay, H. (1971) The kinetics of transfer of plasma amino acids to tissues, and the turnover rates of liver and muscle proteins: the use of continuous infusion technique. *Biochimica et Biophysica Acta* 252, 125–135.

Garlick, P.J., Millward, D.J. and James, W.P.T. (1973) The diurnal response of muscle and liver protein synthesis *in vivo* in meal-fed rats. *Biochemical Journal* 136, 935–945.

Hanigan, M.D., Crompton, L.A., Bequette, B.J., Mills, J.A.N. and France, J. (2002) Modelling mammary metabolism in the dairy cow to predict milk constituent yields, with emphasis on amino acid metabolism and milk protein production: model evaluation. *Journal of Theoretical Biology* 217, 311–330.

Hanigan, M.D., Crompton, L.A., Reynolds, C.K., Wray-Cohen, D., Lomax, M.A. and France, J. (2004) An integrative model of amino acid metabolism in the liver of the lactating dairy cow. *Journal of Theoretical Biology* 228, 271–289.

Heath, D.F. and Barton, R.N. (1973) The design of experiments using isotopes for the determination of the rates of disposal of blood-borne substrates *in vivo* with special reference to glucose, ketone bodies, free fatty acids and proteins. *Biochemical Journal* 136, 503–518.

Henriques, O.B., Henriques, S.B. and Neuberger, A. (1955) Quantitative aspects of glycine metabolism in the rabbit. *Biochemical Journal* 60, 409–423.

Holliday, M.A. (1999) Extracellular fluid and its proteins: dehydration, shock and recovery. *Pediatric Nephrology* 13, 989–995.

Irving, C.S., Thomas, M.R., Malphus, E.W., Marks, L., Wong, W.W., Boulton, T.W. and Klein, P.D. (1986) Lysine and protein metabolism in young women. *Journal of Clinical Investigation* 77, 1321–1331.

Kacser, H. and Burns, J.H. (1973) The control of flux. In: *Rate Control of Biological Processes. Symposia of the Society of Experimental Biology* xxvii. Cambridge University Press, p. 65.

Katz, J., Rostami, H. and Dunn, A. (1974) Evaluation of glucose turnover, body mass and recyclng with reversible and irreversible tracers. *Biochemical Journal* 142, 161–170.

Kreider, M.E., Stumvoll, M., Meyer, C., Overkamp, D., Welle, S. and Gerich, J. (1997) Steady-state and non-steady-state measurements of plasma glutamine turnover in humans. *American Journal of Physiology* 272, E621–627.

Maas, J.A., France, J. and McBride, B.W. (1997) Model of milk protein synthesis. A mechanistic model of milk protein synthesis in the lactating bovine mammary gland. *Journal of Theoretical Biology* 187, 363–378.

Matthews, C.M.E. (1957) The theory of tracer experiments with ^{131}I-labelled plasma proteins. *Physics in Biology and Medicine* 2, 36–53.

Matthews, D.E. and Cobelli, C. (1991) Leucine metabolism in man: lessons from modelling. *Journal of Enteral and Parenteral Nutrition* 15, 86–89S.

Miles, J.M., Nissen, S.L., Rizza, R.A., Gerich, J.E. and Haymond, M.W. (1983) Failure of infused β-hydroxybutyrate to decrease proteolysis in man. *Diabetes* 32, 197–205.

Myhill, J. (1967) Investigation of the effect of data error in the analysis of biological tracer data. *Biophysical Journal* 7, 903–911.

Nissen, S.L. and Haymond, M.W. (1981) Effects of fasting on flux and interconversion of leucine and α-isocaproate *in vivo*. *American Journal of Physiology* 241, E72–75.

Olesen, K., Heilskov, N.C.S. and Schønheyder, F. (1954) The excretion of ^{15}N in urine after administration of ^{15}N-glycine. *Biochimica et Biophysica Acta* 15, 95–107.

Picou, D. and Taylor-Roberts, T. (1969) The measurement of total protein synthesis and catabolism and nitrogen turnover in infants in different nutritional states and receiving different amounts of dietary protein. *Clinical Science* 36, 283–296.

Reilly, P.E.B. and Green, J.R. (1975) Multi-exponential analysis of plasma free amino acid kinetics in the rat. *Biochemica et Biophysica Acta* 381, 424–430.

Reiner, J.M. (1953) The study of metabolic turnover rates by means of isotopic tracers, I and II. *Archives of Biochemistry and Biophysics* 46, 53–99.

Robertson, J.S. (1957) Theory and use of tracers in determining transfer rates in biological systems. *Physiological Reviews* 37, 133–154.

Rodriguez, N., Schwenk, W.F., Beaufrère, B., Miles, J.M. and Haymond, M.W. (1986) Trioctanoin infusion increases *in vivo* leucine oxidation: a lesson in isotope modelling. *American Journal of Physiology* 251, E343–348

Russell, J.A. (1958) The use of isotopic tracers in estimating rates of metabolic reactions. *Perspectives in Biology and Medicine* 1, 138–173.

Schoenheimer, R. (1942) *The Dynamic State of Body Constituents*. Harvard University Press, Cambridge, Massachusetts.

Searle, G.L. and Cavalieri, R.R. (1972) Determination of lactate kinetics in the human analysis of data from single injection vs continuous infusion methods. *Proceedings of the Society of Experimental Biology and Medicine* 139, 1002–1006.

Shipley, R.A. and Clark, R.E. (1972) *Tracer Methods for In Vivo Kinetics* Academic Press, New York.

Siebert, W.M. (1978) Good manners in good modelling: mathematical models and computer simulations of physiological systems. *American Journal of Physiology* 234, R161.

Steele, R. (1971) *Tracer Probes in Steady State Systems*. C.C. Thomas, Springfield, Illinois.

Tessari, P., Inchiostro, S., Zanetti, M. and Barazzoni, R. (1995) A model of skeletal muscle kinetics measured across the human forearm. *American Journal of Physiology* 269, E127–136.

Thomas, M.R., Irving, C.S., Reeds, P.J., Malplus, E.W., Wong, W.W., Boutton, T.W. and Klein, P.D. (1991) Lysine and protein metabolism in the young lactating woman. *European Journal of Clinical Nutrition* 45, 227–242.

Umpleby, A.M., Boroujerdi, M.A., Brown, P.M., Carson, E.R. and Sönksen, P.H. (1986) The effect of metabolic control on leucine metabolism in type I (insulin-dependent) diabetic patients. *Diabetologia* 29, 131–141.

Waterlow, J.C. and Stephen, J.M.L. (1967) The measurement of total lysine turnover in the rat by infusions of L-(U-^{14}C)–lysine. *Clinical Science* 33, 489–506.

Waterlow, J.C. and Stephen, J.M.L. (1968) The effect of low protein diets on the turnover rates of serum, liver and muscle proteins in the rat, measured by continuous infusion of L-^{14}C–lysine. *Clinical Science* 35, 287–305.

Waterlow, J.C., Millward, D.J. and Garlick, P.J. (1978) *Protein Turnover in Mammalian Tissues and in the Whole Body*. North-Holland, Amsterdam.

Welch, G.R. (1986) The organization of cell metabolism: a historical vignette. In: Welch, G.R. and Clegg, J.S. (eds), *The Organization of Cell Metabolism*. Plenum, New York.

Welch, G.R. (1987) The living cells as an ecosystem: hierarchical analogy and symmetry. *Trends in Ecology and Evolution* 2, 305–309.

Westerhoff, H.V. and Welch, G.R. (1992) In: Stadtman, E.R. and Shock, P.B. (eds) *Current Topics in Cellular Regulation,* 33. Academic Press, New York, pp. 361–390.

Wolfe, R.R. (1984) *Tracers in Metabolic Research: Radioisotope and Stable Isotope/Mass Spectrometry Methods*. Alan R. Liss, New York.

Wu, H. and Snyderman, J. (1950) Rate of excretion of N^{15} after feeding N^{15}-labeled L-aspartic acid in man. *Journal of General Physiology* 34, 339–345.

Wu, H., Sendroy, J. and Bishop, C.W. (1959) Interpretation of urinary ^{15}N-excretion data following administration of an N^{15}-labelled amino acid. *Journal of Applied Physiology* 14, 11–21.

Zilversmit, D.B. (1960) The design and analysis of isotope experiments. *American Journal of Medicine* 29, 832–848.

3

Free Amino Acids: Their Pools, Kinetics and Transport

3.1 Amino Acid Pools

3.1.1 Free amino acids

The free amino acid pool is the link between the environment and the proteins of the tissues. The free amino acids are the substrates of protein synthesis and the products of protein breakdown. The inputs to the free pool are from food and protein degradation; the outputs are to protein synthesis and oxidation. Table 3.1 gives some idea of the role of the free amino acid pools in the body's nitrogen economy: tiny in size, but turning over many times in a day.

The table shows that the free leucine pool is renewed every $^1/_2$ h, that of lysine about every 3 h. A doubling of the inward flux over 12 h would, if not compensated, increase the leucine pool 7.5-fold, and double that of lysine. The fact that in general such large changes do not occur shows that input and output must be accurately controlled – a point emphasized by Scornik (1984), who wrote: 'The regulatory role of amino acids is a particularly attractive subject of investigation. Its physiological significance is direct and immediate. The effects correct the cause: if amino acid pools are depleted, slower protein synthesis and faster breakdown tend to replenish them …'

We concentrate here mainly on the essential amino acids (EAAs), because they are the ones primarily concerned with the measurement of protein synthesis and breakdown. The non-essentials (NEAAs) are, of course, as important as the EAAs as components of protein, and some of them also have roles in metabolic pathways that are not concerned with protein. However, the NEAAs cannot be used in the same way as EAAs as markers of protein turnover, because part of

Table 3.1. Turnover rates of free pools of leucine and lysine in human muscle.

	Leucine	Lysine
A. Free pool, mmol kg^{-1} muscle	0.16	0.60
B. Protein-bound, mmol kg^{-1} muscle	122	106
C. Total amino acid entry, mmol kg^{-1} h^{-1}	0.3	0.18
D. Turnover rate of free pool, k, h^{-1}	~2.0	0.3
E. Turnover time, h	0.5	3.3

A: From Bergström *et al.* (1990), recalculated from mmol l^{-1} IC water to $mmol^{-1}$ kg^{-1}, on basis of 650 ml IC water kg^{-1} muscle.
B: Protein content taken as 20% of muscle weight; leucine and lysine taken as 8% of muscle protein, converted to moles.
C: Total amino acid entry as: entry from protein breakdown = 0.001 h^{-1} + entry from plasma, taken from Table 3.11 as 1.44 × rate of breakdown for leucine and 0.73 × rate of breakdown for lysine.
D: Turnover rate k = C/A.
E: Turnover time = 1/k. Alvestrand *et al.* (1988) give much larger values for the turnover time, but they are based only on entry from plasma, and do not include entry from protein breakdown.

their flux is, by definition, derived from *de novo* synthesis in the body.[1]

Jackson (1982, 1991) has suggested that some of the NEAAs should be regarded as 'conditionally essential', where *de novo* synthesis is inadequate to provide as much as is needed under conditions of stress, such as growth in neonates or restricted protein intakes. This group of amino acids would include glycine, arginine, tyrosine and cysteine.

A distinction has also been made on metabolic grounds between amino acids that are transaminated and those that are deaminated, the latter group comprising glycine, serine, threonine, histidine and lysine (Jackson and Golden, 1980).

3.1.2 Free amino acid pools in blood

Plasma

Out of the hundreds of published values of amino acid concentrations in plasma, a representative set of the essentials is shown in Table 3.2 and of the non-essentials in Table 3.3. The extent to which these change under different conditions is considered later.

Red blood cells

The concentrations and enrichments of free amino acids in red blood cells (RBCs) have received relatively little attention. Generally, concentrations in red cell water are higher than in plasma and enrichments lower, so that estimates of whole body protein turnover are higher when based on whole blood than on plasma (Darmaun *et al.*, 1986; Lobley *et al.*, 1996; Savarin *et al.*, 2001).

In a study in humans (Tessari *et al.*, 1996b) enrichments in whole blood and plasma of the general circulation were similar, so that rates of turnover for the whole body agreed well, but in the forearm enrichments were higher in plasma, giving a lower rate of synthesis. Tessari *et al.* were investigating the effect of a meal and concluded that red cells played a key role in 'mediating meal-enhanced protein accretion'. The discrepancy with the whole body results remains unexplained.

It may be helpful to go back to the work of Elwyn (1966). He found, like later workers, that the EAA concentrations in RBCs were about 1.5 × those in plasma. When blood was incubated *in vitro* with ^{14}C-glycine at 37°C, the specific activity in plasma and RBCs became equal after 20 minutes, so that in a study lasting several hours equilibration should not be a problem. Ellory (1987) has discussed the numerous transport systems in RBCs that would allow a concentration gradient to be maintained, but does not explain the greater *dilution* of tracer in the RBC, unless there is continuing protein breakdown, on which we

Table 3.2. Concentrations of essential amino acids in plasma and muscle free pool of healthy men in the post-absorptive state.

	Plasma μmol l^{-1}	Muscle μmol l^{-1} IC water	Concentration ratio muscle:plasma
Histidine	109	505	4.6
Isoleucine	77	133	1.73
Leucine	163	246	1.51
Lysine	194	919	4.74
Methionine	39	90	2.31
Phenylalanine	73	87	1.19
Threonine	189	901	4.77
Valine	277	369	1.33
Total	1121	3250	

No measurements of tryptophan.
After Bergström *et al.* (1990), Table 2.

[1] The term *de novo* synthesis is applied to the carbon skeletons of the amino acids and not to the effects of transamination. Thus the interconversion of leucine and α-ketoisocaproic acid (KIC) as measured by ^{15}N labelling of the amino groups is not regarded as synthesis.

Table 3.3. Concentrations of four free amino acids in muscle in post-absorptive healthy subjects. Comparisons between studies.

	μmol l^{-1} intracellular water			
	Leucine	Lysine	Phenylalanine	Threonine
Bergström *et al.*, 1974	150	1150	70	1030
Milewski *et al.*, 1982	180	800	80	710
Alvestrand *et al.*, 1988	268	–	120	908
Bergström *et al.*, 1990	246	919	87	901
Biolo *et al.*, 1995	232	976	112	–

have no information. For the time being it seems best to treat the RBCs like any tissue, in which the enrichment of an amino acid has no special claim to reflect that of the precursor in other tissues.

3.1.3 Free amino acid pools in muscle

In man muscle is the tissue for which most information is available, since it can be sampled by biopsy. The essential amino acid concentrations in human muscle in the post-absorptive state are shown in Table 3.2. Many data sets could be used for this table, but we have shown an example from the Swedish group who have been making these measurements on human muscle biopsies for many years. There is a certain amount of variability in the measurements, both between those of the same author at different times and between those of different laboratories. Some examples are shown in Table 3.3. Variations of this order could be important for non-steady state calculations, when changes in plasma concentrations are taken as indicative of changes in the free pool of the whole body.

In spite of variations certain points stand out in Table 3.2. In muscle, lysine, threonine and, to a lesser extent, histidine, dominate the picture, accounting for about 70% of the total. All three have high intracellular/extracellular concentration ratios and lysine and threonine are the only amino acids that are not transaminated. Whether there is any connection between these two characteristics has not, as far as we know, been studied.

In animals comparisons can be made between different tissues. Table 3.4 shows a comparison of the EAA free pools in muscle and liver of rats (Lunn *et al.*, 1976). As in man, in both tissues by far the highest concentrations are of lysine, threonine and histidine. Except for these three, the concentrations are somewhat higher in liver than in muscle and some three times higher in rat muscle than human muscle. As is the case so often, one can only speculate about cause and effect. It may be that these higher concentrations are necessary to maintain the much higher rate of protein synthesis in the rat; or, on the contrary, perhaps they result from the more rapid rate of protein breakdown.

Concentrations of the NEAA in plasma and muscle of human subjects are set out in Table 3.5.

Table 3.4. Concentration of free essential amino acids in muscle and liver of rats.

	μmol l^{-1} intracellular water	
	Muscle	Liver
Histidine	2046	907
Isoleucine	83	267
Leucine	128	430
Lysine	2005	1836
Methionine	251	459
Phenylalanine	79	165
Threonine	3403	3792
Valine	261	506
Total	7256	8362

From Lunn *et al.* (1976).

The dominant amino acid by far is glutamine, followed by alanine and glycine in plasma and glutamate and alanine in muscle. The concentration ratios are several times higher than those of most of the essential amino acids, and the ratios are enormously high for aspartate and glutamate. Perhaps these high ratios reflect the intracellular *de novo* synthesis of the amino acids, or they may be related to the great metabolic activity of these amino acids, particularly in transamination reactions. Some notes are given later about the metabolic functions and relationships of these amino acids.

The EAA concentrations in the proteins of the whole body and of muscle are shown in Table 3.6. The second column is the mean of results in three animal species, cattle, sheep and pigs, summarized by Davis *et al.* (1993). There was little variation between the three species, so the results are presented here as an average. The agreement is remarkable between this pattern of essential amino acids and that reported by Widdowson *et al.* (1979) in the whole body of the human fetus. To our knowledge there is no other report of the amino acid composition of the human body at any age. There is fairly good agreement in the amino acid composition of the proteins of different tissues, except skin (MacRae *et al.*, 1993); that of different single proteins varies more widely (e.g. Reeds *et al.*, 1994).

Table 3.7 compares the EAA composition of mixed muscle protein with that of the precursor

Table 3.5. Concentrations of free non-essential amino acids in plasma and muscle of man.

	Plasma μmol l^{-1}	Muscle μmol l^{-1} ICW	Concentration ratio muscle:plasma
Alanine	384	2439	6.35
Arginine	70	504	7.2
Aspargine	85	330	3.9
Aspartate	18	1400	78
Citrulline	48	84	1.75
Glutamate	46	3392	74
Glutamine	655	17215	26
Glycine	306	1719	5.6
Ornithine	114	393	3.4
Proline	296	1420	4.8
Serine	166	994	6.0
Tyrosine	88	227	2.6
Total	2182	29611	

From Bergström *et al.* (1990).

Table 3.6. Essential amino acid concentrations in proteins of the whole body and of muscle.

	g 100 g^{-1} protein		
	Whole body		Muscle
	Human[a]	Animal[b]	Animal[c]
Histidine	2.8	2.6	5.1
Isoleucine	3.8	3.4	4.8
Leucine	8.2	7.5	8.1
Lysine	7.8	7.3	9.8
Methionine	2.1	1.8	2.5
Phenylalanine	4.5	4.1	4.0
Threonine	4.5	4.2	4.7
Valine	5.1	4.9	5.4

[a]Human fetus. From Widdowson *et al.* (1979). Free + bound amino acids, recalculated to g 16 g^{-1} N.
[b]From Davis *et al.* (1993). Mean of pig, calf and sheep.
[c]From Reeds *et al.* (1994).

pool of free amino acids in muscle in terms of moles, rather than grams. The third column gives the ratio between protein-bound and free amino acids, often called the R ratio. It will be seen that R varies quite widely, which means that there must be variation in the proportion of each amino acid's free pool that is taken up in the synthesis of proteins. These rates are shown in the fourth column of the table. They lead to the conclusion that the machinery of protein synthesis is indifferent to the concentration of precursor and behaves like a zero order system. For example, only 0.2 of the lysine pool is taken up per hour, compared with 0.6 of the leucine pool.

It has long been established that proteins are synthesized from amino acids, although there is occasional evidence of synthesis from peptides (Backwell *et al.*, 1994). If the amino acid pattern of a protein is fixed, it follows that all the amino acids that compose it must turn over at rates corresponding to their molar concentrations in the protein. For example, from the data in Table 3.7 the synthesis of 1 g of muscle protein would require 250 μmol of phenylalanine and 570 μmol of leucine, a molar ratio of 0.44:1. This ratio is indeed generally found when measurements of turnover are made simultaneously with these two amino acids (Chapter 6, section 6.9). In 1989 Bier reported a linear relationship between the whole body turnover rates of a number of amino acids and their molar concentrations in body protein (Bier, 1989). The values of the fluxes were probably underestimates, because they were based on enrichments in plasma, but that does not alter the essential relationship, which provided strong evidence of the validity of the precursor method of measuring protein turnover (see Chapter 6).

3.2 Nutritional Effects on the Free Amino Acid Pools

A distinction has to be made between the acute effects of a meal or an infusion of amino acids and the more chronic effects of a continuing diet. In both situations there is a wealth of information about changes in plasma amino acid concentrations, much less about changes in the tissue free pools. For this we have to rely largely on animal experiments.

3.2.1 Acute effects

In response to a protein meal there are substantial increases in amino acid concentrations in portal blood, but these are largely smoothed out in the

Table 3.7. Comparison of free and protein-bound amino acid concentrations in human muscle and their rates of uptake into muscle protein.

	Free μmol g^{-1} muscle	Bound μmol g^{-1} muscle protein	R	Uptake into protein μmol g^{-1} h^{-1}	Uptake/free pool, h^{-1}
	A	B	C	D $\times 10^{-2}$	D/A
Histidine	0.33	66	200	5.28	0.16
Isoleucine	0.086	73	849	5.84	0.68
Leucine	0.160	124	775	9.92	0.62
Lysine	0.597	134	224	10.72	0.18
Methionine	0.058	34	586	2.72	0.47
Phenylalanine	0.057	48	842	3.84	0.67
Threonine	0.586	79	135	6.32	0.11
Valine	0.240	92	383	7.36	0.37

A: per g tissue, from Bergström *et al.* (1990), assuming that intracellular water = 650 ml.kg^{-1} (Bergström *et al.*, 1974);
B: from Reeds *et al.* (1994), assuming that muscle contains 200 g protein kg^{-1};
C: ratio of protein-bound to free amino acid;
D: assuming that fractional synthesis rate of mixed human muscle protein is 0.0008 h^{-1} (1.9% per day).

liver and are much less in the peripheral circulation (Bloxam, 1971). Bergström *et al.* (1990) reported a heroic study in which five biopsies of muscle were obtained at intervals up to 7 h after a meal that either contained protein or was protein-free. With the protein-free meal the EAAs decreased and the NEAAs increased in both plasma and muscle, in each case by about 35% at peak. After the protein-containing meal the changes tended to be in the opposite direction.

An interesting finding in this study was a significant linear relationship, both in plasma and muscle, between the change in concentration of each EAA after the protein meal and its molar concentration in the protein of the meal, which consisted of bovine serum albumin (BSA). Since BSA differs significantly from whole body protein in its amino acid composition, being lower in isoleucine, methionine and phenylalanine, it was suggested that the changes in amino acid concentration depended more on the arterial input than on protein degradation.

In another Swedish study (Lundholm *et al.*, 1987) amino acids were infused for successive 2-h periods at rates of 8.3, 16.7 and 33.2 mg N kg^{-1} h^{-1}, the highest level corresponding to a protein intake of 5 g kg^{-1} day^{-1}. Muscle biopsies were performed at the end of each 2-h period. After the highest rate of infusion the free amino acid concentrations in muscle of methionine and phenylalanine had increased on average to nearly three times their basal levels, whereas those of lysine, threonine, histidine and the NEAAs were unchanged. This finding is very interesting, because the amino acids whose concentrations increased have a high R value and a low tissue/plasma concentration ratio.

The effect of insulin is also relevant, because a meal, particularly a carbohydrate meal, stimulates insulin secretion. Long ago Munro suggested that the fall in plasma amino acid concentration produced by a carbohydrate meal resulted from stimulation of amino acid uptake into protein. More recently it has been shown that insulin infusion decreased the intracellular concentrations of the BCAs and aromatic amino acids by 33% (Alvestrand *et al.*, 1988), presumably because of the effect of insulin in reducing protein breakdown. The impression one gets is that the pool size of this group of amino acids is particularly labile. It may be noted that phenylalanine and leucine are transported by the same carrier.

3.2.2 Chronic changes in protein intake

The responses to different levels of protein intake over days or weeks are broadly similar to the effects of acute changes. In man a protein-free diet fed for 1–2 weeks caused small decreases in the plasma concentrations of EAA with a rise in NEAA (Young and Scrimshaw, 1968; Adibi *et al.*, 1973). In default of studies in man on diet-induced changes in tissue amino acid pools, we again have to rely on animal experiments. In rats on a low protein or protein-free diet for 1–3 weeks the levels of EAAs fell and those of NEAAs rose in plasma, liver and muscle. In starvation the changes were the opposite to those on a low protein diet (Millward *et al.*, 1974, 1976). The changes in plasma seem to predict those in the tissue pools.

It is clear that both immediate food intake and prevailing diet do influence the free amino acid concentrations in plasma and tissue pools, but in few situations do the changes exceed ± 50% and they are usually less. These are superimposed on small daily fluctuations (Lunn *et al.*, 1976). Thus the changes in the free pools are relatively small in relation to the large fluxes through them (Table 3.1). They are most consistent in those amino acids with a high value of the R ratio (concentration in protein/concentration in free pool). It has been suggested that low plasma levels of these amino acids might be diagnostic of protein deficiency. One can conceive that one or other of them, with their small precursor pools, might become limiting for protein synthesis. It is a difficult problem to sort out cause and effect: does the size of the free pool have any effect on rates of protein synthesis and breakdown? Or is it simply determined by the balance of fluxes through it? That is the question posed by Scornik at the beginning of this chapter.

3.3 Kinetics of Free Amino Acids

An apparent rate coefficient, k′, for the turnover of a free amino acid in plasma can be determined from the decay of enrichment in the plasma after a single dose of tracer or from the increase in enrichment with a continuous infusion. In both cases what is obtained is not the true turnover rate, k, of the amino acid but a coefficient, k′, that represents the disposal rate, i.e. that fraction of the free pool that disappears into protein synthesis and oxidation. In the example of Fig. 2.6:

for the plasma pool A: true k
$= V_{BA}/A = 200/100 = 2\ h^{-1}$
apparent $k' = (V_{BA} - V_{AB})/A$
$= (V_{CB} + V_{OB})/A = 100/100 = 1\ h^{-1}$

Thus the apparent k′ underestimates the true k by a factor of 2. A further point, mentioned in the previous chapter, is that k′ is not in reality a single number but the weighted average of all the coefficients of uptake into all the proteins of the body.

There is a difficulty in estimating k′ from decay curves after a single dose, because the first part of the curve is very important and it may be distorted by the time taken for mixing, although consistent results for glycine and alanine were obtained with this method by Nissim and co-workers (Nissim and Lapidot, 1979; Amir *et al.*, 1980). Results can be got more easily from the rising part of the activity curve during a constant infusion, provided that a priming dose of tracer has not been given. k′ can be estimated from the time needed to reach half maximum activity (plateau) according to the relation:

$$k' = \ell n2/t_{1/2}$$

For greater precision Lobley *et al.* (1980) proposed a method in which two tracers were infused, starting at different points of time. This method was elaborated by Dudley *et al.* (1998) in a study of mucosal glycoprotein synthesis. In order to avoid taking multiple mucosal samples they infused no less than six different tracers, leucine labelled with ^{13}C and ^{3}H and 4 isotopomers of phenylalanine, starting at different time-points over a period of 6 h and from these constructed a curve of rise to plateau. The rate constants were identical, whether calculated in the conventional way from multiple samples and a single tracer or from a single sample with multiple tracers.

The values of k′ obtained in a number of studies are shown in Table 3.8. The remarkable thing about this table is the small range of values, except for glutamate, from less than 1 to 3 h^{-1}, regardless of amino acid or species. One would expect k′ to be much higher in rat than in man because of the rat's much greater rate of whole body turnover – about 10 × that of man; and to be higher with leucine than lysine because the

Table 3.8. Apparent turnover rates, k′,of free amino acids in plasma. All measurements by continuous intravenous infusion except where otherwise stated.

Animal	Amino acid	k′, h^{-1}	Reference
Rat	Lysine	1.5	Waterlow *et al.* (1967)
Rat	Tyrosine	0.7	[a]Harney *et al.* (1976)
		2.5	
Rabbit	Proline	2.5	Robins (1979)
Rabbit	Alanine	4.2	[b]Nissim *et al.* (1979)
Rabbit	Tyrosine	< 1	Lobley *et al.* (1980)
Piglet	Phenylalanine + leucine	3.1	Dudley *et al.* (1998)
Pig	Tyrosine	~ 2	Garlick *et al.* (1976)
Pig	Leucine	1.46	Simon *et al.* (1982)
	Lysine	2.0	Simon *et al.* (1982)
Sheep	Phenylalanine + leucine	~ 1	Connell *et al.* (1997)
Man	Glycine	1.7	[b]Amir *et al.* (1980)
neonate			
3–4 weeks	Glycine	4.3	[b]Amir *et al.* (1980)
adult	Tyrosine	1.7	James *et al.* (1976)
	Glutamate	4.8	Darmaun *et al.* (1986)
	Glutamine		
	$2\text{–}^{15}N$	1.7	Darmaun *et al.* (1986)
	$5\text{–}^{15}N$	1.5	Darmaun *et al.* (1986)

[a]In this study specific activities were of CO_2, not of labelled amino acids in plasma.
[b]Single intravenous dose.

free lysine pool is many times larger than that of leucine (Table 3.4).

In another approach, data on liver and muscle derived from constant infusions of [$U^{14}C$]-lysine in the rat were analysed with a five-pool model (Aub and Waterlow, 1970). The analysis provided values for all the rate-coefficients of the system. The true turnover rate of the free lysine pool in liver was 10.9 and in muscle 6.3 h^{-1}. One cannot extrapolate these figures to the whole body, but they may give some indication of the size of the difference between true and apparent turnover rates.[1]

Another point about Table 3.8 is that it leads to a serious discrepancy, and discrepancies are always interesting. If the value of k′ is as shown, and the flux is determined from the plateau, the pool size is given by: pool size = flux/k′. Darmaun *et al.* (1986) infused ^{15}N-glutamate and ^{15}N-glutamine, estimated k′ from the activity curve and flux from the plateau, and obtained the figures shown in Table 3.9. These are 1/25–1/70 of the observed amounts in the free pools of human muscle (Table 3.2). A similar but much smaller discrepancy arises with lysine. These discrepancies are discussed further in the next chapter.

3.4 Amino Acid Transport across Cell Membranes

The rate of transport of amino acids through the cell membrane could be a step limiting their uptake into protein and their exchange between tissues. Much of what is known on this subject comes from the classical work over many years of Christensen and his colleagues, who showed that transport of amino acids into and out of cells is mediated by a complex system of carrier mechanisms (Christensen, 1975; Christensen and Kilberg, 1987). Four main systems were originally identified: A and ASC, covering most of the neutral amino acids; L, covering the branched chain and aromatic amino acids; and Lys, now known as y^+, the basic amino acids. The A and ASC systems are sodium dependent, producing active transport against a gradient. In recent years a number of more selective transporters have been described, such as the cationic amino acid transporter CAT 1 (Hyatt *et al.*, 1997). Christensen has repeatedly emphasized that there is tremendous overlap between the main transporters (see, for example, a useful diagram in Ellory, 1987). Moreover, the relative activity of different transport systems depends not only on the amino acid but also on the type of cell. Grimble (2000) has summarized the present state of knowledge. He has also summarized the evidence for the uptake of peptides, which are then hydrolysed within the cell (Grimble and Silk, 1989).

Some of the transporters, such as A and CAT-1, are adaptively regulated by high or low amino acid concentrations in the medium, being stimulated when the extracellular concentration is low (Christensen and Kilberg, 1987). There is also regulation by many hormones, glucagon in particular being active (Kilberg, 1986). The adaptation of the A system involves a change in Vmax rather than Km, suggesting that there is a change in the number of transporter molecules. This interpretation is confirmed by the finding that adaptation is prevented by cyclohexamide, an inhibitor of protein synthesis (Christensen and Kilberg, 1987). There is evidence also that amino acid starvation leads to an increase in CAT-1

Table 3.9. Comparison between calculated and observed sizes of the free pool of glutamine and glutamate.

	Calculated[a]	Observed[b]
	μmol kg^{-1} wt	
Glutamine	207	5200
Glutamate	17	1200

[a]Based on relation: pool size = flux/k′ (see text).
[b]Estimated from direct measurements, mainly on muscle.
From Darmaun *et al.* (1986).

[1] The difference arises because in tissues we can ignore the return of tracer from protein to tissue free pool, whereas with plasma the return of tracer from tissue pool to plasma cannot be ignored.

mRNA (Hyatt *et al.*, 1997). We therefore have a complex mechanism of homeostatic regulation.

Since the work on amino acid transport described so far was mostly done on isolated cells or tissue slices, the question is, how far is it related to what happens *in vivo*? The first study that we know of in the intact animal was that of Baños *et al.* (1973), who infused [^{14}C]-labelled amino acids into rats with an electronically controlled syringe which brought the radioactivity in the blood to a high level within 10 seconds and then held it constant for up to 40 minutes (Daniel *et al.*, 1975). This was the precursor of priming. Rats were sacrificed after 3 or 10 minutes and radioactivity measured in the free pool of muscle. It was found that for several amino acids the entry rates from plasma varied linearly with the plasma concentration. The entry rates for different amino acids at normal plasma concentration are shown in Table 3.10. These rates do not by themselves mean very much; what is of more interest is to compare them with rates of synthesis, which are also shown in Table 3.10, calculated on the assumption that the fractional synthesis rate of muscle protein in rats of the same size (200 g) is 6% day^{-1} or 0.0025 h^{-1}. Comparison of synthesis (D) and entry rates (B) as a ratio that may be called the 'transport index' shows that for all amino acids, except perhaps histidine, synthesis is unlikely to be limited by the trans-membrane entry rate, since the comparison in the table ignores the contribution to synthesis of amino acids from protein breakdown.

The 'efficiency' of synthesis, F, can be represented as the uptake into synthesis divided by the total amino acid availability – i.e. the sum of entry from plasma and supply from protein breakdown. If a steady state is assumed, so that breakdown = synthesis, the efficiency can be calculated as: $F = 1/(E+1)$.

The conclusion that synthesis is unlikely to be limited by the entry rate is supported by perfusion experiments of Hundal *et al.* (1989) in the rat hind-limb. They studied the transport systems of a range of neutral, acidic and basic amino acids and found that in all cases the Km was many times greater than the normal plasma concentration, so there would be little risk of the transporters being saturated. These authors' conclusions agreed with those of Baños *et al.* (1973): that in muscle the entry of amino acids was greater than their incorporation into protein, so that under normal circumstances transport is not limiting. They cautioned, however, that this conclusion may not hold for other tissues. Thus Salter *et al.* (1986) showed that in liver the transporter that carries the aromatic amino acids may effectively control their catabolism. In the five-pool rat model mentioned above (Aub and Waterlow, 1970) the ratio of entry rate of lysine: uptake into protein was calculated to be 7.1 for muscle but only 1.7 for liver, so transport may come close to being limiting in a rapidly turning over tissue.

Further information about amino acid transport across membranes comes from studies in man on exchanges in the arm or leg. The classical

Table 3.10. Muscle free amino acids in the rat: concentration, rate of entry, turnover, uptake by synthesis, and transport index = entry/synthesis.

	A Concentration nmol g^{-1}	B Entry nmol g^{-1} h^{-1}	C Turnover h^{-1}	D Synthesis nmol g^{-1} h^{-1}	E Transport index
Histidine	135	120	0.9	165	0.73
Isoleucine	31	294	9.5	185	1.59
Leucine	127	642	4.4	310	2.07
Lysine	1971	846	0.4	335	2.53
Methionine	10	144	14.4	85	1.69
Phenylanaline	44	462	10.5	120	3.85
Threonine	602	960	1.6	200	4.80
Valine	136	534	3.9	230	2.32

A and B: from Baños *et al.* (1973), Table 3. Rats of about 200 g.
C: B/A.
D: Fractional synthesis rate of muscle protein in 200 g rat taken as 6% day^{-1}.
E: B/D.

approach of measuring arterio-venous differences, pioneered by Cahill and co-workers (e.g. Pozefsky *et al.*, 1969) has been supplemented by infusion of a tracer amino acid. The original model developed by Cheng *et al.* (1985, 1987) contained a pool of free intramuscular leucine, together with a KIC pool, with inflows from the arterial supply and from protein breakdown, and outflows to protein synthesis, oxidation and venous effluent. Biolo *et al.* (1992), working with dogs, used a simpler model (Fig. 3.1) in which arterio-venous measurements were combined with muscle biopsies to provide direct information about enrichment of tracer in the muscle free pool. Tessari *et al.* (1995) developed a model similar to that of Cheng *et al.*, with separate KIC pools providing indirect estimates of intramuscular precursor enrichment. All these models require measurements of blood flow to the limb in addition to tracee and tracer concentrations. The calculations depend on mass balance of tracer and tracee in the steady state, and the reader is referred to the original papers for details.

The most useful index is the ratio: inflow to the muscle free pool: uptake into protein and oxidation, the transport index (cf. Table 3.10). In the terminology of Biolo and Tessari this ratio is FMA/FOM (see Fig. 3.1). Results are summarized in Table 3.11. The values for leucine in Biolo's studies are rather variable, with an average similar to that of Tessari; lysine has a lower index and phenylalanine a higher one. Increasing the amino acid supply sometimes raises the index, while insulin has little effect. It is remarkable that the transport indices are so similar in Tables 3.10 and 3.11, considering that they were obtained on different animals under very different experimental conditions. An index of 1 or less does not, of course, mean that the supply is limiting the rate of synthesis, because the substrate available for synthesis is derived from protein breakdown as well as from the plasma.

Finally, Hovorka *et al.* (1999) derived an equation for estimating inward transport to the tissues of the whole body:

$$F_{in} = \frac{d}{\varepsilon_A - \varepsilon_B}$$

where F_{in} is the rate of transport from extracellular to intracellular pool, ε_A and ε_B are the enrichments in the plasma and precursor pools, and *d* is the rate

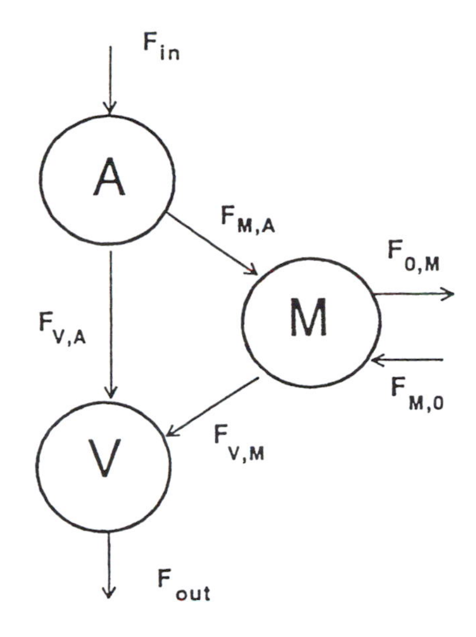

Fig. 3.1. Biolo's model of kinetics in a limb. A, artery; V, vein; M, muscle free pool; O, protein. Fs represent unidirectional fluxes. Reproduced from Biolo *et al.* (1992), by courtesy of the *American Journal of Physiology*.

of tracer dosage. This derivation is slightly in error. They assumed that the backflow, F_{out}, from IC to EC pool is the same as the forward flow. Even in post-absorptive subjects oxidation continues, so that $F_{out} = F_{in}$ – oxidation. Alternatively, the loss from the IC pool by oxidation might be balanced by increased protein breakdown, so that the protein pool is not in a steady state. However, oxidation is small compared with F_{in} and as a first approximation can perhaps be neglected. Hovorka *et al.* (1999) calculated that in post-absorptive men the inward transport rate was 6.7 μmol kg^{-1} min^{-1}, giving a transport index of 3.45. Like Baños *et al.* (1973) they found a significant linear relation between leucine transport and plasma leucine concentration.

3.5 Conclusion

The real problem raised by this work on amino acid transport is whether changes in the amount of amino acid entering the cell have any effect on the rate of protein synthesis. On this opinions seem to differ. The question is complicated by the evidence

Table 3.11. Values of the 'transport index' for selected amino acids in human forearm and hindlimb preparations: transport index = amino acid inflow/disposal rate.

		Leucine	Lysine	Phenylalanine
Cheng *et al.* (1985, 1987)				
	Fasted	1.39	–	–
	Fed	0.82		
Biolo *et al.* (1992)	Post-absorptive	0.75	0.62	2.50
(1995a)	Post-absorptive	1.17	0.67	2.18
Biolo *et al.* (1995b)	Post-absorptive	0.99	0.57	1.82
	+ insulin	0.87	0.74	1.37
Biolo *et al.* (1995c)	Post-absorptive	–	1.11	3.85
(1997)	Post-absorptive	3.06	0.82	3.23
(1997)	+ amino acids	2.15	1.03	6.29
Tessari *et al.* (1995)	Post-absorptive	1.24	–	–
(1996a)	Post-absorptive	1.63	–	–
(1999)	Post-absorptive	1.02	–	–
	+ amino acids and insulin	1.07	–	–
Volpi *et al.* (1998)	Post-absorptive	1.71	0.71	2.31
	+ amino acids	3.18	0.61	4.49
Zanetti *et al.* (1999)	Basal	1.05	–	–
	+ insulin	0.90	–	–
	+ insulin + AAs	1.07	–	–

(see Chapter 5) for compartmentation within the cell, producing some degree of separation between amino acids entering from the plasma and those derived from protein breakdown.

There remains also the evidence, not considered here, that amino acids are transported bound as peptides, which are taken up more efficiently than free amino acids in the gastrointestinal tract, where peptide transporters have been demonstrated in the brush border of the mucosa cells (Grimble, 2000); it has also been shown that peptides are taken up by the lactating mammary gland (Becquette *et al.*, 1998).

3.6 References

Adibi, S.A., Modesto, T.A., Morse, E.L. and Amin, P.M. (1973) Amino acid levels in plasma, liver and skeletal muscle during protein deprivation. *American Journal of Physiology* 225, 408–414.

Alvestrand, A., Defronzo, R.A., Smith, D. and Wahren, J. (1988) Influence of hyperinsulinaemia on intracellular amino acid levels and amino acid exchange across splanchnic and leg tissues in uraemia. *Clinical Science* 74, 155–163.

Amir, J., Reisner, S.H. and Lapidot, A. (1980) Glycine turnover rates and pool sizes in neonates as determined by gas chromatography-mass spectrometry and nitrogen-15. *Pediatric Research* 14, 1238–1244.

Aub, M.R. and Waterlow, J.C. (1970) Analysis of a five-compartment system with continuous infusion and its application to the study of amino acid turnover. *Journal of Theoretical Biology* 26, 243–250.

Backwell, F.R.C., Bequette, B.J., Wilson, D., Calder, A.G., Metcalf, J.A., Wray-Cahen, D., MaCrae, J.C., Beeve, D.E. and Lobley, J.A. (1994) The utilization of dipeptides by the caprine mammary gland for milk protein synthesis. *American Journal of Physiology* 267, R 1–6.

Baños, G., Daniel, P.M., Moorhouse, S.R. and Pratt, E.O. (1973) The movement of amino acids between blood and skeletal muscle in the rat. *Journal of Physiology* 235, 459–475.

Becquette, B.J., Blackwell, F.R.C. and Crompton, L.A. (1998) Current concepts of amino acid and protein metabolism in the mammary gland of the lactating ruminant. *Journal of Dairy Science* 81, 2540–2559.

Bergström, J., Fürst, P. Norée, L.O. and Vinnars, E. (1974) Intracellular free amino acid concentration in human muscle tissue. *Journal of Applied Physiology* 36, 693–697.

Bergström, J., Fürst, P. and Vinnars, E. (1990) Effect of a test meal, with and without protein, on muscle and plasma free amino acids. *Clinical Science* 79, 331–337.

Bier, D.M. (1989) Intrinsically difficult problems: the kinetics of body proteins and amino acids in man. *Diabetes Metabolism Reviews* 5, 111–132.

Biolo, G., Chinkes, D., Zhang, X.-J. and Wolfe, R.R. (1992) A new model to determine *in vivo* the relationship between amino acid transmembrane transport and protein kinetics in muscle. *Journal of Parenteral and Enteral Nutrition* 16, 305–315.

Biolo, G., Fleming, R.Y.D., Maggi, S.P. and Wolfe, R.R. (1995a) Transmembrane transport and intracellular kinetics of amino acids in human skeletal muscle. *American Journal of Physiology* 268, E75–84.

Biolo, G., Fleming, R.Y.D. and Wolfe, R.R. (1995b) Physiologic hyperinsulinaemia stimulates protein synthesis and enhances transport of selected amino acids in human skeletal muscle. *Journal of Clinical Investigation* 95, 811–819.

Biolo, G., Zhang, X.-J. and Wolfe, R.R. (1995c) Role of membrane transport in interorgan amino acid flow between muscle and small intestine. *Metabolism* 44, 719–724.

Biolo, G., Tipton, K.D., Klein, S. and Wolfe, R.R. (1997) An abundant supply of amino acids enhances the metabolic effect of exercise on muscle protein. *American Journal of Physiology* 273, E122–129.

Bloxam, D.L. (1971) Nutritional aspects of amino acid metabolism. 1. A rat liver perfusion method for the study of amino acid metabolism. *British Journal of Nutrition* 26, 393–422.

Cheng, K.N., Dworzak, F., Ford, G.C., Rennie, M.J. and Halliday, D. (1985) Direct determination of leucine metabolism and protein breakdown in humans using L-[1-^{13}C,^{15}N]-leucine and the forearm model. *European Journal of Clinical Investigation* 15, 349–354.

Cheng, K.N., Pacy, P.J., Dworzak, F., Ford, G.C. and Halliday, D. (1987) Influence of fasting on leucine and muscle protein metabolism across the human forearm determined using L-[1-^{13}C,^{15}N] as the tracer. *Clinical Science* 73, 241–246.

Christensen, H.N. (1975) *Biological Transport* 2nd edn. Benjamin, Reading, Massachusetts.

Christensen, H.N. and Kilberg, M.S. (1987) Amino acid transport across the plasma membrane: role of regulation in interorgan flows. In: Yudilevitch, D.L. and Boyd, C.R. (eds) *Amino Acid Transport in Animal Cells.* Manchester University Press, pp. 10–46.

Connell, A., Calder, A.G., Anderson, S.E. and Lobley, G.E. (1997) Hepatic protein synthesis in the sheep: effect of intake as monitored by use of stable-isotope-labelled glycine, leucine and phenylalanine. *British Journal of Nutrition* 77, 255–271.

Daniel, P.N., Donaldson, J. and Pratt, O.E. (1975) A method for injecting substances into the circulation to reach rapidly and to maintain a steady level. *Medical and Biological Engineering* 13, 214–227.

Darmaun, D., Matthews, D.E. and Bier, D.M. (1986) Glutamine and glutamate kinetics in humans. *American Journal of Physiology* 251, E117–126.

Davis, T.A., Fiorotto, M. and Reeds, P.H. (1993) Amino acid compositions of body and milk protein change during the suckling period in rats. *Journal of Nutrition* 123, 947–956.

Dudley, M.A., Burrin, D.G., Wykes, L.J., Toffolo, G., Cobelli, C., Nichols, B.L., Rosenberger, J., Jahoor, F. and Reeds, P.J. (1998) Protein kinetics determined *in vivo* with a multiple tracer, single-sample protocol: application to lactase synthesis. *American Journal of Physiology* 274, G591–598.

Ellory, J.C. (1987) Amino acid transport of systems in mammalian red cells. In: Yudilevitch, D.L. and Boyd, C.A.R. (eds) *Amino Acid Transport in Animal Cells.* Manchester University Press, pp. 106–119.

Elwyn, D.H. (1966) Distribution of amino acids between plasma and red blood cells in the dog. *Federation Proceedings* 25, 854–861.

Garlick, P.J., Burk, T.L. and Swick, R.W. (1976) Protein synthesis and RNA in tissues of the pig. *American Journal of Physiology* 230, 1108–1112.

Grimble, G.K. (2000) Mechanisms of peptides and amino acid transport and their regulation. In: Fürst, P. and Young, V. (eds) *Proteins, Peptides and Amino Acids in Enteral Nutrition,* pp. 63–84.

Grimble, G.K. and Silk, D.B.A. (1989) Peptides in human nutrition. *Nutrition Research Reviews* 2, 87–108.

Harney, M.E., Swick, R.W. and Benevenga, N.J. (1976) Estimation of tissue protein synthesis in rats fed diets labelled with [U-^{14}C] tyrosine. *American Journal of Physiology* 231, 1018–1023.

Hovorka, R., Carrol, P.V., Gowrie, I.J., Jackson, N.C., Russell-Jones, D.L. and Umpleby, A.M. (1999) Surrogate measure of whole body leucine transport across the cell membrane. *American Journal of Physiology* 276, E573–579.

Hundal, H.S., Rennie, M.J. and Watt, P.W. (1989) Characteristics of acidics, basic and neutral amino acid transport in the perfused rat hind limb. *Journal of Physiology* 408, 93–114.

Hyatt, S.L., Aulak, K.S., Malandro, M., Kilberg, M.S. and Hatzoglou, M. (1997) Adaptive regulation of the cationic amino acid transport –1 (Cat-1) in Fao cells. *Journal of Biological Chemistry* 272, 19951–19957.

Jackson, A.A. (1982) Amino-acids, essential and non-essential. *Lancet* 1, 1034–1037.

Jackson, A.A. (1991) The glycine story. *European Journal of Clinical Nutrition* 45, 59–65.

Jackson, A.A. and Golden, M.H.N. (1980) [^{15}N] glycine metabolism in normal man: the metabolic α-amino-nitrogen pool. *Clinical Science* 58, 577–582.

James, W.P.T., Garlick, P.J., Sender, P.M. and Waterlow, J.C. (1976) Studies of amino acid and protein metabolism in normal man with L-[U-^{14}C] tyrosine.

Clinical Science and Molecular Medicine 50, 525–532.

Kilberg, M.S. (1986) System A mediated amino acid transport: metabolic control at the plasma membrane. *Trends in Biochemical Science* 11, 183–186.

Lobley, G.E., Robins, S.P., Palmer, R.M. and McDonald, I. (1980) Measurement of the rates of protein synthesis in rabbits. *Biochemical Journal* 192, 623–629.

Lobley, G.E., Connell, A., Revell, D.K., Bequette, B.J., Brown, D.S. and Calder, A.G. (1996) Splanchnic bed transfers of amino acids in sheep blood and plasma, as monitored through use of a multiple U-^{13}C-labelled amino acid mixture. *British Journal of Nutrition* 75, 217–235.

Lundholm, K., Bennegard, K., Zachrisson, H., Lundgven, F., Eden, E. and Möller-Loswak, A.-C. (1987) Transport kinetics of amino acids across the resting human leg. *American Journal of Clinical Investigation* 80, 763–771.

Lunn, P.G., Whitehead, R.G. and Baker, B.A. (1976) The relative effects of a low-protein-high-carbohydrate diet on the free amino acid composition of liver and muscle. *British Journal of Nutrition* 36, 219–230.

MacRae, J.C., Walker, A., Brown, D. and Lobley, G.E. (1993) Accretion of total protein and individual amino acids by organs and tissues of growing lambs and the ability of nitrogen balance techniques to quantitate protein retention. *Animal Production* 57, 237–245.

Milewski, P.J., Threlfall, C.J., Heath, D.F., Holbrook, I.B., Wilford, K. and Irving, M.H. (1982) Intracellular free amino acids in undernourished patients with or without sepsis. *Clinical Science* 62, 83–91.

Millward, D.J., Nnanyelugo, D.O., James, W.P.T. and Garlick, P.J. (1974) Protein metabolism in skeletal muscle: the effect of feeding and fasting on muscle RNA, free amino acids and plasma insulin concentrations. *British Journal of Nutrition* 32, 127–142.

Millward, D.J., Garlick, P.J., Nnanyelugo, D. and Waterlow, J.C. (1976) The relative importance of muscle protein synthesis and breakdown in the regulation of muscle mass. *Biochemical Journal* 156, 185–188.

Nissim, I. and Lapidot, A. (1979) Plasma amino acid turnover rates and pools in rabbits: *in vivo* studies using stable isotopes. *American Journal of Physiology* 237, E418–427.

Pozefsky, T., Felig, P., Tobin, J.D., Soeldner, J.S. and Cahill, G.F. (1969) Amino acid balance across tissues of the forearm in postabsorptive man. Effects of insulin at two dose levels. *Journal of Clinical Investigation* 48, 2273–2282.

Reeds, P.J., Fjeld, C.R. and Jahoor, F. (1994) Do the differences between the amino acid compositions of acute phase and muscle proteins have a bearing on nitrogen losses in traumatic states? *Journal of Nutrition* 124, 906–910.

Robins, S.P. (1979) Metabolism of rabbit skin collagen. *Biochemical Journal* 181, 75–82.

Salter, M., Knowles, R.G. and Pogson, C.J. (1986) Transport of the aromatic amino acids into isolated liver cells. *Biochemical Journal* 233, 499–506.

Savarin, I.C., Hoskin, S.O., Dennison, N. and Lobley, G.E. (2001) Lysine metabolism across the hindquarters of sheep: effect of intake on transfers from plasma and red cells. *British Journal of Nutrition* 85, 565–573.

Scornik, O. (1984) Role of protein degradation in the regulation of cellular protein content and amino acid pools. *Federation Proceedings* 43, 1283–1288.

Simon, O., Bergner, H., Münchmeyer, R. and Zebrowska, P. (1982) Studies on the range of tissue protein synthesis in pigs: the effect of thyroid hormones. *British Journal of Nutrition* 48, 571–582.

Tessari, P., Inchiostro, S., Zanetti, M. and Barazzoni, R. (1995) A model of skeletal muscle leucine kinetics measured across the human fore-arm. *American Journal of Physiology* 269, E127–136.

Tessari, P., Garibotto, G., Inchiostro, S., Robando, C., Saffioti, S., Vettore, M., Zanetti, M., Russo, R. and Deferrari, G. (1996a) Kidney, splanchnic and leg protein turnover in humans: insight from leucine and phenylalanine kinetics. *Journal of Clinical Investigation* 98, 1481–1492.

Tessari, P., Zanetti, M., Barazzoni, R., Vettore, M. and Michielan, F. (1996b) Mechanisms of postprandial protein accretion in human skeletal muscle. *Journal of Clinical Investiagation* 98, 1361–1372.

Volpi, E., Ferrando, A.A., Yeckel, C.W., Tipton, K.D. and Wolfe, R.R. (1998) Exogenous amino acids stimulate net muscle protein synthesis in the elderly. *Journal of Clinical Investigation* 101, 2000–2007.

Waterlow, J.C. and Stephen, J.M.L. (1967) The measurement of total lysine turnover in the rat by intravenous infusion of L-[U^{14}C]-lysine. *Clinical Science* 33, 489–506.

Widdowson, E.M., Southgate, D.A.T. and Hey, E.N. (1979) Body composition of the fetus and infant. In: Visscher, H.K.A. (ed.) *Nutrition and Metabolism of the Fetus and Infant.* Kluwer, Dordrecht, pp. 169–178.

Young, V.R. and Scrimshaw, N.S. (1968) Endogenous nitrogen metabolism and plasma free amino acids in young adults given a 'protein-free' diet. *British Journal of Nutrition* 22, 9–20.

Zanetti, M., Barazzoni, R., Kiwanuka, E. and Tessari, P. (1999) Effect of branched-chain enriched amino acids and insulin on forearm leucine kinetics. *Clinical Science* 97, 437–448.

4

Metabolism of Some Amino Acids

It is not proposed in this chapter to describe systematically the metabolism of all amino acids, but only to make some points that are relevant to their turnover or otherwise of interest.

4.1 Leucine

For an excellent older review of this subject see Harper and Zapalowski (1981). The branched-chain amino acids (BCAAs) play a special role in nitrogen metabolism, because in muscle they act as donors of nitrogen to alanine and glutamine, which are transported to the rest of the body (Felig, 1975). Their metabolism and its regulation is therefore an excellent example of inter-organ collaboration. Another reason for a special interest in the BCAAs, particularly leucine, is that it is the amino acid most widely used for measuring protein turnover, precisely because, as a first step in its oxidation, it produces a metabolite, α-ketoisocaproic acid (KIC), whose enrichment is a measure of that of intracellular leucine (see Chapter 5). The mechanism and site(s) of leucine oxidation are therefore important. The oxidation occurs in two steps that are distributed between the splanchnic and peripheral tissues: the first is transamination, which is reversible; the second is dehydrogenation and eventual liberation of CO_2, which is irreversible.

4.1.1 Transamination

The same transaminase, BCAT for short, operates for all three BCAAs. It has two iso-enzymes, mitochondrial and cytoplasmic, the former predominating in most tissues except the brain. The mitochondrial location of the BCAT perhaps provides an opportunity for oxidation and acylation with t-RNA, the first step in protein synthesis, to occur in separate compartments (see Chapter 5). Table 4.1 shows the distribution of BCAT in human tissues (Suryawan *et al.*, 1998). The distribution differs in different species: in the rat there is virtually none in the liver, whereas in man there is an appreciable amount. Since transamination precedes oxidation, this difference in distribution could explain the long-held belief, based on animal experiments, that BCAAs are not oxidized by the splanchnic tissues.

The advent of stable isotopes made it possible to determine the rate of transamination *in vivo* in man. In a pioneering study with leucine doubly labelled with ^{15}N and ^{13}C Matthews *et al.* (1981a) obtained the results shown in Table 4.2. The transamination is reversible and in the cycle of deamination/reamination labelled N is lost; the chance of reamination with a labelled N atom from the amino-N pool is virtually nil, whereas the carbon skeleton of the keto-acid, unlike the pyruvate derived from alanine, appears not to exchange rapidly with other metabolites. Thus a large part of the keto acid formed is reaminated. The difference between carbon and nitrogen fluxes represents the rate at which N is added to the flux by reamination. The proportion of the KIC produced that goes on to be oxidized is quite small.

In keeping with the much larger total amount of BCAT in muscle than in liver, only a small proportion of KIC production occurs in the splanchnic region. When labelled leucine was given by intragastric infusion, only 8% of the

Table 4.1. Distribution in human tissues of branched-chain amino acid transaminase (BCAT) and dehydrogenase (BCKD), and the proportion of BKCD in the active state. From Suryawan *et al.* (1998).

	mU g^{-1} wet wt		% of total BCKD active	% of active BCKD in whole body
	BCAT	BCKD		
Muscle	124	4.9	26	54
Liver	248	14.8	28	13
Kidney	880	110	14	8
Small intestine	241	1.6	44	4
Brain	510	10.9	59	20
	Totals: U per total tissue*			
	BCAT	BCKD		
Muscle	3472	137		
Liver	496	30		

*Assuming that in a 70-kg human, liver weight = 2 kg, muscle weight = 28 kg; U, units

dose in the fasted state (Matthews *et al.*, 1993a) and 13% in the fed state (Biolo *et al.*, 1997) was converted to KIC in the splanchnic tissues.

A number of situations have been described in which transamination in both directions is greatly reduced: in normal pregnancy (Kalhan *et al.*, 1998), in the treatment of diabetic patients with insulin (Hutson *et al.*, 1978; Nair *et al.*, 1995; Meek *et al.*, 1998), and with infusion of medium-chain triglycerides (Beaufrère *et al.*, 1985). In all these cases there appears to be an adaptation to economize nitrogen. In pregnancy Kalhan *et al.* (1998) found a significant correlation between the rate of urea synthesis and the rate of amination of KIC, suggesting a sparing of nitrogen. The anabolic effects of insulin are well-known and the effects of triglyceride infused, in the words of Beaufrère *et al.* (1985) 'are consistent with the hypothesis that increased fatty acid oxidation ... may spare essential amino acids'. The rate of transamination may be driven by the plasma leucine concentration. Nissen and Haymond (1986) reported a significant correlation between plasma concentrations of leucine and KIC with their rates of interconversion, and said , '... this is consistent with known characteristics of transamination reactions, in which the rate of reaction is controlled solely by the concentrations of reactants and products'. This statement is not entirely correct, since Krebs and Lund (1977) reported that glutamate shifts the equilibrium away from keto-acid formation. In the studies of Nair *et al.* (1995) and Meek *et al.* (1998) there was a close parallel between the reductions following insulin treatment in both transamination rates and plasma leucine concentrations.

4.1.2 Oxidation

The next step in oxidation is irreversible dehydrogenation by the branched chain keto acid dehydrogenase (BCKD), which is common to all three BCAAs. The data in Table 4.1 show that, if the *in vitro* enzyme assays on human tissues are valid *in vivo*, the concentration of the transami-

Table 4.2. Transamination of leucine. From Matthews *et al.* (1981a).

	μmol kg^{-1} h^{-1}	
	Post-absorptive	Fed
Leucine carbon flux	84	153
Leucine nitrogen flux	199	273
Leucine oxidation	10.5	31.5
Leucine transamination to KIC	122	152
KIC transamination to leucine	111	120

nase is far greater than that of the dehydrogenase, so the changes described above in BCAT activity may not be so relevant.

The BCKD is a complex of three enzymes whose activity is regulated by a kinase and a phosphatase (Harris *et al.*, 1985). The human tissues on which the degree of activation in Table 4.1 was measured were obtained by biopsy in the course of various surgical procedures and probably represent the activation state in postabsorptive subjects at rest. BCKD activity is regulated in two ways. In animal experiments it has been shown to be reduced on low protein diets as a result of increased phosphorylation (Sketcher *et al.*, 1974; Harris *et al.*, 1985). In man an impressive correlation has been observed between the concentration of leucine in plasma and the rate of oxidation (Young *et al.*, 1985). The relationship holds up to leucine concentrations of 150–200 μM. Plasma KIC concentrations are usually about 1/3 of those of leucine, i.e. around 50 μM. In the rat, administration of leucine caused a rapid increase in activation of the BCKD complex (Aftring *et al.*, 1986). The k_M of the BCKD complex in rat liver has been stated to be 10–20 μM (Harper and Zapalowski, 1981), so that with high intakes of leucine or protein the response of enzyme kinetics has to be supplemented by an increase in the activation state of the enzyme. This might be an example of the combination of coarse and fine controls described by Krebs (1972).

A final issue in the metabolism of leucine is the specific effect that has been claimed for this amino acid in stimulating protein synthesis and reducing degradation. These effects are considered in Chapter 9.

4.2 Glycine

Glycine is important in turnover studies because of its role as a tracer, when labelled with [^{15}N], in the end-product method of measuring whole body turnover (Chapter 7).

4.2.1 Metabolism

The metabolism of glycine is comprehensively reviewed by Jackson (1991). When rats were infused with [^{14}C]-glycine the label rapidly appeared in serine and vice versa when they were infused with serine (Fern and Garlick, 1973, 1974). This reversible reaction is catalysed by the mitochondrial enzyme serine hydroxymethyl transferase. Glycine is also irreversibly broken down by the glycine cleavage system (Kikuchi, 1973), catalysed by a multi-enzyme complex similar to the α-keto acid dehydrogenase, which produces ammonia and CO_2 from the carboxyl carbon, while the methyl group combines with tetrahydrofolate to produce methylene tetrahydrofolate. Thus glycine is an important methyl donor. As with the BCDK complex (see section 4.1) a high protein diet stimulates activity of the glycine cleavage system (Ewart *et al.*, 1992). The depletion of the serine pool that would result from catabolism of glycine is counteracted by biosynthesis of serine from phosphoglycerate. The metabolism of serine is described in detail in a review by de Koning *et al.* (2003).

Some work has been done, beginning with that of Ratner *et al.* (1940), on the extent to which glycine transfers its N to that of other amino acids. When rats were given [^{15}N]-glycine and the liver analysed after varying intervals there was general agreement that serine was most highly labelled; after that alanine, glutamic acid, amide-N and arginine were all more or less equally enriched, followed by aspartic acid and tyrosine, with the rest nowhere (Ratner *et al.*, 1940; Aqvist, 1951; Vitti *et al.*, 1963). Similar results were obtained by Matthews *et al.* (1981b) on the distribution of [^{15}N] in the plasma free amino acid pool after feeding labelled glycine for 60 h. ^{15}N ammonium salts give the same kind of pattern but with the enrichments more dispersed: alanine > arginine > glutamic acid > amide N > aspartate > glycine > tyrosine (Aqvist, 1951; Vitti *et al.*, 1963, 1964). This general similarity of pattern could mean that the source of N derived from glycine by the other amino acids is mainly the ammonium produced by the glycine cleavage reaction. On the other hand, Arnstein (1952) found that glycine only sparingly enters into transamination reactions; on infusion of [^{15}N]-glycine no [^{15}N] was found in blood alanine or glutamic acid (Jackson and Golden, 1980). In his 1991 review Jackson concluded that '… the data fit in well with the older literature, which shows that there is specific channelling of amino groups'.

4.2.2 Glycine flux and *de novo* synthesis

Total glycine flux has been measured by constant intravenous infusion (Table 4.3). The extent of *de novo* synthesis has been determined by subtracting from the total flux the component derived from protein breakdown, measured independently with leucine. There is a wide variation in the estimates of flux, with even greater variations in *de novo* synthesis. Neither the administration of glucose (Robert *et al.*, 1982) nor the protein level of the diet (Yu *et al.*, 1985) made any difference. The fluxes were low in diabetes, with or without insulin (Robert *et al.*, 1985). The extremely high flux in the newborns is interesting, in view of the glycine that they need for rapid growth and collagen synthesis. There are a number of situations in which glycine seems to be limiting, so that, as Jackson (1991) has proposed, it should be regarded as a semi-essential amino acid; there is need for more measurements of the capacity for *de novo* synthesis.

4.2.3 Routes of glycine utilization

The main routes of glycine disposal are shown in Table 4.4. Widdowson *et al.* (1979) showed that growth of the newborn infant at birth involved the deposition of about 500 mg glycine a day, whereas 600 ml of breast milk would only provide about 200 mg. They commented: 'In view of its (glycine's) role as part of the molecule of glutathione, of the porphyrin ring of haemoglobin and of the nucleic acids perhaps particular attention should be paid to it in designing food for the low birth weight infant.' Two years later Jackson *et al.* (1981) showed that when pre-term babies fed breast milk were given ^{15}N-glycine, none of the label appeared in urinary urea, suggesting that all available glycine was being retained for synthesis of protein and other metabolites. Collagen in particular contains large amounts of glycine, so that when growth is rapid the need for it is high (Chapter 17).

After protein deposition, the next most important sinks for the irreversible disposal of glycine are creatinine and the purine bases. Creatine is synthesized from glycine + arginine; creatinine is formed by irreversible non-enyzmic dehydration of creatine phosphate and is excreted in a normal adult at the rate of 1.5–2 g day^{-1}. The rate is very constant in any individual and depends on the muscle mass. Uric acid excreted in the urine, at a rate of about 500 mg per day, is a degradation product of the purines. Out of four nitrogen atoms in the molecule, one is derived from glycine. Glycine also provides the nitrogen for the synthesis of haem. A normal rate of haemoglobin production, assuming a lifetime for the red cell of 100 days, would amount to about 150 mg Hb kg^{-1} day^{-1}, so this route of glycine consumption is not large. Glycine is also a scavenger of toxic substances, with which it forms conjugates such as hippuric acid when conjugated with benzoic acid.

Table 4.3. Estimates of glycine flux and *de novo* synthesis.

	Flux	*De novo* synthesis μmol kg^{-1} h^{-1}	*De novo* synthesis % of flux	Method
Normal adults				
Robert *et al.*, 1982	240	195	81	Constant infusion glycine + leucine
Yu *et al.*, 1985				
High protein diet	189	64	33.5	Constant infusion glycine + leucine
Low protein diet	165	59	35	Constant infusion glycine + leucine
Gersovitz *et al.*, 1980	458	351	77	60 h feeding; synthesis from protein breakdown
Lapidot and Nissim, 1980	260–430	–	–	[^{15}N] glycine single dose IV
Neonates				
Amir *et al.*, 1980	3020			Single dose IV

Table 4.4. Routes of irreversible loss of glycine.

	μmol kg^{-1} h^{-1}	
Net deposition of protein		
Premature infant	~200	a
Newborn infant	~100	b
Pregnancy, 3rd trimester	7	c
Lactation	11	c
Other routes of loss in adult		
Creatine	8–12	
Uric acid	~0.5	d
Haem	0.5	e
Erythrocyte glutathionine	~3	f
Urinary losses, ? as conjugates	3	g
Acute phase protein production	33 +	h

[a]Assuming protein gain = 2.6 g kg^{-1} day^{-1}.
[b]From Widdowson *et al.*, 1979.
[c]Protein gains from WHO, 1985.
[d]Assuming that one out of five N atoms is derived from glycine.
[e]Assuming that red cells have a life of 100 days.
[f]From Jackson, 2004.
[g]From Documenta Geigy, 1962: 529.
[h]From Reeds, 1994; routine surgery (probably much higher with severe injury).

Infection may also increase the requirement for glycine, because of massive synthesis of acute-phase proteins. Thus Reeds *et al.* (1994) calculated that during a typical response to uncomplicated surgery about 33 μmol kg^{-1} h^{-1} might be incorporated into acute phase proteins, and with severe injury or infection the amount will be much larger. Rough estimates of the rates of glycine consumption by these various routes are shown in Table 4.4.

Glycine is also a component, with glutamic acid and cysteine, of the tripeptide glutathione, which is present in all cells and is an important anti-oxidant. Most measurements of the synthesis rate of GSH have been made on erythrocytes. Jackson *et al.* (2004) and Lyons *et al.* (2000) report absolute synthesis rates (ASR) of ~2 mmol l^{-1} of erythrocytes day^{-1}. This reduces to an uptake of glycine of ~3 μmol kg^{-1} h^{-1}. GSH concentrations in red blood cells were found to be low in children with severe oedematous malnutrition (Jackson, 1986; Golden and Ramdath, 1987) and the synthesis rates are also reduced in these cases (Reid *et al.*, 2000). They are low too in burns injuries (Yu *et al.*, 2002), and even in normal people on moderately low protein intakes (Jackson *et al.*, 2004).

These findings in children led Jackson to the hypothesis that the low levels of GSH and its synthesis might result from glycine insufficiency, and a test for it was devised (Jackson *et al.*, 1987). The rationale is shown in Fig. 4.1. If glycine is lacking the enzyme glutathione synthase will be less active: γ-glutamylcysteine accumulates, is degraded to L-5-oxoproline and excreted. An increase in oxoproline excretion therefore indicates insufficiency of glycine. Such an increase compared with controls has been found in many situations, listed in Table 4.5. Young infants and people with infections or trauma are particularly vulnerable.

In one experimental study on volunteers a glycine-free or a sulphur-amino-acid-free diet produced barely significant increases in oxoproline excretion (Metges *et al.*, 2000).

Glycine fluxes were not measured in the conditions listed in Table 4.5, but it may seem paradoxical that a flux that in normal subjects is ~200 μmol kg^{-1} h^{-1}, should be so reduced that it cannot support glutathione synthesis at a rate of ~3 μmol kg^{-1} h^{-1}. The solution seems to come from an old study of Henriques *et al.* (1955) on rabbits, in which it was found that the synthesis rate of GSH in liver was ~600 μmol kg^{-1} h^{-1}, and in muscle about 30. If the rates are of the same order of magnitude in man, it is clear that

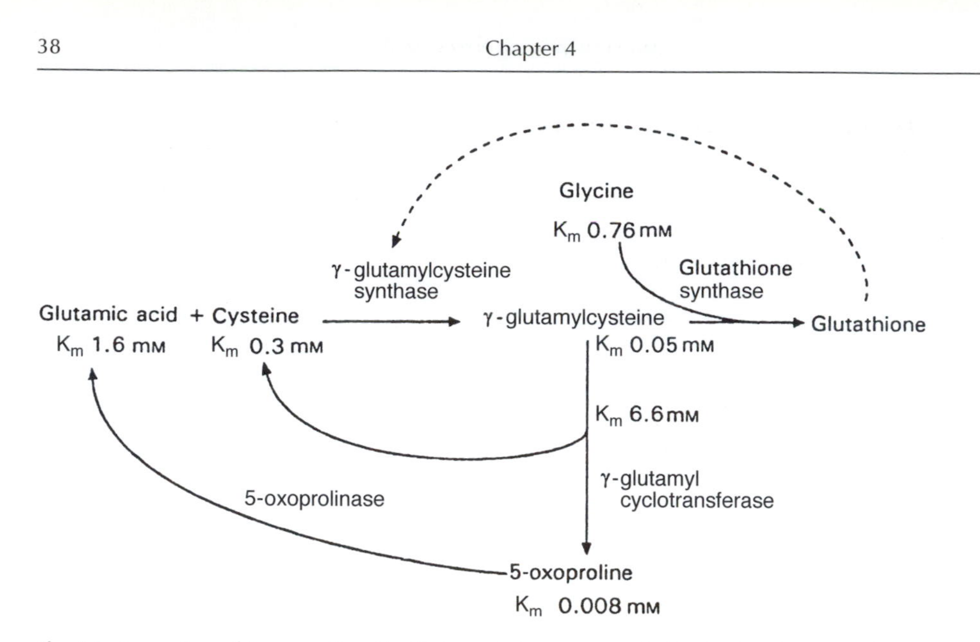

Fig. 4.1. Formation of 5-oxoproline. If sufficient glycine is not available for glutathione synthesis, γ-glutamylcysteine accumulates and is converted to 5-oxoproline, which is excreted in the urine. Reproduced from Jackson *et al.* (1997b), by courtesy of the *British Journal of Nutrition.*

glutathione synthesis is a major drain on glycine flux and supply, unless the glycine released in the production of oxo-proline is re-utilized.

4.3 Alanine

4.3.1 Metabolism

Alanine is at the cross-roads of protein and energy metabolism. Studies by arterio-venous cannulation showed that in the post-absorptive state PA, although there is a net release of virtually all amino acids from muscle, alanine and glutamine predominate; alanine, like glutamine, is a nitrogen carrier (Felig, 1975). Most of the alanine released is synthesized *de novo*; the nitrogen from the branched-chain amino acids, the carbon-skeleton from pyruvate. Some results in humans are shown in Table 4.6.

According to Perriello *et al.* (1995), in humans about half the alanine flux is derived from muscle. Experiments by Ben Galim *et al.* (1980) on dogs showed that ~50% of the leucine N flux was transferred to alanine. About 20% of the alanine N flux was derived from leucine; if the same proportion applies to isoleucine and valine, it follows that almost all the N of newly synthesized alanine comes from the branched-chain amino acids.

Studies with [^{13}C]-alanine have shown that about 90% of alanine carbon is oxidized, about half after conversion to glucose (Consoli *et al.*, 1990; Battezatti *et al.*, 1999). The splanchnic uptake of alanine on first pass is very high, about 70% of the

Table 4.5. 5-L–oxoproline excretion in various conditions.

Subjects	Oxoproline excretion as per cent of controls	References
Normal adults, vegetarians on low protein diets	220	Jackson *et al.*, 1996
Premature infants	153	Jackson *et al.*, 1997a
Infants (6 weeks) in Jamaica	275	Persaud *et al.*, 1997
Infants (6 weeks) in Trinidad	222	Lenton *et al.*, 1998
Late pregnancy, Southampton	310	Jackson *et al.*, 1997b
Jamaica	595	
Type II diabetics	380	Forrester *et al.*, 1990

Table 4.6. Alanine flux and *de novo* rate of synthesis.

		Flux	*De novo* synthesis µmol kg^{-1} h^{-1}	*De novo* synthesis as per cent of flux	Tracer
Robert *et al.*, 1982	PA	381	310	81	^{2}H
	+ glucose	470	408	87	–
	PA	317	219	69	^{15}N
	+ glucose + insulin	344	280	81	–
Royle *et al.*, 1982	PA	351	–		^{13}C
	+ glucose	457			^{13}C
Miles *et al.*, 1983	PA	329	228	69	? ^{15}N or ^{2}H
	+ βOHB	462	354	77	
Robert *et al.*, 1985	Diabetics PA	345	258	75	^{2}H
	+ insulin	330	264	80	
Yu *et al.*, 1985	PA 1.5 g protein	274	160	58	^{15}N
	0.6 g protein	264	164	62	
Hoffer *et al.*, 1985	PA	402	320	80	^{2}H
	Fed	994	883	89	
Pierello *et al.*, 1995	PA	274	–		^{13}C
Yang *et al.*, 1986	PA high protein	256	136	53	^{15}N
	low protein	385	294	76	and ^{2}H
	high energy	369	264	72	
	low energy	195	98	50	

flux, according to Battezatti *et al.* (1999). Thus *de novo* alanine production in muscle serves not only nitrogen transport but also fuel transport from the periphery to the splanchnic region.

As for the disposal of alanine-N, Jeevanandam *et al.* (1979) concluded from a single dose given IV, that alanine-N is 'rapidly transaminated to other amino acids … ' and that the excretory data of urea and NH_3 after 1 h '… represents the characteristics of the overall metabolic pool of N and does not reflect the metabolism of alanine only'. Thus they confirmed the finding of Aqvist (1951) that 3 days after giving [^{15}N] alanine there was substantial enrichment of the amino acids of liver protein, that of glutamic acid being nearly equal to that of alanine, followed in descending order by aspartic acid, the BCAAs, arginine, glycine and serine. On this basis Jeevanandam (1981) suggested that [^{15}N] alanine would be better than [^{15}N] glycine as a tracer for measuring whole body turnover by the end-product method (Chapter 7).

4.3.2 Flux and *de novo* synthesis

The extent of *de novo* synthesis can be determined, as with glycine, by subtracting the amount derived from protein breakdown, measured with labelled leucine, from the total alanine flux.

Early measurements of alanine flux encountered a difficulty: substantial differences were found between fluxes measured with [^{15}N]-, [^{13}C]- and [^{2}H]-alanine (Matthews *et al.*, 1985). The authors reasoned that 'upon transamination the alanine N becomes incorporated into glutamate and the carbon skeleton becomes a pyruvate … we conclude that … the flow of carbon through pyruvate is considerably faster than the flow of nitrogen through the transaminating pool.' The high values given by [^{2}H]-alanine were attributed to rapid removal of the deuterium atoms during or after transamination, as had been demonstrated by studies on transaminating enzymes *in vitro*. It seems, therefore, that an estimate of *de novo* synthesis will depend in part on the tracer used for measuring alanine flux.

There is a little information about factors that modify *de novo* alanine synthesis. Yang *et al.* (1986) measured it in the post-absorptive state (12 h after the last meal) in subjects who had previously consumed for 1 week diets providing different levels of protein (1.5 to 0 g kg^{-1} day^{-1}) or energy (140–240 kJ kg^{-1} day^{-1}). Alanine flux, measured with both [^{15}N] and [^{2}H]-alanine, and *de novo* synthesis were higher with protein intakes below the level of requirement, but the energy intake had an even greater effect; *de novo* synthesis in the low energy intake group was only about 40% of that on the high intake. In general, with low fluxes the proportion of *de novo* synthesis was lower. Regardless of the tracer used, there was a highly significant linear correlation between the rate of *de novo* synthesis and the plasma alanine concentration. This, perhaps, was the signal that carried a message persisting after 12 h fasting.

There is also more direct information about the effect of increasing energy supply. Alanine flux and *de novo* synthesis increased when a mixed meal was fed every hour (Hoffer *et al.*, 1985) or with infusion of glucose (Robert *et al.*, 1982; Royle *et al.*, 1982) or of ketone bodies (Miles *et al.*, 1983), but not by the hyperglycaemia of diabetes (Robert *et al.*, 1985). Flux and *de novo* synthesis were also increased by cortisol infusion (Darmaun *et al.*, 1988). It is particularly interesting that infusions of glucose, while increasing alanine flux decreased the production of urea (Royle *et al.*, 1982), an observation that has been confirmed in sheep (Obitsu *et al.*, 2000).

This recalls the long established finding, that increasing the energy supply improves nitrogen balance and decreases urinary N excretion by 1–2 mg per extra kilocalorie (Calloway and Spector, 1954) (see Chapter 10). Royle *et al.* (1982) showed that reduced urea production during infusion of glucose was not due to decreased consumption of alanine for gluconeogenesis. As far as I know this sparing effect of energy has not been explained in biochemical terms. Alanine indeed stands at a crossroads.

4.4 Glutamine

4.4.1 Metabolism

In 1980 Hans Krebs wrote 'Most amino acids have multiple functions, but glutamine appears to be the most versatile' (Krebs, 1980). It is also by far the most abundant free amino acid in the body, with a concentration in the free pool of human muscle of about 17 mmol l^{-1}, some 30 times greater than in plasma (Bergström *et al.*, 1990). As much glutamine is stored in muscle as there is glycogen in liver (Newsholme and Calder, 1997). Perhaps these two attributes are linked: if a substance is likely to be needed at short notice in an emergency, there must be an adequate store of it.

Thirty years ago it was shown by Cahill and co-workers, by measurement of arterio-venous differences, that in the post-absorptive state muscle releases large amounts of glutamine and alanine, which have received their nitrogen by transamination, mainly from the branched chain amino acids (Marliss *et al.*, 1971; Felig, 1975). Quantitatively, glutamine is more important than alanine (Nurjhan *et al.*, 1995). Adipose tissue also releases glutamine, where, as in muscle, it is synthesized from glutamate (Elia, 1991). These nitrogen carriers are transported to the splanchnic bed, where they are metabolized as sources of fuel, glutamine mainly in the gastrointestinal mucosa, alanine in the liver (Souba *et al.*, 1985). When glutamine is given by nasogastric tube, 55–75% of it is sequestered in first pass by the splanchnic tissues (Matthews *et al.*, 1993b; Mittendorfer *et al.*, 2001); about 80% of this uptake is oxidized in the splanchnic region (Matthews *et al.*, 1993b; Haisch *et al.*, 2000). Heitman and Bergmann (1978) in a classical study showed that glutamine production by the viscera is even greater than by muscle; the difference between them is that there is net consumption by the viscera and net production by muscle (Table 4.7). The kidneys also produce glutamine, and in acidosis consumption exceeds production, because glutamine is the main source of the increased output of urinary ammonia (Pitts and Pilkington, 1966). Nevertheless, the extra ammonia generated only takes up a small proportion (2–4%) of the glutamine flux (Golden *et al.*, 1982; Waterlow *et al.*, 1994).

Newsholme *et al.* (1985) were the first to point out the important role of glutamine in the metabolism of cells of the immune system, lymphocytes and macrophages, which are said to account for 1.4 kg in weight in the average human (Newsholme and Calder, 1997). Since these cells use the carbon skeleton of glutamine as a fuel, and the nitrogen for the synthesis of RNA and DNA (Martin, 1987; Gate *et al.*, 1999) as well as of protein, their proliferation and production of antibodies must depend on an adequate supply of this substrate. These observations have stimulated a vast number of clinical studies in recent years, and it is natural that there should be an increasing interest in the kinetics of glutamine.

Table 4.7. Uptake and release of glutamine in tissues of the sheep. From Heitmann and Bergmann (1978).

	Uptake	Release	Net
Viscera + liver			
Normal, fed	9.4	6.4	+3.0
Fasted	8.0	4.6	+3.4
Peripheral tissues			
Fed	2.4	4.7	–2.3
Fasted	2.3	6.0	–3.7
Kidneys			
Fed	0.5	1.2	–0.7
Fasted	1.2	0.9	+0.3

Data are mmol h^{-1}, per whole animal (50 kg).

Liver contains both glutaminase and glutamine synthetase, the former being mitochondrial, the latter cytoplasmic. In the liver there is a particularly interesting situation because the two enzymes not only occupy separate compartments within a cell but are expressed in different cells. Häussinger (1987) has shown that the periportal cells of the liver lobule contain glutaminase as well as the enzymes of the urea cycle; this represents a low affinity–high capacity system that deals with most of the ammonia derived from glutamine coming to the liver. Downstream are the perivenous cells, about 7% of total hepatocytes, which contain glutamine synthetase, a high affinity system that acts as a scavenger of any ammonia escaping urea synthesis. Thus there is a cooperative activity of these two enzymes.

4.4.2 Flux and *de novo* synthesis

In 1986 Darmaun *et al.* published a study in which they infused for 4 h glutamine labelled with ^{15}N in either the amide (5–N) or the amino (2–N) nitrogen; [^{15}N] glutamate was also infused. With both glutamine tracers a plateau seemed to have been achieved after about 4 h. These were unprimed infusions and plasma was sampled every 15 minutes during the rise to plateau. This allowed calculation both of the flux, from the plateau labelling, and of the apparent rate coefficient, k′, from the rising part of the curve. Since Q (flux) = k′ P, the pool size could be determined. The results, already shown in Table 3.9, are repeated in Table 4.8.

It is evident that the calculated pool sizes are far smaller than the measured ones, and Darmaun *et al.* (1986) suggested that there must be intracellular compartmentation, with a large pool that does not mix readily with plasma. Very similar results were obtained by Kreider *et al.* (1997), with plateaux being reached in 2–3 hours. In a later study, with improved methods of measurement, it was found that even after 11 h of infusion the enrichments of [^{15}N]- and [^{13}C]-glutamine in muscle were slowly rising, at rates of 0.040 and 0.028 per hour (van Acker *et al.*, 1998). These increases were too small to be distinguishable from a zero slope with the gas chromatograph–mass spectrometers available earlier. Moreover, in these latter experiments enrichment of glutamine in the muscle free pool was very low and

Table 4.8. Estimated and measured sizes of free glutamine and glutamate pools. Data of Darmaun *et al.* (1986).

μmol kg^{-1}	Glutamine		Glutamate	
	EC	IC	EC	IC
Measured	110	5200[a]	7	1240[b]
Estimated	207		17	

[a]Muscle only.
[b]Muscle + liver.
Estimated pool sizes calculated from the observed rate coefficients (see text).

rose steadily throughout the 11 h of infusion, by which time it was still only 1/5 of that in plasma. This work confirms the evidence for a large very slowly mixing pool. As Matthews has said, the glutamine flux is an operational flux, a measure of the extent of inter-organ transfer. Perhaps a more accurate term for it would be glutamine *release* rather than flux, and the glutamine pool could be regarded as a store.

Since that early study glutamine flux has been measured many times with a variety of tracers. The results (Table 4.9) are reasonably consistent and show the same small differences with different labels in different positions in the molecule as have been found with alanine (q.v.), with 5-^{15}N giving the lowest rate and ^{3}H the highest. The reasons are discussed in detail by Matthews *et al.* (1985).

As with glycine and alanine, *de novo* synthesis has been determined by subtracting from the total flux the amount contributed by protein breakdown. In normal subjects *de novo* synthesis accounts, by this method, for 60–80% of the glutamine flux (Darmaun *et al.*, 1988; Matthews *et al.*, 1992, 1993b; Hankard *et al.*, 1995). With a different approach, based on arterio-venous differences in the leg and muscle biopsies, it was found that about 60% of the total glutamine flux through muscle was provided by *de novo* synthesis (Biolo *et al.*, 1995; Mittendorfer *et al.*, 2001), so the two methods agree quite well. *De novo* synthesis was not affected by the level of protein intake (Matthews *et al.*, 1992) but, as might be expected, was decreased by infusions of glutamine (Hankard *et al.*, 1995, 1998; Claeyssens *et al.*, 2000). Infusions of cortisol in normal humans increased glutamine flux and *de novo* synthesis (Darmaun *et al.*, 1988), but in patients with burns and hypercortisolaemia there were large decreases in both intramuscular glutamine concentration and *de novo* synthesis (Biolo *et al.*, 2000).

These results raise two questions: the first is the relationship between the huge IC pool in muscle, the much smaller EC pool and the flux between them. Because the EC pool is only about 1/50 of the IC pool, even large changes in the EC glutamine concentration will have little effect on the IC concentration. Input to the IC pool from plasma is small (Biolo *et al.*, 1995) and it is difficult to raise the glutamine concentration in it by infusions of glutamine (Zachwieja *et al.*, 2000) in spite of the existence of a very active inward transporter described by Rennie *et al.* (1994). Therefore it seems probable that the main determinant of plasma glutamine concentration is, as Newsholme suggested, the rate of efflux from the IC pool. This efflux is enhanced by glucocorticoids (Rennie *et al.*, 1986), which increase the glutamine concentration in plasma and decrease it in muscle (Muhlbacher *et al.*, 1984). Whether glucocorticoids have a direct action on the outward transporter or its synthesis seems not to be known.

In muscle the major input of glutamine to the

Table 4.9. Estimates of glutamine flux in humans with different tracers.

Flux, μmol kg^{-1} h^{-1}			
Tracer	Mean	No.	Range
5-[^{15}N]	234	2	185–283[a,b]
2-[^{15}N]	318	5	280–348[a,c,d,e,f]
^{13}C or ^{14}C	352	7	285–373[f,g,h,i]
^{3}H	380	2	368–393[f,g]

[a]Darmaun *et al.*, 1986
[b]Van Acker *et al.*, 1998
[c]Darmaun *et al.*, 1988
[d]Matthews *et al.*, 1992
[e]Matthews *et al.*, 1993b
[f]Kreider *et al.*, 1997
[g]Haisch *et al.*, 2000
[h]Hankard *et al.*, 1995
[i]Nurjhan *et al.*, 1995
Note: Studies by Golden *et al.* (1982) and Waterlow *et al.* (1994) with 5-[^{15}N] glutamine gave fluxes of 743 and 600 μmol kg^{-1} h^{-1}, but are not included here because, uniquely, glutamine was estimated enzymically.

IC pool is by *de novo* synthesis, which depends on the balance between glutamine synthase and glutaminase. According to Rennie *et al.* (1994) in rat muscle, in contrast with liver and kidney, the synthase is about 50 times as active as the glutaminase. There is some evidence that the synthesis is affected by outside factors; when Hankard *et al.* (1995) gave enteral infusions of large amounts of glutamine, plasma glutamine concentrations were doubled and *de novo* synthesis depressed by 40%. On the other hand, in the study of Muhlbacker *et al.* (1984) cortisol decreased the glutamine concentration in muscle with no effect on glutamine synthase or glutaminase. The free glutamine pool in muscle is greatly depleted in sepsis, trauma and other disease states (Rennie *et al.,* 1986), presumably as a result of glucocorticoid action, so the first question is whether the effect is primarily on the outward transporter or on glutamine synthase. The answer of Newsholme *et al.* (1990) is quite clear; they pointed out that the outward glutamine transporter operated at a K_m far below the normal intracellular glutamine concentration, so that it was always saturated and the concentration would be irrelevant. In terms of metabolic control theory, Newsholme regarded efflux as the flux-generating step in glutamine metabolism. The hypothesis, therefore, is that adaptations of glutamine metabolism in conditions of stress depend neither on changes in enzyme activities nor in plasma or muscle concentrations but on regulation of the outward transporter by nutrients and hormones.

The second question is whether the depletion of the muscle glutamine pool matters. In rats on diets containing different amounts of protein or with endotoxaemia (Jepson *et al.*, 1986) and in perfusions of the hind-limb with different concentrations of glutamine (MacLennan *et al.*, 1987) correlations were found between the free glutamine concentration in muscle and the rate of muscle protein synthesis, and also the concentration of RNA. Jepson admitted that the relation might be coincidental rather than causal and this indeed seems to be the case. Adding glutamine or alanyl-glutamine to total parenteral nutrition was claimed to reduce depletion of the glutamine pool and the fall in muscle protein synthesis (Hammerqvist *et al.*, 1989, 1990). However, when Olde Damink *et al.* (1999) gave an inhibitor of glutamine synthase to rats they found that in spite of severe depression of plasma and muscle glutamine concentrations, there was no effect on whole body or muscle protein synthesis.

In the last decade there has been intense activity to determine whether supplements of glutamine have any beneficial clinical effects, from two points of view: whether they help to support the activities of the immune system and the gut; and whether by repletion of the glutamine pool it is possible to reduce nitrogen loss and hasten recovery in injured or septic patients (see Chapter 13). Useful reviews are those of Griffiths *et al.* (1997), Wilmore (2002), Neu *et al.* (2002) and the American Society for Nutritional Sciences (2001).

4.5 Glutamic acid

Krebs' remark about glutamine applies, perhaps even more forcibly, to glutamic acid. Young and Ajami (2000) described glutamate as 'An amino-acid of particular distinction'. This description has been borne out by network analysis applied to the metabolism of *Escherichia coli*, which places glutamic acid at the centre of the metabolic web, with the highest degree of connectivity to all other reactants (Fell and Wagner, 2000; Wagner and Fell, 2001).

Young and Ajami (2000) show how the remarkable reactivity of glutamic acid is related to its structure; five carbon atoms give it properties which are denied to aspartic acid, with only four. The main reactions are: dehydrogenation to α-ketoglutaric acid and NH_3; transamination to many amino acids, whereas aspartate transaminates only with glutamate; a precursor of glutamine and an intermediate in the citric acid cycle. Young and Ajami also say that glutamic acid is the most abundant amino acid in the α-helices of proteins and '... modulates their hydration state ... in step with the physiological processes that require proteins to do work by folding and unfolding repeatedly'. This repeated change of state could be an important factor in the breakdown of proteins (see Chapter 2).

The concentration in plasma and the size of the free pool are much smaller, by a factor of 5–10, for glutamate than for glutamine (Matthews *et al.*, 1992, 1993b). As with glutamine the miscible pool of glutamate is very much smaller than the measured pool; in fact, the discrepancy is greater for glutamate (Table 4.8). The calculations of Darmaun *et al.* (1986) show that the miscible pool, at 17 μmol kg^{-1}, is nearly half the size of the IC pool of liver, but only about 1.4% of the IC pool of muscle. In a later paper Matthews *et al.* (1993b) concluded that 'glutamate is very tightly compartmentalized within cells and that most glutamine is synthesized from glutamate, which mixes very slowly with extracellular glutamate'. It seems likely that by far the greatest part of this tightly compartmented pool is in muscle. The liver pool is much smaller, and probably exchanges freely with glutamate entering from plasma. That such exchanges occur very rapidly is shown by the experiments of Cooper *et al.* (1988) with the very short half-life isotope ^{13}N. They showed that 60 seconds after injection of [^{13}N]-glutamate into the portal vein of rats, 21% of the tracer was found in aspartate, 14% in glutamine and 10% in alanine. Similarly, Matthews *et al.* (1993b) showed that 88% of [^{15}N] glutamate given by nasogastric tube was sequestered on first pass through the splanchnic bed, and that substantial amounts of the label appeared in the BCAs, alanine, proline and glutamine. The reverse reaction, glutamine → glutamate, was also occurring in the liver, in agreement with Haüssinger's demonstration (1990) that glutamine synthetase and glutaminase both operate in the liver, although in separate cells. It has also been found with [^{13}C] glutamate that almost 80% of the sequestered glutamate was oxidized (Battezzati *et al.*, 1995).

The *de novo* synthesis of glutamate can also be calculated in the same way as for alanine and glycine (*vid.sup*). Some results are shown in Table 4.9. The effect of a high protein diet was to decrease the rate of *de novo* synthesis.

The glutamate flux in plasma is smaller than that of glutamine and much of it is derived by transamination in tissues other than muscle; thus whereas we can regard the glutamine flux as being mainly concerned with inter-organ transfer of N, that of glutamate can be seen as the end-result of a vast number of metabolic exchanges in a variety of tissues.

4.6 Phenylalanine

The first step in the catabolism of phenylalanine is hydroxylation to tyrosine. The hydroxylase occurs mainly in the liver, but it is also active in the kidney (Møller *et al.*, 2000). Numerous studies of phenylalanine metabolism were made in the 1960s and 1970s in patients with the genetic condition phenylketonuria (PKU), usually with loading doses of phenylananine and bolus doses of tracer. In homozygous PKU the hydroxylase is deficient or absent. Clarke and Bier (1982) were the first to apply the constant infusion method to determine phenylalanine fluxes and hydroxylation rates in normal subjects. They measured the phenylalanine flux with ring [$^2H_5$] phenylalanine and showed that the rate of hydroxylation to tyrosine could be calculated as:

$$Qp \rightarrow t = Q_t \times \frac{\varepsilon[^2H_4]\ tyr}{\varepsilon\ [^2H_5]\ phe}$$

where Q_t is the tyrosine flux and the enrichment ratio gives the proportion of the tyrosine flux derived from the deuterated phenylalanine. The method therefore requires an independent measure of the tyrosine flux, for which Clarke and Bier (1972) used [1–^{13}C] tyrosine, and later workers [$^2H_2$] tyrosine (Marchini *et al.*, 1993; Price *et al.*, 1994; Sanchez *et al.*, 1995, 1996). Another way round the problem is simply to assume that the flux of tyrosine bears the same relation to phenylalanine as their concentrations in body protein (Thompson *et al.*, 1989; Cortiella *et al.*, 1992; Price *et al.*, 1994). Shortland *et al.* (1994) found good agreement between this indirect estimate and the directly measured tyrosine flux.

A further difficulty is that the phenylalanine and tyrosine fluxes are measured from their enrichment in plasma, with no allowance for intracellular dilution; it was therefore suggested that the same dilution factor should be applied as for KIC:leucine (Thompson *et al.*, 1989; Price *et al.*, 1994). The correction gives flux ratios of phenylanaline to tyrosine that correspond better with their molar ratios in body proteins (Price *et al.*, 1994).

Fluxes obtained with different phenylalanine tracers – [$^2H_5$], [^{13}C] and [^{15}N] – agreed well (Krempf *et al.*, 1992; Marchini *et al.*, 1993). However, this agreement does not extend to hydroxylation, as shown in Table 4.10. It was suggested that the five deuterium atoms in ring-[$^2H_5$] may interfere with the activity of the hydroxylase enzyme; extensive references are given in Marchini's paper to the possible effects of deuterated substrates on other enzyme reactions.

Table 4.10. Phenylalanine hydroxylation measured with $[^2H_5]$ and $[1\text{-}^{13}C]$ phenylalanine. From Marchini *et al.* (1993).

		μmol kg^{-1} h^{-1}	
		Flux	Hydroxylation
Fasted:	$[^2H_5]$	45	5.1
	$[^{13}C]$	47.5	11.1
Fed:	$[^2H_5]$	56	6.8
	$[^{13}C]$	60.5	12.7

Measurements of phenylalanine turnover have been useful in two practical situations. The first is the estimation of human requirements for the aromatic amino acids (Zello *et al.*, 1990; Sánchez *et al.*, 1995, 1996; Basil-Filho *et al.*, 1998). (See Chapter 10.)

The second derives from the fact that phenylalanine is not oxidized in muscle, so that in studies in the forearm or leg net disposal of phenylalanine represents synthesis into protein, with no necessity to take account of oxidation. Calder *et al.* (1992) also made the point that ring-$[^2H_5]$phenylalanine can be measured by gas chromatography–mass spectrometry (GCMS) with such sensitivity that enrichments in plasma and muscle protein can be determined with the same instrument.

4.7 Arginine

Arginine is considered to be a semi-indispensable amino acid for humans, and its synthesis is a good example of inter-organ cooperation. It is synthesized from citrulline, mainly in the kidney (Dhanakoti *et al.*, 1990), and citrulline is synthesized in the liver and gastro-intestinal tract by a bio-synthetic pathway that starts with glutamate (Bender, 1985). Arginine is of particular interest for two reasons: it plays a major role in the ornithine cycle of urea formation, and it is a substrate for the production of nitric oxide (NO), which is important as a smooth muscle relaxant, a neurotransmitter and a product of immune reactions (Moncada, 1999).

The kinetics of arginine turnover have been worked out in an impressive series of studies by Young's group at MIT. They labelled arginine in three different ways: [guanidino-^{13}C]; [guanidino-$^{15}N_2$]; and [guanidino-$^{15}N_2$, 5,5-$^2H_2$]. In the third case, the deuterium atoms are attached to the carbon atoms of the non-guanidino part of the molecule. These three tracers all gave similar estimates of the plasma arginine flux, which means that there is no separate recycling of the guanidine part of the molecule.

The plasma arginine flux in the fed state is about 70–80 μmol kg^{-1} h^{-1} (Castillo *et al.*, 1996). *De novo* synthesis from citrulline in the kidney is only about 10% of this flux, and thus makes a much smaller contribution than *de novo* synthesis does to the fluxes of other non-essential amino acids such as glycine and alanine. Moreover, the rate of urea production, at ~300 μmol kg^{-1} h^{-1}, is some four times higher than the plasma arginine flux. Only 5% of urea produced is derived from plasma arginine, which, therefore, is almost totally excluded from the hepatic cells that are the site of urea synthesis. This barrier is attributed to the low activity in hepatic cells of the CAT (formerly y+) system for the transmembrane transport of cationic amino acids (Castillo *et al.,* 1993a). In the face of this, it seems paradoxical that the first-pass uptake of arginine by the splanchnic tissues was 33–48% (Castillo *et al.*, 1993b), considerably higher than that of leucine or phenylananine (Chapter 6); moreover, a study of Lobley *et al.* (1996a) in sheep showed hepatic uptake of arginine to be similar to that of other amino acids.

When an arginine-rich and an arginine-free diet were compared, there was no difference between the endogenous arginine fluxes but the rates of conversion of arginine to ornithine and of ornithine to citrulline were greatly reduced on the arginine-free diet, at least in the fed state, as was the rate of ornithine oxidation (Castillo *et al.*, 1993c, 1994a,b). The conclusion drawn was that adaptation to the arginine-free diet did not involve any increase in *de novo* synthesis but was achieved by suppression of arginine oxidation, perhaps triggered by a reduced plasma arginine concentration.

Arginine is also the substrate of a reaction that produces nitric oxide (NO) and citrulline in equimolar amounts (Moncada, 1999). From measurement of the rate of conversion it was estimated that NO was produced from arginine at the rate of ~1 μmol kg^{-1} day^{-1} (Castillo *et al.*, 1996). There was excellent agreement with an independent method, measurement of the urinary excretion of nitrite/nitrate. This synthesis accounted for ~1% of the plasma arginine flux.

This body of work, although described only in outline here, is an excellent example of how tracers can provide quantitative estimates of flows through various pathways of amino acid metabolism.

4.8 Methionine

Methionine, apart from being an essential amino acid, is interesting for other reasons. It is a methyl group donor and according to Lobley *et al.* (1996b) over 100 methylation reactions have been described, the methylation of DNA being obviously of particular importance. It is a precursor of homocysteine, which is attracting much interest as a risk factor for cardiovascular disease: and it plays an essential role in the biosynthesis of cysteine, a component of the anti-oxidant tripeptide glutathione.

An outline of methionine metabolism is shown in Fig. 4.2. There are two branch-points: the first is where the methionine flux is partitioned either to protein synthesis or, by transmethylation, to homocysteine (pathway TM). At the second branch-point homocysteine may be remethylated to methionine (pathway RM) or converted by condensation with serine to cystathionine. The final irreversible step is splitting of cystathionine to cysteine + ketobutyrate. This step is referred to either as oxidation or transsulphuration (TS).

In a pioneer study Storch *et al.* (1988) infused $[1-^{13}C]$ and [methyl $^{2}H_3$] methionine to obtain values for the methionine carbon flux and the methionine methyl flux. Oxidation/transsulphuration (TS) was derived from the output of $^{13}CO_2$. From these data the extent of transmethylation (TM) and remethylation (RM) could be calculated.

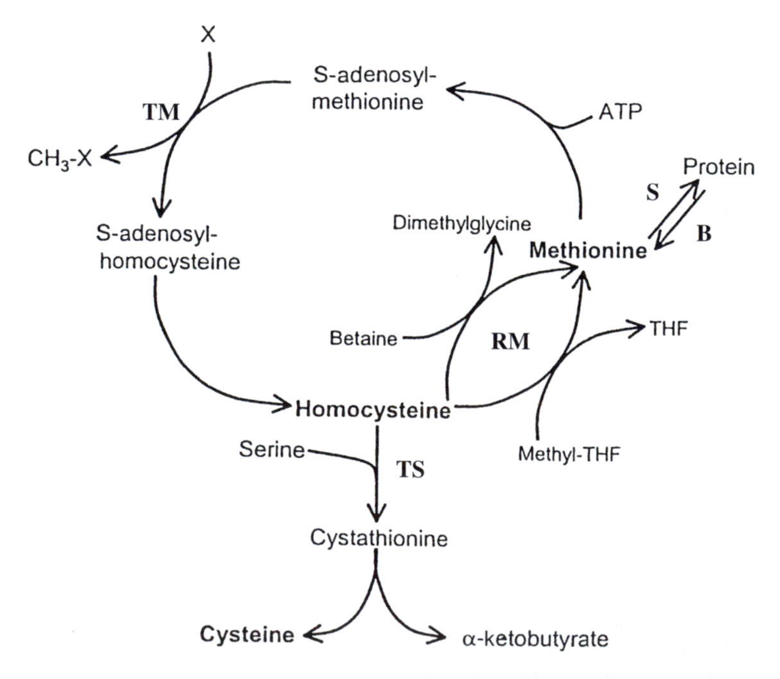

Fig. 4.2. Pathways of methionine metabolism in humans. Methionine may be taken up into protein (S) or undergo transmethylation (TM) to homocysteine + a methyl-group acceptor (TM). Homocysteine may be remethylated to methionine (RM) by two different pathways; or its sulphur atom may be transferred to form cysteine and its carbon skeleton oxidized (TS). Reproduced from MacCoss *et al.* (2001), by courtesy of the *American Journal of Physiology*.

If Q_c is the carbon flux and Q_m is the methyl flux,

$$Q_c = S + TS; Q_m = S + TM;$$

$$Q_m - Q_c = TM - TS = RM; RM + TS = TM$$

Storch *et al.* (1988) assumed, by analogy with leucine, that the precursor enrichment of methionine could be taken as 0.8 × the plasma enrichment. In a subsequent study MacCoss *et al.* (2001), using the same tracers, determined the precursor:plasma enrichment ratio from the enrichment of [^{13}C]-homocysteine or [^{13}C]-cystathionine, argued that these metabolites bore the same relation to their progenitor as KIC does to leucine. The ratio they found was 0.58; the values obtained are shown in Table 4.11. Such measurements, in subjects with raised plasma homocysteine concentrations, would reveal whether the cause is increased production or decreased disposal of homocysteine.

Some disturbances of these fluxes have been recorded. The transmethylation pathway is inhibited by supplements of creatine and choline (Lobley *et al.*, 1996b). On a methionine-free diet transmethylation was reduced to ~ $^1/_4$ of the rate on a normal intake, and of the small amount of homocysteine formed only ~ $^1/_3$ went on to the TS pathway (Storch *et al.*, 1990). There was thus considerable conservation of methionine.

On diets containing no sulphur amino acids, the synthesis of glutathione is reduced (Lyons *et al.*, 2000) as it is also in some pathological states – malnourished children (Jackson, 1986), patients with burns (Yu *et al.*, 2001) or critically ill patients (Hammarqvist *et al.*, 1997).

4.9 References

Aftring, R.P., Block, K.P. and Buse, M.G. (1986) Leucine and isoleucine activate skeletal muscle branched-chain α-keto acid dehydrogenase *in vitro*. *American Journal of Physiology* 250, E599–604.

American Society for Nutritional Sciences (2001) Glutamine metabolism: nutritional and clinical significance. *Journal of Nutrition* 131, 2447S–2596S.

Amir, J., Reisner, S.H. and Lapidot, A. (1980) Glycine turnover rates and pool sizes in neonates as determined by gas chromatography–mass spectrometry and nitrogen 15. *Pediatric Research* 14, 1238–1244.

Åqvist, S.E.G. (1951) Metabolic relationships among amino acids studied with isotopic nitrogen. *Acta Chemica Scandinavica* 5, 1046–1064.

Arnstein, H.R.V. (1952) The metabolism of glycine. *Advances in Protein Chemistry* 9, 1–91.

Basil-Filho, A., Baumier, L. and El Khoury, A. (1998) Twenty-four-hour L-[1-^{13}C] tyrosine and L[3,3-$^2H_2$] phenylalanine and tracer studies at generous, intermediate and low phenylalanine intakes to estimate aromatic amino acid requirements in adults. *American Journal of Clinical Nutrition* 67, 640–659.

Battezzati, A., Brillon, D.J. and Matthews, D.E. (1995) Oxidation of glutamic acid by the splanchnic bed in humans. *American Journal of Physiology* 269, E269–276.

Battezatti, A., Haisch, M., Brillon, D.J. and Matthews, D.E. (1999) Splanchnic utilization of enteral alanine in humans. *Metabolism* 48, 915–921.

Beaufrère, B., Tessari, P., Cattalini, M., Miles, J. and Haymond, M.W. (1985) Apparent decreased oxidation and turnover of leucine during infusion of medium-chain triglycerides. *American Journal of Physiology* 249, E175–182.

Bender, D.A. (1985) *Amino Acid Metabolism* 2nd edn. John Wiley, Chichester, UK.

Ben Galim, E., Houska, K., Bier, D.M., Matthews, D.E. and Haymond, M.W. (1980) Branched-chain amino acid nitrogen transfer to alanine *in vivo* in dogs. *Journal of Clinical Investigation* 66, 1295–1304.

Bergström, J., Fürst, P. and Vinnars, E. (1990) Effect of a test meal, with and without protein, on muscle and plasma free amino acids. *Clinical Science* 79, 331–337.

Biolo, G. and Tessari, P. (1997) Splanchnic versus whole-body production of α-ketoisocaproate from leucine in the fed state. *Metabolism* 46, 164–167.

Table 4.11. Turnover of methionine and homocysteine in healthy adults. Data of MacCoss *et al.* (2001).

	μmol kg^{-1} h^{-1}
Methionine flux[a]	37.3
Methionine transmethylation (TM)	9.7
Homocysteine remethylation (RM)	4.4
Homocysteine transulphuration/oxidation (TS)	5.4
Methionine into protein	32.0

[a]The value for flux is that obtained with [1-^{13}C]-methionine.
For TM, RM, TS: see Fig. 4.2.

Biolo, G., Fleming, R.Y.D., Maggi, S. and Wolfe, R.R. (1995) Transmembrane transport and intracellular kinetics of amino acids in human skeletal muscle. *American Journal of Physiology* 268, E75–84.

Biolo, G., Fleming, R.Y., Maggi, S.P., Nguyen, T.T., Hemdon, D.N. and Wolfe, R.R. (2000) Inhibition of muscle glutamine formation in hypercatabolic patients. *Clinical Science* 99, 189–194.

Buse, M.G. and Reid, S.S. (1975) Leucine: a possible regulator of protein turnover in muscle. *Journal of Clinical Investigation* 56, 1250–1261.

Calder, A.G., Anderson, S.E., Great, I., McNurlan, M.A. and Garlick, P.J. (1992) The determination of low d_5-phenylalanine enrichment (0.002–0.09 atom per cent excess) after conversion to phenylethylamine, in relation to protein turnover studies by gas chromatography/electron ionization mass spectrometry. *Rapid Communications in Mass Spectrometry* 6, 421–424.

Calloway, D.H. and Spector, H. (1954) Nitrogen balance as related to caloric and protein intake in young men. *American Journal of Clinical Nutrition* 2, 405–411.

Castillo, L., de Rojas, T.C., Chapman, T.E., Vogt, J., Burke, J.F., Tannenbaum, S.R. and Young, V.R. (1993a) Splanchnic metabolism of dietary arginine in relation to nitric oxide synthesis in normal adult man. *Proceedings of the National Academy of Sciences USA* 90, 193–197.

Castillo, L., Chapman, T.E., Yu, Y.-M., Ajami, A., Burke, J.F. and Young, V.R. (1993b) Dietary arginine uptake by the splanchnic region in adult humans. *American Journal of Physiology* 265, E532–539.

Castillo, L., Chapman, T.E., Sánchez, M., Yu, Y.-M., Burke, J.F., Ajami, A., Vogt, J. and Young, V.R. (1993c) Plasma arginine and citrulline kinetics in adults given adequate and arginine-free diets. *Proceedings of the National Academy of Sciences USA* 90, 7749–7753.

Castillo, L., Sanchez, M., Chapman, T.E., Ajami, A., Burke, J.F. and Young, V.R. (1994a) The plasma flux and oxidation rate of ornithine adaptively decline with restricted arginine intake. *Proceedings of the National Academy of Sciences USA* 91, 6393–6397.

Castillo, L., Ajami, A., Branch, S., Chapman, T.E., Yu, Y.-M., Burke, J.F. and Young, V.R. (1994b) *Metabolism* 43, 114–122.

Castillo, L., Beaumier, L., Ajami, A.M. and Young, V.R. (1996) Whole body nitric oxide synthesis in healthy men determined from [^{15}N] arginine-to-[^{15}N] citrulline labeling. *Proceedings of the National Academy of Sciences, USA* 93, 11460–11465.

Claeyssens, S., Bouteloup-Demange, C., Gachon, P., Hecketsweler, B., Lerebours, S., Lavoinne, A. and Dechelotte, P. (2000) Effect of enteral glutamine on leucine, phenylalanine and glutamine metabolism in hypercortisolemic subjects. *American Journal of Physiology* 278, E817–824.

Clarke, J.T.R. and Bier, D.M. (1982) The conversion of phenylalanine to tyrosine in man. Direct measurement by continuous intravenous tracer infusions of L-[ring $^2H_5$] phenylalanine and L-[1-^{13}C] tyrosine in the post-absorptive state. *Metabolism* 31, 999–1005.

Consoli, A., Nurjhan, N., Reilly, J.J., Bier, D.M. and Gerich, J.E. (1990) Contribution of liver and skeletal muscle to alanine and lactate metabolism in humans. *American Journal of Physiology* 259, E677–684.

Cooper, A.J.L., Nieves, E., Rosenspive, C.E., Filc-DeRicco, S., Gelbard, A.S. and Brusilow, S.W. (1988) Short-term metabolic fate of ^{13}N-labeled glutamate, alanine and glutamine (amide) in rat liver. *Journal of Biological Chemistry* 263, 12268–12273.

Cortiella, J., Marchini, J.S., Branch, S., Chapman, T.E. and Young, V.R. (1992) Phenylalanine and tyrosine kinetics in relation to altered protein and phenylalanine and tyrosine intakes in healthy young men. *American Journal of Clinical Nutrition* 56, 517–525.

Darmaun, D., Matthews, D.E. and Bier, D.M. (1986) Glutamine and glutamate kinetics in humans. *American Journal of Physiology* 251, E117–126.

Darmaun, D., Matthews, D.E. and Bier, D.M. (1988) Physiological hypercortisolemic increases proteolysis, glutamine, and alanine production. *American Journal of Physiology* 255, E366–373.

de Koning, P.J., Snell, K., Duran, M., Berger, R., Poll-Thee, B.T. and Surtees, R. (2003) L-serine in disease and development. *Biochemical Journal* 371, 653–661.

Dhanakoti, S.N., Brosnan, J.T., Herzberg, G.R. and Brosnan, M.E. (1990) Renal arginine synthesis studies *in vitro* and *in vivo*. *American Journal of Physiology* 259, E437–442.

Documenta Geigy (1962) *Scientific Tables* 6th edn. Geigy Pharmaceutical Co., Manchester, UK.

Elia, M. (1991) The inter-organ flux of substrates in fed and fasted man, as indicated by arterio-venous balance studies. *Nutrition Research Reviews* 4, 3–31.

Ewart, H.S., Jois, M. and Brasnan, J.T. (1992) Rapid stimulation of the hepatic glycine-cleavage system in rats fed a single high protein meal. *Biochemical Journal* 283, 441–442.

Felig, P. (1975) Amino acid metabolism in man. *Annual Reviews of Biochemistry* 44, 933–955.

Felig, P. (1982) Inter-organ amino acid exchange. In: Waterlow, J.C. and Stephen, J.M.L. (eds) *Nitrogen Metabolism in Man*. Applied Science Publishers, London, pp. 45–62.

Fell, D.A. and Wagner, A. (2000) The small world of metabolism. *Nature Biotechnology* 18, 1121–1122.

Fern, E.B. and Garlick, P.J. (1973) The specific radioactivity of the precursor pool for estimates of the rate of protein synthesis. *Biochemical Journal* 134, 1127–1130.

Fern, E.B. and Garlick, P.J. (1974) The specific radioactivity of the tissue free amino acid pool as a basis for measuring the rate of protein synthesis of the rat *in vivo*. *Biochemical Journal* 142, 413–419.

Forrester, T.E., Badaloo, V., Bennett, F.I. and Jackson, A.A. (1990) Excessive excretion of 5-oxoproline and decreased levels of blood glutathione in type II diabetes mellitus. *European Journal of Clinical Nutrition* 44, 847–850.

Gate, J.J., Parker, D.S. and Lobley, G.E. (1999) The metabolic fate of the amino-N group of glutamine in the tissues of the gastrointestinal tract in 24 h-fasted sheep. *British Journal of Nutrition* 81, 297–306.

Gersovitz, M., Bier, D., Matthews, D., Udall, J., Munro, H.N. and Young, V.R. (1980) Dynamic aspects of whole body glycine metabolism: influence of protein intake in young adult and elderly males. *Metabolism* 29, 1087–1094.

Golden, M.H.N. and Ramdath, D. (1987) Free radicals in the pathogenesis of kwashiorkor. *Proceedings of the Nutrition Society* 46, 53–68.

Golden, M.H.N., Jahoor, P. and Jackson, A.A. (1982) Glutamine production rate and its contribution to urinary ammonia in normal man. *Clinical Science* 62, 299–305.

Griffiths, R.D., Jones, C. and Palmer, T.E.A. (1997) Six-month outcome of critically ill patients given glutamine-supplemented parenteral nutrition. 13, 295–302.

Haisch, M., Fukegawa, N.K. and Matthews, D.E. (2000) Oxidation of glutamine by the splanchnic bed in humans. *American Journal of Physiology* 278, E593–602.

Hammarqvist, F., Wernerman, J., Ali, M.R., von der Decken, A. and Vinnars, E. (1989) Addition of glutamine to total parenteral nutrition after elective abdominal surgery spares free glutamine in muscle, counteracts the fall in muscle proteins synthesis, and improves nitrogen balance. *Annals of Surgery* 209, 455–461.

Hammarqvist, F., Wernerman, J., von der Decken, A. and Vinnars, E. (1990) Alanyl-glutamine counteracts the depletion of free glutamine and the post-operative decline in proteins synthesis in skeletal muscle. *Annals of Surgery* 212, 637–644.

Hammarqvist, F., Luo, J.L., Cotgreave, I.A., Anderson, K. and Wernerman, J. (1997) Skeletal muscle glutathione is depleted in critically ill patients. *Critical Care in Medicine* 25, 78–84.

Hankard, R.G., Darmaun, D., Sager, B.K., D'Amore, D., Parsons, W.R. and Haymond, M. (1995) Response of glutamine metabolism to exogenous glutamine in humans. *American Journal of Physiology* 269, E663–670.

Hankard, R.G., Hammond, D., Haymond, M.W. and Darmaun, D. (1998) Oral glutamine slows down whole body protein breakdown in Duchenne muscular dystrophy. *Pediatric Research* 43, 222–226.

Harper, A.E. and Zapalowski, C. (1981) Metabolism of branched chain amino acids. In: Waterlow, J.C. and Stephen, J.M.L. (eds) *Nitrogen Metabolism in Man.* Applied Publishers, London, pp. 97–115.

Harris, R.A., Paxton, R. and Jenkins, P. (1985) Nutritional control of branched chain α-ketoacid dehydrogenase in rat hepatocytes. *Federation Proceedings* 44, 2463–2468.

Häussinger, D. (1987) Structural-functional organization of hepatic glutamine and ammonium metabolism. *Biochemical Society Transactions* 15, 369–372.

Heitman, R.N. and Bergman, E.N. (1978) Glutamine metabolism, inter-organ transport and glucogenicity in the sheep. *American Journal of Physiology* 234, E197–203.

Henriques, O.B., Henriques, S.P. and Neuberger, A. (1955) Quantitative aspects of glycine metabolism in the rabbit. *Biochemical Journal* 60, 409–424.

Hoffer, L.J., Yang, R.D., Matthews, D.E., Bistrian, B.R., Bier, D.M. and Young, V.R. (1985) Effects of meal consumption on whole body leucine and alanine kinetics in young adult men. *British Journal of Nutrition* 53, 31–38.

Hutson, S.M., Cree, K. and Harper, A.E. (1978) Regulation of leucine and of α-ketoisocaproate metabolism in skeletal muscle. *Journal of Biological Chemistry* 253, 8126–8133.

Jackson, A.A. (1986) Blood glutathione in severe malnutrition in childhood. *Transactions of the Royal Society of Tropical Medicine and Hygiene* 80, 911–913.

Jackson, A.A. (1991) The glycine story. *European Journal of Clinical Nutrition* 45, 59–65.

Jackson, A.A. and Golden, M.H.N. (1980) [^{15}N] glycine metabolism in normal man: the metabolic α-amino-nitrogen pool. *Clinical Science* 58, 517–522.

Jackson, A.A., Shaw, J.C.L., Barber, A. and Golden, M.H.N. (1981) Nitrogen metabolism in pre-term infants fed human donor breast milk: the possible essentiality of glycine. *Pediatric Research* 15, 1454–1461.

Jackson, A.A., Badaloo, A.V., Forrester, T., Hibbert, J.M. and Persaud, C. (1987) Urinary excretion of 5-oxoproline (pyroglutamic aciduria) as an index of glycine insufficiency in normal man. *British Journal of Nutrition* 58, 207–214.

Jackson, A.A., Persaud, C., Meakins, T.S. and Bundy, R. (1996) Urinary excretion of 5-L-oxoproline (pyroglutamic acid) is increased in normal adults consuming vegetarian or low protein diets. *Journal of Nutrition* 126, 2813–2822.

Jackson, A.A., Persaud, C., Hall, M., Smith, S., Evans, N. and Rutter, N. (1997a) Urinary excretion of 5-L-oxoproline (pyroglutamic acid) during early life in term and preterm infants. *Archives of Disease in Childhood* 76, F152–157.

Jackson, A.A., Persaud, C., Werkmeister, G., McClellan, I.S., Badaloo, A. and Forrester, T. (1997b) Comparison of urinary 5-L-oxoproline (L-pyroglutamate) during normal pregnancy in women in England and Jamaica. *British Journal of Nutrition* 77, 183–196.

Jackson, A.A., Gibson, N.R., Lu, Y. and Jahoor, F. (2004) Synthesis of erythrocyte glutathione in healthy adults consuming the safe amount of dietary protein. *American Journal of Clinical Nutrition* 80, 101–107.

Jeevanandam, M. (1981) 'Metabolic pool' and the use of ^{15}N-labelled amino acids. *Clinical Science* 61, 349–351.

Jeevanandam, M., Long, C.L. and Kinney, J.M. (1979) Kinetics of intravenously administered ^{15}N-L-alanine in the evaluation of protein turnover. *American Journal of Clinical Nutrition* 32, 975–980.

Jepson, M.M., Bates, P.C., Broadbent, P., Pell, J.M. and Millward, D.J. (1986) Relationship between glutamine concentration and protein synthesis in rat skeletal muscle. *American Journal of Physiology* 255, E166–172.

Kalhan, S.C., Rossi, K.Q., Gruca, L.L., Super, D.M. and Savin, S.M. (1998) Relation between transamination of branched-chain amino acids and urea synthesis: evidence from human pregnancy. *American Journal of Physiology* 275, E423–431.

Kikuchi, G. (1973) The glycine cleavage system: composition reaction mechanism and physiological significance. *Molecular and Cellular Biology* 1, 169–187.

Krebs, H.A. (1972) Some aspects of the regulation of fuel supply in omnivorous animals. *Advances in Enzyme Regulation* 10, 397–420.

Krebs, H.A. (1980) Glutamine metabolism in the animal body. In: Moran, J. and Palacios, R. (eds) *Glutamine: Metabolism, Enzymology and Regulation*. Academic Press, New York, pp. 319–329.

Krebs, H.A. and Lund, P. (1977) Aspects of the regulation of the metabolism of the branched-chain amino acids. In: Weber, G. (ed.) *Advances in Enzyme Regulation* Vol. 15. Pergamon Press, Oxford, UK.

Kreider, M.E., Stumroll, M., Meyer, C., Overkempf, D., Welte, S. and Gerich, J. (1997) Steady-steady state and non-steady-state measurements of plasma glutamine turnover in humans. *American Journal of Physiology* 272, E621–627.

Krempf, M., Hoerr, R.A., Marks, L. and Young, V.R. (1990) Phenylalanine flux in adult men: estimates with different tracers and routes of administration. *Metabolism* 39, 560–562.

Lapidot, A. and Nissim, I. (1980) Regulation of pool sizes and turnover rates of amino acids in humans: ^{15}N-glycine and ^{15}N-alanine single-dose experiments using gas chromatography–mass spectrometry analysis. *Metabolism* 29, 230–239.

Lenton, C., Ali, Z., Persaud, C. and Jackson, A.A. (1998) Infants in Trinidad excrete more 5-L-oxoproline (L-pyroglutamic acid) in urine than infants in England: an environmental not ethnic difference. *British Journal of Nutrition* 80, 51–55.

Lobley, G.E., Connell, A., Revell, D.K., Bequette, B.J., Brown, D.S. and Calder, A.G. (1996a) Splanchnic bed transfers of amino acids in sheep blood and plasma, as monitored through use of a multiple U-^{13}C-labelled amino acid mixture. *British Journal of Nutrition* 75, 217–235.

Lobley, G.E., Connell, A. and Revell, D. (1996b) The importance of transmethylation reactions to methionine metabolism in sheep; effects of supplementation with creatine and choline. *British Journal of Nutrition* 75, 47–56.

Lyons, J., Rauh-Pfeiffer, A., Yu, Y.M., Lu, X.-M, Zurakowski, D., Tompkins, R.G., Ajami, A.M. and Young, V.R. (2000) Blood glutathione synthesis rates in healthy adults receiving a sulfur amino acid-free diet. *Proceedings of the National Academy of Sciences* 97, 5071–5076.

MacCoss, M.J., Fukegawa, N.K. and Matthews, D.E. (2001) Measurement of intracellular sulfur amino acid metabolism in humans. *American Journal of Physiology* 280, E947–955.

MacLennan, P.A., Brown, R.A. and Rennie, M.J. (1987) A positive relationship between protein synthetic rate and intracellular glutamine concentration in perfused rat skeletal muscle. *FEBS Letters* 215, 187–191.

Marchini, J.S., Castillo, L., Chapman, T.E., Vogt, J.A., Ajami, A. and Young, V.R. (1993) Phenylalanine conversion to tyrosine: comparative determination with L-[ring $^{2}H_{5}$] phenylalanine and L-[1–^{13}C] phenylalanine as tracers in man. *Metabolism* 42, 1316–1322.

Marliss, E.B., Aoki, T.T., Pozefsky, T., Most, A.S. and Cahill, G.F. (1971) Muscle and splanchnic glutamine and glutamate metabolism in post-absorptive and starved man. *Journal of Clinical Investigation* 50, 814–817.

Martin, D.W. (1987) Metabolism of purine and pyrimidine nucleotide. In: Martin, D.W., Mayes, P.A. and Rodwell, V.W. (eds) *Harper's Review of Biochemistry*. Langes, New York, pp. 331–348.

Matthews, D.E. and Campbell, R.G. (1992) The effect of dietary protein intakes on glutamine and glutamate metabolism in humans. *American Journal of Clinical Nutrition,* 55, 963–970.

Matthews, D.E., Bier, D.M., Rennie, M.J., Edwards, R.H.T., Halliday, D., Millward, D.J. and Clugston, G.A. (1981a) Regulation of leucine metabolism in man: a stable isotope study. *Science* 214, 1129–1131.

Matthews, D.E., Conway, J.M., Young, V.R. and Bier, D.M. (1981b) Glycine nitrogen metabolism in man. *Metabolism* 30, 886–893.

Matthews, D.E., Yang, R.D., Bier, D.M. and Young, V.R. (1985) What is whole body alanine metabolism when studied with labelled alanine tracers? In: Garrow, J.S. and Halliday, D. (eds) *Substrate and Energy Metabolism.* John Libbey, London, pp. 187–194.

Matthews, D.E., Marano, M.A. and Campbell, R.G. (1993a) Splanchnic bed utilization of leucine and phenylalanine in humans. *American Journal of Physiology* 264, E109–118.

Matthews, D.E., Marano, M.A. and Campbell, R.G. (1993b) Splanchnic bed utilization of glutamine and glutamic acid in humans. *American Journal of Physiology* 264, E848–854.

Meek, S.E., Persson, M., Ford, C.G. and Nair, K.S. (1998) Differential regulation of amino exchange and protein dynamics across splanchnic and skeletal muscle beds by insulin in healthy human subjects. *Diabetes* 47, 1824–1835.

Metges, C.C., Yu, Y.-M., Cai, W., Lu, X.-M., Wong, S., Regan, M.M., Ajami, A. and Young, V.R. (2000) Oxoproline kinetics and oxoproline urinary excretion during glycine- or sulfur amino acid-free diets in humans. *American Journal of Physiology* 278, E676–686.

Miles, J.M., Nissen, S.L., Rizza, R.A., Gerich, J.E. and Haymond, M.W. (1983) Failure of infused β-hydroxybutyrate to decrease proteolysis in man. *Diabetes* 32, 197–205.

Mittendorfer, B., Volpi, E. and Wolfe, R.R. (2001) Whole body and skeletal muscle glutamine metabolism in healthy subjects. *American Journal of Physiology* 280, E323–333.

Møller, N., Meek, S., Bigelow, M., Andrews, J. and Nair, K.S. (2000) The kidney is an important site for *in vivo* phenylalanine-to-tyrosine conversion in adult humans: a metabolic role of the kidney. *Proceedings of the National Academy of Sciences* 97, 1242–1246.

Moncada, S. (1999) Nitric oxide: discovery and impact on clinical medicine. *Journal of the Royal Society of Medicine* 92, 164–169.

Muhlbacher, F., Kapedia, C.R., Colpoys, F., Smith, R.J. and Wilmore, D.W. (1984) Effects of glucocorticoids on glutamine metabolism in skeletal muscle. *American Journal of Physiology* 247, E75–83.

Nair, K.S., Ford, G.C., Ekberg, K., Fernqvist-Forbes, E. and Wahren, J. (1995) Protein dynamics in whole body and in splanchnic and leg tissues in type 1 diabetic patients. *Journal of Clinical Investigation* 95, 2926–2937.

Neu, J., DeMarco, V. and Li, N. (2002) Glutamine: clinical applications and mechanisms of action. *Current Opinion in Clinical Nutrition and Metabolic Care* 5, 69–75.

Newsholme, E.A. and Calder, P.C. (1997) The proposed role of glutamine in some cells of the immune system and speculated consequences for the whole animal. *Nutrition* 13, 728–730.

Newsholme, E.A. and Parry-Billings, M. (1990) Properties of glutamine release from muscle and its importance for the immune system. *Journal of Parenteral and Enteral Nutrition* 14, 635–685.

Newsholme, E.A., Crabtree, B. and Ardewi, M.S. (1985) Glutamine metabolism in lymphocytes in biochemical, physiological and clinical importance. *Quarterly Journal of Experimental Physiology* 70, 473–489.

Nissen, S. and Haymond, M.W. (1986) Changes in leucine kinetics during meal absorption: effect of dietary leucine availability. *American Journal of Physiology* 250, E695–701.

Nurjhan, N., Bucci, A., Periello, G., Stumvoll, M., Dailey, G., Bier, D.M., Toft, I., Jenssen, T.G. and Gerich, J.E. (1995) Glutamine: a major gluconeogenic precursor and vehicle for carbon transport in man. *Journal of Clinical Investigation* 95, 272–277.

Obitsu, T., Bremner, D., Milne, E. and Lobley, G.E. (2000) Effect of abomasal glucose infusion on alanine metabolism and urea production in sheep. *British Journal of Nutrition* 84, 157–163.

Olde Damink, S.W.M., de Blaauw, I., Dentz, N.E.P. and Soeters, P.B. (1999) Effects *in vivo* of decreased plasma and intracellular muscle glutamine concentration on whole body and hindquarter protein kinetics in rats. *Clinical Science* 96, 634–646.

Persaud, C., Pate, E., Forrester, T. and Jackson, A.A. (1997) Urinary 5-L-oxoproline (pyroglutamic acid) excretion is greater in infants in Jamaica than in infants in England. *European Journal of Clinical Nutrition* 51, 510–513.

Pierello, G., Jorde, R., Nurjhan, N., Stumroll, M., Dailey, G., Jenssen, T., Bier, D.M. and Gerich, J.E. (1995) Estimation of glucose-alanine-lactate-glutamine cycles in postabsorptive humans: role of skeletal muscle. *American Journal of Physiology* E269, 443–450.

Pitts, R.F. and Pilkington, R.A. (1966) The relation between plasma concentration of glutamine and glycine and utilization of their nitrogen as sources of urinary ammonia. *Journal of Clinical Investigation* 42, 263–276.

Price, G.M., Halliday, D., Pacy, P.J., Quevedo, M.R. and Millward, D.J. (1994) Nitrogen homoeostasis in man: influence of protein intake on the diurnal cycling of body nitrogen. *Clinical Science* 86, 91–102.

Ratner, S., Rittenberg, D., Keston, A.S. and

Schoenheiner, R. (1940) Studies on protein metabolism. XIV. The chemical interaction of dietary glycine and body protein in rats. *Journal of Biological Chemistry* 134, 665–676.

Reeds, P.J., Fjeld, C.R. and Jahoor, F. (1994) Do the differences between the amino acid compositions of acute-phase and muscle proteins have a bearing on nitrogen loss in traumatic states? *Journal of Nutrition* 124, 906–910.

Reid, M., Badaloo, A., Forrester, T., Morlese, J.F., Frazer, M., Heird, W.C. and Jahoor, F. (2000). *In vivo* rates of glutathione synthesis in children with severe protein-energy malnutrition. *American Journal of Physiology* 278, E405–412.

Rennie, M.J., Hundal, M.S., Babij, P., MacLennan, P., Taylor, P.M., Watt, P.W., Jepson, M.M. and Millward, D.J. (1986) Characteristics of a glutamine carrier in skeletal muscle have important consequences for nitrogen loss in injury, infection and chronic disease. *Lancet,* 2, 1008–1012.

Rennie, M.J., Tadros, L., Khogali, S., Ahmed, A. and Taylor, P.M. (1994) Glutamine transport and its metabolic effects. *Journal of Nutrition* 124, 1503–1508S.

Robert, J.J., Bier, D.M., Zhao, X.H., Matthews, D.E. and Young, V.R. (1982) Glucose and insulin effects on the *de novo* synthesis in young men: studies with stable isotope-labelled alanine, glycine, leucine and lysine. *Metabolism* 31, 1210–1218.

Robert, J.J., Beaufrère, B., Desjeux, F., Bier, D.M. and Young, V.R. (1985) Whole body *de novo* alanine synthesis in type I (insulin-dependent) diabetes studied with stable isotope-labelled leucine, alanine and glycine. *Diabetes* 34, 67–73.

Royle, G.T., Molnar, J.A., Wolfe, M.H., Wolfe, R.R. and Burke, J.P. (1982) Urea, glucose and alanine kinetics in man: effects of glucose infusion. *Clinical Science* 62, 553–556.

Sánchez, M., El-Khoury, A.E., Castillo, L., Chapman, T.E. and Young, V.R. (1995) Phenylalanine and tyrosine kinetics in young men throughout a continuous 24-h period, at a low phenylalanine intake. *American Journal of Clinical Nutrition* 61, 555–570.

Sánchez, M., El-Khoury, A.E., Castillo, L., Chapman, T.E., Basil-Filho, A., Beaumier, L. and Young, V.R. (1996) Twenty-four-hour intravenous and oral tracer studies with L-[1-^{13}C] phenyalanine and L[3,3-$^{2}H_2$] tyrosine at a tyrosine-free generous phenylalanine intake in adults. *American Journal of Clinical Nutrition* 63, 532–545

Shortland, G.J., Walter, G.H., Fleming, P.J. and Halliday, D. (1994) Phenylalanine kinetics in sick preterm neonates with respiratory distress syndrome. *Pediatric Research* 36, 713–718.

Sketcher, R.D., Fern, E.B. and James, W.P.T. (1974) The adaptation in muscle oxidation of leucine to dietary protein and energy intake. *British Journal of Nutrition* 31, 333–342.

Souba, W.W., Smith, R.J. and Wilmore, D.W. (1985) Glutamine metabolism by the intestinal tracer. *Journal of Parenteral and Enteral Nutrition* 9, 608–617.

Storch, K.J., Wagner, D.A., Burke, J.F. and Young, V.R. (1988) Quantitative study *in vivo* of methionine cycle in humans using [methyl-$^{2}H_3$] – and [1-^{13}C] methionine. *American Journal of Physiology* 255, E322–331.

Storch, K.J., Wagner, J.A., Burke, J.F. and Young, V.R. (1990) [1-^{13}C: methyl-$^{2}H_3$] methionine kinetics in humans: methionine conservation and cystine sparing. *American Journal of Physiology* 258, E790–798.

Suryawan, A., Hawes, J.W., Harris, R.A., Shimomura, Y., Jenkins, A.E. and Hutson, S.E. (1998) A molecular model of human branched-chain amino acid metabolism. *American Journal of Clinical Nutrition* 68, 72–81.

Thompson, G.L., Pacy, P.J., Merritt, H., Ford, G.C., Reed, M.A., Cheng, K.N. and Halliday, D. (1989) Rapid measurement of whole body and forearm protein turnover using a [$^{2}H_5$] phenylalanine model. *American Journal of Physiology* 256, E631–639.

van Acker, B.A.C., Hulsewé, K.E.W., Wagenmakers, A.J.M., Deutz, N.E.P., van Kreel, B.K., Halliday, D., Matthews, D.E., Soeters, P. and van Meyenfelet, M. (1998) Absence of glutamine isotopic steady state: implications for the assessment of whole-body glutamine production rate. *Clinical Science* 95, 339–346.

Vitti, T.G. and Gaebler, O.H. (1963) Effects of growth hormone on metabolism of nitrogen from several amino acids and ammonia. *Archives of Biochemistry and Biophysics* 101, 292–298.

Vitti, T.G., Vukmivovich, R. and Gaebler, O.H. (1964) Utilization of ammonia nitrogen, administered by intragastric, intrapectoral and subcutaneous routes: effects of growth hormone. *Archives of Biochemistry and Biophysics* 106, 475–482.

Wagner, A. and Fell, D.A. (2001) The small world inside large metabolic networks. *Proceedings of the Royal Society* B 268, 1803–1810.

Waterlow, J.C., Jackson, A.A., Golden, M.H.N., Jahoor, F., Sutton, G. and Fern, E.B. (1994) The effect of acidosis on the labelling of urinary ammonia during infusion of [amide ^{15}N] glutamine in human subjects. *British Journal of Nutrition* 72, 83–92.

WHO (1985) Energy and Protein Requirements: report of a Joint FAO/WHO/UNU Consultation. *Technical Report Series* no. 724. World Health Organization, Geneva.

Widdowson, E.M., Southgate, D.A.T. and Hey, E.H. (1979) Body composition of the fetus and infant. In: Visser, H.A.K. (ed.) *Nutrition and Metabolism of the Fetus and Infant.* Martinus Nijhoff, The Hague, Netherlands, pp. 169–178.

Wilmore, D.W. (2002) The effect of glutamine supplementation in patients following elective surgery and accidental injury. *Journal of Nutrition* 131, 2543S–2549S.

Yang, R.D., Matthews, D.E., Bier, D.M., Wen, Z.M. and Young, V.R. (1986) Response of alanine metabolism in humans to manipulation of dietary protein and energy intakes. *American Journal of Physiology* 250, E39–46.

Young, V.R. and Ajami, A.M. (2000) Glutamate: an amino acid of particular distinction. *Journal of Nutrition* 130, 892S–900S.

Young, V.R., Meredith, C., Hoerr, R., Bier, D.M. and Matthews, D.E. (1985) Amino acid kinetics in relation to protein and amino acid requirements: the primary importance of amino acid oxidation. In: Garrow, J.S. and Halliday, D. (eds) *Substrate and Energy Metabolism in Man.* John Libbey, London, pp. 119–134.

Yu, Y.-M., Yang, R.D., Matthews, D.E., Zhi, M.W., Burke, J.F., Bier, D.M. and Young, V.R. (1985) Quantitative aspects of glycine and alanine nitrogen metabolism in postabsorptive young men: effects of level of nitrogen and dispensable amino acid intake. *Journal of Nutrition* 115, 399–410.

Yu, Y.-M., Ryan, C.M., Fei, Z-W., Lu, X-M., Castillo, L., Schultz, J.T., Tompkins, R.G. and Young, V.R. (2001) Plasma L-5-oxoproline kinetics and whole body glutathione synthesis rates in severely burned human adults. *American Journal of Physiology* 282, E247–258.

Zachwieja, J.J., Witt, T.L. and Yaresheki, K.E. (2000) Intravenous glutamine does not stimulate mixed muscle protein synthesis in healthy young men and women. *Metabolism* 49, 1555–1560.

Zello, G.A., Pencharz, P.B. and Ball, R.O. (1990) Phenylalanine flux, oxidation and conversion to tyrosine in humans studied with L-[1-^{13}C] phenylalanine. *American Journal of Physiology* 259, E835–843.

5

The Precursor Problem

Rates of protein synthesis, whether in the whole body or in tissues, cannot be determined with labelled amino acids without knowledge of the specific radioactivity or enrichment of the precursor. In the 1960s and '70s there were many kinetic studies of the incorporation of amino acids into proteins, on preparations ranging from bacteria and single-cell organisms to cell suspensions, tissue slices, intact muscles and perfused organs. This work was reported in some detail by Waterlow *et al.* (1978) and will not be discussed here at length except for one important point that emerged: the possibility that the precursors for synthesis and for oxidation are different (Rosenberg *et al.*, 1963; Schneible *et al.*, 1981). We return to this question later.

5.1 Transfer-RNA as the Precursor for Synthesis

Transfer-RNA (tRNA) is the undoubted precursor of synthesis, but it needs tissues for its isolation, and measurement of its labelling is difficult because it is extremely labile. Nevertheless, valuable work has been done, mainly on animals, on the relation of its enrichment to that of the EC and IC pools. Methods for its isolation and determination of its activity have been described in some detail by Wallyn *et al.* (1974), Martin *et al.* (1977), Ljungkvist *et al.* (1997) and Ahlman (2001).

In an important paper Mortimore *et al.* (1972), working on rat livers perfused with [^{3}H] valine, proposed a model (Fig. 5.1) based on kinetic considerations in which there was an expandable pool whose valine content varied directly with that of the perfusate and which was the precursor for synthesis; and a non-expandable pool whose valine content did not vary and which received the input from protein breakdown. The non-expandable pool was tentatively located in the lysosomes.

Shortly afterwards Airhart *et al.* (1974) proposed a model (Fig. 5.2) with a different type of compartmentation. This is the first study we know of in which the enrichment of tRNA was measured. After injection of [^{3}H] valine into rats the enrichment of valyl-tRNA in the liver was intermediate between that of valine in the EC and IC pools because of dilution of the tracer from plasma by the unlabelled product of protein breakdown. It was suggested that the precursor pool was located within the transport system of the cell membrane and that the composition of this pool would be determined by the concentration gradient across the membrane. In a later paper (Vidrich *et al.*, 1977) this group injected labelled valine at different times after rats had been allowed food for a fixed period of 3 h and measured the specific radioactivity of valine in plasma, liver and valyl-tRNA. The hypothesis that the composition of the membrane pool should be proportional to the concentration gradient across the membrane allowed them to generate an equation to predict the activity of valyl-tRNA:

$$\varepsilon_{tRNA} = \frac{\varepsilon_E[E] + \varepsilon_I[I]}{[E] + [I]}$$

where [ ] denote concentrations and ε activities.

There was excellent agreement between predicted and observed activities of tRNA. The study shows that ε_{tRNA} is much less variable in

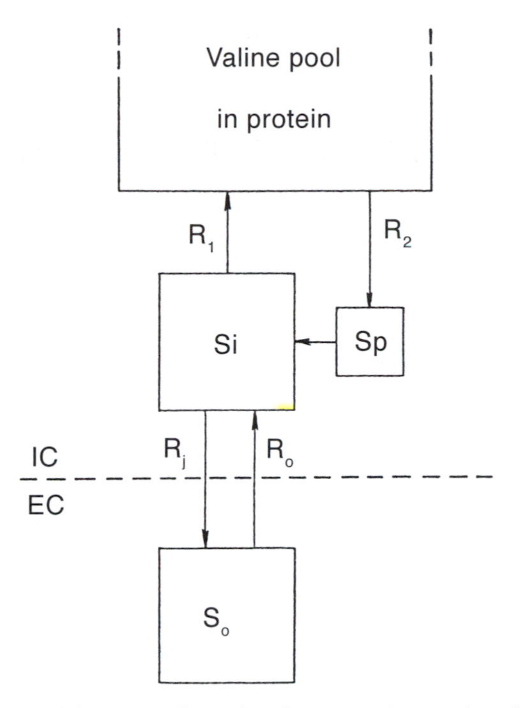

Fig. 5.1. Hypothetical scheme of the major relationships between valine pools in the perfused liver. R_1, R_2 are flows into and out of protein; R_i and R_o are transport fluxes across the cell membrane; S_i is the expandable free valine pool, S_p the non-expandable pool. Reproduced from Mortimore *et al.* (1972) by courtesy of the *Journal of Biological Chemistry*.

relation to feeding and fasting, than either the EC or the IC activities. To our knowledge no alternative model has been proposed.

Out of 12 studies *in vivo* published since the work of Airhart and Vidrich all show that the activity of aminoacyl-tRNA lies between that of tracer amino acid in the EC and tissue pools (Table 5.1). In eight cases the activity in tissue fluid was some 2–14% higher than in tRNA. The reason may be that usually, except in the case of Vidrich, no correction was made for the ~20% of EC fluid that is included in the tissue pool, so that its activity is higher than that of the true IC tracer. Most studies were of muscle, but the pattern in liver is similar. That of Watt *et al.* (1991) is the pioneer study in man. It is noteworthy that in most cases (Vidrich is again an exception) the ratio ε tissue fluid: εtRNA was not significantly changed either by a meal (Ljungqvist *et al.*, 1997; Davis *et al.*, 1999) or by insulin (Young *et al.*, 1994; Davis *et al.*, 1999; Caso *et al.*, 2001). Since both food and insulin decrease protein breakdown, one would expect the precursor activity to increase; this must be counteracted by decreased labelling of amino acids entering from the plasma.

5.2 A 'Reciprocal' Metabolite as Precursor

The 'reciprocal' approach is based on the principle that if two products are derived from the same precursor and only that precursor, their level of labelling will be the same. Arends and Bier (1991) examined four amino acids, glycine, leucine, lysine and phenylalanine, and concluded that only the first two had metabolites that would be useful as reciprocals in measurements of protein synthesis.

5.2.1 Glycine-hippurate

In the early days, when measurements of isotope abundance in small quantities of material such as

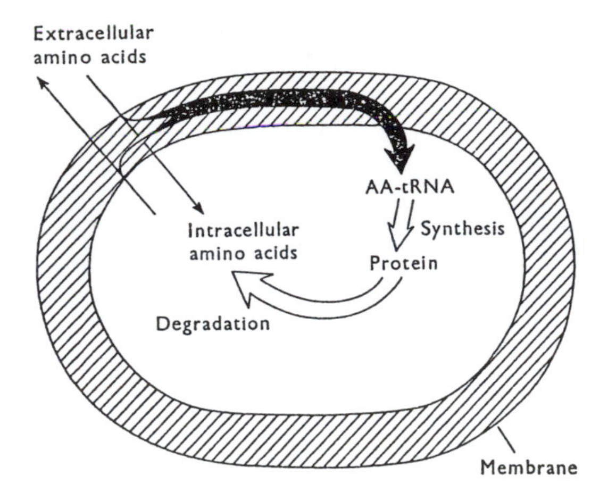

Fig. 5.2. Model showing compartmentation of amino acid for protein synthesis within the transport system of the cell membrane. Reproduced from Airhart *et al.* (1974), by courtesy of the *Biochemical Journal*.

plasma were limited by the instruments available, advantage was taken of the fact that benzoic acid, a potentially toxic substance, is detoxicated by combination with glycine to form hippuric acid. With a sufficient dose of benzoic acid quite large amounts of hippuric acid are excreted in the urine. For early results by this method see Waterlow *et al.* (1978). Since then Connell *et al.* (1997) in sheep have found excellent agreement between the activities of free glycine in liver and urinary hippurate. The method has been used to measure the synthesis rates of various plasma proteins: albumin in premature infants (Yudkoff *et al.*, 1987); fibrinogen (Stein *et al.*, 1978; Thompson *et al.*, 1989); fibronectin in prematures (Polin *et al.*, 1989) and in burned or traumatized patients (Thompson *et al.*, 1989). It was found that when labelled glycine was infused its activity in plasma at plateau was substantially higher than that of its metabolite – hippurate in plasma or urine – that had presumably been formed intracellularly (Arends *et al.*, 1991, 1995; Thompson *et al.*, 1989). There was therefore agreement with the experimental finding that the activity of an infused tracer was always lower in the intracellular pool than in plasma (see Chapter 3). However, there are also contradictory results: in two sets of experiments with rat liver perfused with labelled glycine the activity of hippurate in the perfusate was 30–40% higher than that of free glycine in the liver (Matsushima *et al.*, 1989; Marsh and Diffenderfer, 1991). The idea of differential labelling in periportal and perivenous hepatocytes was not considered likely, because the same difference was found by retrograde perfusion; the authors concluded that there must be intracellular compartmentation of glycine transport. Probably as a result of these contradictions hippurate has fallen out of use in recent years.

5.2.2 Glycine-serine

Serine is another metabolite of glycine that could be used as a precursor. Fern and Garlick (1973) showed that when U-^{14}C-glycine was infused in rats, at plateau the specific activity ratio of serine:glycine in liver and muscle protein was much closer to the ratio in the tissue free amino acid pool than to the ratio in plasma, by a factor of about 2. In a second paper (Fern and Garlick, 1974) they found that when either serine or glycine was infused rates of protein synthesis calculated from the labelling in the free tissue pool were very similar, whether the precursor was taken to be the infused amino acid or its reciprocal. These experiments made it very clear that the labelling of the infused amino acid in plasma is not representative of that at the site of the precursor. There is no reason why this method should not be applied to studies in man, using either infusions of glycine with measurements on plasma serine, or vice versa. It would be interesting to compare the two combinations, but this has not been done. Fern and Garlick (1976) also found that when two different proteins were isolated, albumin and

Table 5.1. Enrichment relative to tRNA in plasma and tissue fluid in whole body studies.

Author	Animal	Method	Relative activity	
			Plasma tRNA	Tissue fluid tRNA
Muscle				
Watt *et al.*, 1991	Rat	CI-leu	1.20	0.85
Young *et al.*, 1994	Rat	CI-phe	1.42	0.82
Baumann *et al.*, 1994	Pig	CI-leu	1.33	1.26
		CI-phe	1.57	1.09
Davis *et al.*, 1999	Pig	FD-phe	1.05	1.02
Caso *et al.*, 2001	Dog	CI-phe	1.84	1.14
		+ phe	1.31	0.95
		CI-leu	1.72	1.02
Caso *et al.*, 2002	Dog	CI-leu	1.70	1.06
		FD-phe	1.06	1.11
Watt *et al.*, 1991	Man	CI-leu	1.28	0.92
Ljungqvist *et al.*, 1997	Man	CI-leu	1.55	0.89
Smith and Sun, 1995	Rat	FD-val	1.52	1.37
Vidrich *et al.*,	Rat	SD-val	2.45	0.44
Liver				
Watt *et al.*, 1991	Rat	CI-leu	2.98	0.58
Baumann *et al.*, 1994	Pig	CI-leu	1.57	1.06
Davis *et al.*, 1999	Pig	FD-phe	1.06	0.97
Ahlman *et al.*, 2001	Pig	CI-leu	1.33	0.85
		CI-phe	1.41	0.89
Smith and Sun, 1995	Rat	FD-val	1.09	1.08

CI, constant infusion; FD, flooding dose; SD, single dose. Most measurements in the post-absorptive state. In a number of studies food or amino acids with or without insulin were given (Ljungqvist *et al.*, 1997; Davis *et al.*, 1999; Caso *et al.*, 2001, 2002) but they had no significant effect on the relative activities.

ferritin, the specific activity ratios of serine:glycine were significantly different in the two. This again was evidence of compartmentation.

5.2.3 Urea

Urea was introduced by McFarlane *et al.* (1963) for measuring the synthesis rates of plasma proteins. The idea, following Swick (1958), was to label the guanidino-carbon (C_6) of arginine with a single dose of [^{14}C]-carbonate; this arginine on the one hand is taken up into protein, on the other produces urea by the action of arginase. The method depends, as does the leucine-KIC method (see below) on the assumption of a common pool for synthesis and oxidation, and some doubts were expressed on this score. However, in a small number of studies in which synthesis of albumin was measured by this method and breakdown with [^{131}I] albumin, good agreement was found (Tavill *et al.*, 1968; Wöchner *et al.*, 1968) (see Chapter 16). Labelling of the C6 of arginine has been more widely used as a method of measuring decay rate (see Chapter 13).

A somewhat similar approach was used by Gersovitz *et al.* (1980). They gave frequent doses of ^{15}N-glycine, which produced a plateau in urea labelling after 45 h. The synthesis rate of albumin was then calculated from the increase in labelling of albumin arginine between 45 and 60 h and the enrichment of urea at plateau. This is too slow to be useful, even if the urea pool had been labelled with a priming dose.

5.2.4 Leucine-α-ketoisocaproate (KIC) (also denoted 4-methyl-2-oxopentanoate (MOP))

Leucine was first used for measuring protein turnover in man by O'Keefe *et al.* (1974) and Golden and Waterlow (1977). This amino acid was chosen because the small size of its free pool should cause plateau activity to be achieved within 1–2 h, which was important in clinical studies. The first step in the metabolism of leucine is reversible transamination to KIC, followed by irreversible decarboxylation (Chapter 4, section 4.2). Matthews *et al.* (1982) were the first to propose that the enrichment of KIC in plasma would be a good measure of intracellular enrichment, and introduced the useful term 'reciprocal pool'. Matthews found that the enrichment ratio of plasma KIC to plasma leucine was about 0.7–0.8 and varied little with the level of protein intake or whether the subjects were in the fed or fasted state. This ratio has been fully confirmed in numerous other studies (Table 5.2.). It is clearly remarkably stable. More important is the ratio of activity of plasma KIC to that of t-RNA. This has not been measured very often, but Table 5.3 gives such results as there are. There is quite a wide scatter, but the suggestion is that KIC will overestimate the precursor enrichment and underestimate the synthesis rate by 15–20%.

Further experiments were done in which ^{3}H-leucine and ^{14}C-KIC were infused together (Schwenk *et al.*, 1985; Chinkes *et al.*, 1996). The first paper gives the impression that it is immaterial which couple one chooses as primary/reciprocal, as would be expected if the transamination is reversible and rapid. However, the paper of Chinkes *et al.* (1996) reached a different conclusion: they found that when KIC was infused, enrichment of its reciprocal, leucine, in plasma agreed better with that of tissue leucine, than the reciprocal, KIC, when leucine was infused. They therefore recommended that a 'reverse' infusion of KIC gave a better estimate of precursor activity. This suggestion has not been generally followed up. It is worth mentioning that the relationship of leucine and KIC to the true precursor leucyl-tRNA is not symmetrical.

If leucine is infused (a) the sequence is:

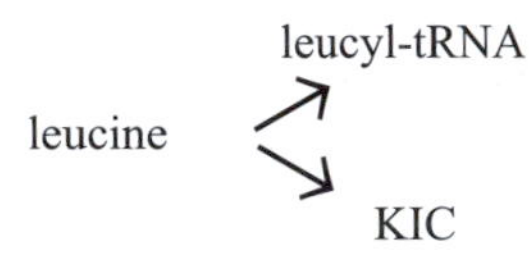

i.e. two products of the same precursor.

Table 5.2. Ratio of enrichment of plasma KIC to enrichment of plasma leucine after infusions of labelled leucine.

	ε KIC/ε plasma leucine		
	No. of studies	Mean	Range
Post-absorptive	29	0.79	0.63–0.91
Fed	9	0.78	0.60–0.92
+ insulin	5	0.77	0.62–0.88

Table 5.3. Relative enrichment of plasma KIC to that of tRNA.

Author	Animal	ε KIC/ε tRNA
Muscle		
Rennie *et al.*, 1991	Rat	0.96
Watt *et al.*, 1991	Rat	1.33
Caso *et al.*, 2002	Dog	1.26
Baumann *et al.*, 1994	Pig	0.89
Watt *et al.*, 1991	Man	1.18
Ljungqvist *et al.*, 1997	Man	1.53
	Mean, unweighted	1.19
Liver		
Watt *et al.*, 1991	Rat	1.57
Baumann *et al.*, 1994	Pig	0.89
Boirie *et al.*, 2001	Pig	1.04
Ahlman *et al.*, 2001	Pig	1.07
	Mean, unweighted	1.14

If KIC is infused (b) the sequence is:

KIC → leucine → leucyl-tRNA

which, in spite of the speed of transamination, may not come to the same thing.

5.2.5 Glucose metabolites

Jahoor *et al.* (1994) infused U-^{14}C-glucose in infant pigs and compared the activity of alanine derived from glucose in the rapidly turning over protein apolipoprotein B-100 (see below) with that of alanine in plasma and found perfect agreement. When leucine was infused the agreement was less good; the ratio of ε-leucine in apoB: εKIC in plasma was only 0.74. This approach, using alanine as a 'reciprocal' of glucose, seems promising.

An ingenious application of the same principle was published by Hellerstein *et al.* (1986, 1987) and Hellerstein and Munro (1987). UDP-glucose[1] formed in the liver is the precursor of both glucuronic acid and of the galactose moiety of α_1-acid glycoprotein, an important acute-phase protein in the response to inflammation. Because the glycoprotein is synthesized relatively slowly, it required 40 hours of glucose infusion to bring it to a plateau, but when this had been achieved activity in the glycoprotein was the same as that in urinary glucuronic acid, demonstrating that there was no compartmentation of the precursor. In a subsequent paper Hellerstein and Munro (1987) used the activity in glucuronic acid to measure the rise and fall of glycoprotein synthesis produced by an injection of turpentine.

5.3 A Rapidly Synthesized Protein as Proxy for Precursor

The most recently proposed solution to the precursor problem has been to find a protein that turns over so rapidly that within the period of observation its activity has come to a plateau, which may be taken to be the true precursor activity. The pancreatic enzyme described by Bennet *et al.* (1993) would be a good candidate, since its activity reached plateau in about 6 h from the start of infusion, but obviously this protein would not be suitable for general use because of the difficulty of sampling it. Studies on the turnover rates of the very low density lipoproteins showed that apolipoprotein B100 (apo-B for short) reached plateau labelling in 8–15 h (Cryer *et al.*, 1986). Since apo-B can be sampled in the plasma it is suitable for the practical measurement of precursor activity, but obviously it is only applicable to proteins synthesized in the liver, including export proteins.

A number of studies have compared the labelling of apo-B at plateau with that of the infused amino acid (Reeds *et al.*, 1992; Motil *et al.*, 1994; Jahoor *et al.*, 1994; Cayol *et al.*, 1996; Connell *et al.*, 1997). These results are not of great interest since it is very clear, from the evidence cited earlier, that the activity of the free amino acid in plasma does not give an adequate measure of precursor labelling; therefore this comparison is no test of the validity of apo-B as precursor. A better approach is given by a reciprocal of the infused amino acid, of which three examples have been discussed above: serine from glycine, KIC from leucine and alanine from glucose.

Arends *et al.* (1995) infused ^{15}N-glycine in human subjects and found that after 2 h the enrichments in apo-B-glycine and apo-B-serine were identical, confirming the findings of Fern and Garlick (1973); by 8 h, when apo-B was approaching plateau, labelling in apo-B glycine and apo-B serine were very close to that of plasma serine. Table 5.4 shows the enrichment ratio apo-B/plasma KIC at plateau with infusion of leucine. With the exception of Halliday *et al.*'s study (1993), shown in Fig. 5.3, many of the ratios are low, suggesting that the apo-B had perhaps not reached plateau labelling (Fig. 5.3). The low level of apo-B labelling in the fed state in Reeds' experiment perhaps reflects dilution of the precursor in the liver by amino acids absorbed from the gut, which would affect apo-B more than KIC. Similarly one might explain the high labelling of apo-B with Cayol's IG infusion of tracer by first-pass uptake in the splanchnic bed (Chapter 6).

There is unfortunately only one study in which enrichments of apo-B and tRNA have both been measured – that of Ahlman *et al.* (2001). Both leucine and phenylalanine gave a ratio εapo-B: εtRNA of ~0.5.

[1] UDP–glucose is an activated form of glucose, the substrate for glycogen synthase.

Table 5.4. Relative enrichment of apolipoprotein B-100 to that of KIC.

Author	Animal	Tracer infusion	ε apo-B/ε KIC
Connell *et al.*, 1997	Sheep	IV fasted	1.12
		IV fed	0.84
Jahoor *et al.*, 1994	Pig	IV ?	0.74
Ahlman *et al.*, 2001	Pig	IV basal	0.63
		IV + insulin and AAs	0.72
Halliday *et al.*, 1993	Man	IV post-absorptive	1.00
Reeds *et al.*, 1992[a]	Man	IV post-absorptive	I 0.96
		post-absorptive	II 1.01
		fed	I 0.72
		fed	II 0.79
Cayol *et al.*, 1996	Man	IV post-absorptive	0.80
		IG[b] post-absorptive	1.56

[a]In this study leucine was infused for 48 h in the consecutive 12 h periods – fed-fasting-fed-fasting. There undoubtedly was recycling (Fig. 5.3), but both apo-B and KIC should be equally affected.
[b]By nasogastric tube.

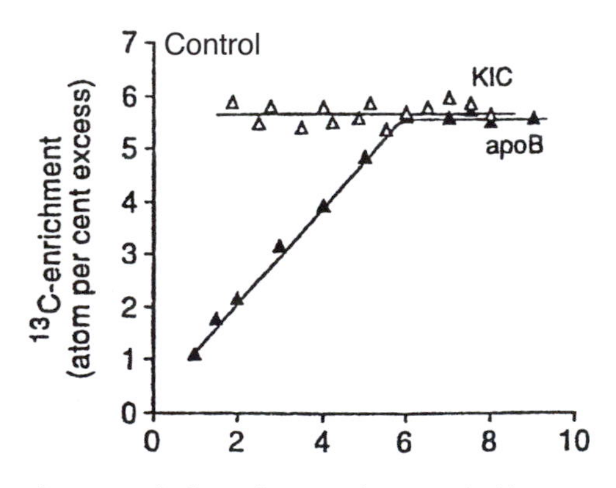

Fig. 5.3. Labelling of KIC in plasma and of leucine in apo B-100 during a 10-h infusion of [^{13}C]-leucine. Reproduced by permission from Halliday *et al.* (1993), by courtesy of the *American Journal of Clinical Nutrition* © Am. J. Clin. Nutr. American Society for Nutrition.

5.4 Conclusion

The most useful reciprocals seem to be leucine/KIC and glycine/serine. The former has the great advantage that in whole body studies it enables oxidation to be measured as well as synthesis. It is logical, therefore, that leucine/KIC should be adopted as the gold standard, at least for measurements on the whole body. This approach has been criticized on the ground that enrichment of KIC in plasma reflects mainly that of the precursor in muscle (Garlick *et al.*, 1997; Dangin *et al.*, 2001); however, most tissues produce KIC (see Table 4.1) and its activity in plasma must be the average of that in different tissues, weighted according to its production rate in those tissues. Moreover, as Table 5.3 shows, the relative enrichment in liver is little different from that in muscle.

The fact that KIC overestimates the precursor labelling (Table 5.3) and hence underestimates synthesis is probably not important, since most studies are comparative.

A problem still unsolved is whether the precursor pool for oxidation is the same as that for synthesis; i.e. whether acylation and transamination occur at the same site. The experiments of Schneible *et al.* (1981) on cultured muscle cells led them to conclude that internally derived amino acids are funnelled directly to protein synthesis while the oxidative machinery uses mainly amino acids from external sources. This is in agreement with the model of Mortimore *et al.* (1972) (Fig. 5.1), but in contrast to that of Airhart *et al.* (1974), who took the position that 'amino acids from an extracellular pool funnel directly into the synthetic machinery'. The very small amount of available information summarized in Table 5.3 appears to favour Schneible's view, since KIC is on average, more highly labelled than tRNA. There is clearly a need for more measurements in the same study of the enrichments of KIC, apo-B and tRNA in muscle and in liver.

It has been said that whenever there is a discrepancy, attribute it to 'compartmentation'. Fern and Garlick's paper (1976) is important because it extends the concept of compartmentation from different sites of synthesis and oxidation to differences in the sites of synthesis of individual proteins. They discuss one possible explanation: that albumin is synthesized by membrane-bound ribosomes attached to the endoplasmic reticulum, while ferritin is thought to be synthesized by free ribosomes. The word 'compartmentation' may give a wrong impression; it seems more likely that there is a gradient or series of gradients, with labelling of the tracer amino acid highest at points near the cell surface and lowest at points near the sites of degradation. Moreover, as has often been pointed out, the cell is not just a watery bag of enzymes but has a highly complex internal structure. It seems entirely possible that the gradients differ locally, according to the rates at which amino acids penetrate through the crevices of the cytoplasmic gel.

5.5 References

Ahlman, B., Charlton, M., Fu, A., Berg, C., O'Brien, P. and Nair, K.S. (2001) Insulin's effect on synthesis rates of liver proteins: a swine model comparing various precursors of protein synthesis. *Diabetes* 50, 947–954.

Airhart, J., Vidrich, A. and Khairallah, E.A. (1974) Compartmentation of free amino acids for protein synthesis in rat liver. *Biochemical Journal* 140, 539–548.

Arends, J. and Bier, D.M. (1991) Labelled amino acid infusion studies of *in vivo* protein synthesis with stable isotope tracers and gas chromatography – mass spectrometry. *Analytica Chimica Acta* 247, 255–263.

Arends, J., Schäfer, G., Schauder, P., Bircher, J. and Bier, D.M. (1995) Comparison of serine and hippurate as precursor equivalents during infusion of [^{15}N] glycine for measurement of fractional synthesis rates of apolipoprotein B of very-low-density apolipoprotein. *Metabolism* 44, 1253–1258.

Baumann, P.K., Stirewalt, W.S., O'Rourke, B.O., Howard, D. and Nair, K.S. (1994) Precursor pools of protein synthesis: a stable isotope study in a swine model. *American Journal of Physiology* 267, E203–209.

Bennet, W.M., O'Keefe, S.J.D. and Haymond, M.W. (1993) Comparison of precursor pools with leucine, α-ketoisocaproate and phenylalanine tracers used to measure splanchnic protein synthesis in man. *Metabolism* 42, 691–695.

Boirie, Y., Short, K.R., Ahlman, B., Charlton, M. and Nair, K.S. (2001). Tissue specific regulation of mitochondrial and cytoplasmic protein synthesis rates by insulin. *Diabetes* 50, 2652–2658.

Caso, G., Ford, G.C., Nair, S.K., Vosswinkel, J.A., Garlick, P.J. and McNurlan, M.A. (2001) Increased concentration of tracee affects estimates of muscle protein synthesis. *American Journal of Physiology* 280, E937–946.

Caso, G., Ford, C., Nair, K.S., Garlick, P.J. and McNurlan, M.A. (2002) Aminoacyl-tRNA enrichment after a flood of labeled phenylalanine: insulin effect on muscle protein synthesis. *American Journal of Physiology* 282, E1029–1038.

Cayol, M., Boirie, Y., Prugnaud, J., Gachon, P., Beaufrère, B. and Obled, C. (1996) Precursor pool for hepatic protein synthesis in humans: effect of tracer route infusion and dietary proteins. *American Journal of Physiology* 270, E980–987.

Chinkes, D., Klein, S., Zhang, X.-J. and Wolfe, R.R. (1996) Infusion of KIC is more accurate than labeled leucine to determine human muscle protein synthesis. *American Journal of Physiology* 270, E67–71.

Connell, A., Calder, A.G., Anderson, S.E. and Lobley, G.E. (1997) Hepatic protein synthesis in the sheep: effect of intake as monitored by use of stable-isotope-labelled glycine, leucine and phenylalanine. *British Journal of Nutrition* 77, 255–271.

Cryer, D.R., Matsushima, T., Marsh, J.B., Yudkoft, M. and Coates, P.M. (1986) Direct measurement of apolipoprotein B synthesis in human very low density lipoprotein using stable isotopes and mass spectrometry. *Journal of Lipid Research* 27, 508–516.

Dangin, M., Boirie, Y., Garcia-Rodenas, C., Gachon, P., Fauquant, J., Callier, P., Ballèvere, O. and Beaufrère, B. (2001) The digestion rate of protein is an independent regulating factor of postprandial protein retention. *American Journal of Physiology* 280, E340–348.

Davis, T.A., Fiorotto, M.L., Nguyen, H.V. and Burrin, D.G. (1999) Aminoacyl-tRNA and tissue free amino acid pools are equilibrated after a flooding dose of phenylalanine. *American Journal of Physiology* 277, E103–109.

Fern, E.B. and Garlick, P.J. (1973) The specific radioactivity of the precursor pool for estimates of the rate of protein synthesis. *Biochemical Journal* 134, 1127–1130.

Fern, E.B. and Garlick, P.J. (1974) The specific radioactivity of the tissue free amino acid pool as a basis for measuring the rate of protein synthesis in the rat *in vivo*. *Biochemical Journal* 142, 423–425.

Fern, E.B. and Garlick, P.J. (1976) Compartmentation of albumin and ferritin synthesis in rat liver *in vivo*. *Biochemical Journal* 156, 189–192.

Garlick, P.J., McNurlan, M.A. and Caso, G. (1997)

Critical assessment of methods used to measure protein synthesis in human subjects. *Yale Journal of Biology and Medicine* 70, 1–12.

Gersovitz, M., Munro, H.N., Udall, J. and Young, V.R. (1980) Albumin synthesis in young and elderly subjects using a new stable isotope methodology: response to level of protein intake. *Metabolism* 29, 1075–1086.

Golden, M.H.N. and Waterlow, J.C. (1977) Total protein synthesis in elderly people: a comparison of results with [^{15}N] glycine and [^{14}C] leucine. *Clinical Science Molecular Medicine* 53, 277–288.

Halliday, D., Venkatesan, S. and Pacy, P. (1993) Apolipoprotein metabolism: a stable isotope approach. *American Journal of Clinical Nutrition* 57, 726S–731S.

Hellerstein, M.K., Greenblatt, D.J. and Munro, H.N. (1986) Glycoconjugates as non-invasive probes of intrahepatic metabolism: pathways of glucose entry into compartmentalized hepatic UDP-glucose pools during glycogen accumulation. *Proceedings of the National Academy of Sciences, USA.* 83, 7044–7048.

Hellerstein, M.K., Greenblatt, D.J. and Munro, H.N. (1987) Glycoconjugates as non-invasive probes of intrahepatic metabolism: I. Kinetics of label incorporation with evidence of a common precursor UDP-glucose pool for secreted glycoconjugates. *Metabolism* 36, 988–994.

Hellerstein, M.K. and Munro, H.N. (1987) Glycoconjugates as non-invasive probes of intrahepatic metabolism: II. Application to measurement of plasma α-1-acid glycoprotein turnover during inflammation. *Metabolism* 36, 995–1000.

Horber, F.F., Horber-Feyder, F.M., Krayer, S., Schwenk, W.F. and Haymond, M.W. (1989) Plasma reciprocal pool specific activity predicts that of intracellular free leucine for protein synthesis. *American Journal of Physiology* 257, E385–E399.

Jahoor, F., Burrin, D.G., Reeds, P.J. and Frazer, M. (1994) Measurement of plasma protein synthesis rate in infant pig: an investigation of alternative tracer approaches. *American Journal of Physiology* 267, R221–227.

Ljungqvist, O.H., Persson, M., Ford, C. and Nair, S.K. (1997) Functional heterogeneity of leucine pools in human skeletal muscle. *American Journal of Physiology* 273, E564–570.

Marsh, J.B. and Diffenderfer, M.R. (1991) Use of [^{15}N] glycine in the measurement of apolipoprotein B synthesis in perfused rat liver. *Journal of Lipid Research* 32, 2019–2024.

Martin, A.F., Rabinowitz, M., Blough, R., Prior, G. and Zak, R. (1977) Measurements of half-life of rat cardiac myosin heavy chain with leucyl-tRNA used as precursor pool. *Journal of Biological Chemistry* 252, 3422–3429.

Matsushima, T., Cryer, D.R., Winkler, K.E., Marsh, J.B. and Cortner, J.A. (1989) Measurement of apolipoprotein B synthesis in perfused rat liver using stable isotopes: [^{15}N] hippurate as a measure of the intracellular [^{15}N] glycine precursor enrichment. *Journal of Lipid Research* 30, 841–846.

Matthews, D.E., Schwartz, H.P., Yang, R.D., Motil, K.G., Young, V.R. and Bier, D.M. (1982) Relationship of plasma leucine and α-ketoisocaproate during a L-[^{13}C] leucine infusion in man: a method for measuring human intracellular tracer enrichment. *Metabolism* 31, 1105–1112.

McFarlane, A.S. (1963) Measurement of synthesis rates of liver produced plasma proteins. *Biochemical Journal* 89, 277–290.

Mortimore, G.E., Woodside, K.H. and Henry, J.E. (1972) Compartmentation of free valine and its relation to protein turnover in rat liver. *Journal of Biological Chemistry* 247, 2776–2784.

Motil, K.J., Opekun, A.R., Montandon, C.M., Berthold, H.R., Davis, T.A., Klein, P. and Reeds, P.J. (1994) Protein oxidation changes rapidly after dietary protein intake is altered in adult women but lysine flux is unchanged as is lysine incorporation into VLDL-apolipoprotein B-100. *Journal of Nutrition* 124, 41–51.

O'Keefe, S.J.D., Sender, P.M. and James, W.P.T. (1974) 'Catabolic' loss of nitrogen in response to surgery. *Lancet* ii, 1035–1037.

Polin, R.A., Yoder, M.C., Douglas, S.D., McNelis, W., Nissim, I. and Yudkoff, M. (1989) Fibronectin turnover in the premature neonate measured with [^{15}N] glycine. *American Journal of Clinical Nutrition* 49, 314–319.

Reeds, P.J., Hachey, D.L., Patterson, B.W., Motil, K.J. and Klein, P.D. (1992) VLDL apoliproprotein B-100, a potential indicator of the isotopic labeling of the hepatic protein synthetic precursor pool in humans: studies with multiple stable isotopically labeled amino acids. *Journal of Nutrition* 122, 457–466.

Rennie, M.J., Bennett, W.M., Watt, P.W., Connacher, A.A., Smith, K. and Jung, R.T. (1991) Protein turnover and regulation of muscle mass in man. In: Rothwell, N.J. and Stock, M.J. (eds) *Obesity and Cachexia.* John Wiley, New York, pp. 49–62.

Rosenberg, L.E., Berman, M. and Segal, S. (1963) Studies of the kinetics of amino acid transport, incorporation into protein and oxidation in kidney cortex slices. *Biochemica et Biophysica Acta* 71, 664–675.

Schwenk, W.F., Beaufrère, B. and Haymond, M.W. (1985) Use of reciprocal pool specific activities to model leucine metabolism in man. *American Journal of Physiology* 249, E646–650.

Schneible, P.A., Airhart, J. and Low, R.B. (1981) Differential compartmentation of leucine for oxidation and for protein synthesis in cultured skeletal muscle. *Journal of Biological Chemistry* 256, 4888–4894.

Smith, C.B. and Sun, Y. (1995) Influence of valine flooding on channeling of valine into tissue pools and on protein synthesis. *American Journal of Physiology* 268, E735–744.

Stein, T.P., Leskin, M.J. and Wallace, H.W. (1978) Measurement of half-life of human plasma fibrinogen. *American Journal of Physiology* 234, E504–510.

Swick, R.W. (1958) Measurement of protein turnover in rat liver. *Journal of Biological Chemistry* 231, 751–763.

Tavill, A.S., Craigie, A. and Rosenoer, V.M. (1968) The measurement of the synthetic rate of albumin in man. *Clinical Science* 34, 1–28.

Thompson, C., Blumenstock, F.A., Saba, T.M., Feuster, P.J., Kaplan, J.E., Fortune, J.B., Hough, L. and Gray, V. (1989) Plasma fibronectin synthesis in normal and injured humans as determined by stable isotope incorporation. *Journal of Clinical Investigation* 84, 1226–1235.

Vidrich, A., Airhart, J., Bruno, M.K. and Khairallah, E.A. (1977) Compartmentation of free amino acids for protein synthesis. *Biochemical Journal* 162, 257–266.

Wallyn, C.S., Vidrich, A., Airhart, J. and Khairallah, E.A. (1974) Analysis of the specific radioactivity of valine isolated from aminoacyl-transfer ribonucleic acid of rat liver. *Biochemical Journal* 140, 545–548.

Waterlow, J.C., Garlick, P.J. and Millward, D.J. (1978) *Protein Turnover in Mammalian Tissues and in the Whole Animal.* North Holland, Amsterdam.

Watt, P.W., Lindsay, Y., Scrimgeour, C.M., Chien, P.A., Gibson, J.N.A., Taylor, D.J. and Rennie, M.J. (1991) Isolation of aminoacyl-tRNA and its labeling with stable-isotope tracers: use in studies of human protein synthesis. *Proceedings of the National Academy of Sciences, USA* 88 5892–5896.

Wöchner, R.D., Weissman, S.M., Waldmann, T.G., Houston, D. and Berlin, N.I. (1968) Direct measurement of the rates of synthesis of plasma proteins in control subjects and patients with gastrointestinal protein loss. *Journal of Clinical Investigation* 47, 971–982.

Young, L.H., Shrewalt, W., McNulty, P.H., Revkin, J.H. and Barrett, E.J. (1994). Effect of insulin on rat heart and skeletal muscle phenylalanine – tRNA labeling and protein synthesis *in vivo. American Journal of Physiology* 267, E337–342.

Yudkoff, M., Nissin, I., McNellis, W. and Polin, R.A. (1987) Albumin synthesis in premature infants: Determination of turnover with [^{15}N] glycine. *Pediatric Research* 21, 49–53.

6

Precursor Method: Whole Body Protein Turnover Measured by the Precursor Method

6.1 Background

According to Wagenmakers (1999) 'the L-[1-^{13}C] leucine method has survived two decades of intense scrutiny and is now considered the reference method to obtain fair estimates of whole-body protein metabolism in most physiological conditions'. This judgement is justified for most purposes, but it took many years for the method to achieve this position.

The starting point was the simultaneous but independent publication of three papers on continuous infusion of labelled amino acids in the rat (Loftfield and Harris, 1956; Gan and Jeffay, 1967; Waterlow and Stephen, 1967). We used ^{14}C-lysine and Gan also used ^{14}C-tyrosine. They showed that the specific radioactivity of free lysine in plasma approached a plateau in 1–3 h; both groups observed that the activities of free lysine in liver and muscle also reached plateaus, which were at lower levels than that in plasma (Gan and Jeffay, 1967, 1971; Waterlow and Stephen, 1968). From these observations it was possible to calculate rates of synthesis of liver and muscle proteins which were more accurate than previous observations derived from decay curves. We were particularly interested in deriving estimates of whole body protein turnover (WBPT) from the plateau activity of plasma free lysine, according to the formula: $Q = {}^{d}\!/_{\varepsilon}$, described in Chapter 2, and applied the method in three human subjects (Waterlow, 1967). A list of earlier measurements of WBPT by compartmental analysis is given in that paper. Because of the limits of permissible radioactive dosage, counts in plasma leucine were as low as 2 dpm, so the results cannot be considered very accurate. The problems with radioactive isotopes turned our minds towards the use of ^{15}N-glycine, and in the decade of the 1970s only three papers appeared describing measurements of WBPT with radioactive tracers. Studies were made with ^{14}C-leucine in surgical patients (O'Keefe *et al.*, 1974) and in the elderly (Golden and Waterlow, 1977), and with U^{14}C-tyrosine in normal male volunteers (James *et al.*, 1976).

The mid-1970s saw a breakthrough with the development of gas chomatography-mass spectrometry (GCMS) that made possible the measurement of stable isotope concentrations in the small samples of free amino acids that could be isolated from plasma. This methodology was first applied to ^{15}N-lysine (Halliday and McKeran, 1975; Bier and Christopherson, 1979; Conway *et al.*, 1980) and then to [1-^{13}C]-leucine (Matthews *et al.*, 1980). Descriptions of the instrumentation available at that time are given by Bier and Matthews (1982) and Halliday and Rennie (1982). Hard on the heels of these developments came the proposal by Matthews *et al.* (1981) to use α-ketoisocaproic acid (KIC) as a 'reciprocal' of leucine to provide a better measure of the precursor activity (see Chapter 5). Other amino acids labelled with stable isotopes have indeed been used for measuring WBPT, in particular $^{2}H_{2}$-leucine and $^{2}H_{5}$-ring phenylalanine. A number of other amino acids were used from 1986 onwards by Young's group at MIT, specifically for measuring oxidation in carbon balance studies of amino acid requirements. These included lysine (Meredith *et al.*, 1986), threonine (Zhao *et al.*, 1986) and valine (Meguid *et al.*, 1986), but the reciprocal method with ^{13}C-leucine remained, as Wagenmakers (1999) has said, the gold standard.

The only point that has not yet been agreed is whether the enrichment of KIC should be used for calculating both flux and oxidation, or oxidation only (Cortiella *et al.*, 1988; Garlick *et al.*, 1997; Dangin *et al.*, 2001) (see Chapter 4, section 4.1).

6.2 Outline of the Method

The bare bones of the procedure with leucine are as follows: after an initial blood sample to measure base-line leucine concentration and enrichment an intravenous infusion of [1-^{13}C] leucine is set up and continued for 3–8 h. Beyond 8 h recycling (see below) may become a problem. The infusion is into an antecubital vein, and samples are taken from a hand vein on the opposite side, arterialized by warming. It is usual to give priming doses of both ^{13}C-leucine and ^{13}C-bicarbonate to hasten the rise to plateau, which is generally achieved in 1–2 h. At plateau several (four or more) blood samples are taken for measuring enrichment of leucine and KIC and to check that a reasonable plateau with zero slope has been achieved. Breath CO_2 is collected by frequent sampling to determine VCO_2 and its enrichment. If the experiment involves feeding, care has to be taken that the natural abundance of ^{13}C in the food protein is not too different from that in body protein. In the calculations of flux and oxidation, since the stable isotope tracer is not massless, a correction has to be made to the standard flux formula: $Q = d/\varepsilon_p$ (Chapter 4) by subtracting the mass of tracer infused; the corrected flux is:

$$Q_{corr} = d\left(\frac{\varepsilon_i}{\varepsilon_p} - 1\right)$$

where d is the rate of infusion of tracer (μmol h^{-1}), ε_i its enrichment and ε_p the enrichment in plasma.

Enrichment is calculated as:

$\frac{[^{13}C]}{[^{12}C]+[^{13}C]}$ rather than as the molar ratio $\frac{[^{13}C]}{[^{12}C]}$

(Slater *et al.*, 2001).

This short description of the method raises many points that need to be discussed in more detail (see below).

6.3 Variability of Whole Body Synthesis Rates in Healthy Adults by the Precursor Method

A widely used protocol is for the infusion of leucine to be started with the subject in the post-absorptive state, having eaten nothing overnight. After 2½–3 h, when it is expected that enrichment will have reached a plateau, blood samples are taken and feeding or infusion of amino acids is begun. The tracer infusion is continued for another 3–5 h, when further samples are taken. This protocol is referred to as a fasting/feeding study. The subjects for the results in Table 6.1 were all healthy, young or in a few cases middle-aged. Elderly people have not been included. No distinction has been made between men and women; usually when people of both sexes have been included in a study the results have not been reported separately except for Volpi *et al.* (1998). The results reported here are mainly for the post-absorptive state, whether or not it was followed by a period of feeding, since the purpose is to examine the variability of the method under standardized conditions. They are for synthesis rather than flux, in order to avoid the complicating factor of oxidation, which even in the post-absorptive state varies a little with the protein content of the preceding diet (Waterlow, 1996).

Two series have been analysed: the first includes 65 studies from America, Britain, France, Italy and India, containing in all 502 subjects. In fasting/feeding studies, when two or more levels of protein were fed, each preceded by a separate baseline post-absorptive measurement, they have been regarded as separate studies. A sub-series is the mammoth experiment of Marchini *et al.* (1993) (Group E) in which three diets with different amino acid patterns were fed. All the post-absorptive measurements are included here. The results of both series are shown in Table 6.1.

In the whole series (A) the variation, both between studies and between individuals within each study, is high. There is clearly a tendency for some laboratories to produce high results and others low ones. It might be that the variation would be larger in earlier studies when workers were less experienced, but no chronological effect was found.

The purpose of separating out sub-series E is to show what can be achieved by a rigidly

Table 6.1. Whole body uptake of leucine to protein synthesis[a] in healthy normal adults. Marchini's study analysed separately.

	No. of groups[b]	No. of subjects	Synthesis μmol leucine kg^{-1} h^{-1} [c]	SD group[d]	SD individuals[e]
Main series					
Post-absorptive (PA)					
A. Whole series	65	502	103.5	22.9	18.1
B. PA without feeding	22	197	105.3	24.8	20.9
C. PA before feeding	32	223	99.4	16.7	11.8
D. PA before amino acids intravenously	11	82	102.3	27.6	24.4
On feeding					
E[f].	32	171	97	18.9	17.5
Marchini's series					
F. Post-absorptive	8	52	82.1	3.6	9.7
G. Fed	8	52	82.2	3.0	12.2

[a]Synthesis rates were calculated with KIC as precursor. In a few cases where the precursor enrichment was that of plasma leucine the values have been multiplied by 1.25.
[b]A group represents subjects who received the same treatment. There might be several groups in the same study.
[c]Weighted mean.
[d]Calculated as $\sqrt{\sum [(x_i - x)^2 . n_i]/N - 1}$.
[e]Calculated as $\sqrt{\sum [(SD_i^2 . (n_i - 1)]/N}$.
[f]Four studies from C were omitted.
References*:* Schwenk *et al.*, 1985b; Castellino *et al.*, 1987; Gelfand and Barrett, 1987; Schwenk and Haymond, 1987; Young *et al.*, 1987; Conway *et al.*, 1988; Cortiella *et al.*, 1988; Gelfand *et al.*, 1988; Pacy *et al.*, 1988, 1994; Beaufrère *et al.*, 1989; Fukagawa *et al.*, 1989; Melville *et al.*, 1989, 1990; Thompson *et al.*, 1989; Bruce *et al.*, 1990; Louard *et al.*, 1990; Hoerr *et al.*, 1991; McHardy *et al.*, 1991; Tessari *et al.*, 1991, 1995, 1996b,c,d, 1999; 1992; Heslin *et al.*, 1992; Reaich *et al.*, Beshiyah *et al.*, 1993; Marchini *et al.*, 1993; Fong *et al.*, 1994; Russell-Jones *et al.*, 1994; El-Khoury *et al.*, 1995; Nair *et al.*, 1995; Tauveron, 1995; Gibson *et al.*, 1996; Giordano *et al.*, 1996; Balagopal *et al.*, 1997; Boirie *et al.*, 1997; Bouteloupe-Demange *et al.*, 1998; Fereday *et al.*, 1998; Forslund *et al.*, 1998; Macallan *et al.*, 1998; Volpi *et al.*, 1998; Zanetti *et al.*, 1999; Millward *et al.*, 2000; Kurpad *et al.*, 2001.

standardized technique in a single experienced laboratory (Massachusetts Institute of Technology). It is not clear why the mean synthesis rate is substantially lower than in the other sub-groups. However, the variation between studies was reduced to a very low level. On the other hand, the between-subject variability had a coefficient of variation of about 10%. Thus, Welle and Nair (1990) in a study on 47 men and women, found a CV of 10% in leucine flux when related to total body potassium. This will not remove all variability, because there may be variations in the inter-organ distribution of lean body mass. Garby and Lammert (1994) showed that between-subject variation in energy expenditure could be largely explained by variations in organ size.

There is also the contribution of within-subject variation from one day to another. There is little information about this, but in one study fluxes were compared in the same six subjects on two separate occasions, with a mean difference of only 3.8% (Welle and Nair, 1990).

6.4 Sites of Administration and of Sampling

This subject has raised some difficult problems and the extensive literature is quite confusing. Katz (1982, 1992) produced a detailed experimental and mathematical study of it. He distinguished between infusion into a vein with sampling from an artery – the so-called VA mode – and infusion into an artery with sampling from a vein – the AV mode. He favoured the latter, and stated that 'the ideal AV mode is to supply tracer uniformly to all tissues and to sample the mixed effluent from all parts of the body. Injections into the aorta and sampling from the vena cava appear to be the most practical (!) approach to this goal. However, alternative methods, providing that the capillary bed is the site of administration and sampling, *may be adequate*'.

The classification into AV and VA modes is not helpful. For example, are infusions into the pulmonary artery (Pell *et al.*, 1983) arterial or venous? The important points surely are as follows: (i) sampling must be downstream from the infusions; and (ii) a distinction has to be made between sampling before and sampling after the blood has passed through the tissues (Fig. 6.1).

The flux that we are concerned to measure is the disposal rate, which is determined by the extent to which the tracer has been taken up by the tissues and its enrichment at the site of disposal. This uptake is the difference between the amount of tracer delivered in the arterial supply and the amount removed in the mixed venous effluent, and in the isotopic steady state is equal to the rate of tracer dosage. Thus:

$d = V_A.\varepsilon_A - V_V.\varepsilon_V$ where V_A and V_V are the fluxes into and out of the tissues, and ε_A and ε_V are enrichments in arterial and mixed venous blood. V_A and V_V are not equal, as can be seen in Fig. 2.6, Chapter 2, where they correspond to V_{BA} and V_{AB}. These fluxes are not easily measured, but Layman and Wolfe (1987) got round the difficulty by substituting plasma flow for flux, since it could be assumed that the flow rates into and out of the tissues are the same. The equation then becomes:

$d = (F.C_A + I)\ \varepsilon_A - F.C_V\varepsilon_V$ where F = plasma flow, I is the mass of infused tracer, and C is the tracee concentration in plasma, arterial and venous. Layman and Wolfe (1987) assumed that C_A and C_V were equal, which is unlikely to be the case. Nevertheless, with this assumption, in experiments on dogs they calculated the expected value of ε_V for a number of metabolites, including leucine and alanine, and found good agreement with the observed values.

The drawback to this approach is the need to measure plasma flow as well as tracer concentrations. The alternative is to abandon any attempt to make measurements on the venous side, and to concentrate on getting the best estimate that we can of the enrichment of the precursor at the intracellular site of protein synthesis and oxidation. As discussed in Chapter 5, the enrichment

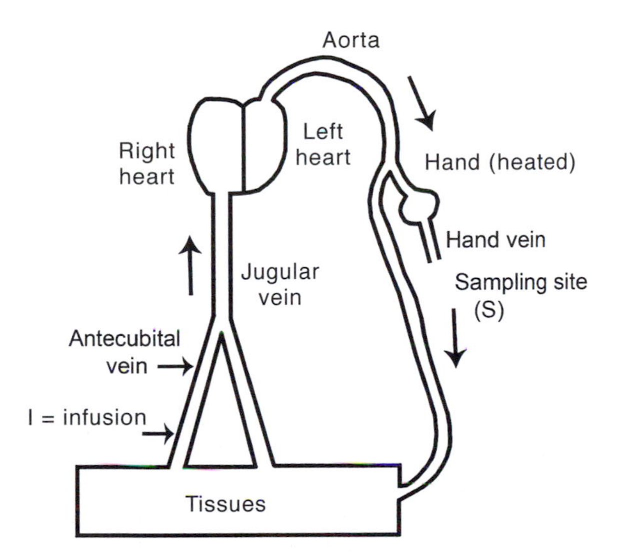

Fig. 6.1. Model to illustrate the sampling site. The infusion of tracer at I is upstream from the sampling site of a heated vein, S, which is equivalent to an arterial site. Arrows denote direction of blood flow. Using KIC as a precursor is a 'proxy' for sampling the venous effluent from the tissues.

of KIC in plasma is probably the best approximation that is available so far. In the study of Layman and Wolfe (1987) ε_V was not significantly different from ε_{KIC}; it was about 20% lower than ε_A and about 15% higher than the enrichment of free leucine in liver, muscle, gut and kidney.

This is the approach that is used in the technique proposed by Abumrad *et al.* (1981), now widely adopted and part of the standard method. The tracer is infused into an ante-cubital vein, passes into the vena cava and thence through the right side of the heart and the pulmonary circulation to the aorta. In passing through the lungs the blood loses only a small amount of tracer, estimated by Pell *et al.* (1983) as about 4%. At the sampling site in the hand, which is kept in a box heated to about 65°C, the blood flows through very rapidly and perfuses only a very small amount of tissue, so that it is in effect arterial blood. Copeland *et al.* (1992) have shown good agreement in leucine enrichment between blood from the aorta and from a heated hand vein. Thus, contrary to Katz, the method is anatomically in the V-A mode, but the key point is that the infusion is upstream from the sampling site, and far enough up to be well mixed before it reaches that site.

6.5 Priming

The prime is a bolus dose of tracer given at the beginning of the infusion, the object being to reach a plateau as nearly as possible instantaneously. It is intuitively obvious that, for this to be achieved, the curve of falling enrichment after the bolus should be the mirror image of the rising enrichment with the infusion (Fig. 6.2). This requires that the initial enrichment from the bolus should be equal to the plateau enrichment produced by the infusion, which is easily proved.

On the assumption of instantaneous mixing, the initial enrichment, ε_{bo}, produced by the bolus = p/M, where p is the priming dose of tracer and M is the pool (mass) of tracee (i.e. volume × concentration) in which it is diluted. Likewise, with the infusion,

$$\varepsilon_{max} = d/Q$$

where Q is the flux and d the rate of infusion of tracer;

$$\therefore p/M = d/Q$$

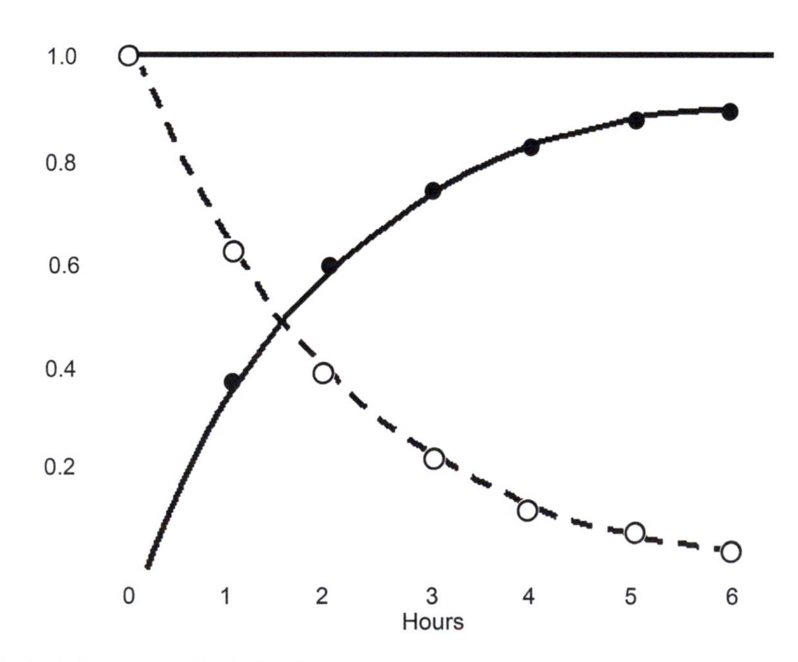

Fig. 6.2. Ideal priming curve with a bolus dose of tracer at the beginning of a continuous infusion. S_B = activity produced by the prime ○; S_I = activity produced by the infusion ●; at any point of time the sum of the two activities = plateau activity $S_{I\,max}$.

but Q = kM, where k is the fractional turnover rate of the pool. Thus:

$$p/\mathrm{M} = d/k\mathrm{M} \text{ and } p/d = 1/\mathrm{k}$$

k, of course, is not known, but it might be estimated from experience as between about 0.01 and 0.02 min^{-1} (Waterlow and Stephen, 1968). This would mean that a prime would be 50–100 times the amount infused in 1 minute. Very often a prime is chosen to be equal to 1 hour of infusion. It is impossible, without independent assessment of k, to choose a priming dose that exactly fits the criteria, but an approximation to 'ideal' k will speed up the attainment of plateau and reduce the error in defining where the plateau starts.

6.6 The First-pass Effect

When the tracer is given by mouth or naso-gastric tube (IG) it reaches first the gut and portal-drained viscera, where some of it is taken up ('sequestered'), while the remainder goes on to the peripheral circulation. As a consequence the enrichment of tracer at the sampling site (heated hand vein) is lower, and the calculated flux greater with intragastric than with intravenous administration of tracer. This is known as the 'first-pass effect'.

The proportion taken up in the first pass,

$$F = I - (\varepsilon_{IG}/\varepsilon_{IV}), \text{ or } I - Q_{IV}/Q_{IG}$$

where ε are enrichments of the tracer in plasma, Q are fluxes and the subscripts indicate the route of administration. F is measured by giving by the two routes tracers labelled with two different isotopes, e.g. ^{13}C- and ^{2}H-leucine or ^{2}H- and ^{15}N-lysine (Hoerr *et al.*, 1991, 1993). The formula based on fluxes is preferable to that based on enrichments: since $Q = d/\varepsilon$, it automatically compensates for any difference in the dose rates of the two tracers. It is also, of course, necessary to be certain that the body treats the two tracers in exactly the same way quantitatively; it is not clear that this is always the case.

An early measurement of the first-pass effect was by Cortiella *et al.* (1988) as part of a study on the effect of different levels of leucine intake. Having previously established the equivalence of ^{2}H- and ^{13}C-leucine they infused the ^{2}H tracer IV in the fasting state for 8 h; during the last 5 h the ^{13}C tracer was infused IG with food. This protocol has been widely used since then; results are shown in Table 6.2. As the table shows, studies have also been made with other essential amino acids – phenylalanine (Matthews *et al.*, 1993a) and lysine (Hoerr *et al.*, 1993) and with non-essential amino acids – glutamate/glutamine (Matthews, 1993b) and arginine (Castillo *et al.*, 1993).

Table 6.2. The 'first-pass effect': estimates of splanchnic uptake of amino acids in the fed state in normal adults.

	Number of studies	Mean uptake by splanchnic viscera (% of dose)	Range	References
Leucine	17	23	14–34	a
Phenylalanine	6	32	21–50	b
Tyrosine	1	40		c
Lysine	2	23		d
Glutamine	1	54		e
Glutamate	1	88		e
Arginine	1	37		f

a, Cortiella *et al.*, 1988; Gelfand *et al.*, 1988; Hoerr *et al.*, 1991, 1993; Biolo *et al.*, 1992; Castillo *et al.*, 1993, 1997; Krempf *et al.*, 1993; Matthews *et al.*, 1993a; Collin-Vidal *et al.*, 1994; Tessari *et al.*, 1994; Boirie *et al.*, 1997; Cayol *et al.*, 1997; Sánchez *et al.*, 1995; Arnal *et al.*, 2000; Metges *et al.*, 2000.
b, Biolo *et al.*, 1992; Krempf *et al.*, 1993; Tessari *et al.*, 1994; Sánchez *et al.*, 1995; Volpi *et al.*, 1999.
c, Basil-Filho *et al.*, 1998.
d, Hoerr *et al.*, 1993.
e, Matthews, 1993b; Reeds *et al.*, 1996.
f, Castillo *et al.*, 1993.

Matthews and Hoerr designed singularly complex protocols to examine the effect in more detail. Matthews (1993a) studied two groups of six subjects on separate days, all in the fasted state. ^{2}H- and ^{13}C-leucine were infused simultaneously, one IV and the other IG. The infusions were continued for 7 h, and halfway through the tracers were switched. In the second group of subjects the same protocol was used, except that the order of the tracers was reversed. It is thus possible to generate two sets of results, one with the same combination of tracers in different subjects, the other with different tracers infused in the same subjects.

The results of the first comparison are shown in Table 6.3A. It indicates that when the same tracer is given IV and IG, e.g. tests 1 and 2 versus 3 and 4, splanchnic uptake (SU, final column) is similar with the two tracers, even though IV and IG infusions are made in different subjects on different days. The other comparison is shown in Table 6.3B. It is clear that the combination ^{13}C-leu IV + ^{2}H-leu IG, the one most frequently used in later work, gives quite different results from the other combination. It can also be seen from the table that for comparable measurements, IV or IG, the enrichments are greater, and the fluxes lower with ^{13}C than with ^{2}H.

The protocols used by Hoerr *et al.* (1991) and by Castillo *et al.* (1993) were similar to that of Matthews but included studies in the fed state. Infusions were made on two consecutive days, on the first day with ^{2}H-leucine IV and ^{13}C-leu IG, and on the second day with the tracers reversed.

In Hoerr *et al.*'s study ^{13}C consistently gave higher fluxes than ^{2}H, with a difference of about

Table 6.3. Leucine enrichment with intravenous and intragastric infusions.

A. Results for the *same* tracer in different subjects on *different* days, normalized to equal dose rates.

Day	Test	Period	Tracer	Route	Enrichment	Mean	Splanchnic uptake %
1	1	A	^{13}C-leucine	IV	4.65	4.90	21
2	2	B	^{13}C-leucine	IV	5.16		
1	3	B	^{13}C-leucine	IG	3.78	3.88	
2	4	A	^{13}C-leucine	IG	3.98		
1	5	B	^{2}H-leucine	IV	3.57	3.52	16
2	6	A	^{2}H-leucine	IV	3.47		
1	7	A	^{2}H-leucine	IG	2.63	2.95	
2	8	B	^{2}H-leucine	IG	3.28		

B. Results with *different* tracers infused simultaneously in the *same* subjects by intravenous and intragastric routes.

Day	Period	Tracer	Route	Enrichment	Splanchnic uptake, %
1	A	^{13}C-leucine	IV	4.65	43
		^{2}H-leucine	IG	2.63	
2	B	^{13}C-leucine	IV	5.16	36
		^{2}H-leucine	IG	3.28	
1	B	^{2}H-leucine	IV	3.57	Negative
		^{13}C-leucine	IG	3.78	
2	A	^{2}H-leucine	IV	3.47	Negative
		^{13}C-leucine	IG	3.98	

Data of Matthews *et al.* (1993a).
The studies were made on two separate days in different groups of six subjects. Intravenous and intragastric infusions were given for 7 h in the fasted state. Halfway through each infusion the routes of administration were switched, so that each infusion has two periods, A and B.

15%, a point to which the authors drew attention in their paper. The difference was in the opposite direction to that in Matthews' data, and so was presumably due to technical rather than biological factors. This work illustrates some of the difficulties and uncertainties of these studies. Even when tracers agree, since SU is a measurement by difference there is inevitably much variation.

An interpretation often given to the first pass effect is that the actual flux with tracer IG is greater than that with tracer IV, and that it might be desirable, when a flux has been measured IV, to 'correct' it by adding 20–25% to compensate for splanchnic uptake (El Khoury *et al.*, 1995). We believe that this interpretation is not correct. A person, whether in the fed or fasted state, has a certain rate of amino acid flux – of synthesis, breakdown and oxidation. The body, however, has no means of sensing the tracer; because it is a tracer, the flux in reality remains the same, whatever the route of tracer administration. When tracer is given IG, the viscera take more than their fair share of it; one might say that the estimate of whole body flux is 'biased' towards the viscera, and when tracer is given IV the estimate is 'biased' towards the peripheral tissues. What it comes down to is that the two routes of dosage provide two different *estimates* of flux, neither of which is correct for the whole body. The position is exactly analogous to the two estimates of whole body flux given by the two end-products, urea and ammonia, in the end-product method of measuring flux (Chapter 7). Since the fluxes or disposal rates in the two regions, visceral and peripheral, are roughly the same, as shown by Yu *et al.* (1992 in Table 6) in dogs and Tessari *et al.* (1996b) in humans, it would be logical to get the best estimate of whole body flux by averaging the estimates from the two routes of infusion.

The average ratio from many studies of plasma leucine enrichment with IG to that with IV infusions comes to about 0.8, conventionally indicating a splanchnic uptake of 20%. If εKIC is used as a better measure of precursor activity, the ratio from the available data comes to 0.88, closer to unity.

It has been suggested that KIC will not give a representative picture of what is happening in the body as a whole, because it is formed mainly in muscle. However, this appears not to be the case. In the dogs studied by Yu *et al.* (1990) formation of KIC accounted for 40% of the metabolic uptake of leucine by the splanchnic region, and Cortiella *et al.* (1988) calculated that 20% of the KIC entering the plasma was derived from this region. On the other hand, Biolo *et al.* (1997) found that only 10% of the first-pass uptake of leucine was deaminated to KIC and concluded that 'the splanchnic region is relatively less efficient than the whole body in KIC production'. Until this question is resolved, it seems sensible to use plasma εKIC to provide the best estimate of the overall flux, whether tracer is given IV or IG. The ideal would be to give tracers by both routes and take the average.

We do not suggest that the splanchnic sequestration of tracer has no significance; it must give some information about the disposal rate in the splanchnic area, but without further information the rate cannot be quantified. Differences in splanchnic uptake may well be interesting. For example, the data in Table 6.2 suggest that, under the same conditions, the uptake is greater with phenylalanine than with leucine. This may be because proteins synthesized in the liver are richer in aromatic amino acids than the average body protein, as was pointed out by Reeds *et al.* (1994) for the acute-phase proteins. Again, the splanchnic uptake of labelled leucine was found to be almost twice as high in premature infants and in the elderly as in young adults (Beaufrère *et al.*, 1992; Boirie *et al.*, 1997). This may reflect the higher proportion of splanchnic tissue in the old and the very young.

6.7 Recycling

A distinction must be made between recycling and reutilization: the one relates to the tracer, the other to the tracee. Reutilization describes the process by which amino acids liberated by protein breakdown are reincorporated into protein, either in the same cell (internal reutilization) or in different cells or tissues (external reutilization). The same distinction can be made for recycling of tracer.

The problem of recycling is particularly serious for measurements of degradation rates from decay curves of tracer. We attempted to quantify the effect of both internal and external recycling by injecting rats with [$U^{14}C$] arginine and [^{14}C] sodium carbonate, as suggested by Swick (1958), followed by measurements on liver and serum proteins (Stephen and Waterlow, 1966). The tissues were treated with arginase + urease to determine the activity of the non-recycled guanidine carbon (C6) and with arginine decarboxylase to determine that of the recycled carboxyl-carbon (C1). With both liver and serum proteins the decay rates of the carboxyl-carbon were twice those of the guanidine-carbon, showing clearly the extent of the error produced by recycling.

In short-term measurements of synthesis, before most proteins have had time to become highly labelled, recycling nevertheless can have an important effect on the calculation of flux. If it is an infusion study recycling will convert the supposedly zero slope of the pseudoplateau into a positive slope. As a result, fluxes calculated at the end of a long infusion are lower than those from a shorter one. If it is a single-dose experiment recycling means that it will take longer for the plasma to be cleared of tracer, which again reduces the estimate of flux.

In theory, if the fractional rate of whole body protein turnover is 2.5% day^{-1} or 0.001 h^{-1}, this is the rate at which recycling will occur.[1] Since even after a whole day of infusion only 2.5% of protein will be labelled, 0.001 of this is a very small amount. However, the rate will increase as time goes on, as tissue proteins become more and more highly labelled. Gibson *et al.* (1996) calculated that if 1.25% of tracer per hour is added to the metabolic pool, the enrichment would be increased by 3.75% at 6 h and 7.5% at 9 h, leading to corresponding underestimates of the true flux.

In practice, both higher and lower rates of recycling have been observed, as summarized in Table 6.4. These rates are quite variable; to get more accurate estimates would require longer infusions, and this presents practical difficulties: the subject must be in a steady state throughout, but either fasting or feeding continuously for more than 24 h is likely to produce a non-steady state with adaptive changes in flux and plasma enrichment. Reeds *et al.* (1992) tried to get over this problem by infusing labelled amino acids for 48 h, divided into alternating 12 h periods of feeding and fasting. The percentage increases in enrichment between periods of each pair are shown in Table 6.4. These rates, like many of those in the table, are some tenfold higher than would be expected on the basis of the average turnover rate of whole body protein, and have been attributed to rapid turnover of a small pool of protein. The question then arises whether such a small pool, however fast it turned over, could produce the observed effect, or whether it would be swamped by the much larger mass of slowly turning over protein. One way to answer this question was to construct a simple model with two protein pools communicating independently with a small central pool. The fixed parameters of this model are given in Fig. 6.3 (Slevin and Waterlow, to be published). With these parameters the model predicts that a pseudoplateau of enrichment in the central pool would be reached in about 5 h, whereas it would take some 50,000 h to achieve a truly steady state in which enrichment of all protein-bound and free amino acids is equal, and equal to that of the infusate. Enrichment in the central pool at pseudoplateau is only about 1/10 of that in the state of complete equilibrium. The enrichment at plateau depends entirely on the flux out of the pool.

The total protein pool was then divided into two parts, fast and slow, varying in size but each

Table 6.4. Estimates of recycling from the literature.

Author	Period of observation	Rate of recycling (% h^{-1})
Tsalikian *et al.*, 1984	14 h	1.1
Schwenk *et al.*, 1985a	20 h	1.0
Melville *et al.*, 1989	12 h	0.7
Thompson *et al.*, 1989	26 h	0.36
Carraro *et al.*, 1991	120 days	0.12
Matthews *et al.*, 1993a	3.5 h	2.9
Reeds *et al.*, 1992	48 h	0.4

Recycling defined as rate of increase in enrichment in plasma, expressed as percentage of the enrichment at the beginning of the pseudoplateau.

[1] Observed turnover = ~5 g protein per kg body weight per day containing 20% protein.

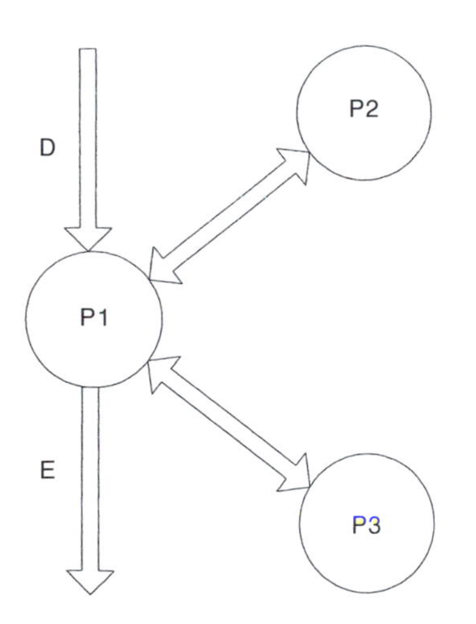

Fig. 6.3. Three-pool model for the calculation of recycling. P1 is a precursor pool containing 100 units. P2 and P3 are fast and slow protein pools with a combined content of 100,000 units; E is excretion at a rate of $0.1 \times P1\ h^{-1}$; D is material + tracer entering to balance E. Reproduced from K. Slevin, personal communication.

maintaining a constant proportion of the flux out of the central pool. The results are shown in Fig. 6.4 and Table 6.5. Case A shows that a fast pool containing only 2% of body protein could lead to a recycling rate of ~1% h^{-1}. Such a small but fast pool, composed of lipoproteins, enzymes, etc., is easily envisaged.

To ignore the effect of recycling can have quite a large influence on estimates of changes in synthesis and breakdown. Thus in a fasting-feeding study lasting 9 h, Gibson *et al.* (1996) found that, if recycling was ignored, feeding produced an increase of 11% in synthesis and a decrease of 47% in breakdown; but if recycling was assumed to occur at 1.5% h^{-1}, the increase in synthesis would rise to 27% and the decrease in breakdown would fall to 36%.

The quantitative effect of recycling on estimates of synthesis and breakdown under various conditions has been largely ignored.

6.8 Regional Turnover

6.8.1 Studies on the arm and leg

Halliday's group was the first to measure protein turnover in a single limb as opposed to the whole body. Their two papers (Cheng *et al.*, 1985, 1987) covered the fed and fasted states. The method, used by all subsequent workers, combines mea-

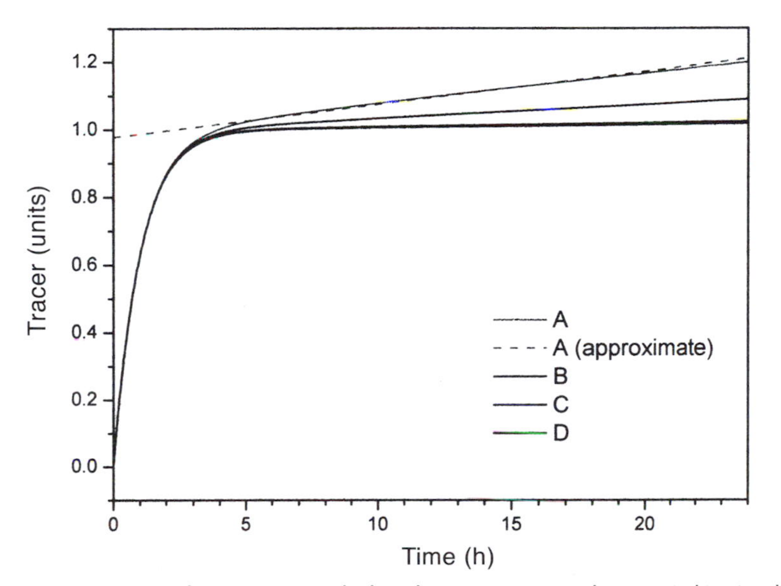

Fig. 6.4. Activity curves in the precursor pool when the two protein pools are varied in size, but have equal fluxes into and out of P1. The relative sizes of the two protein pools that were tested are given in Table 6.5. Reproduced from K. Slevin, personal communication.

Table 6.5. Slopes of post-plateau *increase in* enrichment (*recycling*) in the central pool when the total protein pool is divided into two parts of different sizes with different turnover rates, but each contributing equal amounts to the total flux.

	Protein pool size % of total		Turnover rate, $h^{-1} \times 10^3$		Slope of pseudo-plateau (% h^{-1}) (sum of fast and slow components)
	Fast	Slow	Fast	Slow	
A	2	98	22.5	0.46	0.93
B	5	95	9.0	0.47	0.38
C	5	75	1.8	0.60	0.097
D	50	50	0.9	0.90	0.073

Parameters of the model: central pool: 100 units; turnover of central pool: protein pool(s):100,000; rate 1.0 h^{-1}: excretion 0.1 h^{-1} In the above example the fast and slow protein pools each generate half the total flux, excluding excretion, i.e. 45 units. Slope of pseudo-plateau calculated from the approximate equation: slope = $[d \cdot k_{21} k_{12} \cdot t]/k_{21} + k_{12} + k_{31} + e]^2$. Reference: Slevin and Waterlow, to be published. See Fig. 6.4.

surement of blood flow and of arterio-venous (A-V) differences in amino acid concentrations across the limb – the classical method before the isotopic era – with infusion of a labelled amino acid to give A-V differences in enrichment. The model had four pools, protein, leucine, KIC and CO_2. For details of the calculations the original paper (Cheng *et al.*, 1985) should be consulted. The tracer was doubly labelled with $[1\text{-}^{13}C,^{15}N]$ leucine, from which rates of KIC formation and reamination could be calculated. Oxidation was determined from A-V differences in CO_2 concentration and enrichment, with KIC as precursor.

The results of the combined studies of Cheng *et al.* (1985, 1987) are shown in Table 6.6. There was a negative leucine balance in the fasted state, which was converted to a positive balance with infusion of amino acids, mainly through increased synthesis. The rates of leucine-KIC interchange agree well with those of Bier and Matthews (1982) in the whole body (Chapter 4).

Later workers have developed different models, e.g. Biolo *et al.* (1992) and Tessari *et al.* (1995), but they all rely on achieving a steady state and determining transport rates of tracee and tracer from A-V differences in concentration and enrichment (see Chapter 3). Biolo's model, which has been widely used, is shown in Fig. 3.1. The details of the calculation are somewhat complicated, and are set out in the original paper. Further studies on the arm, leg or splanchnic region, describing the effects of amino acids and insulin, are summarized in Chapter 9.

Studies on arm and leg have paid dividends in several ways. First, they have enabled calculation of the rate of transport of amino acids across the muscle cell membrane, described in Chapter 3. Secondly, in some cases, in addition to the infusions muscle biopsies have been taken, which made it possible to compare the direct estimate of fractional synthesis rate (FSR) of muscle from the biopsy with the indirect estimate from amino acid

Table 6.6. Leucine metabolism in the forearm.

	μmol h^{-1} per 100 ml arm	
	Fasted[a]	Fed[b]
Leucine C flux	6.2	8.4
Leucine N flux	17.8	25.0
Deamination	12.5	23.3
Reamination	11.9	19.8
Oxidation	0.3	2.6
Protein synthesis	4.2	7.6
Protein breakdown	5.0	5.2

Data of Cheng *et al.*, [a]1987, [b]1985.

uptake by the limb. Excellent agreement was found between the two methods (Biolo *et al.*, 1995a). In order to calculate the FSR from the data on the whole limb, the volume of the limb was determined by anthropometry (Jones and Pearson, 1969) and it was assumed that 60% of this volume was muscle. In limb studies the contributions of skin and bone are often ignored, and since these tissues have high rates of turnover (see below), the turnover rate of muscle protein will be inflated if no account is taken of them. Biolo *et al.* (1994) tackled this problem by experiments on dogs in which biopsies were taken of skin as well as of muscle. The relative masses of the two tissues were assumed to be given by the relative rates of blood flow to them. On this basis it was calculated that skin would not contribute more than about 10% to total turnover in the limb.

A third dividend from the limb studies is that they provide comparisons between metabolism in the peripheral tissues, mainly muscle, and in the whole body. Heslin *et al.* (1992) calculated that in the basal state skeletal muscle contributed about one quarter of whole body synthesis and breakdown. The estimates of Tessari *et al.* (1996a) are slightly higher – about one third.

6.8.2 Turnover in the splanchnic region

Great interest naturally attaches to protein turnover in the splanchnic region, where metabolism is so active. The region includes the gastrointestinal tract, liver and spleen and to study its protein turnover requires sampling from a catheter in the hepatic vein. To separate the contributions of liver and GI tract it would be necessary to catheterize the portal vein, which cannot easily be done in man, but which was achieved by Yu *et al.* (1990, 1992) in their very important experiments on dogs.

Tessari *et al.* (1996a) estimated total protein synthesis in the splanchnic area in the basal state as about 70 g day^{-1}. On the basis of the nitrogen content of the splanchnic organs, this would give a fractional synthesis rate (FSR) for the splanchnic region of about 15% day^{-1}, or about 5 × the FSR of the whole body (Chapter 6). The figure fits in reasonably well with such information as is available for man on the FSR of the proteins of different organs (Chapter 9).

Table 6.7 gives information from five studies on the percentage contributions of splanchnic protein synthesis and breakdown to that in the whole body in the basal state. It seems that Tessari's estimate may be a little low, but comparisons are difficult because different models have been used for the calculations.

Protein turnover in the human kidney was measured for the first time in humans by Tessari *et al.* (1996a) by separately catheterizing the renal vein. They found that protein synthesis in the kidneys accounted for about 9% of whole body synthesis. Assuming that the kidneys contain 25% protein, the FSR would be about 36% day^{-1}, more than twice that of the splanchnic viscera as a

Table 6.7. Percentage contribution of the splanchnic region to whole body protein turnover. Measurements with labelled leucine. Comparison of results in the basal state with those after giving amino acids, a meal or insulin.

	Per cent of whole body turnover				
	In the basal state		With intervention		
	Synthesis	Breakdown	Synthesis	Breakdown	
Man					
Gelfand *et al.*, 1988	58	50	52	68	+ amino acids
Nair *et al.*, 1995	42	26	28	22	+ insulin
Tessari *et al.*, 1996a	27	22	–	–	
Cayol *et al.*, 1997	40	77	68	62	+ meal
Dog					
Yu *et al.*, 1992	22	18	73	30	+ amino acids
Sheep					
Lobley *et al.*, 1996	24	–	–	–	–

whole. Oxidation in the kidney was also very high, achieved by increasing net deamination of leucine and net extraction of KIC from the arterial blood. Holliday *et al.* (1967) give a figure of 120 kcal day^{-1} for the oxygen uptake of the kidneys, which at a BMR of 6.3 MJ day^{-1} would be 8% of total oxygen uptake. Thus the kidneys contribute almost identical proportions of whole body protein and oxygen turnover. The paper of Tessari *et al.* (1996a) contains a huge amount of valuable information. It is a pity that we do not have similar data for the human brain.

6.9 Measurement of Protein Turnover with Amino Acids other than Leucine

In 1989, in an important paper (Bier, 1989), Bier published a diagram, reproduced by Matthews (1993a) (Fig. 6.5A) showing the relation between the whole body turnover rates of different amino acids and their concentrations in muscle or gut protein (Fig. 6.5B). The data for the essential amino acids (EAA) all fell on a straight line, suggesting that if the fluxes obtained with any EAA are corrected for their concentration in whole

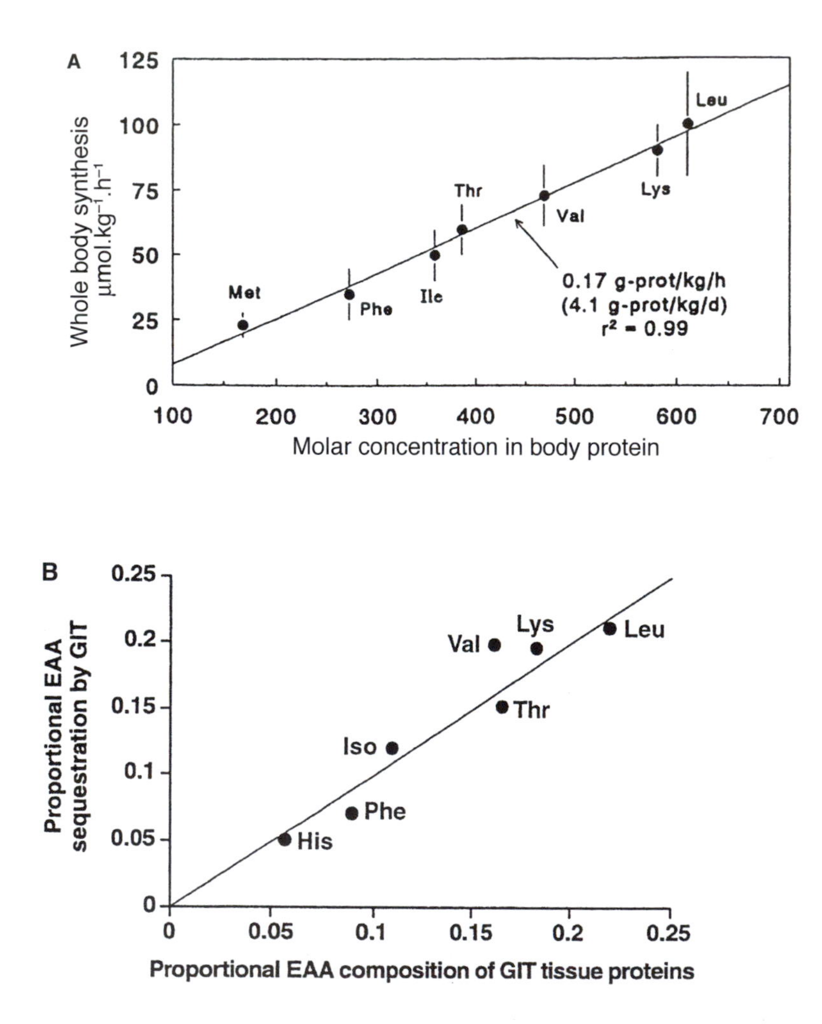

Fig. 6.5. Three-pool model for studies in forearm or leg. A: Relationship between post-absorptive fluxes of essential amino acids in man, calculated from their enrichment in plasma and their molar concentrations in body proteins. Reproduced, by permission, from Matthews (1993b). B: Relation between proportional uptake of essential amino acids by the gastrointestinal tract (GIT) and the composition of GIT proteins. Reproduced from MacRae *et al.* (1997), by courtesy of the *American Journal of Physiology*.

body protein, they would all be the same. A similar diagram was published by MacRae *et al.* (1997) showing a linear relation between the sequestration of EAA in the gastrointestinal tract and their proportions in GIT protein (Fig. 6.5B).

This relationship is to be expected but it faces two difficulties. The first is that the estimates of flux cited by Bier were based on enrichments of the amino acids in plasma; can it be assumed that the dilution of the intracellular precursor, which is the key to the estimation of flux, is the same for all amino acids? This would involve the further assumption that the rate of amino acid transport into the cell, relative to its rate of entry from protein breakdown, is the same for all EAA – an assumption that is not tenable because the rate of entry is determined in part by the plasma amino acid concentration.

Some information on this point can be obtained from the studies of Biolo on the leg. Data from three of his experiments are shown in Table 6.8. The intracellular dilutions of the plasma enrichment are similar for the different amino acids, but by no means the same, and there is quite a large variation between studies.

The second difficulty is that amino acids may differ in their concentrations in the proteins of different tissues. It can easily be shown with an arithmetic model that if there are two protein pools, one with a fast and the other with a slow turnover rate, and if amino acid X has a relatively higher concentration than Y in the fast pool and a lower concentration in the slow pool, an estimate of whole body flux based on X will be greater than one based on Y. Data on the amino acid concentrations in the proteins of different tissues are surprisingly meagre: some are shown in Table 6.9.

6.9.1 Whole body fluxes and synthesis rates

Phenylalanine is the amino acid that has been most widely used in parallel with leucine, mainly because it is oxidized via hydrolysis to tyrosine, so that if tyrosine flux is measured as well, there is no need for collection and analysis of respiratory CO_2. A further advantage is that when biopsies of tissues are taken a method developed by Calder *et al.* (1992) enables measurement of the relatively high enrichment in plasma and the low enrichment in tissue proteins to be determined in the same instrument.

Data on the phenylalanine flux in the whole body, relative to leucine, the flux ratio, are shown in Table 6.10. The expected ratio, from the concentrations in whole body protein (Table 6.9) is ~0.40. However, the leucine fluxes in Table 6.10 are calculated from the enrichment in KIC, because that is how most authors have reported them, whereas the phenylalanine fluxes are based on plasma enrichment. To restore them to the same base the 'leucine' fluxes should be multiplied by a correction factor of 0.8 or the flux ratios divided by 0.8 (Chapter 4). After this correction an average flux ratio of 0.38 becomes 0.47: there is reasonable agreement with the theoretical ratio (Table 6.9), particularly since proteins in the parts of the body with a high rate of turnover – e.g. the liver – have a greater relative concentration of phenylalanine. In keeping with this, the first-pass uptake of phenylalanine is twice as great as that of leucine

Table 6.8. Contributions to the intracellular pool of amino acids entering from the plasma.

	FMA/(FMA + FMO)		
	Leucine	Lysine	Phenylalanine
Biolo *et al.*, 1995a			
basal	0.51	0.35	0.61
Biolo *et al.*, 1995b			
basal	0.47	0.47	0.57
+ insulin	0.61	0.44	0.50
Biolo and Tessari, 1997			
basal	0.71	0.55	0.90
mean	0.575	0.45	0.645

FMA = rate of entry of amino acid from plasma into the intracellular pool; FMO = rate of entry of amino acid from protein breakdown into the intracellular pool.

Table 6.9. Ratio of molar concentrations of phenylalanine and lysine to that of leucine in animal tissues.

		[Phe]/[Leu]	[Lys]/[Leu]
Whole body			
Widdowson *et al.*, 1979	Man (fetus)	0.43	0.86
Davis *et al.*, 1993	Rat	0.40	0.84
Carcass			
MacRae *et al.*, 1993	Lamb	0.40	0.83
Muscle			
Reeds *et al.*, 1994	Sheep, pig, cattle	0.30	1.09
Adibi *et al.*, 1973	Rat	0.45	–
Munro (1969)	Pig	0.51	1.11
Liver			
MacRae *et al.*, 1993	Lamb	0.485	0.69
Adibi *et al.*, 1973	Rat	0.51	–
Munro and Fleck, 1969	Pig	0.57	0.85
Skin			
Seifter and Gallop, 1966	Man	0.49	1.11
Collagen			
Bornstein and Traub, 1979	Man	0.50	1.12
Birk and Lande, 1981	Rabbit	0.61	1.00

(Matthews *et al.*, 1993a; Sánchez *et al.*, 1995) (Table 6.2). The ratio of the fluxes does not appear to be significantly affected by feeding state, nor would one expect it to be.

Lysine is the only other essential amino acid that has been compared concurrently with leucine in measurements of whole body turnover in man. The results of two studies are shown in Table 6.10B. The expected flux ratio, based on whole body amino acid concentrations, is ~0.85.

Unique experimental studies were carried out by Obled *et al.* (1989, 1991) in rats, in which

Table 6.10. A: ratio of whole body fluxes, phenylalanine/leucine (KIC), measured by IV infusion; B: ratio of whole body fluxes, lysine/leucine.

A.	Mean ratio unweighted	N[a]
Normal adults, post-absorptive	0.385	22
fed	0.365	11
Children with cancer	0.39[b]	1

[a]N = number of studies.
[b]This study included 32 children (Daley *et al.*, 1996).
Expected ratio from tissue composition (Table 6.9): 0.4–0.5 (phenylalanine/leucine, not KIC). References: Thompson *et al.*, 1989; Bennet *et al.*, 1990; Tessari *et al.*, 1991, 1994, 1996a,b,c,d, 1999; Biolo *et al.*, 1992; Cortiella *et al.*, 1992; Pacy *et al.*, 1994; Sánchez *et al.*, 1995, 1996; Lee, 1996; Connell *et al.*, 1997 (sheep); Meek *et al.*, 1998.

B.	Ratio
Man, IV infusion, normal adults	
Robert *et al.*, 1982	0.82
Hoerr *et al.*, 1993	
high protein diet	0.72
low protein diet	0.61
Rat, flooding dose, Obled *et al.*, 1989, 1991	0.83

Expected ratio (Table 6.9): 0.85–1.10.

whole body synthesis rates were measured by three different methods, with six different amino acids. The data in Table 6.11 illustrate the difficulty of comparing methods. The results for lysine and threonine are remarkably consistent across methods, those for leucine inconsistent. However, the FSRs with lysine and threonine differ significantly, which is hard to explain.

6.9.2 Measurements on forearm and leg

In a number of studies on the limb, particularly by Tessari and Biolo and their colleagues, phenylalanine has been used as well as leucine, and sometimes also lysine. Since phenylalanine is not oxidized in the muscle and skin of the limb, whereas leucine is, release (= breakdown) will reflect the relative concentrations in the tissue better than uptake (= disposal).

Considering the variability of the data, the mean ratio of phenylalanine/leucine degradation, 0.37, is not too far from that expected from the relative concentrations in skin and muscle proteins (Gelfand *et al.*, 1987; Bennet *et al.*, 1990; Jessari *et al.*, 1991, 1996a,b; Newman *et al.*, 1994; Biolo *et al.*, 1995a). With lysine the position is different. The mean ratio of breakdown rates from all the data was 1.40. If two very aberrant values are omitted, it becomes 1.16. This is close to the concentration ratio in muscle (Reeds *et al.*, 1994) and skin (Seifter and Gallop, 1966) and well above that in the whole body (Table 6.10).

6.9.3 Tissue biopsies

With tissues incorporation into protein is measured directly, and there is no need to correct for the concentration of tracer amino acid in the protein. Measurements of fractional synthesis rates (FSR) in muscle with different amino acids have shown good agreement; lysine with phenylalanine (Biolo *et al.*, 1944); valine with leucine (Smith *et al.*, 1992; Tjader *et al.,* 1996); leucine with threonine (Smith *et al.*, 1998); leucine with phenylalanine and valine (Rocha *et al.*, 1993) (see Chapter 13). Agreement between different amino acids was also found in the FSR of skin (Rocha *et al.*, 1993; Biolo *et al.*, 1994). Jahoor *et al.* (1994) obtained very similar FSRs of albumin and fibrinogen, with leucine, lysine and alanine, based on their enrichments in apolipoprotein B100 as precursor. The alanine was derived from infusion of [$U^{13}C$]-glucose.

6.10 Conclusion

Simultaneous measurements of whole body fluxes, when related to the relative molar concentrations of amino acids in whole body protein, seem, within the limits of error, to confirm the theoretical expectation. This may appear an obvious and trivial conclusion, but it is not; a serious discrepancy would undermine the validity of all the results obtained with a single amino acid such as leucine. There are some deviations in the results of the limb studies, but these can probably be explained by the amino acid composition of the limb proteins, much of which is collagen in skin, being different from that of the mixed proteins of the whole body. However, this possible explanation needs to be confirmed.

There does not seem to be any point in continuing to do turnover studies with more than one amino acid: if two or more are used and the results

Table 6.11. Fractional synthesis rates (FSR, % day^{-1}) of the whole body proteins of young rats measured by different methods and with different amino acids.

	Constant infusion	Flooding dose	Carcass analysis
Leucine	55	31.5	49
Lysine	26	25.5	25
Threonine	37	39	33
Tyrosine	40	–	36
Phenylalanine	–	31	–
Histidine	–	23	–

Data of Obled *et al.*, 1989, 1991.

agree, that is only to be expected; if they disagree, usually no attempt is made to investigate why.

6.11 References

Abumrad, N.N., Rabin, D., Diamond, M.P. and Lacy, W.W. (1981) Use of a superficial hand vein as an alternative site for the measurement of amino acid concentrations and for the study of glucose and alanine kinetics in man. *Metabolism* 30, 836–840.

Adibi, S.A., Modesto, T.A., Morse, E.L. and Amin, P.F. (1973) Amino acid levels in plasma, liver and skeletal muscle during protein deprivation. *American Journal of Physiology* 225, 408–414.

Arnal, M.A., Mosoni, L., Boirie, Y., Gachon, P., Genest, M., Boyle, G., Grizard, J., Arnal, M., Antoine, J.M., Beaufrère, B. and Mirand, P.P. (2000) Protein turnover modifications induced by the protein feeding pattern still persist after the end of the diets. *American Journal of Physiology* 278, E902–909.

Balagopal, P., Rooyackers, O.E., Adey, D.B., Ades, P.A. and Nair, K.S. (1997) Effects of aging on *in vivo* synthesis of skeletal muscle myosin heavy chain and sarcoplasmic protein in humans. *American Journal of Physiology* 273, E790–800.

Beaufrère, B., Horber, F.F., Schwenk, F.W., Marsh, H.M., Matthews, D.E., Gerich, J.E. and Haymond, M.H. (1989) Glucocorticoids increase leucine oxidation and impair leucine balance in humans. *American Journal of Physiology* 257, 712–721.

Beaufrère, B., Fournier, V., Salle, B. and Putet, G. (1992) Leucine kinetics in fed low birthweight infants: importance of splanchnic tissues. *American Journal of Physiology* 263, E214–220.

Bennet, W.M., Connacher, A.A., Scrimgeour, C.M., Jung, R.T. and Rennie, M.J. (1990) Euglycemic hyperinsulinemia augments amino acid uptake by human leg tissues during hyperaminoacidemia. *American Journal of Physiology* 259, E185–194.

Beshiyah, M.A., Sharp, P.S., Gelding, S.V., Halliday, D. and Johnston, D.G. (1993) Whole-body leucine turnover in adults on conventional treatment for hypopituitarism. *Acta Endocrinologica* 129, 158–164.

Bier, D.M. (1989) Intrinsically difficult problems: the kinetics of body proteins and amino acids in man. *Diabetes/Metabolism Reviews* 5, 111–132.

Bier, D.M. and Christopherson, H.L. (1979) Rapid micromethod for determination of ^{15}N enrichment in plasma lysine: application to measurement of whole body protein turnover. *Analytical Biochemistry* 94, 242–248.

Bier, D.M. and Matthews, D.E. (1982) Stable isotope methods for *in vivo* investigations. *Federation Proceedings* 41, 2679–2685.

Biolo, G., Tessari, P., Inchiostro, S., Brutomesso, D., Fongher, C., Sabatin, L., Fratton, M.G., Valerio, A. and Tengo, A. (1992) Leucine and phenylalanine kinetics during mixed meal ingestion: a multiple tracer approach. *American Journal of Physiology* 262, E455–463.

Biolo, G., Gastadelli, A., Zhang, X.-J. and Wolfe, R.R. (1994) Protein synthesis and breakdown in skin and muscle: a leg model of amino acid kinetics. *American Journal of Physiology* 267, E467–474.

Biolo, G., Fleming, R.Y.D., Maggi, S.P. and Wolfe, R.R. (1995a) Transmembrane transport and intracellular kinetics of amino acids in human skeletal muscle. *American Journal of Physiology* 268, E75–84.

Biolo, G., Fleming, R.Y.D. and Wolfe, R.R. (1995b) Physiologic hyperinsulinaemia stimulates protein synthesis and enhances transport of selected amino acids in skeletal muscle. *Journal of Clinical Investigation* 95, 811–819.

Biolo, B. and Tessari, P. (1997) Splanchnic versus whole-body production of α-ketoisocaproate from leucine in the fed state. *Metabolism* 46, 164–167.

Birk, D.E. and Lande, M.A. (1981) Corneal and scleral collagen fiber formation *in vitro*. *Biochimica et Biophysica Acta* 670, 362–369.

Boirie, Y., Gachon, P. and Beaufrère, B. (1997) Splanchnic and whole-body leucine kinetics in young and elderly men. *American Journal of Clinical Nutrition* 65, 489–495.

Bornstein, P. and Traub, W. (1979) The chemistry and biology of collagen. In: Neurath, H. and Hill, R.L. (eds) *The Proteins*, 3rd edn. Vol. IV. Academic Press, New York, pp. 412–530.

Bouteloup-Demange, C., Boirie, Y., Déchelotte, P., Gachon, P. and Beaufrère, B. (1998) Gut mucosal protein synthesis in fed and fasted humans. *American Journal of Physiology*, 274, E541–546.

Bruce, A.C., McNurlan, M.A., McHardy, K.C., Broom, J., Buchanan, K.D., Calder, A.G., Milne, E., McGaw, B.A., Garlick, P.J. and James, W.P.T. (1990) Nutrient oxidation patterns and protein metabolism in lean and obese subjects. *International Journal of Obesity* 14, 631–646.

Calder, A.G., Anderson, S.E., Grant, I., McNurlan, M.A. and Garlick, P.J. (1992) The determination of low d_5-phenylalanine, in relation to protein turnover studies by gas chromatography/electron ionization mass spectrometry. *Rapid Communications in Mass Spectrometry* 6, 421–424.

Carraro, F., Rosenblatt, J. and Wolfe, R.R. (1991) Isotopic determination of fibronectin synthesis in humans. *Metabolism* 40, 553–561.

Castellino, B., Luzi, L., Simonson, D.C., Haymond, M., and Defronzo, R.A. (1987) Effect of insulin and plasma amino acid concentrations on leucine metabolism in man. *Journal of Clinical Investigation* 80, 1784–1793.

Castillo, L., de Rojas, T.C., Chapman, T.E., Vogt, J., Burke, J.F., Tannenbaum, S.R. and Young, V.R. (1993) Splanchnic metabolism of dietary arginine in relation to nitric oxide synthesis in normal adult man. *Proceedings of the National Academy of Sciences, USA* 90, 193–197.

Cayol, M., Boirie, Y., Rambourdin, F., Prugnaud, J., Gachou, P., Beaufrère, B. and Obled, C. (1997) Influence of protein intake on whole body and splanchnic leucine kinetics in humans. *American Journal of Physiology* 272, E584–591.

Cheng, K.N., Dworzak, F., Ford, G.C., Rennie, M.J. and Halliday, D. (1985) Direct determination of leucine metabolism and protein breakdown in humans using L-[1-^{13}C, ^{15}N] leucine and the forearm model. *European Journal of Clinical Investigation* 15, 349–354.

Cheng, K.N., Pacy, P.J., Dworzak, F., Ford, G.C. and Halliday, D. (1987) Influence of fasting on leucine and muscle protein metabolism across the human forearm determined using L-[1-^{13}C, ^{15}N] as the tracer. *Clinical Science* 73, 241–246.

Collin-Vidal, C., Cayol, M., Obled, C., Ziegler, F., Bommelaer, G. and Beaufrère, B. (1994) Leucine kinetics are different during feeding with whole protein or oligopeptides. *American Journal of Physiology* 267, E907–914.

Connell, A., Calder, A.G., Anderson, S.E. and Lobley, G.E. (1997) Hepatic protein synthesis in the sheep: effect of intake as monitored by use of stable-isotope-labelled glycine, leucine and phenylalanine. *British Journal of Nutrition* 77, 255–271.

Conway, J.M., Bier, D.M., Motil, K.J. and Burke, J.F. (1980) Whole body lysine flux in young adult men: effects of reduced total protein and of lysine intake. *American Journal of Physiology* 239, 192–200.

Conway, J.M., Marable, N.L. and Bodwell, C.E. (1988) Whole–body leucine and energy metabolism in adult women. *European Journal of Clinical Nutrition* 42, 661–669.

Copeland, K.C., Kenney, F.A. and Nair, K.S. (1992) Heated dorsal hand vein sampling for metabolic studies: a reappraisal. *American Journal of Physiology* 263, E1010–1014.

Cortiella, J., Matthews, D.E., Hoerr, R.A., Bier, D.M. and Young, V.R. (1988) Leucine kinetics at graded intakes in young men: quantitative fate of dietary leucine. *American Journal of Clinical Nutrition* 48, 998–1009.

Cortiella, J., Marchini, J.S., Branch, S., Chapman, T.E. and Young, V.R. (1992) Phenylalanine and tyrosine kinetics in relation to altered protein and phenylalanine and tyrosine intakes in healthy young men. *American Journal of Clinical Nutrition* 56, 517–525.

Daley, S.E., Pearson, A.D.J., Craft, A.W., Kernahan, J., Wyllie, R.A., Price, L., Brock, C., Hetherington, C., Halliday, D. and Bartlett, K. (1996) Whole body protein metabolism in children with cancer. *Archives of Disease in Childhood* 75, 273–281.

Dangin, M., Boirie, Y., Garcia-Rodenas, C., Gachon, P., Fauquant, J., Callier, P., Ballèvre, Q. and Beaufrère, B. (2001) The digestion rate of protein is an independent regulating factor of post-prandial protein retention. *American Journal of Physiology* 280, E340–348.

Davis, T.A., Fiorotto, M. and Reeds, P.J. (1993) Amino acid compositions of body and milk protein change during the suckling period in rats. *Journal of Nutrition* 123, 947–956.

El Khoury, A.E., Sánchez, M., Fukagawa, N.K. and Young, V.R. (1995) Whole body protein synthesis in healthy adult humans: $^{13}CO_2$ technique vs. plasma precursor approach. *American Journal of Physiology* 268, E174–184.

Fereday, A., Gibson, N.R., Cox, M., Pacy, P.J. and Millward, D.J. (1998) Variation in the apparent sensitivity of the insulin-mediated inhibition of proteolysis to amino acid supply determines the efficiency of protein utilization. *Clinical Science* 96, 725–733.

Fong, Y., Matthews, D.E., He, W., Marana, M.A., Moldawer, L.L. and Lowry, S.F. (1994) Whole body and splanchnic leucine, phenylalanine and glucose kinetics during endotoxaemia in humans. *American Journal of Physiology* 266, R419–425.

Forslund, A.M., Hambraeus, L., Olsson, R.M., El-Khoury, A.E., Yu, Y.-M. and Young, V.R. (1998) The 24-h whole body leucine and urea kinetics at normal and high protein intakes with exercise in healthy men. *American Journal of Physiology* 275, E310–320.

Fukagawa, N.K., Minaker, K.L., Young, V.R., Matthews, D.E., Bier, D.M. and Rowe, J.W. (1989) Leucine metabolism in aging humans: effect of insulin and substrate availability. *American Journal of Physiology* 256, E288–294.

Gan, J.C. and Jeffay, H. (1967) Origins and metabolism of the intracellular amino acid pools in rat liver and muscle. *Biochimica et Biophysica Acta* 148, 448–459.

Gan, J.C. and Jeffay, H. (1971) The kinetics of transfer of plasma amino acids to tissues and the turnover rates of liver and muscle proteins: the use of continuous infusion technique. *Biochimica et Biophysica Acta* 252, 125–135.

Garby, L. and Lammert, O. (1994) Between-subjects variation in energy expenditure: estimation of the effect of variations in organ size. *European Journal of Clinical Nutrition* 48, 376–378.

Garlick, P.J., McNurlan, M.A. and Caso, G. (1997) Critical assessment of methods used to measure protein synthesis in human subjects. *Yale Journal of Biological Medicine* 70, 65–76.

Gelfand, R.A. and Barrett, E.J. (1987) Effect of physiologic hyperinsulinaemia on skeletal muscle protein synthesis and breakdown in man. *Journal of Clinical Investigation* 80, 1–6.

Gelfand, R.A., Glickman, M.G., Castellino, P., Louard, R.J. and De Fronzo, R.A. (1988) Measurement of L-[1-^{14}C] leucine kinetics in splanchnic and leg tissues in humans. Effect of amino acid infusion. *Diabetes* 37, 1365–1372.

Gibson, N.R., Fereday, A., Cox, M., Halliday, D., Pacy, P.J. and Millward, D.G. (1996) Influences of dietary energy and protein on leucine kinetics during feeding in healthy adults. *American Journal of Physiology* 270, E282–291.

Giordano, M., Castellino, P. and De Fronzo, R.A. (1996) Differential responsiveness of protein synthesis and degradation to amino acid availability in humans. *Diabetes* 45, 493–499.

Golden, M.H.N. and Waterlow, J.C. (1977) Total protein synthesis in elderly people: a comparison of results with [^{15}N] glycine and [^{14}C] leucine. *Clinical Science and Molecular Medicine* 53, 277–288.

Halliday, D. and McKeran, O. (1975) Measurement of muscle protein synthesis rate from serial muscle biopsies and total body protein turnover in man by continuous intravenous infusion of L-[α-^{15}N] lysine. *Clinical Science and Molecular Medicine* 49, 581–590.

Halliday, D. and Rennie, M.J. (1982) The use of stable isotopes for diagnosis and clinical research. *Clinical Science* 63, 485–496.

Heslin, M.J., Newman, E., Wolf, R.W., Pisters, P.W.T. and Brennan, M.F. (1992) Effect of hyperinsulinaemia on whole body and skeletal muscle leucine carbon kinetics. *American Journal of Physiology* 262, E911–918.

Hoerr, R.A., Matthews, D.E., Bier, D.M. and Young, V.R. (1991) Leucine kinetics from [$^{2}H_3$] and [^{13}C] leucine infused simultaneously by gut and vein. *American Journal of Physiology* 260, E111–117.

Hoerr, R.A., Matthews, D.E., Bier, D.M. and Young, V.R. (1993) Effects of protein restriction and acute refeeding on leucine and lysine kinetics in young men. *American Journal of Physiology* 264, E567–575.

Hoffer, L.J., Yang, R.D., Matthews, D.E., Bistrian, B.R., Biuer, D.M. and Young, V.R. (1985) Effects of meal consumption on whole body leucine and alanine kinetics in young men. *British Journal of Nutrition* 53, 31–38.

Holliday, M., Potter, D., Jarrah, A. and Bearg, S. (1967) The relation of metabolic rate to body weight and organ size. *Pediatric Research* 1, 185–195.

Jahoor, F., Burrin, D.G., Reeds, P.J. and Frazer, M. (1994) Measurement of plasma protein synthesis rate in infant pig: an investigation of alternative tracer approaches. *American Journal of Physiology* 267, R221–227.

James, W.P.T., Garlick, P.J., Sender, P.M. and Waterlow, J.C. (1976) Studies of amino acid and protein metabolism in normal man with L-[U-^{14}C] tyrosine. *Clinical Science* 50, 525–532.

Jeevandam, M., Brennan, N.F., Horowitz, G.D., Rose, D., Mihranian, M.H., Daly, J. and Lowry, S. (1985) Tracer priming in human protein turnover studies. *Biochemical Medicine* 34, 214–225.

Jones, P.M.P. and Pearson, J. (1969) Anthropometric determination of leg fat and muscle plus bone volumes in young male and female adults. *Journal of Physiology (London)* 204, 63–66.

Katz, J. (1982) Importance of sites of tracer administration and sampling in turnover studies. *Federation Proceedings* 41, 123–128.

Katz, J (1992) On the determination of turnover *in vivo* with tracers. *American Journal of Physiology* 263, E417–424.

Krempf, M., Hoerr, R.A., Pelletier, V.A., Marks, L.M., Gleason, R. and Young, V.R. (1993) An isotopic study of the effect of dietary carbohydrate on the metabolic fate of dietary leucine and phenylalanine. *American Journal of Clinical Nutrition* 57, 161–169.

Kurpad, A.V., Raj, T., El-Khoury, A.E., Kuriyan, R., Maruthy, K., Borgonha, S., Chandukudlu, D., Regan, M. and Young, V.R. (2001) Daily requirement for and splanchnic uptake of leucine in healthy adult Indians. *American Journal of Clinical Nutrition* 74, 747–755.

Layman, D.K. and Wolfe, R.R. (1987) Sample site selection for tracer studies applying a unidirectional circulatory approach. *American Journal of Physiology* 253, 173–178.

Lobley, G.E., Connell, A., Revell, D.K., Bequette, B.J., Brown, D.S. and Calder A.G. (1996) Splanchnic bed transfers of amino acids in sheep blood and plasma, as monitored through use of a multiple U-^{13}C-labelled amino acid mixture. *British Journal of Nutrition* 75, 217–235.

Loftfield, R.B. and Harris, A. (1956) Participation of free amino acids in protein synthesis. *Journal of Biological Chemistry* 219, 151–159.

Louard, R.J., Barrett, E.J. and Gelfand, R.A. (1990) Effect of branched-chain amino acids on muscle and whole-body amino acid metabolism in man. *Clinical Science* 79, 457–466.

Macallan, D.C., McNurlan, M.A., Kurpad, A.V., de Souza, G., Shetty, P.S., Calder, A.G. and Griffin, G.E. (1998) Whole body protein metabolism in human pulmonary tuberculosis and undernutrition: evidence for anabolic block in tuberculosis. *Clinical Science* 94, 321–331.

MacRae, J.C., Walker, A., Brown, D. and Lobley, G.E. (1993) Accretion of total protein and individual amino acids by organs and tissues of growing lambs and the ability of nitrogen balance techniques to quantitate protein retention. *Animal Production* 57, 237–245.

MacRae, J.C., Bruce, L.A., Brown, D.S. and Calder, A.G. (1997) Amino acid use by the gastrointestinal tract of sheep given lucerne forage. *American Journal of Physiology* G 1158–1165.

Marchini, J.S., Cortiella, J., Hiramatsu, T., Chapman, T.E. and Young, V.R. (1993) Requirements for indispensable amino acids in adult humans: longer-term amino acid kinetic study with support for the adequacy of the Massachusetts Institute of Technology amino acid requirement pattern. *American Journal of Clinical Nutrition* 58, 670–683.

Matthews, D.E., Mohl, K.J., Rohrbaugh, D.K., Burke, J.F., Young, V.R. and Bier, D.M. (1980) Measurement of leucine metabolism in man from a primed continuous infusion of L-[1-^{13}C] leucine. *American Journal of Physiology* 238, 473–479.

Matthews, D.E., Bier, D.M., Rennie, M.J., Edwards, R.H.T., Halliday, D., Millward, D.J. and Clugston, G.A. (1981) Regulation of leucine metabolism in man: a stable isotope study. *Science* 214, 1129–1131.

Matthews, D.E., Marano, M.A. and Campbell, R.G. (1993a) Splanchnic bed utilization of leucine and phenylalanine in humans. *American Journal of Physiology* 264, E104–118.

Matthews, D.E. (1993b) Stable isotope methodologies in studying human amino acid and protein metabolism. *Italian Journal of Gastroenterology* 25, 72–78.

McHardy, K.C., McNurlan, M.A., Milne, E., Calder, A.G., Fearns, L.M., Broom, J. and Garlick, P.J. (1991) The effect of insulin suppression on post-prandial nutrient metabolism: studies with infusion of somatostatin and insulin. *European Journal of Clinical Nutrition* 45, 515–526.

Meek, S.E., Persson, M., Ford, G.C. and Nair, K.S. (1998) Differential regulation of amino acid exchange and protein dynamics across splanchnic and skeletal muscle beds by insulin in healthy human subjects. *Diabetes* 47, 1824–1835.

Meguid, M.M., Matthews, D.E., Bier, D.M., Meredith, C.N. and Young, V.R. (1986) Valine kinetics at graded valine intakes in young men. *American Journal of Clinical Nutrition* 43, 781–786.

Melville, S., McNurlan, M.A., McHardy, K.C., Broom, J., Milne, E., Calder, A.G. and Garlick, P.J. (1989) The role of degradation in the acute control of protein balance in adult man: failure of feeding to stimulate protein synthesis as assessed by L-[1-^{13}C] leucine infusion. *Metabolism* 38, 248–255.

Melville, S., McNurlan, M.A., Calder, A.G. and Garlick, P.J. (1990) Increased protein turnover despite normal energy metabolism and responses to feeding in patients with lung cancer. *Cancer Research* 50, 1125–1131.

Meredith, C.M., Wen, Z.-W., Bier, D.M., Matthews, D.E. and Young, V.R. (1986) Lysine kinetics at graded lysine intakes in young men. *American Journal of Clinical Nutrition* 43, 787–794.

Metges, C.C., El-Khoury, A.E., Selvaraj, A.B., Tsay, R.H., Atkinson, A., Regan, M.M. and Young, V.R. (2000) Kinetics of L-[1-(13)C] leucine when ingested with free amino acids, unlabelled or intrinsically labelled casein. *American Journal of Physiology* 278, E1000–1009.

Millward, D.J., Fereday, A., Gibson, N.R. and Pacy, P.J. (2000) Human adult amino acid requirements: [1-^{13}C] leucine balance evaluation of the efficiency of utilization and apparent requirements for wheat protein and lysine compared with those for milk protein in healthy adults. *American Journal of Clinical Nutrition* 72, 112–121.

Munro, H.N. and Fleck, A. (1969) Analysis of tissues and body fluids for nitrogenous constituents. In: Munro, H.N. (ed.) *Mammalian Protein Metabolism* Vol. III, Chapter 30. Academic Press, New York, pp. 423–525.

Nair, K.S., Ford, G.C., Ekberg, K., Fernqvist-Forbes, E. and Wahren, J. (1995) Protein dynamics in whole body and in splanchnic and leg tissues in type 1 diabetic patients. *Journal of Clinical Investigation* 95, 2926–2937.

Newman, E., Heslin, M.J., Wolf, R.F., Pisters, P.W.T. and Breman, M.F. (1994) The effect of systemic hyperinsulinaemia with concomitant amino acid infusion on skeletal muscle turnover in the human forearm. *Metabolism* 43, 70–78.

Obled, C., Barre, F., Millward, D.J. and Arnal, M. (1989) Whole body protein synthesis: studies with different amino acids in the rat. *American Journal of Physiology* 257, E639–646.

Obled, C., Barre, F. and Arnal, M. (1991) Flooding dose of various amino acids for measurement of whole-body protein synthesis in the rat. *Amino Acids* 1, 17–27.

O'Keefe, S.J.D., Sender, P.M. and James, W.P.T. (1974) 'Catabolic' loss of body nitrogen in response to surgery. *Lancet* ii, 1035–1039.

Pacy, P.J., Garrow, J.S., Ford, G.C., Merritt, H. and Halliday, D. (1988) Influence of amino acid administration on whole-body leucine kinetics and resting metabolic rate in postabsorptive normal subjects. *Clinical Science* 75, 225–231.

Pacy, P.J., Price, G.H., Halliday, D., Quevedo, and Millward, D.J. (1994) Nitrogen homoeostasis in man: the diurnal responses of protein synthesis and degradation and amino acid oxidation to diets with increasing protein intake. *Clinical Science* 86, 103–118.

Pell, J.M., Caldarone, E.M. and Bergman, E.N. (1983) Importance of sites of tracer administration and blood sampling in relation to leucine metabolism. *Biochemical Journal* 214, 1015–1018.

Reaich, D., Channon, S.M., Simgeour, C.M. and Goodship, Th.J. (1992) Ammonium chloride-induced acidosis increases protein breakdown and amino acid oxidation in humans. *American Journal of Physiology* 263, E735–739.

Reeds, P.J., Hachey, D.L., Patterson, B.W., Mahl, K.J. and Klein, P.D. (1992) VLDL Apolipoprotein B-100, a potential indicator of the isotopic labeling of the hepatic protein synthetic precursor pool in humans: studies with multiple stable isotopically labeled amino acids. *Journal of Nutrition* 122, 457–466.

Reeds, P.J., Fjeld, C.R. and Jahoor, F. (1994) Do the differences between the amino acid composition of acute-phase and muscle proteins have a bearing on nitrogen loss in traumatic states? *Journal of Nutrition* 124, 906–910.

Reeds, P.J., Burrin, D.G., Jahoor, F., Wykes, L., Henry, J. and Frazer, E.M. (1996) Enteral glutamate is almost completely metabolized in first pass by the gastro-intestinal tract of infant pigs. *American Journal of Physiology* 270, E413–418.

Robert, J.-J., Bier, D.M., Zhao, X.H., Matthews, D.E. and Young, V.R. (1982) Glucose and insulin effects on de novo amino acid synthesis in young men: studies with stable isotope labelled alanine, glycine, leucine and lysine. *Metabolism* 31, 1210–1218.

Rocha, H.J.G., Nash, J.E., Connell, A. and Lobley, G.E. (1993) Protein synthesis in urine muscle and skin: sequential measurements with three different amino acids based on the large dose procedure. *Comparative Biochemistry and Physiology* 105B, 301–307.

Russell-Jones, D.L., Umpleby, A.M., Hennessy, T.R., Bowes, S.B., Shojaee-Moradie, F., Hopkins, K.D., Jackson, N.C., Kelly, J.M., Jones, R.H. and Sönksen, P.H. (1994) Use of a leucine clamp to demonstrate that IGF-I actively stimulates protein synthesis in normal humans. *American Journal of Physiology* 267, 591–598.

Sánchez, M., El-Khoury, A.E., Castillo, L., Chapman, T.E. and Young, V.R. (1995) Phenylalanine and tyrosine kinetics in young men throughout a continuous 24 h period, at a low phenylalanine intake. *American Journal of Clinical Nutrition* 61, 555–570.

Sánchez, M., El-Khoury, A.E., Castillo, L., Chapman, T.E., Basil-Filho, A., Beaumier, L. and Young, V.R. (1996) Twenty-four-hour intravenous and oral tracer studies with L-[1-^{13}C] phenylalanine and L-[3,3$-^{2}H_2$] tyrosine at a tyrosine-free, generous phenylalanine intake in adults. *American Journal of Clinical Nutrition* 63, 532–545.

Schwenk, W.F., Tsalikian, E., Beaufrère, B. and Haymond, M.W. (1985a) Recycling of an amino acid label with prolonged isotope infusion: implications for kinetic studies. *American Journal of Physiology* 248, E482–487.

Schwenk, W.F., Beaufrère, B. and Haymond, M.W. (1985b) Use of reciprocal pool specific activities to model leucine metabolism in humans. *American Journal of Physiology* 249, E646–650.

Schwenk, W.F. and Haymond, M.W. (1987) Effects of leucine, isoleucine or threonine infusion on leucine metabolism in humans. *American Journal of Physiology* 253, E428–E434.

Seifter, S. and Gallop, P.M. (1966) The structure proteins. In: Neurath, H. (ed.) *The Proteins – Composition, Structure and Function.* Academic Press, New York, pp. 155–257.

Slater, C., Preston, T. and Weaver, L.T. (2001) Stable isotopes and the international system of units. *Rapid Communications in Mass Spectrometry* 15, 1270–1273.

Smith, K., Barua, J.M., Walt, P.W., Scrimgeour, P.M. and Rennie, M.J. (1992) Flooding with L–[1-^{13}C] leucine stimulates human muscle incorporation of continuously infused L-[1-^{13}C] valine. *American Journal of Physiology* 262, E372–376.

Smith, K., Reynolds, N., Downie, S., Patel, A. and Rennie, M.J. (1998) Effects of flooding amino acids on incorporation of labelled amino acids into human muscle protein. *American Journal of Physiology* E73–78.

Stephen, J.M.L. and Waterlow, J.C. (1966) Use of carbon-14-labelled arginine to measure the catabolic rate of serum and liver proteins and the extent of amino-acid recycling. *Nature (London)* 211, 978–980.

Swick, R.W. (1958) Measurement of protein turnover in rat liver. *Journal of Biological Chemistry* 231, 751–764.

Tessari, P., Inchiostro, S., Biolo, G., Vincenti, E., Sabadin, L. and Vettore, M. (1991) Effects of systemic hyperinsulinaemia on forearm muscle proteolysis in healthy man. *Journal of Clinical Investigation* 88, 27–32.

Tessari, P., Inchiostro, S., Barazzoni, R., Zanetti, M., Orlando, R., Bido, G., Sergi, G., Pino, A. and Tiengo, A. (1994) Fasting and postprandial phenylalanine and leucine kinetics in liver cirrhosis. *American Journal of Physiology* 267, E140–149.

Tessari, P., Inchiostro, S., Zanetti, M. and Barazzoni, R. (1995) A model of skeletal muscle leucine kinetics measured across the human forearm. *American Journal of Physiology* 269, E127–136.

Tessari, P., Zanetti, M., Barazzoni, R., Vettore, M. and Michielan, F. (1996a) Mechanisms of post-prandial protein accretion in human skeletal muscle: insight from leucine and phenylalanine forearm kinetics. *Journal of Clinical Investigation* 98, 1361–1372.

Tessari, P., Garibolto, G., Inchiostro, S., Robando, C., Saffioti, S., Vettore, M., Zanetti, M., Russo, R. and Deferrari, G. (1996b) Kidney, splanchnic and leg protein turnover in humans: insight from leucine and phenylalanine kinetics. *Journal of Clinical Investigation* 98, 1431–92

Tessari, P., Zanetti, M., Barazzoni, R., Biolo, G., Orlando, R., Vettore, M., Inchiostro, S., Perini, P. and Tiengo, A. (1996c) Response of phenylalanine and leucine kinetics to branched-chain-enriched amino acids in patients with cirrhosis. *Gastroenterology* 111, 127–137.

Tessari, P., Barazzoni, R., Zanetti, M., Vettore, M., Normand, S., Bruttomesso, D. and Beaufrère, B. (1996d) Protein degradation and synthesis measured with multiple amino acid tracers *in vivo*. *American Journal of Physiology* 271, E733–741.

Tessari, P., Barazzoni, R. and Zanetti, M. (1999) Differences in estimates of forearm protein synthesis between leucine and phenylalanine tracers following unbalanced amino acid infusion. *Metabolism* 48, 1564–1570.

Thompson, G.N., Pacy, P.J., Merritt, H., Ford, G.C., Read, M.A., Cheng, K.N. and Halliday, D. (1989) Rapid measurement of whole body and forearm protein turnover using a [$^2H_5$] phenylalanine model. *American Journal of Physiology* 256, E631–639.

Tjäder, I., Essen, P., Wernerman, J., McNurlan, M.A., Garlick, P.J., Smith, K. and Rennie, M.J. (1996) Comparison of constant infusion and flooding dose methods for measuring muscle protein synthesis: effect of conventional and total parenteral nutrition. *Proceedings of the Nutrition Society* 56, 151A.

Tsalikian, E., Howard, C., Gerich, J.E. and Haymond, G.E. (1984) Increased leucine flux in short-term fasted human subjects: evidence for increased proteolysis. *American Journal of Physiology* 247, E323–327.

Volpi, E., Lucichi, P., Bolli, G.B., Santeusanio, F. and De Feo, P. (1998) Gender differences in basal protein kinetics in young adults. *Journal of Clinical Endocrinology and Metabolism* 83, 4363–4367.

Volpi, E., Mitterndorfer, B., Wolfe, S.E. and Wolfe, R.R. (1999) Oral amino acids stimulate muscle protein anabolism in the elderly despite higher first-pass splanchnic extraction. *American Journal of Physiology* 277, E513–520.

Wagenmakers, A.M. (1999) Tracers to investigate protein and amino acid metabolism in human subjects. *Proceedings of the Nutrition Society* 58, 987–1000.

Waterlow, J.C. (1967) Lysine turnover in man measured by intravenous infusion of L-[U-^{14}C] lycine. *Clinical Science* 33, 507–515.

Waterlow, J.C. (1996) The requirements of adult man for indispensable amino acids. *European Journal of Clinical Nutrition* 50, Suppl. 1, S151–179.

Waterlow, J.C. and Stephen, J.M.L. (1967) The measurement of total lysine turnover in the rat by intravenous infusion of L-[U-^{14}C] lysine. *Clinical Science* 33, 489–506.

Waterlow, J.C. and Stephen, J.M.L. (1968) The effect of low protein diets on the turnover rates of serum liver and muscle protein in the rat, measured by continuous infusion of L-[^{14}C] lysine. *Clinical Science* 35, 287–305.

Welle, S. and Nair, K.S. (1990) Relationship of resting metabolic rate to body composition and protein turnover. *American Journal of Physiology* 258, E990–998.

Widdowson, E.M., Southgate, D.A.T. and Hay, E.N. (1979) Body composition of the fetus and infant. In: Visser, H.K.A. (ed.) *Nutrition and Metabolism of the Fetus and Infant.* Martinus Nijhoff, The Hague, Netherlands, pp. 169–178.

Young, V.R., Gucalp, C., Rand, W.M., Matthews, D.E. and Bier, D.M. (1987) Leucine kinetics during three weeks at submaintenance-to-maintenance intakes of leucine in men: adaptation and accommodation. *Human Nutrition: Clinical Nutrition* 41C, 1–18.

Yu, Y.-M., Wagner, D., Tredget, E.E., Walaszewski, J.A., Burke, J.F. and Young, V.R. (1990) Quantitative role of splanchnic region in leucine metabolism: 1-[^{13}C, ^{15}N] leucine and substrate balance studies. *American Journal of Physiology* 259, E31–51.

Yu, Y.-M., Burke, J.F., Vogt, J.A., Chambers, L. and Young, V.R. (1992) Splanchnic and whole body L-[1-^{13}C, ^{15}N] leucine kinetics in relation to enteral and parenteral amino acid supply. *American Journal of Physiology* 262, E687–694.

Zanetti, M., Barazzoni, R., Kiwanuke, E. and Tessari, P. (1999) Effects of branched-chain-enriched amino acids and insulin on forearm leucine kinetics. *Clinical Science* 97, 437–448.

Zhao, X.-L., Wen, Z.-W., Meredith, C.M., Matthews, D.E., Bier, D.M. and Young, V.R. (1986) Threonine kinetics or graded threonine intakes in young men. *American Journal of Clinical Nutrition* 43, 795–802.

7

Measurement of Whole Body Protein Turnover by the End-product Method

7.1 History

The first measurement ever made of whole body protein turnover in man was by Sprinson and Rittenberg in 1949. They gave a single dose of ^{15}N-glycine and measured the excretion of ^{15}N in the urine over 72 h. A two-pool model was assumed, with a metabolic pool, a protein pool and no re-entry of label from the protein pool (Sprinson and Rittenberg, 1949). They showed empirically that the cumulative excretion of isotope could be represented by the expression:

$$e_t/d = A\,[1 - \exp(-Bt)]$$

where e_t is the cumulative excretion of isotope up to time t, d is the dose and A and B are constants. As t approaches infinity the exponential term becomes zero and e_t reaches a plateau, at which A = e_t/d. It was assumed that this fraction, the proportion of the tracer excreted, was the same as the proportion of the flux excreted, E/Q, and so the flux was determined. Sprinson and Rittenberg's paper is important because it was the foundation of all later work on the end-product method of estimating the rate of protein synthesis.

In the years that followed a few studies appeared that used Sprinson and Rittenberg's single dose approach. Wu and Snyderman (1950) obtained the cumulative excretion at infinite time by extrapolating the enrichment curve of urinary N between 6 and 30 h to zero abundance, and showed that the synthesis rate in children was twice as high as in the adult. Bartlett and Gaebler (1952) found that in dogs treated with growth hormone the synthesis rate was 50% higher than in untreated dogs. Sharp *et al.* (1957) showed that the rate of protein turnover was lower in elderly people. The main interest of this paper is that ^{15}N-labelled yeast was used, and the synthesis rate in young adults was almost identical with that obtained with ^{15}N-glycine by Sprinson and Rittenberg. However, in a more recent study this agreement was not confirmed (Fern *et al.*, 1985b).

A new approach was proposed by San Pietro and Rittenberg (1953), based on the precursor-product relationship between the metabolic N pool and the urea N pool. Theory predicts that after a single dose the enrichment of tracer will be equal in the two pools at the point of time when enrichment in the product pool is at its maximum (Zilversmit, 1960). If at that time the enrichment of the metabolic pool is determined from the urea data, the rate of protein turnover can then be calculated. For details of the calculation the reader is referred to the original paper. The approach was criticized by Tschudy *et al.* (1959); the principal defect of it is the difficulty of estimating the time of peak labelling, considering the slow turnover rate of urea and the delay before it appears in the urine.

This method was quite extensively used, e.g. in studies of the effects of low and high intakes of energy and protein (Tschudy *et al.*, 1959), of myxoedema (Crispell *et al.*, 1956) and of pituitary disease, with results that were quite variable. In the end it proved to be a wrong turning which almost certainly led to clinical studies on protein turnover being largely abandoned for the next 10 years or so.

There were two other lines of work during this period which are still of interest today because they were concerned with methods. The first is the study of Olesen *et al.* (1954), in which full-scale compartmental analysis was applied to the excretion curve of ^{15}N over a period of 15 days after a dose of ^{15}N-glycine. Their model was designed to demonstrate the existence of two protein pools, one turning over rapidly, the other slowly. However, from my calculations it would take at least 10 days of measurement to establish the existence of these separate pools. This study was followed by a heroic experiment in which subjects were almost totally encased in a plaster cast for 6 days in order to examine the effects of immobilization (Schønheyder *et al.*, 1954). During that time there was a substantial negative nitrogen balance which resulted from a fall in protein synthesis rather than an increase in breakdown.

The second major contribution of this era was the work of Wu and co-workers (Wu and Bishop, 1959; Wu and Sendroy, 1959; Wu *et al.*, 1959), who studied the pattern of ^{15}N excretion in the first few hours after giving a single dose of a variety of labelled compounds – glycine, phenylalanine, aspartic acid and ammonium citrate. They found that the shape of the early excretion curve was complex and must represent the sum of several exponentials which could not be separated. They concluded that it is virtually impossible to interpret the curve over the first 24 h after the dose and wrote: 'Any theoretical interpretation of the ^{15}N excretion data following the administration of an N^{15}-labelled compound by mathematical analysis, e.g. Olesen *et al.*, is a matter of conjecture' (Wu *et al.*, 1959). They went on to say: 'The synthesis rate can be calculated from the cumulative excretion at infinite time, without the necessity of fitting the isotope concentration curve to an equation.' Thus they had moved from the compartmental to the stochastic approach. It is interesting that the estimates of synthesis obtained in this way were the same, whether the tracer amino acid was given orally or intravenously. The work of Wu and his colleagues was not wasted; their great contribution was the comparison of different amino acids, from which they concluded that glycine is not a very good tracer, because its abundance never becomes equal to that of ammonia-N, which they took to be representative of that of the amino-N of the metabolic pool. This subject is discussed in more detail below (section 7.4).

Until the end of the 1960s almost no further work on protein turnover was published. One reason for the loss of interest in what seemed a very promising field of research was probably the cumbersome and complex mathematical treatment. This contrasted with the treatment of the subject by those who were writing at that time on the general theory of tracer kinetics, e.g. Reiner (1953), Robertson (1957), Russell (1958), Zilversmit (1960) and Sheppard (1962). These and other theoretical papers were virtually never cited by the practical workers. Certainly the relatively difficult mathematics of compartmental analysis influenced us in the choice of method when our interest was first stimulated by a purely clinical problem, the high mortality of severely malnourished children and their failure to respond to treatment. The hypothesis was that protein deficiency was an important causal factor, which perhaps led to an inability to synthesize protein; and since the synthetic enzymes are themselves proteins, the damage would be irreversible. The challenge, then, was to find a method of measuring whole body protein turnover (WBPT) that was both simple mathematically and minimally invasive, and the answer was constant infusion of tracer.

7.2 Theory

The hypothesis, as originally formulated by Rittenberg, is that newly synthesized protein and excreted end-product are derived from the same metabolic pool of amino-N. Absent from Rittenberg's treatment was the central point that products of the same labelled precursor have the same activity, provided that they are derived only from that precursor (Zilversmit, 1960). On that basis:

(1) $Q/d = S/s = E_{ep}/e_{ep}$

where Q is the flux, S is the amount of amino-N synthesized into protein, E_{ep} is the amount of N excreted as end-product, d is the dose of tracer and s and e_{ep} are, respectively, the amounts of tracer going to synthesis and to end-product. Since e_{ep}/E_{ep} is the enrichment, ε_{ep}, of the end-product:

(2) $Q = d/\varepsilon_{ep}$

Thus the calculation of flux does not involve the amount of end-product. With the stochastic approach (Chapter 2, section 2.3) this simple expression can be used in two ways. Either a single dose of tracer is given and the end-product collected over a given time, in which case Q is the amount of flux during that time, so the rate of flux Q = Q/t. With the single dose method the collection of urine over the chosen time period must be complete, because ε will be varying throughout that period. Alternatively the tracer is given by continuous infusion or multiple dosage, until ε_{ep} reaches a plateau, at which $Q = d/\varepsilon_{ep}$. In this case there is no need for the collection to be complete.

As with the precursor method, once the flux is known synthesis can be calculated as:

(3) $S = Q - E_T$

where E_T is the total N excretion, which is not the same as $E_{ep,}$ the amount of end-product excreted. It is customary to equate E_T with total urinary N, and to disregard endogenous faecal and other N losses, which are small compared with the other components of equation 3.

7.3 Alternative End-products (EP)

7.3.1 Urea

The initial studies with continuous infusion of [^{15}N]-glycine, either by the oral or intravenous route, were based on urea as EP. It was found that it took about 30 h to reach plateau labelling of urea in children (Picou *et al.*, 1968; Picou and Taylor-Roberts, 1969) and up to 60 h in adults (Steffee *et al.*, 1976). This is unacceptably long for practical purposes. There were several ways round the problem. The first was to return to the single dose–cumulative excretion method. A decision had to be made on the point in time at which clearance of tracer into synthesis and excretion could be taken as effectively complete. If the time is too short the flux will be underestimated, if too long there will be re-entry of tracer from labelled protein. The solution was to collect urine for 9 or 12 h; at the end of that time to take a blood sample and add to the ^{15}N-urea excreted the amount retained in the body's urea pool (Fern *et al.*, 1981). It was assumed that urea is uniformly distributed throughout body water and total body water was estimated by the formula of Watson *et al.* (1980). At 9 h about 50% of ^{15}N in urea was still in the body pool, and by 24 h about 20% (Fern *et al.*, 1981). This addition of a single blood sample is now an essential feature of the method. A disadvantage is that it requires accurate collection of urine over the chosen period. This is particularly important in the first few hours, when the excretion of isotope is at its maximum. Thus cumulative excretion after a single dose, with measurement of urea retention at 9 or 12 h, seemed to be the method of choice. Alternatively, with a constant infusion the time to reach plateau could be reduced by priming with both glycine and urea (Jahoor and Wolfe, 1987), but this method never came into general use. In studies in the fed state, when tracer is given by mouth, it is more convenient than an infusion to give it with food in small amounts at frequent intervals, which vary with different workers from 20 minutes to 1 h. This is referred to as the 'multiple' dose method.

7.3.2 Ammonia

In the 1970s it was suggested that a way of shortening the time needed for measuring the flux was to use ammonia instead of urea as end product (Waterlow *et al.*, 1978).

The ammonia pool is small and turns over faster than that of urea, so that with a single dose excretion of tracer tends more rapidly to zero (Fig. 7.1). It was found, however, that the two end-products did not give the same value for flux and the difference varied with different conditions of measurement, whether in the fed or fasted state and whether the tracer was given orally or intravenously (see below).

The estimate of flux based on ammonia may be designated Q_A, that on urea Q_U. It is important to recognize that these are not separate fluxes but different estimates of the whole body amino-N flux. It was clearly, then, necessary to abandon the original concept of a single homogeneous 'metabolic' pool and to postulate two pools, geographically and metabolically separate, with fluxes through them leading predominantly to different end-products, although it would be unrealistic to regard these pools as totally separate (Fern *et al.*, 1981, 1984, 1985a,b; Fern and Garlick, 1983). It was suggested that Q_A may be 'biased' towards the peripheral tissues, since urinary ammonia is derived mainly from glutamine (Pitts and Pilkington, 1966) which is synthesized

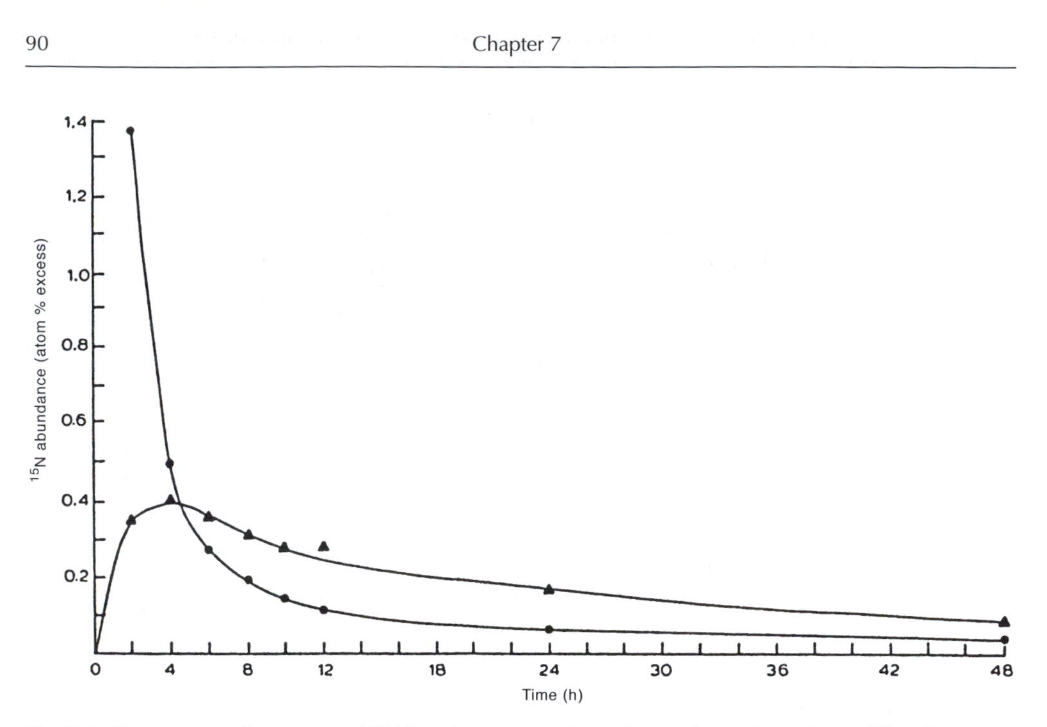

Fig. 7.1. Time-course of excretion of ^{15}N in urinary ammonia and urea after a single dose of ^{15}N-glycine. Reproduced from Grove and Jackson (1995), by courtesy of the *British Journal of Nutrition*.
▲, urea; ●, ammonia.

in muscle; and Q_U is 'biased' towards the viscera, since urea is synthesized exclusively in the liver. This interpretation would be an example of Fern's 'geographical' separation. The model so derived is shown in Fig. 7.2 and our concept of the metabolic pathways in Fig. 7.3.

Fern *et al.* (1985b) also pointed out that Q_A and Q_U tended to vary inversely. This finding suggested that the best estimate of the 'true' flux is the average of the two different estimates. The arithmetic average, $(Q_A + Q_U)/2$, implies that the dose of tracer is equally distributed between the two pools or pathways. The harmonic average:

$$\frac{2}{1/Q_A + 1/Q_U}$$

implies that the fluxes through the two pools are equal. In practice these two averages differ very little.

In considering the metabolic meaning of these fluxes, it is important to recognize that what is measured is the enrichment of the end-product, and the flux is the inverse of the enrichment. A low flux implies a high level of labelling of the end-product and of the metabolites that give rise to it. This high level could result either from greater 'access' of the tracer to the metabolite – i.e. a preferred pathway between them; or from lesser dilution by unlabelled metabolite – i.e. that the metabolite is a less active player in the game. At present it is not possible to distinguish between these two possibilities.

7.3.3 Respiratory carbon dioxide

A third end-product that has been used occasionally is respiratory CO_2 when a carbon-labelled amino acid such as [1-^{13}C]-leucine is the tracer (Golden and Waterlow, 1977; Yagi and Walser, 1990; El-Khoury *et al.*, 1995). The activity of expired CO_2 cannot be used because it is diluted by CO_2 derived from sources other than leucine. However, the total amount of labelled CO_2, e_{CO_2}, can be measured and related to the amount of leucine oxidized, E_{leu}, which in the steady state is assumed to be equal to the leucine intake.

7.4 Measurement of Flux with a Single End-product

Originally many studies, including our own, were made with urea as a single end-product. This is

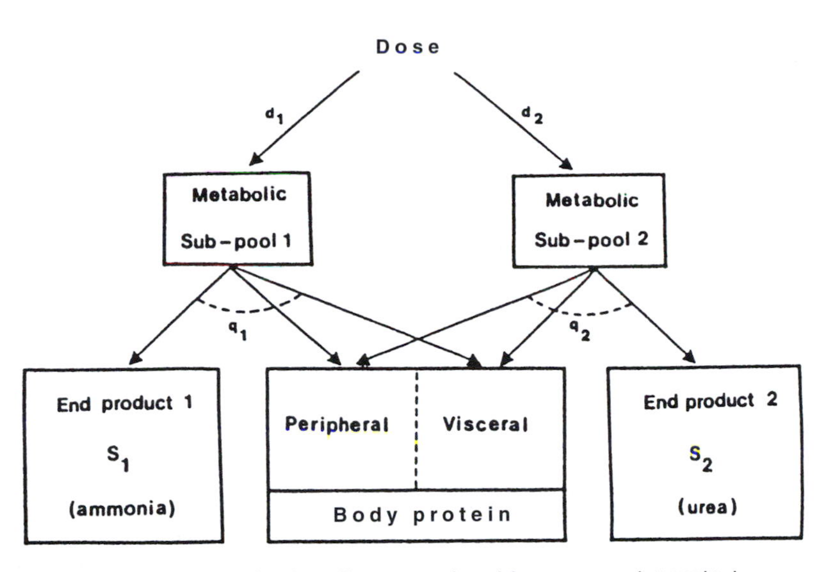

Fig. 7.2. Fern's hypothesis of metabolic chanelling. Reproduced from Fern *et al.* (1985b), by courtesy of *Human Nutrition: Clinical Nutrition*. d_1, d_2, partiition of dose; q_1, q_2, fluxes through the two channels; S_1, S_2, synthesis rates of the two end products.

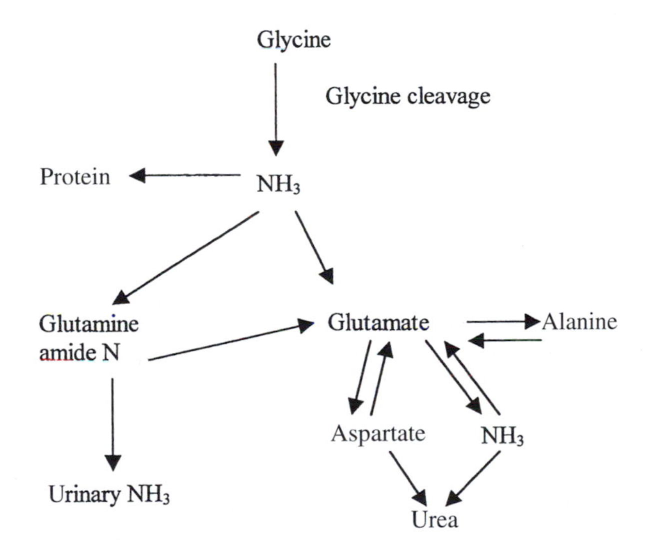

Fig. 7.3. Metabolic scheme of channelling of ^{15}N from glycine to end products.

no longer a viable option. Continuous dosage has to be carried on for so long to achieve a plateau that, in addition to the inconvenience, recycling of tracer almost certainly becomes a problem. Single dose requires a blood sample and assumptions about the urea space.

On the other hand ammonia as sole end-product has been very useful. Since the flux ratio, Q_A/Q_U, is generally in the region of 0.7, the error compared with Q_{av}, in using ammonia only will not be very large, unless there is a gross distortion of the metabolic pathways. Ammonia as EP has been particularly useful in field studies under difficult conditions. An example is a study by

Tomkins *et al.* (1983) in Nigeria in children with varying degrees of infection and malnutrition, which showed that healthy children with an acute infection responded with a large increase in protein turnover, whereas there was no response in children who were severely malnourished. Another advantage of ammonia as EP is that, because the method is completely non-invasive, numerous measurements can be made over a short period of time. An example here is a study of the effects of dietary change in obese patients, illustrated in Fig. 7.4.

7.5 Behaviour of Different Amino Acids in the End-product Method: Choice of Glycine

The most extensive study of this problem is that by Fern *et al.* (1985b). Their results are shown in Table 7.1. By far the highest rates of flux and synthesis are given by lysine. This fits with the data of Aqvist and others quoted above (Chapter 3), which show minimal exchanges of lysine-N with other amino acids. The high value of Q_A/Q_U indicates that most of the N that is exchanged goes into the urea pathway.[1] Next comes leucine, which, according to the data in Table 7.1, exchanges only modest amounts of label with the end products; the very low level of Q_A/Q_U shows that more of it goes into ammonia than into urea.

Aspartate and glutamate, the transaminators *par excellence*, form a couple. The remarkable thing about these two amino acids is the difference in their responses to oral and intravenous tracer. When they are given orally transfer of label to urea strongly predominates; when given IV there is greater transfer to ammonia. Alanine and glutamine, the two major transporters of N between organs, also behave very much alike. The low values of flux and synthesis indicate that transfer of ^{15}N from precursors to end-products is particularly active, whether the tracer is given orally or IV. As with the other non-essential amino acids, oral dosage favours the urea

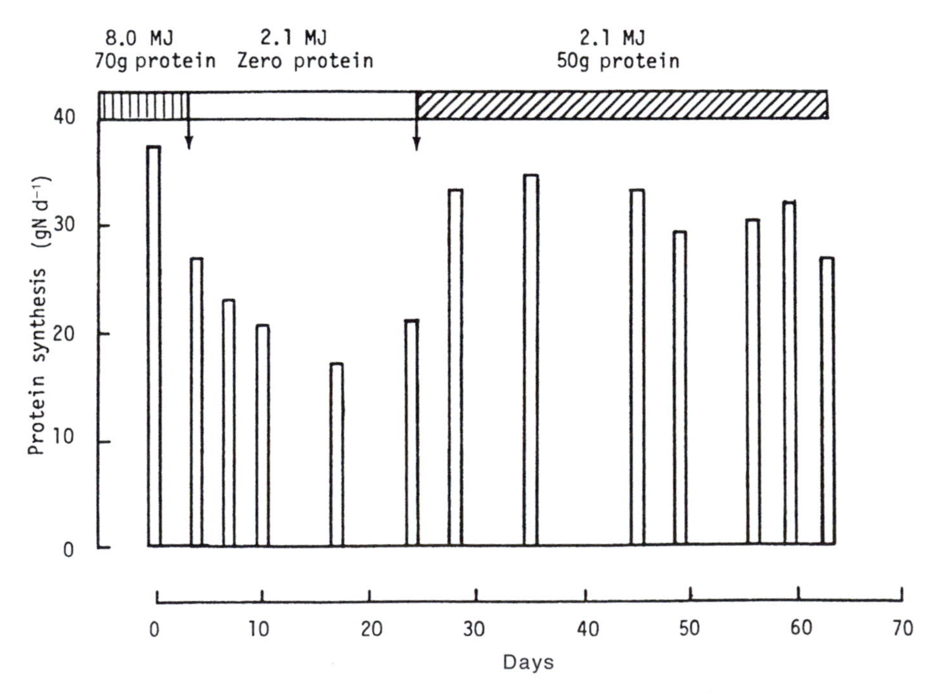

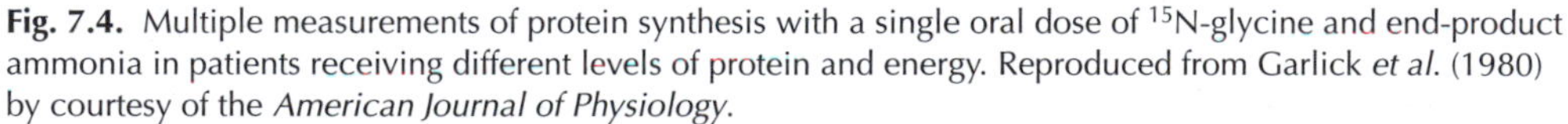
Fig. 7.4. Multiple measurements of protein synthesis with a single oral dose of ^{15}N-glycine and end-product ammonia in patients receiving different levels of protein and energy. Reproduced from Garlick *et al.* (1980) by courtesy of the *American Journal of Physiology*.

[1] Since flux is the inverse of enrichment, a *high* value of Q_A means ammonia is *poorly* enriched.

Table 7.1. Estimates of flux and flux ratio with different labelled amino acids.

		Q_{av} g N kg^{-1} per 12 h	Q_A/Q_U	S_{av} mg N kg^{-1} h^{-1}
Glycine	Oral	27.1	1.09	140
	IV	28.0	0.73	147
Alanine	Oral	16.5	0.94	70
	IV	18.7	0.55	83
Glutamine	Oral	16.8	1.20	60
	IV	17.2	0.51	70
Aspartate	Oral	31.5	2.71	179
	IV	19.9	0.77	94
Glutamate	Oral	33.7	2.29	189
	IV	25.7	0.68	133
Leucine	Oral	42.2	0.49	243
	IV	37.0	0.57	242
Lysine	Oral	126.7	3.06	847
	IV	152.7	2.30	1028
Wheat	Oral	27.5	1.68	135
Yeast	Oral	44.0	1.93	255

From Fern *et al.*, 1985a
Measurements by single dose of tracer in the fed state.

pathway, the IV route the ammonia pathway. Taruvinga *et al.* (1979), working with rats, found similar differences between essential amino acids (valine and leucine) and non-essentials (glycine and aspartic acid) in fluxes derived from urea and ammonia in liver and kidney (Table 7.2). Uniformly labelled wheat, fed orally, gave results very close to those of glycine, except for a greater distortion of the flux ratio. Yeast, on the other hand, gave substantially higher values for flux and synthesis; one can only speculate that this may be because it contains relatively high amounts of the essential amino acids which have rather limited exchange with the end-products.

It would seem from the results in Table 7.1 that if consistency is the criterion of the most suitable amino acid – consistency between oral and IV dosage and between the two end-products – then glycine is the one to choose. It is purely fortuitous – and fortunate – that glycine is the amino acid that has been most used for measuring protein turnover in the whole body, initially because it was very cheap. But is consistency really important? It must be taken as true that the fluxes to the two end-products are 'geographically' and meta-

Table 7.2. Fluxes with different amino acids in the rat, derived from enrichments of urea and ammonia in liver and kidney.

		Flux, g protein day^{-1} kg^{-1} End-product		
Amino acid		Urea	NH_3	EPA
[^{15}N] glycine	Liver	38	36	37
	Kidney	35	37	36
[^{15}N] aspartate	Liver	30	26	28
	Kidney	29	21	25
[^{15}N] valine	Liver	74	73	73.5
	Kidney	71	55	63
[^{15}N] leucine	Liver	61	70	65.5
	Kidney	79	36	57.5

EPA: end-product average.
Data on six rats (Taruvinga *et al.*, 1979).

bolically different. However, it has been shown that, starting from the hypothetical 'ideal' case that the dose of tracer is equally divided between the two pathways and the enrichments of the end-products are the same, a fourfold variation in the partition of the dose and a twofold variation in the ratio of end-product enrichments produces less than 10% variation in the average estimate of flux, the harmonic average being slightly better than the arithmetic average (Fern *et al.*, 1985a).

Even if the distorting effect of two pathways is virtually eliminated, the basic assumption still has to be satisfied: that the proportions of the tracer dose going to end-product and to protein are the same as the proportions of the amino-N flux going down these two pathways. It is obvious that this assumption cannot hold with, on the one hand, glutamate/aspartate, which 'flood' the end-product; or, on the other hand with lysine, which is well taken up into protein but has virtually no exchange with the end-product. A step towards satisfying the assumption was a little-quoted paper by Matthews *et al.* (1981). After oral dosage of ^{15}N-glycine for 60 h, by which time it may be presumed that plateau had been reached, they measured the quantitative pattern (enrichment x amino acid concentration) of the tracer in plasma amino acids. From this they calculated the expected enrichment in urea, and found quite reasonable agreement with the observed enrichment – 0.38 *vs.* 0.44 atom % excess. The final link in the chain, although rather a tenuous one, was the demonstration by Bier and Matthews (1982) that the distribution of ^{15}N in the amino acids isolated from plasma albumin was similar to that of the free amino acids of plasma. One would not expect the distribution to be identical, since the pattern of uptake into protein is not the same as the pattern in plasma or intracellular fluid. It is strange that there have not been more studies of this kind. That of Bier and Matthews (1982) was cited as the first example of the use of gas-chromatograph-mass spectrometry (GCMS) for analysis of enrichment in individual amino acids.

7.6 Comparisons of Different Protocols

The most useful comparison is when the measurements have been made with different protocols in the same subjects.

7.6.1 Oral *vs.* intravenous dosage of tracer

Fern *et al.* (1981) compared the results with a single dose of [^{15}N]-glycine given orally or IV in the same four subjects. There was no difference in either the fed or fasted state but the fasted level was only 60% of that in the fed state. In another study (Fern *et al.*, 1984), when measurements were repeated on the same individual at different times, there was again no difference between oral and intravenous routes of dosage.

7.6.2 Single *vs.* multiple dosage or constant infusion

The best comparison of these two protocols is one by Grove and Jackson (1995) in which they were both carried out in the same person in 11 out of 13 subjects. In the single dose method urine was collected for 9 h, with corrections for ^{15}N retained in the urea pool; in the multiple dose method tracer was given orally every 3 h for 15 h. It is essential that with single dose collection of urine over the designated period should be complete, particularly at early times; with multiple dose, where a plateau is reached, this is not important. With both methods the subjects were fed every 3 h. The multiple dose method gave values for flux that on average were 25% lower than those with the single dose method. Two factors may have contributed to this discrepancy: with the multiple dose method a rather large priming dose of ^{15}N glycine was given 6 h before the multiple doses began, in order that plateau labelling should be achieved in 24 h. This dose may not all have been disposed of by the end of the study; at the same time the 15 h period of dosage may have allowed some recycling of tracer from protein breakdown. The difference between the two protocols is a technical matter: it shows that we have not yet achieved a 'gold standard' for the EP method.

7.7 Summary of Measurements of Protein Synthesis in Normal Adults by the End-product Method

A literature search by S. Duggleby identified 77 studies that met the criteria for inclusion in our analysis: that fluxes should be based on both ammonia and urea, and that in studies in the fed state the protein intake should be within a reasonably normal range of 0.5–1.5 g kg^{-1} day^{-1}. As well as healthy adults the analysis includes data on premature infants, pregnant women, the elderly and patients in various pathological states. The results in these groups are considered in Chapters 12 and 17.

Table 7.3 shows the results for synthesis rather than flux, in order to avoid the confounding effect on the flux of differences in protein intake. The table gives results in 427 healthy subjects in 57 studies. No distinction of gender has been made. By far the largest group is that given a single dose orally. The difference between the fed and fasted rates is significant (P = 0.03). The multiple dose (MD) gives significantly lower results (P = 0.05), in agreement with the finding of Grove and Jackson (1995) discussed above. The intravenous single dose (SD) gives rates that are not significantly different from the oral SD, suggesting that there is no first-pass effect. However, it may perhaps be argued that when a single dose is given and the flux aggregated over the next 12–24 h there has been plenty of time for an equilibrium to be established between the contributions of visceral and peripheral tissues to the flux. The situation is quite different when splanchnic uptake is measured by continuous infusion of tracer. The low rates by constant IV infusion are possibly because in most such studies the tracer was infused for 2–3 days, so that recycling would significantly increase the activity of the end-product.

Synthesis rates at various ages, measured by the EP method, are shown in Table 7.4 (see also Chapter 12). The table is approximate because some results were obtained with ammonia only as EP, some with urea only and some with the EP average. Nevertheless, it does show the expected increase in synthesis at early ages. More detail is in Chapter 12.

7.8 Variability

A study by Fern *et al.* (1984) allows an estimate of the intrinsic variability of the method when measurements were repeated on the same two subjects over a period of 3–4 years (Table 7.5). The coefficients of variation are of the order of 5–6%, with, as mentioned above, no difference between oral and intravenous dosage. In five other subjects measured on two occasions at intervals of 25–66 weeks, the mean difference

Table 7.3. Whole body synthesis rates by the end-product method (mg protein kg^{-1} h^{-1}).

	Method	n	N	S	SD(g)	SD(i)
A	Oral, single dose, fed	22	146	179	49.1	37.3
B	Oral, single dose, fasted	13	130	145	29.7	33.0
C	Oral, multiple dose, fed	6	48	137	27.4	17.6
D	IV, single dose, fed	5	16	183	19.8	56.5
E	IV, single dose, fasted	5	36	167	31.8	49.4
F	Constant IV infusion, fed	3	9	96	6.0	20.1
G	Constant IV infusion, fasted	3	42	63	11.9	12.5

Statistics	A *vs.* B	A *vs.* C
Difference	37	47
SD of difference	15.7	20.9
95% confidence intervals	5.4 to 68.4	5.0 to 91
F	0.1>P>0.05	0.05>P>0.02

Calculation of standard deviations:

SD(g) = $\sqrt{\Sigma[(x_i - x)^2 \times n_i]/N\text{-}1}$

SD(i) = $\sqrt{\Sigma[SD_i \times (n_i\text{-}1)]/N}$

n, number of groups studied; N, number of individuals studied; S, weighted mean synthesis rate, mg protein kg^{-1}. h^{-1}; SD(g), standard deviation of weighted group means; SD(i), weighted mean SD of individuals in all groups. Data of Duggleby and Waterlow (2005); the references for this table are given in that paper.

Table 7.4. Synthesis rates at various ages, measured by the EP method.

	S (mg protein kg^{-1} day^{-1})	n (studies)	N (subjects)
Premature infants			
Oral, multiple dose, fed	442	6	60
Neonates			
IV, constant infusion, TPN	322	2	24
Infants			
Oral, single or multiple dose, fed[a]			
Mid-recovery from malnutrition	255	2	21
Recovered from malnutrition	241	3	37
Pregnancy			
Oral, single or multiple dose, fed			
Early	169	2	15
Mid	162	4	46
Late	163	4	46
Elderly			
Oral, single dose or constant infusion, fed[b]	151	9	85

[a]No difference in rates of synthesis with single or multiple dosage – in separate subjects.
[b]In one study (Pannemans *et al.*, 1995) results given separately for men and women: S for women 114, for men 183 mg protein kg^{-1} day^{-1}.

between the two measurements was 10%. Of course, there may have been real temporal changes. Glynn *et al.* (1988) made two studies, at an interval of a month, on a patient with Guillaine-Barré syndrome. The tracer was given IV over 1 h and the patient was fed by naso-gastric tube. Synthesis rates differed by only 4%. This patient is of interest because her muscle mass was only 17% of body weight; as might be expected, the whole body synthesis rate was very high – at 310 mg protein kg^{-1} h^{-1}, almost twice that in normal subjects.

The variability both between and within groups in the overall analysis (Table 7.3) is high. The between-group SD presumably reflects technical as well as true variation. In the studies in Group F of the table, the between-group SD is very low, but the within-group SD quite high.

7.9 Comparison of Synthesis Rates Measured by the End-product and Precursor Methods

A few comparisons of the two methods have been made in the same subjects at the same time. Golden and Waterlow (1977) fed six elderly patients by naso-gastric tube with [^{15}N]-glycine added to the feed. In three patients [^{14}C]-leucine was infused intravenously; in the other three it was added to the naso-gastric infusion. No distinction was made in the paper between the two routes of leucine administration. Some corrections should be made to the results of this study as originally published. With leucine, flux and synthesis were determined from the enrichment of plasma leucine; if KIC had been used the synthesis rate would be increased by ~ 20%. With glycine, urea

Table 7.5. Variability of synthesis rates measured on five occasions (oral) or three occasions (IV) over periods of 3.5–4 years.

Subject		N	Synthesis, mg protein kg^{-1} day^{-1}	CV %
1	Oral	5	150	6.6
	IV	3	158	5.5
2	Oral	5	157	7.7

Synthesis rates are the arithmetic average of rates derived from Q_U and Q_A, by single dose of [^{15}N] glycine in the fed state. Recalculated from Fern *et al.* (1984).

Table 7.6. Comparison of whole body protein synthesis rates by precursor and end-product methods in the same subjects. A: Pacy *et al.* (1994); young adults; leucine by continuous intravenous infusion, glycine by intravenous dosage over 1 h; both in fed state. B: Pannemans *et al.* (1997); elderly women; leucine by continuous intravenous infusion, glycine by single oral dose; both in fasted state.

A.

Previous protein intake, g kg^{-1} day^{-1}	Protein synthesis, mg protein kg^{-1} h^{-1} (SD)					
	Leucine			Glycine		
	Mean		n	Mean		n
0.36	170	(26)	5	128	(31)	5
0.77	174	(17)	5	176	(33)	5
1.59	193	(7)	6	209	(78)	5

B.

Protein/energy ratio of previous diet	Mean protein synthesis, mg protein, kg^{-1} h^{-1} (SD)	
	Leucine (n, 6)	Glycine (n, 6)
10.6	129 (21)	75 (17)
19.6	129 (8)	108 (25)

reached an estimated 90% of its plateau by 30 h, when the infusions were ended, so the enrichment should be increased by 10% to be that at plateau. With these corrections the synthesis rate with glycine (S_{av}) was 121 and with leucine 133 mg protein kg^{-1} h^{-1}, a difference of 10%.

Pacy *et al.* (1994) compared synthesis rates in the same subjects (Table 7.6A) and found good agreement at the two higher levels of protein intake but not at the lowest. Pannemans *et al.* (1997) made a similar comparison in elderly women (Table 7.6B). Interpretation of these results is complicated by differences in the route of dosage. Of interest is a study by Nissim *et al.* (1983), who gave a pulse dose of ^{15}N glycine and from compartmental analysis of the plasma decay curve calculated a synthesis rate of 147 mg protein kg^{-1} h^{-1}, compared with a rate of 122 mg from the end-product average. Although this study involved only a single tracer, it is a true comparison between the two methods.

Another comparison in the same subjects was made by De Benoist *et al.* (1984) in premature infants. ^{13}C-leucine was infused for 60 h (!) and its activity was measured in urine, which agreed well with that in plasma (see also Chapter 12). Protein synthesis by N balance and from the leucine data agreed within 3%. However, if KIC had been taken as the precursor, synthesis from leucine would have exceeded synthesis from N balance by 25%.

The decisive comparison of the precursor and end-product methods is of the average results obtained with them, even though they relate to different populations. To make this comparison is not straightforward, because the most widely used protocols have not been the same. For the EP method the preferred protocol has been a single dose orally in the fed state (Table 7.3); for the precursor method continuous dosage, by vein, in both fed and fasted states. The comparison is shown in Table 7.7. The results are again given in terms of synthesis rather than flux in order to eliminate variations in intake and oxidation. Moreover, it is synthesis in which we are generally interested, flux being a step on the way to determining it. Because of the first-pass effect (Chapter 6) one would expect the rate to be higher by the EP method with oral dosage than with IV tracer.

The other important difference between the two series is that the variability is greater by the EP method; the reason is probably that details of the method have been less strictly standardized.

7.10 Comparison of Oxidation Rates by the Two Methods

In a number of studies the oxidation of protein, calculated from that of [^{13}C] leucine on the

Table 7.7. Comparison of synthesis in normal adults by end-product and precursor methods. EP data by oral, single dose; precursor data by constant IV infusion.

	Synthesis, mg protein $kg^{-1} h^{-1}$	
	End-product[a]	Precursor
Post-absorptive		
Synthesis	145	167
n	12	65
N	121	502
SD_g	29.7	37.6
SD_i	33.0	29.7
Fed		
Synthesis	179	155[b]
n	20	28
N	126	197
SD_g	49.1	30.5
SD_i	37.3	20.5

n, number of groups; N, number of individuals; SD_g, between groups; SD_i, between individuals.
[a]S, average of estimates from ammonia and urea.
[b]From Table 6.1, group F. (Note that in Table 6.1 units are μmol leucine $kg^{-1} h^{-1}$ Protein assumed to contain 8% leucine.)
References: EP data: Duggleby and Waterlow (2005); precursor data: see Chapter 6.

assumption that the protein turning over in the body contains 8% leucine, has been compared with that derived from urinary N excretion. It might be thought more correct to make the comparison with urea production, which exceeds excretion by some 30% because of uptake of urea into the colon (see Chapter 8). The urea 'salvaged' in the colon is hydrolysed to ammonia, which is returned to the metabolic N pool. A small fraction of this N is recycled back to urea and to include it in the comparison will represent double counting. The appropriate comparison with CO_2 is of *net* oxidative N loss, by excretion in the urine and to a small extent by other routes.

In pigs Reeds *et al.* (1980) found an almost exact correspondence between the excretion of urea + ammonia and that calculated from the oxidation of leucine. In studies on humans by the group at MIT the mean oxidation ratio, leucine/N, was 1.05 (range 0.92–1.25) (El-Khoury *et al.*, 1994, 1995, 1996, 1997; Forslund *et al.*, 1998). On the other hand in the study of Price *et al.* (1994) the relation was consistently closer to 0.8 than to 1. These comparisons are not as straightforward as they seem. Price introduced a correction for the leucine content of their food protein: if it was greater than the 8% assumed for body protein, the excess would be oxidized. Another difficulty is that leucine oxidation was usually measured over 2-h periods, whereas N excretion was measured over 12 h, so that the two are not strictly simultaneous. Lastly, there is some doubt about the extent to which losses of N by the faeces and skin represent oxidation of protein. On the whole, however, these results validate the measurements of carbon balance that are so important for estimates of amino acid requirements (see Chapter 10).

7.11 The Flux Ratio

The results for the flux ratio (Q_A/Q_U) in normal subjects are summarized in Table 7.8. The outstanding feature is the higher ratio in the fasted than in the fed state with the single oral dose. This was not found when the tracer was given intravenously. On the hypothesis proposed earlier, this difference could imply that in fasting protein metabolism is better conserved in peripheral tissues.

It had been hoped that the flux ratio (Table 7.9) would provide useful information about differences in the partition of protein metabolism according to age, habitual protein intake, nutri-

Table 7.8. Flux ratio (Q_A/Q_U) in normal adults by different methods.

Method	$Q_A/Q_U \times 100$	n	N	SD[a]
Oral				
A Single dose, fed	73.5	16	96	21.2
B Single dose, fasted	108.9	6	47	19.8
C Multiple dose, fed	84.2	6	48	18.6
D Multiple dose, fasted		No studies		
IV				
E Single dose, fed	64.0	5	16	9.9
F Single dose, fasted	58.5	5	36	11.0
G Multiple dose, fed	70.5	3	9	5.0
H Multiple dose, fasted		No studies		

[a]SD of group means of Q_A/Q_U.
The inter-individual SDs could not be calculated because in many papers individual values for Q_A/Q_U are not given. Reference: Duggleby and Waterlow (2005).

tional state, etc. This hope has so far not been fulfilled. The ratio tends to be a little low in prematures, in malnourished children and in pregnancy – all conditions in which protein is being rapidly deposited. An outstanding change is a high ratio in trauma, implying on our hypothesis a relatively greater flux through peripheral than visceral tissues. In trauma this is unexpected since the patients should be producing in the liver large amounts of acute phase proteins (e.g. Preston *et al.*, 1995). Perhaps this result indicates enhanced activity of the immune system throughout the body. There are no studies meeting our criteria on subjects with infection rather than trauma.

Table 7.9. Flux ratio ($Q_A/Q_U \times 100$) measured by the end-product method in various conditions.

Subjects	n	N	$Q_A/Q_U \times 100$	SD_g	References
Oral tracer, single or multiple dose, fed state					
Infants					
Premature	5	52	63	15	1–3
Malnourished	2	20	65	9	4–5
Recovered	4	69	81	145	4–6
Adults					
Pregnant, all stages	8	97	61	12	7–9
Elderly	2	21	89	12	10, 11
Undernourished, males	4	21	60	1	1–14
Undernourished, females	2	14	86	6.5	14
Undernourished, elderly	2	14	81	5	15
IV tracer, single dose or constant infusion, fasting or TPN					
Adults					
After surgery	7	50	108	25	16–19
Multiple trauma	2	16	94	–	20
With gastrointestinal disease, or TPN	4	24	112	25	21, 22

n, number of groups; N, number of individuals; mean Q_A/Q_U and SD weighted for study size.
SD_g between groups. SD for individuals not available.
References: 1. van Goudoever *et al.*, 1995; 2. Pencharz *et al.*, 1989; 3. Catzeflis *et al.*, 1985; 4. Waterlow *et al.*, 1978; 5. Golden and Golden, 1992; 6. Jackson *et al.*, 1983; 7. de Benoist *et al.*, 1984; 8. Willomet *et al.*, 1992; 9. Duggleby, 1999; 10. Arnal *et al.*, 1999; 11. Golden and Waterlow, 1977; 12. Soares *et al.*, 1991; 13. Soares *et al.*, 1994; 14. Vaisman *et al.*, 1992; 15. Bos *et al.*, 2000; 16. Yoshida *et al.*, 1996; 17. Ma and Jiang, 1990; 18. Taggart *et al.*, 1991; 19. Powell-Tuck *et al.*, 1984; 20. Jeevanandam *et al.*, 1991; 21. Powell-Tuck and Glynn, 1985; 22. Glynn *et al.*, 1987.

7.12 Kinetic Findings by the End-product Method

Measurements of urinary ammonia enrichment have suggested the existence of a pool of protein turning over by non-random kinetics. Jackson observed that when ^{15}N-glycine was given together with food by continuous intragastric infusion, plateau labelling of NH_3 was reached in 18–24 h, but instead of a smooth curve there was a clearly marked 'step' at 6–12 h (Jackson and Golden, 1980). When a priming dose of tracer was given to shorten the time to plateau, a step was found at 2–3 h. Indications of such a step could be seen in figures reproduced by other authors (Jeevanandam *et al.*, 1985; Catzeflis *et al.*, 1985).

No combination of pools turning over by first order kinetics could produce such a step. It was therefore suggested that it resulted from a pool of protein that turned over by life-cycle kinetics. Such a system, modelled by computer, produced curves that agreed quite well with those observed experimentally (Slevin *et al.*, 1991). It seemed possible that the life-cycle pool might consist of protein of the villous cells of the gastrointestinal mucosa. Their turnover could not be related in any simple way to feeding, since food was infused throughout the experiment. In a further study of this effect subjects received hourly feeds + tracer for 36 h (Jackson *et al.*, 1997b). They were divided into three groups, starting the experiment at 6 am, noon or 6 pm. Enrichment of NH_3 reached a plateau in about 4 h from the start; then at about midnight, regardless of the time at which the experiment started, the enrichment began to rise to a second plateau, which was achieved in about 4 h and maintained until the evening (Fig. 7.5). At the second plateau the calculated flux was 71% of that at the first plateau.

This pattern suggests a circadian rhythm in which the first phase lasts from about 6 am to 6 pm and the second from 6 pm to 6 am. In the second phase there is an increase in the enrichment of ammonia which could arise in several ways: it could imply a decreased flux through peripheral tissues; alternatively there might be increased breakdown of highly labelled protein, as observed in the evidence for recycling (q.v.), where it was shown that very rapid turnover of even a small part of the total protein pool can have a significant effect on the labelling of the precursor. This breakdown could be occurring in a pool turning over by life-time kinetics. It would be interesting to see if a similar pattern was found with a 24-h infusion of leucine in which feeding was continued throughout.

Circadian rhythms have been described in the activity of liver enzymes, e.g. Wurtman and Axelrod (1967); in the secretion of some proteins by the liver (Marckmann *et al.*, 1993) and the exocrine pancreas (Maouyo *et al.*, 1993); and in the turnover of bone (Eastell, 1992). This is an area that deserves further investigation.

7.13 Conclusion

The end-product method, producing a value for flux which is the average of two separate and different estimates, may seem to be something of a fudge. Nevertheless, there is a rational metabolic justification for this separation. There is still no gold standard for the EP method. To achieve one requires standardizing the protocol as regards route and type of dosage and applying it with great attention to detail. The data of Fern *et al.* (1984) show the precision that can be achieved by the same worker using exactly the same method.

At the end of the day the question of the validity of the results for whole body protein turnover by either method is still an open one. There has been no experiment with the EP method such as that in which Obled *et al.* (1989, 1991) compared the precursor method with the ultimate standard, carcass analysis (see Chapter 6). Therefore we have to rely on the argument that reasonable agreement between two sets of results based on different assumptions, implies that they are both measurements of a real biological function – whole body protein synthesis.

The precursor and EP methods are complementary; the choice between them will depend on the purpose of the exercise and on practical considerations. The practical pros and cons of the two methods are summarized in Table 7.10. There are some results that cannot be obtained by the EP method, such as studies of carbon balance with different labelled amino acids, which are major contributions of Young's group at MIT. On the other hand the EP method may be the one of choice for population studies aimed at determining whole body rates of synthesis and their func-

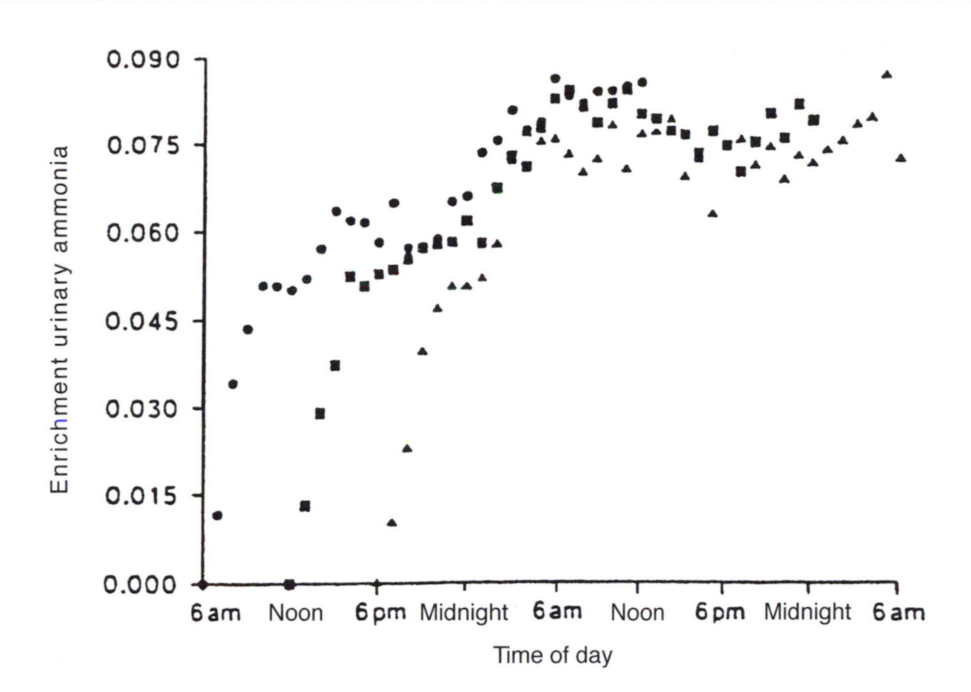

Fig. 7.5. Enrichment of urinary NH_3 when [^{15}N]-glycine was given every hour, with food, for 36 h. Reproduced from Jackson *et al.* (1997b), by courtesy of *Clinical Science*.
●, studies started at 06.00 h; ■, studies started at 12.00 h; ▲ studies started at 18.00 h.

Table 7.10. Pros and cons of precursor and end-product methods.

	Precursor	End-product
Invasiveness		
Blood	Two in-dwelling cannulae, for infusion and blood samples Multiple samples at plateau	For SD, one blood sample, at end, unless urine sample is collected between 9 and 24 h For MD, no blood samples
Urine	No samples	For SD, two samples minimum: the collection must be complete. For MD, at least four samples to establish plateau
Breath	Continuous or multiple sampling of CO_2 output at plateau	No samples needed
Freedom of movement	Restricted	Unrestricted except for urine collection
Equipment	Complex: gas chromatograph–mass spectrometry	Relatively simple: isotope ratio mass spectrometry
Cost	High	Relatively low
Multiple studies	Difficult	Easy (see Fig. 7.5)
Facilities needed	Metabolic ward	Can be free-living
Optimum groups for study	Small groups: basic research	Larger groups: population studies, gender differences.

SD, single dose method; MD, multiple dose method.

tional correlates in genetically or environmentally different groups, particularly under difficult field conditions, or in repeated studies in patients on different diets, as shown in Fig. 7.4. It may also be useful under difficult clinical and field conditions, as in the work of Tomkins *et al.* (1983) in Nigeria on malnourished infected children.

Bier, in an important paper published in 1989, says: '… the approach of "validating" nitrogen flux calculations against related values derived from "primary" and "reciprocal" pool methods is equally unconvincing … Demonstrating the identity of two parameters from incomplete or incorrect models validates neither the parameter nor the models.' (Bier, 1989). We cannot agree with this conclusion. The idea of a 'structurally and mathematically correct model as a reference' is a chimera. These two methods certainly give only approximations to what 'really' happens; but science advances by successive approximations, and the fact that they are in broad agreement hopefully indicates that we are on the right road.

7.14 References

Arnal, M.A., Mosoni, L., Boirie, Y., Houlier, M.L., Morin, L., Verdier, E., Ritz, P., Antoine, J.M., Prugnaud, J., Beaufrère, B. and Mirand, P.P. (1999) Protein pulse feeding improves protein retention in elderly women. *American Journal of Clinical Nutrition* 69, 1202–1208.

Bier, D.M. (1989) Intrinsically difficult problems: the kinetics of body proteins and amino acids in man. *Diabetes/Metabolism Reviews* 5, 111–132.

Bier, D.M. and Matthews, D.E. (1982) Stable isotope methods for *in vivo* investigations. *Federation Proceedings* 41, 2679–2685.

Boirie,Y., Gachon, P., Corny, S., Fauquant, J., Maubois, J.-L. and Beaufrère, B. (1966) Acute post-prandial changes in leucine metabolism as assessed with an intrinsically labelled milk protein. *American Journal of Physiology* 271, E1083–1091.

Bos, C., Benamouzig, R., Bruhat, A., Roux, C., Mahe, S., Valensi, P., Gaudichon, C., Ferriere, F., Rautureau, J. and Tome, D. (2000) Short-term protein and energy supplementation activates nitrogen kinetics and accretion in poorly nourished elderly subjects. *American Journal of Clinical Nutrition* 71, 1129–1137.

Catzeflis, C., Schütz, Y., Micheli, J.L., Welsch, C., Arnaud, M.J. and Jéquier, E. (1985) Whole body protein synthesis and energy expenditure in very low birth weight infants. *Pediatric Research* 19, 679–687.

Crispell, K.R., Parson, W. and Hollifield, G. (1956) A study of the rate of protein synthesis before and during administration of L-tri-iodothyronine to patients with myxoedema and healthy volunteers using N^{15}-glycine. *Journal of Clinical Investigation* 35, 154.

De Benoist, B., Abdulrazzak, Y., Brooke, O.G., Halliday, D. and Millward, D.J. (1984) The measurement of whole body protein turnover in the preterm infant with intragastric infusion of L-[1-^{13}C] leucine and sampling of the urinary leucine pool. *Clinical Science* 66, 154–164.

Duggleby, S.L. (1999) Protein turnover and urea kinetics during pregnancy; maternal body composition and fetal growth. PhD Thesis, University of Southampton, UK.

Duggleby, S.L. and Waterlow, J.C. (2005) The end-product method of measuring whole body protein turnover: a review of published results and comparison with those obtained by leucine infusion. *British Journal of Nutrition*, 94, 141–153.

Eastell, R., Simmons, P.S., Colwell, A., Assiri, M.A., Burritt, M.F., Russell, R.G.G. and Riggs, B.L. (1992) Nyctohemeral changes in bone turnover assessed by serum bone GLa-protein concentration and urinary deoxypyridinoline excretion: effects of growth and ageing. *Clinical Science* 83, 375–382.

El-Khoury, A.E., Fukagawa, J.K., Sanchez, M., Tsay, R.H., Gleason, E., Chapman, T.E. and Young, V.R. (1994) Validation of the tracer-balance concept with reference to leucine 24-h intravenous tracer studies with L-[1-^{13}C] leucine and [^{15}N-^{13}N] urea. *American Journal of Clinical Nutrition* 59, 1000–1011.

El-Khoury, A.E., Sanchez, M., Fukegawa, N. and Young, V.R. (1995) Whole body protein synthesis in healthy adult humans: $^{13}CO_2$ technique vs plasma precursor approach. *American Journal of Physiology* 268, E174–184.

El-Khoury, A.E., Ajami, A.A., Fukagawa, N.K., Chapman, T.E. and Young, V.R. (1996) Diurnal pattern of the interrelationships among leucine oxidation, urea production, and hydrolysis in humans. *American Journal of Physiology* 271, E563–573.

El-Khoury, A.E., Forslund, A., Olsson, R., Branth, S., Sjödin, A., Andersson, A., Atkinson, A., Selvaraj, A., Hambraeus, L. and Young, V.R. (1997) Moderate exercise and energy balance does not affect 24-h leucine oxidation or nitrogen retention in healthy men. *American Journal of Physiology* 273, E394–407.

Fern, E.B. and Garlick, P.J. (1983) The rate of nitrogen metabolism in the whole body of man measured with [^{15}N]-glycine and uniformly labelled [^{15}N]-wheat. *Human Nutrition: Clinical Nutrition* 37C, 91–107.

Fern, E.B., Garlick, P.J., McNurlan, M.A. and Waterlow, J.C. (1981) The excretion of isotope in urea and ammonia for estimating protein turnover in man with [^{15}N] glycine. *Clinical Science* 61, 217–228.

Fern, E.B., Garlick, P.J., Sheppard, H. and Fern, M. (1984) The precision of measuring the rate of whole-body nitrogen flux and protein synthesis in man with a single dose of [^{15}N]-glycine. *Human Nutrition: Clinical Nutrition* 38C, 63–73.

Fern, E.B., Garlick, P.J. and Waterlow, J.C. (1985a) Apparent compartmentation of body nitrogen in one human subject: its consequences in measuring the rate of whole body protein synthesis with ^{15}N. *Clinical Science* 68, 271–282.

Fern, E.B., Garlick, P.J. and Waterlow, J.C. (1985b) The concept of the single body pool of metabolism nitrogen in determining the rate of whole-body nitrogen turnover. *Human Nutrition: Clinical Nutrition* 39C, 85–99.

Forslund, A.H., Hambraeus, L., Olsson, R.M., El-Khoury, A.E., Yu, Y.-M. and Young, V.R. (1998) The 24-h whole body leucine and urea kinetics at normal and high protein intakes with exercise in healthy adults. *American Journal of Physiology* 275, E310–320.

Garlick, P.J., Clugston, G.A. and Waterlow, J.C. (1980) Influence of low-energy diets on whole-body protein turnover in obese subjects. *American Journal of Physiology* 238, E235–244.

Glynn, M.J., Metzne, S., Halliday, D. and Powell-Tuck, J. (1987) Whole body protein metabolism in parenterally fed patients. Glucose versus fat as the predominant energy source. *Clinical Nutrition* 6, 91–96.

Glynn, M.J., Powell-Tuck, J. and Halliday, D. (1988) Reproducibility of whole body protein turnover measurements in an 'ideal' metabolic subject. *European Journal of Clinical Nutrition* 42, 273–275

Golden, B.E. and Golden, M.H.N. (1992) Effect of zinc on lean tissue synthesis during recovery from malnutrition. *European Journal of Clinical Nutrition* 46, 697–706.

Golden, M.H.N. and Waterlow, J.C. (1977) Total protein synthesis in elderly people: a comparison of results with [^{15}N] glycine and [^{14}C] leucine. *Clinical Science and Molecular Medicine* 53, 277–288.

Grove, G. and Jackson, A.A. (1995) Measurement of protein turnover in normal man using the end-product method with oral [^{15}N]-glycine: comparison of single dose and intermittent dose régimes. *British Journal of Nutrition* 74, 491–507.

Halliday, D. and Rennie, M.J. (1982) The use of stable isotopes for diagnosis and clinical research. *Clinical Science* 63, 485–496.

Jackson, A.A. and Golden, M.H.N. (1980) [^{15}N]-glycine in normal man: the metabolic α-amino-nitrogen pool. *Clinical Science* 58, 577–582.

Jackson, A.A., Golden, M.H., Byfield, R., Jahoor, F., Royes, J. and Soutter, L. (1983) Whole-body protein turnover and nitrogen balance in young children at intakes of protein and energy in the region of maintenance. *Human Nutrition: Clinical Nutrition* 37, 433–446.

Jackson, A.A., Persaud, C., Werkmeister, G., McClelland, I.S.M., Badaloo, A. and Forrester, T. (1997a) Comparison of urinary 5-L-oxoproline (L-pyroglutamate) during normal pregnancy of women in England and Jamaica. *British Journal of Nutrition* 77, 183–196.

Jackson, A.A., Soares, M.J., Grove, G. and Waterlow, J.C. (1997b) Enrichment in urinary ammonia and urea with hourly doses of [^{15}N] glycine: evidence for a step function and a circadian rhythm in protein turnover. *Clinical Science* 93, 265–271.

Jahoor, F. and Wolfe, R.R. (1987) Reassessment of primed constant infusion tracer method to measure urea kinetics. *American Journal of Physiology* 252, E557–564.

Jeevanandam, M., Brennan, M.F., Horowitz, G.D., Rose, D., Mihiranian, M.F., Daly, J. and Lowry, S.F. (1985) Tracer priming in human protein studies with [^{15}N] glycine. *Biochemical Medicine* 34, 214–225.

Jeevanandam, M.D., Shamos, R.F., Casabo, S.F. and Schiller, W.R. (1991) Glucose infusion improves endogenous protein synthesis efficiency in multiple trauma victims. *Metabolism* 40, 1199–1206.

Jeeranandam, M., Young, D.H. and Schiller, N.R. (1991b) Obesity and the metabolic response to severe multiple trauma in man. *Journal of Clinical Investigation* 87, 262–269.

Ma, E.L. and Jiang, Z.M. (1990) Determination of protein turnover changes in preoperative patients by the ^{15}N-glycine constant infusion method. *Proceedings of the Chinese Academy of Medical Sciences, Peking Union Medical College* 5, 97–101.

Maouyo, D., Sarfati, P., Guan, D., Morisset, J. and Adelson, J.W. (1993) Circadian rhythm of exocrine pancreatic secretion in rats: major and minor cycles. *American Journal of Physiology* 264, G792–800.

Marckmann, P., Sandstrom, B. and Jesperson, J. (1993) Dietary effects on circadian fluctuations in human blood coagulation factor VII and fibrinolysis. *Atherosclerosis* 101, 225–234.

Matthews, D.E., Conway, J.M., Young, V.R. and Bier, D.M. (1981) Glycine nitrogen metabolism in man. *Metabolism* 30, 886–893.

McClelland, I.S.M., Persaud, C. and Jackson, A.A. (1997) Urea kinetics in healthy women during normal pregnancy. *British Journal of Nutrition* 77, 165–181.

Nissim, I., Yudkoff, M. and Segal, S. (1983) A model for determination of total body protein synthesis based upon compartmental analysis of the plasma

[15N] glycine decay curve. *Metabolism* 32, 646–653.

Obled, C., Barre, F., Millward, D.J. and Amal, M. (1989) Whole body protein synthesis studies with different amino acids in the rat. *American Journal of Physiology* 257, E639–646.

Obled, C., Barre, F. and Arnal, M. (1991) Flooding-dose of various amino acids for measurement of whole-body protein synthesis in the rat. *Amino Acids* 1, 17–27.

Olesen, K., Heilskov, N.C.S., Schønheyder, F. (1954) The excretion of ^{15}N in urine after administration of ^{15}N-glycine. *Biochimica et Biophysica Acta* 15, 95–107.

Pacy, P.J., Price, G.M., Halliday, D., Quevedo, M.R. and Millward, D.J. (1994) Nitrogen homoeostasis in man: the diurnal responses of protein synthesis and degradation and amino acid oxidation to diets with increasing protein intakes. *Clinical Science* 86, 103–118.

Pannemans, D.L.E., Halliday, D. and Westerterp, K.S. (1995) Whole body protein turnover in elderly men and women: responses to two protein studies. *American Journal of Clinical Nutrition* 61, 33–38.

Pannemans, D.L.E., Wagenmakers, A.J.M., Westerterp, K.R., Schaafsma, G. and Halliday, D. (1997) The effect of an increase of protein intake on whole-body protein turnover in elderly women is tracer dependent. *Journal of Nutrition* 127, 1788–1794.

Pencharz, P.B., Clarke, R., Papageorgiou, A. and Farri, L. (1989) A reappraisal of protein turnover values in neonates fed human milk or formula. *Canadian Journal of Physiology and Pharmacology* 67, 282–286.

Picou, D. and Taylor-Roberts, T. (1969) The measurement of total protein synthesis and catabolism and nitrogen turnover in infants in different nutritional states and receiving different amounts of dietary protein. *Clinical Science* 36, 283–296.

Picou, D., Taylor-Roberts, T. and Waterlow, J.C. (1968) Measurement of total protein synthesis and nitrogen flux in man by constant infusion of [^{15}N] glycine. *Journal of Physiology* 200, 52–53P.

Pitts, R.F. and Pilkington, L.A. (1966) The relation between plasma concentrations of glutamine and glycine and utilization of their nitrogen as sources of urinary ammonia. *Journal of Clinical Investigation* 45, 86–93.

Powell-Tuck, J. and Glynn, M.J. (1985) The effect of insulin infusion on whole-body protein metabolism in patients with gastrointestinal disease fed parenterally. *Human Nutrition: Clinical Nutrition* 39, 181–191.

Powell-Tuck, J., Fern, E.B., Garlick, P.J. and Waterlow, J.C. (1984) The effect of surgical trauma and insulin on whole-body protein turnover in parenterally-fed undernourished patients. *Human Nutrition: Clinical Nutrition* 38, 11–22.

Preston, T., Fearon, K.C.H., McMillan, D.C., Winstanley, F.P., Slater, C., Shenkin, A. and Carter, D.C. (1995) Effect of ibuprofen on the acute-phase response and protein metabolism in patients with cancer and weight loss. *British Journal of Surgery* 82, 229–234.

Price, G.M., Halliday, D.H., Pacy, P.J., Quevedo, M.R. and Millward, D.J. (1994) Nitrogen homoeostasis in man: influence of protein intake on the amplitude of diurnal cycling of body nitrogen. *Clinical Science* 86, 91–102.

Reeds, P.J., Cadenhead, A., Fuller, M.F., Lobley, G.E. and McDonald, J.D. (1980) Protein turnover in growing pigs. Effects of age and food intake. *British Journal of Nutrition* 43, 445–455.

Reiner, J.M. (1953) The study of metabolic turnover rates by means of isotopic tracers. I. Fundamental relations. *Archives of Biochemistry and Biophysics* 46, 53–79.

Robertson, J.S. (1957) Theory and use of tracers in determining transfer rates in biological systems. *Physiological Reviews* 37, 133–154.

Russell, J.A. (1958) The use of isotopic tracers in estimating rates of metabolic reactions. *Perspectives in Biology and Medicine* 1, 138–173.

San Pietro, A. and Rittenberg, D. (1953) A study of the rate of protein synthesis in humans. II. Measurement of the metabolic pool and the rate of protein synthesis. *Journal of Biological Chemistry* 201, 457–473.

Schønheyder, F., Heilskov, N.S.C. and Olesen, K. (1954) Isotopic studies on the mechanism of negative nitrogen balance produced by immobilization. *Scandinavian Journal of Clinical and Laboratory Investigation* 6, 178–189.

Sharp, G.S., Lassen, S., Shankman, S., Hazlet, J.W. and Kendis, M.S. (1957) Studies of protein retention and turnover using nitrogen-15 as a tag. *Journal of Nutrition* 63, 155–162.

Sheppard, C.W. (1962) *Basic Principles of the Tracer Method.* John Wiley, New York.

Slevin, K., Jackson, A.A. and Waterlow, J.C. (1991) A model for the measurement of whole-body protein turnover incorporating a protein pool with lifetime kinetics. *Proceedings of the Royal Society, Series B,* 243, 87–92.

Soares, M.J., Piers, L.S., Shetty, P.S., Robinson, S., Jackson, A.A. and Waterlow, J.C. (1991) Basal metabolic rate, body composition and whole body protein turnover in Indian men with different nutritional status. *Clinical Science* 81, 419–425.

Soares, M.J., Piers, L.S., Shetty, P.S., Jackson, A.A. and Waterlow, J.C. (1994) Whole body protein turnover in chronically undernourished individuals. *Clinical Science* 86, 441–446.

Sprinson, D.B. and Rittenberg, D. (1949) The rate of interaction of the amino acids of the diet with the tissue proteins. *Journal of Biological Chemistry* 180, 715–726.

Steffee, W.P., Goldsmith, R.S., Pencharz, P.B., Scrimshaw, N.S. and Young, V.R. (1976) Dietary protein intake and dynamic aspects of whole-body nitrogen metabolism in adult humans. *Metabolism* 25, 281–290.

Taggart, D.P., McMillan, D.C., Preston, T., Shenkin, A., Wheatley, D.J. and Burns, H.J.G. (1991) Effect of surgical injury and intraoperative hypothermia on whole-body protein metabolism. *American Journal of Physiology* 260, E118–125.

Taruvinga, M., Jackson, A.A. and Golden, M.H.N. (1979) Comparison of ^{15}N-labelled glycine, aspartate, valine and leucine for measurement of whole-body protein turnover. *Clinical Science* 57, 281–283.

Tomkins, A.M., Garlick, P.J., Schofield, W.N. and Waterlow, J.C. (1983) The combined effect of infection and malnutrition on protein metabolism in children. *Clinical Science* 65, 313–324.

Tschudy, D.P., Bacchus, H., Weissman, S., Watkin, D.M., Eubanks, M. and White, J. (1959) Studies of the effect of dietary protein and caloric levels on the kinetics of N metabolism using N^{15}-L-aspartic acid. *Journal of Clinical Investigation* 38, 892.

Vaisman, N., Clarke, R., Rossi, M., Goldberg, E., Zello, G.A. and Pencharz, P.B. (1992) Protein turnover and resting energy expenditure in patients with undernutrition and chronic lung disease. *American Journal of Clinical Nutrition* 55, 63–69.

Van Goudoever, J.B., Sulkers, E.J., Halliday, D., Degenhart, H.J., Carnielli, V.P., Wattimena, J.L. and Sauer, P.J. (1995) Whole-body protein turnover in preterm appropriate for gestational age and small for gestational age infants: comparison of [^{15}N] glycine and [1-^{13}C] leucine administered simultaneously. *Pediatric Research* 37, 381–388.

Waterlow, J.C., Golden, M.H.N. and Garlick, P.J. (1978) Protein turnover in man measured with ^{15}N: comparison of end-products and dose régimen. *American Journal of Physiology* 235, E165–174.

Watson, P.E., Watson, S.P. and Batt, R.D. (1980) Total body water volumes for adult males and females estimated from simple anthropometric measurements. *American Journal of Clinical Nutrition* 33, 27–39.

Willommet, L., Schutz, Y., Whitehead, R., Jequier, E. and Fern, E.B. (1992) Whole body protein metabolism and resting energy expenditure in pregnant Gambian women. *American Journal of Physiology* 263, E624–631.

Wu, H. and Bishop, C.W. (1959) Pattern of N^{15}-excretion in man following administration of an N^{15}-labeled glycine. *Journal of Applied Physiology* 14, 1–5.

Wu, H. and Sendroy, J. (1959) Pattern of N^{15}-excretion in man following administration of an N^{15}-labeled L-phenylalanine. *Journal of Applied Physiology* 14, 6–10.

Wu, H., Sendroy, J. and Bishop, C.W. (1959) Interpretation of urinary N^{15}-excretion data following administration of an N^{15}-labeled amino acid. *Journal of Applied Physiology* 14, 11–21.

Wurtman, R.J. and Axelrod, J. (1967) Daily rhythmic changes in tyrosine transaminase activity of rat liver. *Proceedings of the National Academy of Sciences* 57, 1594–1598.

Yagi, M. and Walser, M.N. (1990) Estimation of whole body protein synthesis from oxidation of infused [1-^{14}C] leucine. *American Journal of Physiology* 258, E151–157.

Yoshida, S., Noake, T., Tanaka, K., Ishibashi, N., Shirouzu, Y., Shirouzu, K. and Stein, T.P. (1996) Effect of fentanyl citrate anaesthesia on protein turnover in patients with esophagectomy. *Journal of Surgical Research* 64,120–127.

Zilversmit, D.B. (1960) The design and analysis of isotope experiments. *American Journal of Medicine* 29, 832–848.

8

Amino Acid Oxidation and Urea Metabolism

Carbon dioxide and urea are two irreversible end-products of amino acid metabolism.

8.1 Amino Acid Oxidation

Oxidation is a component of amino acid disposal or irreversible loss; it is necessary to measure it in order to determine the rate of synthesis by subtraction from the flux: $S = Q - O$. Since oxidation is seldom more than about 20% of flux, errors in its measurement will not have such an important effect on synthesis as errors in the estimate of flux. Expired CO_2 has also been used as an end-product in the determination of amino acid flux (Golden and Waterlow, 1977; Yagi and Walser, 1990; El-Khoury *et al.*, 1995a,b). Oxidation is also important in its own right, as the basis of the carbon balance method of determining amino acid requirements (see Chapter 10). Lastly, the extent of dilution of tracer in expired CO_2 has been developed as a method of estimating total energy expenditure over short periods (Elia *et al.*, 1992, 1995). Good reviews of amino acid oxidation have been published by Leijssen and Elia (1996) and by van Hall (1999).

The measurement of oxidation as part of whole body protein turnover is apparently quite straightforward when ^{13}C-leucine is the tracer: the bicarbonate pool is primed; ^{13}C-bicarbonate is infused and expired air collected during the plateau to give the rate of CO_2 output, VCO_2, and its enrichment. With priming plateau enrichment is achieved in about an hour. If KIC is taken as the precursor, the total rate of leucine oxidation is: $(VCO_2 \times \varepsilon_{CO_2})/\varepsilon_{KIC}$. Fig. 8.1 is an example of oxidation measured over a 24-h period of fasting and feeding.

There are, however, some underlying problems.

8.1.1 Recovery of CO_2

It has been recognized for a long time in metabolic studies that when labelled carbon is oxidized, not all the label is recovered in expired CO_2. Numerous measurements have therefore been made of the recovery in breath of labelled CO_2 after a bolus dose or a continuous infusion of labelled bicarbonate. All the results in human subjects that were reported before 1994 were tabulated by El-Khoury *et al.* (1994b) and are summarized in Table 8.1. Similar results were recorded in a review by Leijssen and Elia (1996). In most of the studies the collection period was for 6 h or less, but in a few it was longer, up to 12 or 14 h. With unprimed infusion a plateau is reached in about 6 h, but priming with labelled bicarbonate reduces this time to 2 h or less. There is a tendency for the recovery to be greater the longer the period of infusion; thus Fuller *et al.* (1990), with an unprimed infusion obtained the results shown in Table 8.2. Elia *et al.* (1995) infused $NaH^{14}CO_3$ subcutaneously for 5 days, which included periods of feeding and spells of exercise. After the first 24 h the recovery of $^{14}CO_2$ in breath was 95% and was very consistent between subjects. There seems also to be a direct relation between the amount of CO_2 that is produced and the fraction that is recovered. Recovery is consistently higher in the fed than in the fasted state and higher in the day than in the night. There

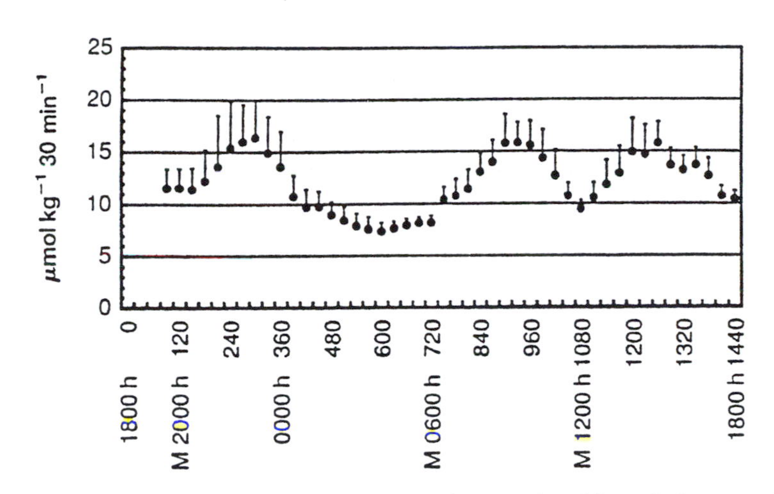

Fig. 8.1. 24 h pattern of leucine oxidation in relation to meals. Reproduced from El-Khoury *et al.* (1995b), with permission of the *American Journal of Clinical Nutrition.* © Am. J. Clin. Nutr. American Society for Nutrition.

is no difference between intravenous and intragastric infusions (Hoerr *et al.*, 1989). Recovery is also higher when subjects are exercised during the period of CO_2 collection (Coggan *et al.*, 1993; Bowtell *et al.*, 1994; Elia *et al.*, 1995). Both food and exercise of course increase CO_2 production.

During exercise the total body pool of CO_2 increases severalfold (Barstow *et al.*, 1990; Leese *et al.*, 1997). There is an immediate increase in the fractional recovery of labelled CO_2 to about twice the rate at which it is being infused (Wolfe *et al.*, 1982; Barstow *et al.*, 1990; van Hall, 1999), but the peak is not maintained: within an hour during exercise it falls to the pre-exercise level. It is tempting to suppose that this pattern represents mobilization of CO_2 from a pool identified with muscle (see below), which fills up again when the exercise stops.

There is no clear evidence about the site or mechanism of CO_2 sequestration. A common view is that it is taken up into a relatively labile fraction of bone carbonate, but the amount of bone mineral available is not enough for the amounts sequestered. Ram *et al.* (1999) infused sheep for 24 h with labelled bicarbonate and measured the acid-fast radioactivity in various tissues. The largest amounts were found in muscle, fat, skin and bone, but the total amount in these tissues was only 4% of the dose in the fed state and 6% in fasted animals. It may be presumed that this carbon is fixed in protein, since ^{14}C-bicarbonate has long been used for labelling body proteins *in vivo* (Swick, 1958; McFarlane, 1963; Millward, 1970). It is possible also that some CO_2 is fixed temporarily in intermediates of the Krebs cycle, and is released when CO_2 production is increased by exercise. This possibility leads on to a consideration of CO_2 kinetics.

Table 8.1. Summary of results on recovery of CO_2 in healthy adults.

Administration of tracer	Number of studies	Mean recovery, %
Bolus injection		
Fasting	14	70
Fed	2	73
Constant infusion		
Fasting	13	77
Fed	8	84.5

Summarized from El-Khoury *et al.* (1994b).

Table 8.2. Percentage recovery of CO_2 over different periods of collection.

Hours of infusion	% of dose	
	Recovery in breath	Recovery in urea
0–3	50	0.35
3–6	81	0.81
6–12	87	1.44
12–36	96	1.85

Data of Fuller *et al.* (1990).

8.1.2 Kinetics of CO_2 metabolism

Several groups have studied the kinetics of CO_2 by compartmental analysis (Fowler *et al.*, 1964; Irving *et al.*, 1983, 1984; Barstow *et al.*, 1990; Cobelli *et al.*, 1991; Leese *et al.*, 1997). Irving's paper of 1983 provides, together with new data, an excellent account of earlier work. This approach requires that the tracer be given as a bolus dose. In most cases the enrichment curve has been followed for only 4–6 h, which is probably not long enough. All agree that the activity curves can yield three exponentials, suggesting a mammillary model with a central pool and two peripheral pools, one fast and one slow. The total pool size of bicarbonate calculated from these models ranges from about 9 to 15 mmol kg^{-1}, of which, according to Irving *et al.* (1983) and Barstow *et al.* (1990), about 80% is in the slow pool, exchanging with the central pool with rate coefficients of 1.5–2 h^{-1}. It was suggested that the fast pool represents the extracellular fluid and the slow pool the intracellular fluid. A simple physiological calculation is reasonably consistent with this idea, at least as regards total pool size.

ECF 280 ml kg^{-1}: HCO_3 25 mmol l^{-1} = 7 mmol kg^{-1}
ICF 340 ml kg^{-1}: HCO_3 10 mmol l^{-1} = 3.4 mmol kg^{-1}
Total 10.4 mmol kg^{-1}.

Irving *et al.* (1983) compared their results by compartmental analysis with those of a 'composite model' based on data on blood flow rates and organ levels of HCO_3 (Fahri and Rahn, 1960). This calculation resulted in a total pool size of HCO_3 of 11 mmol kg^{-1}, of which ~60% was in muscle. CO_2 production from organs other than muscle was estimated to occur at a fractional rate of 3.2 h^{-1}, while the fractional production from muscle was only 0.24 h^{-1}. Even this relatively slow turnover rate, corresponding to a half-life of some 3 h, is too fast to explain the fact that it takes at least 12 h for recovery of CO_2 to approach 100% (Table 8.2).

These laborious compartmental studies are interesting, but have little practical application to measurements of protein turnover. As usual with this approach there was great variability in the estimates of pool sizes and transfer rates between comparable subjects and even in the same subject when measured on more than one occasion (Irving *et al.*, 1983). Cobelli *et al.* (1991) admitted that their analysis of the 'carbon dioxide system', which added four pools to the six of their model of amino acid metabolism, did not contribute anything useful to their studies of protein turnover.

8.1.3 Labelled CO_2 as an end-product for measurement of flux

When leucine is the tracer, labelled CO_2 is an end-product. The assumption (see Chapter 7) is that with an essential amino acid the partition of flux between end-product and synthesis is the same as the partition of tracer. Thus: $Q/d = S/s = E/e$, where the capital letters stand for amounts of tracee and the lower case letters for amounts of tracer. e is measured directly as the rate of expiration of ^{13}C, but E cannot be taken as the total rate of CO_2 production, since the greater part of expired CO_2 is derived from oxidation of substrates other than leucine. Therefore an estimate of E has been based either on the N intake (Golden and Waterlow, 1977) or on N excreted (Yagi and Walser, 1990; El-Khoury *et al.*, 1995a). It has been claimed of the method that the proportion of dose oxidized is independent of assumptions about the precursor, but this is not really true (see Chapter 5). However, Yagi and Walser (1990) obtained excellent agreement in rats between synthesis by this method and by the conventional method based on the enrichment of plasma KIC.

8.1.4 Dilution of labelled CO_2 as a measure of energy expenditure

In theory the dilution of labelled CO_2 during infusion of labelled bicarbonate should provide a measure of the total CO_2 production rate, from which the energy expenditure can be calculated. The method had been used in animals, but Elia (1991) was the first to suggest that it could be applied to short-term measurements of energy expenditure in man. This idea was taken up by El-Khoury *et al.* (1994b). In this approach two problems are encountered: the first is incomplete recovery, which has already been discussed; the second is more subtle. Infusions of labelled HCO_3 are made into the blood, which is part of the extracellular pool; but CO_2 derived from the oxidation of substrates is produced intracellularly. Is it justifiable to assume that the CO_2 from these two different sources encounters the same degree of dilution?

To examine this question use was made of the carbon of urea, as a marker of intracellular CO_2 enrichment, because it is introduced into the urea molecule by the mitochondrial enzyme carbamoyl phosphate synthetase 1. Marsolais *et al.* (1987) found in perfused livers equal specific activities in the carbon of urea and infused $NaH^{14}CO_3$, and also in $^{14}CO_2$ generated from ^{14}C-KIC or ^{14}C-pyruvate. However, this equality was not confirmed by studies in humans. Hamel *et al.* (1993) infused ^{14}C-bicarbonate for 14 h. In spite of a huge priming dose it took 6 h for a plateau of urea labelling to be reached, probably because of the large size and slow turnover of the urea pool. At plateau the specific radioactivity of urea was only 83% of that of breath CO_2, a figure very close to that found by Elia *et al.* (1992, 1995). Thus if urea-carbon and breath-CO_2 are representative of intra- and extra-cellular pools it seems that both can be in a steady state at different levels of labelling – a situation exactly parallel to that of KIC and leucine. Total CO_2 production, calculated as the dose divided by the specific radioactivity of urea, corrected for the experimentally determined retention factor of 0.83, gave results that agreed extremely well with those obtained by standard calorimetry. From the point of view of studies of energy expenditure, this was very satisfactory since it meant that accurate measurements could be made over 2–4 days more quickly and more cheaply than with doubly labelled water, and without the inconvenience of any collections of expired air. As in measurements of protein turnover, urea can be regarded as the reciprocal of CO_2.

8.2 Metabolism of Urea

In the words of Krebs *et al.* (1973): 'Experience shows that virtually every detail of living matter serves a physiological purpose.' From a teleological point of view it must be agreed that the 'purpose' of urea production in mammals is to dispose of excess nitrogen, a result that is achieved in different ways in different animal families; e.g. in birds by the production of uric acid. It has been contended that the function of urea production is to maintain normal pH by disposing of bicarbonate generated in the metabolism of protein (Atkinson and Bourke, 1984). Since, however, acidosis and alkalosis were not found to have any effect on urea production *in vivo* (Halperin *et al.*, 1986) this function may, in the present context, be regarded as a side-line.

8.2.1 Regulation of urea synthesis

Nearly half a century ago Duda and Handler (1958) concluded from their pioneer studies on rats with ^{15}N-ammonium that '... the rate of urea synthesis over the entire range of dosage was dependent solely on the available substrate, and neither the necessary enzymes nor coenzymes seemed to limit the synthetic rate'. Later workers showed in man linear correlations between urea appearance or production and plasma amino acid concentrations (Rafoth and Onstad, 1975; Vilstrup, 1980, 1989) (Fig. 8.2). These findings seem to fit in with the biochemistry of the Krebs-Henseleit cycle. NH_3 rather than NH_4^+ is the substrate for the first enzyme in the cycle, carbamoyl-phosphate synthetase 1 (CPS-1) (Cohen *et al.*, 1985), the ammonia being derived, via the transaminases, from glutamate by the action of glutamate dehydrogenase (Krebs *et al.*, 1973). These reactions occur in the liver virtually instantaneously, as was shown by Cooper *et al.*

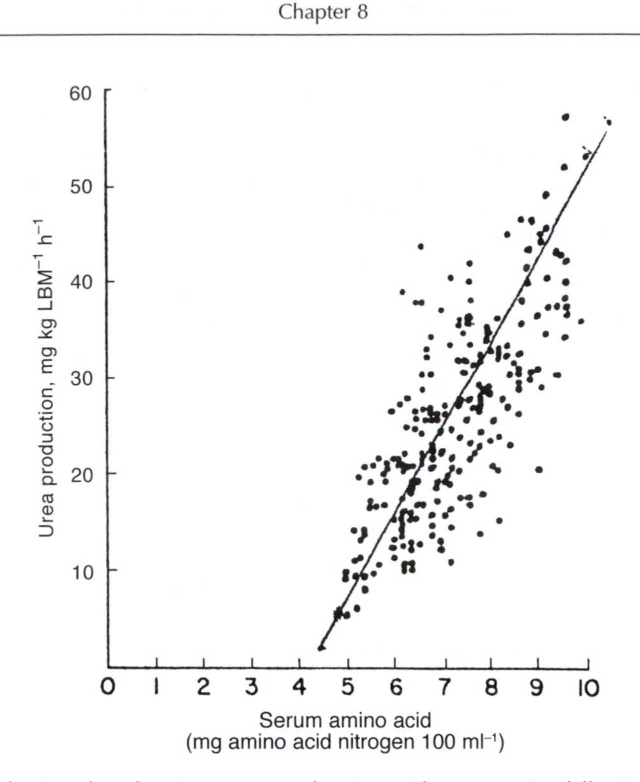

Fig. 8.2. Urea production plotted against serum total amino acid concentration following different protein loads. Reproduced from Rafoth and Onstad (1975), by courtesy of the *Journal of Clinical Investigation.*

(1987, 1988) in experiments with the very short-lived isotope ^{13}N. CPS-1 is not reversible, operates far from equilibrium and is capable of handling a wide range of NH_3 supply, so that it is usually operating at only a fraction of its capacity (Krebs *et al.*, 1973; Meijer *et al.*, 1990). It is reasonable, therefore, to regard the regulation of urea production as the automatic and instantaneous response to changes in input. However, CPS-1 has for its activity an absolute requirement for a co-factor, N-acetylglutamate (NAG), which complexes with CPS-1 to form the functional catalytic unit. There has been much debate about whether NAG should be regarded as a regulator or simply an essential co-factor.

There is a further problem. In 1962 Schimke published two classical papers showing that all the urea cycle enzymes in the liver increased or decreased coordinately in response to changes in dietary protein intake (Schimke, 1962a,b). The time-course of these changes was examined in more detail by Das and Waterlow (1974) (see Chapter 10), and the subject reviewed by Morris (1992) and M. Jackson *et al.* (1986). As a physiologist one can but ask: what is the purpose or reason for the 2–3-fold increase in amount and activity of the enzymes with quite a modest increase in protein intake, when from the biochemical evidence it seems that the ornithine cycle could easily handle such an increase in load without any help from enzyme induction? Perhaps it is a hangover from the time when the hunter-gatherers' diet was much higher in protein than it is now (Speth, 1984).

As was said by Morris in 1992: 'Sixty years after its initial description the physiologic role of the urea cycle continues to be the subject of new ideas and new controversies.' As will be seen below, there are still mysteries about how nitrogen balance is so exquisitely regulated that output matches intake over a range of intakes and long periods of time (Waterlow, 1999).

8.2.2 Urea kinetics

The pioneer paper in the field of urea kinetics was that of Walser and Bodenlos (1959), establishing three points on which all later workers are agreed: first, that after correction for changes in the size of the urea pool, only part of the urea produced is excreted in the urine; secondly, that the 'missing' urea passes into the colon or lower ileum where it is hydrolysed to ammonia by microbial urease; thirdly, that part of the ammonia produced in the gut is recycled back to urea. The end result of these three processes is that nitrogen balance is maintained over a wide range of intakes, but exactly how this is achieved is far from clear. It was also shown that in normal man the urea pool turns over rather slowly, with a half-life of 8–12 h.

In the years that followed Walser and Bodenlos's paper clinicians began to be interested in urea kinetics in conditions such as sepsis (Long *et al.*, 1978), renal failure (Varcoe *et al.*, 1975) and cirrhosis of the liver (Rypins *et al.*, 1980). A great deal of work has also been done on normal subjects by Jackson and his colleagues in Southampton, stimulated by his initial experiences with malnourished children. Six points have to be considered: the most reliable and convenient method of measuring urea production; the amount produced in relation to nitrogen intake; the amount and proportion of urea metabolized or salvaged, to use the term introduced by Jackson; the proportion of N derived from metabolism in the colon that is recycled back to urea; the fate of the salvaged N that is retained; and the regulation of salvage.

8.2.3 Urea production

The method mostly used in the early years was to give a single IV bolus of urea labelled with various tracers, and to follow the enrichment in urine for times ranging from 6 to 16 h. Urea production was calculated from the slope of the semi-log plot of enrichment, after correcting for mixing time. This work was valuable in showing no difference between results with ^{15}N-, ^{13}C-, ^{14}C-, or ^{18}O-labelled urea. A good deal of attention was also given to the problem of 'delay time', which led to enrichment in urine over a given period being higher than in plasma (Walser and Bodenlos, 1959).

A further development was the primed constant intravenous infusion of labelled urea, usually $^{15}N^{15}N$. Matthews and Downey (1984) investigated the effect of the priming dose; they concluded that because of the slow turnover of urea, it was critical to get the priming dose right, and that the correct prime should be 12.5 times the hourly infusion rate. Jahoor and Wolfe (1987), however, obtained good plateaus with primes equivalent to only 5–10 h infusion. The protocol adopted by Young's group at MIT is to give a prime equivalent to 15 h of infusion, followed by IV infusion for 24 h, during which time the subjects are partly fasted, partly fed (El-Khoury *et al.*, 1994a, 1996; Forslund *et al.*, 1998; Young *et al.*, 2000). In England, on the other hand, Jackson's group have consistently given the tracer by mouth, either as a single dose with output in the urine being followed for 48 h; or as a 15 h prime, followed by intermittent doses every 3 h for 15–18 h. This method gives acceptable plateaus; urea production is determined from the plateau enrichment of $^{15}N^{15}N$ urea in the urine.

Relationship of urea production to N intake

This relationship is of great importance, because it has a key bearing on the regulation of nitrogen balance. On this there is serious controversy. The classical view, as we have seen, is that production (P) responds linearly to changes in intake. Krebs *et al.* (1973) concluded that flux through the urea cycle responds *automatically* and *immediately* to changes in N supply. It was therefore not unexpected that Young's group obtain a linear relationship between intake and production over a range of intakes from 0 to nearly 400 mg N kg^{-1} day^{-1} (Fig. 8.3). It should be pointed out that the only true measure of urea production at zero protein intake is that of Young *et al.* (2000), because their subjects had been on a protein-free diet for 5 days.

In contrast to Young's data the results by oral dosage are very different. Jackson, summarizing the extensive work of his group over a number of years, concluded that P varies little with N intake. The findings in 23 groups of subjects, including pregnant women and adolescent boys, are also shown in Fig. 8.3. However, the intakes do not cover a very wide range; half the studies were in the intake range 140–180 mg N kg^{-1} day^{-1}, and gave average values for P varying from 158 to

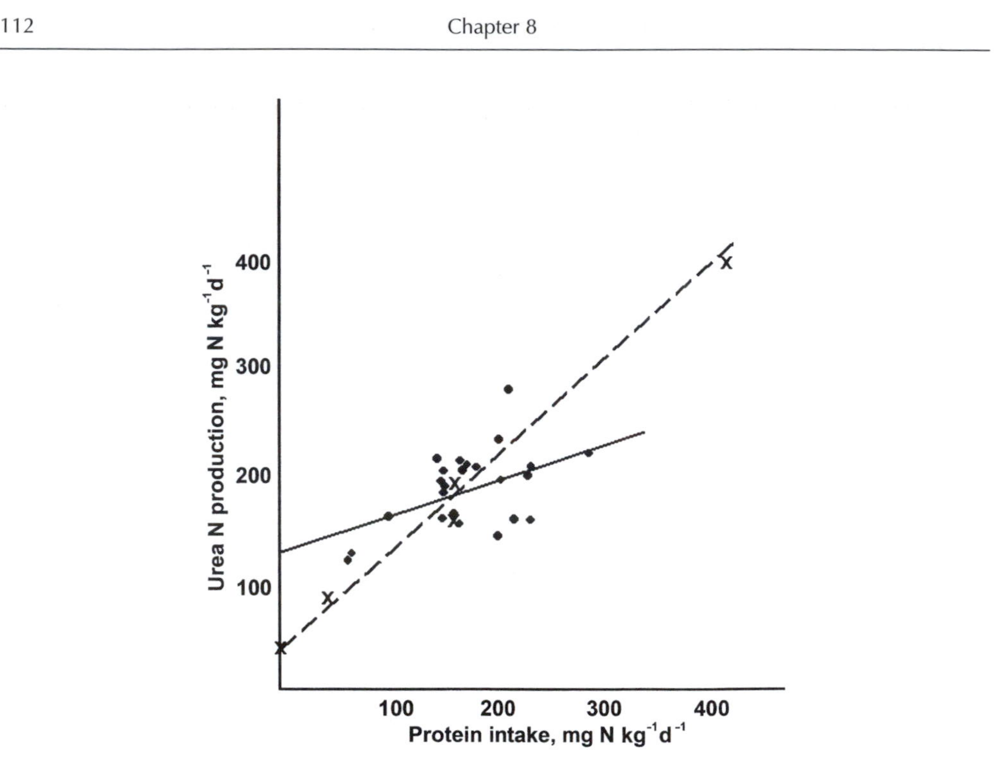

Fig. 8.3. Urea N production *vs.* protein N intake, mg kg^{-1} day^{-1}.
x - - - x, data of Young *et al.* (2000) by constant intravenous infusion of $^{15}N^{15}N$ urea; •——•, data of Jackson and co-workers by single or multiple oral dosage of $^{15}N^{15}N$ urea. References for data of Jackson *et al.*: Jackson *et al.*, 1984, 1993; Hibbert and Jackson, 1991; Danielson and Jackson, 1992; Hibbert *et al.*, 1992, 1995; Langran *et al.*, 1992; Bundy *et al.*, 1993; Forrester *et al.*, 1994; McClelland *et al.*, 1996, 1997; Meakins and Jackson, 1996; Child *et al.*, 1997.

218 mg N kg^{-1} day^{-1}. Nevertheless, a regression of the means of Jackson's data does show a significant relationship: production = 0.33 intake + 124, r = 0.53, $p < 0.01$. This contrasts with the data of Young *et al.*: P = 0.89 intake + 35, r = 0.99, $p < 0.01$. It has been suggested that a more appropriate comparison is of production versus (intake + hydrolysis), since both contribute to the N available for urea production (McClelland *et al.*, 1997); this gives a linear relationship with a steeper slope. A logical solution would be to relate urea production to the nitrogen flux rather than to the intake. Since flux does not vary very much with intake (see Chapter 9), it is not surprising that urea production should be relatively independent of and sometimes greater than intake. One could perhaps reconcile the divergent results of Jackson and Young on the basis that the urea cycle has prior access, through a first-pass effect, to the food-intake component of the flux, and this effect would be most marked at the two extremes, not represented in the Jackson studies, of zero and very high protein intakes.

Variability is a major problem. Hibbert and Jackson (1991) showed that urea production measured on five separate occasions in the same subject on the same intake was very constant. Nevertheless, there is considerable variation, not only between the mean values of P in different studies with similar intakes, but also between individuals in the same study. The mean coefficient of variation of P between subjects in the four studies of Young *et al.* (2000) was 18%; in the four studies of Danielson and Jackson (1992) and Langran *et al.* (1992) it was 21%. This complicates interpretation of the data.

We must conclude that urea production varies with N flux rather than N intake. The relationship is less close with the oral than with the intravenous route of dosage; it is not clear whether the difference is an artefact of the route by which the tracer is given or of some other difference in the protocols of the two sets of studies.

8.2.4 Salvage of urea

The process by which some of the urea produced fails to be excreted in the urine has been described by various names – transfer, metabolism, hydrolysis, salvage. We prefer the term ‘salvage’, introduced by Jackson, since it suggests an active purposeful process, but it seems better to denote it by the symbol T for transfer, rather than S, to avoid confusion with synthesis. Thus T = P – E, where E is the urea excreted. Mammalian species, from rat to camel, salvage a large proportion of the urea that they produce (Wrong *et al.*, 1985). Therefore this process must have some evolutionary advantage, in keeping with Jackson’s view, which I share, that it has a physiological function in economizing nitrogen and regulating nitrogen balance.

In man it is generally supposed, as originally postulated by Walser and Bodenlos (1959), that salvage depends on transfer of urea from plasma to the colon, where it is broken down to ammonia by bacterial urease. It is probable that this process occurs also in the lower ileum, since substantial hydrolysis has been found in patients without a functioning colon (Gibson *et al.*, 1976). The lumen of the colon may not be the most important site of hydrolysis; there is evidence that much of the urease activity has a mucosal or juxta-mucosal location. However, whatever the exact site, bacteria must be responsible, since hydrolysis is greatly reduced when the gut is sterilized by antibiotics.

Data on the proportion of urea production that is salvaged (T/P) are shown in Table 8.3A. There is no relation between T/P and production. At similar production rates there are large variations in rates of salvage both between and within studies. The coefficient of variation between individuals in the same study is sometimes as high as 50%, presumably because salvage is a difference measurement.

Table 8.3B shows that in neonates and infants the proportion is higher than in adults, in keeping with Jackson’s hypothesis that salvage is high when the demand for N is large in relation to intake. In malnourished infants the proportion salvaged, although not the amount, is higher on low protein diets. No consistent increase in T/P was observed in pregnancy (Forrester *et al.*, 1994; McClelland *et al.*, 1997).

An interesting observation is diurnal variation. It is well known that urea excretion is higher in the day than the night (Steffee *et al.*, 1981; Parsons *et al.*, 1983; El-Khoury *et al.*, 1994a); production, however, is relatively constant, so salvage varies inversely with excretion, being less by day than by night (Table 8.4). However, the day/night division does not correspond exactly to the divide between fasting and feeding; in the experiments of Table 8.4 and in earlier observations there is a lag of 3–6 h between the beginning of feeding and the peak of urea excretion, which continues at a high rate for some hours after the feeding has stopped. Presumably this lag results from the slow turnover rate of the urea pool. Another factor that may contribute to it is the transit time of food to reach the lower ileum. It has been shown that food stimulates the urease activity of the microbes in the gut.

If it is true that P is not very sensitive to intake, at least within the usual range, whereas E is regulated to match the intake and maintain nitrogen balance, T must play an important role in this regulation.

Fate of salvaged urea

Our concept is that $^{15}NH_3$, derived by hydrolysis of $^{15}N^{15}N$ urea by the microbes of the lower bowel, is re-absorbed and transported to the liver, where it joins the circulating amino-N pool. That pool is constantly being distributed to protein synthesis, synthesis of other compounds, and urea. The labelled ^{15}N must be distributed in the same metabolic pathways and in the same proportions as the total non-essential amino-N flux, so that some of it is recycled back to urea. The recycled urea will be $^{15}N^{14}N$ urea, since the probability of two ^{15}N atoms coming together is extremely small. Thus the extent of recycling can be determined by measuring the excretion of $^{15}N^{14}N$ urea in the urine. The calculations, which are somewhat complex, are set out in Jackson *et al.* (1984) and in simplified form in Jackson *et al.* (1993). This group is the only one to have made an extensive study of recycling. They found in ten studies on adults that only 10–20% of salvaged N was recycled back into urea. This is

Table 8.3. Proportion of urea produced (P) that is salvaged (T).

		Production mg N kg^{-1} day^{-1}	T/P, %
A. Adults			
Oral dosage[a]			
Intake range mg N kg^{-1} day^{-1}	Number of subjects		
< 100	4	169	42
100–150	3	187	46
150–200	12	178	39
> 200	5	205	41
Intravenous dosage[b]			
Intake, mg N kg^{-1} day^{-1}			
0		35	8
44		83	22
161		157	23
157		184	46
392		386	22
B. Infants			
Neonates (Wheeler *et al.*, 1991)			80
Breast fed, 1–3 months (Steinbrecher *et al.*, 1996)			76
Infants with malnutrition (Badaloo *et al.*, 1999)			
Malnourished,	high protein		59
	low protein		64
Rapidly growing,	high protein		47
	low protein		54
Recovered,	high protein		47
	low protein		59

[a]For references, see Fig. 8.3. This group contains healthy subjects on normal diets, vegetarians, pregnant women, adolescent boys. There are no substantial differences between these subjects.
[b]Data of Young *et al.* (2000), one subject on each intake.

about the proportion of the total amino-N flux that on normal diets goes to the synthesis of urea. The low rate of recycling derived from measurements of $^{14}N^{15}N$ contrasts with Young's view that almost all salvaged N is recycled.

There has been controversy about what happens to the salvaged nitrogen that is not recycled to urea. Picou and Phillips (1972) originally recognized two destinations for it, into protein synthesis (S) and into synthesis of other metabolites (X), but later workers have not maintained this distinction. X in any case was a small proportion. Walser (1981) has argued strongly that salvage does not produce any *net* gain of N to the body, but merely exchanges with or replaces other amino-N. In the adult in nitrogen balance this clearly must be correct. When labelled N from urea is found, for example, in plasma albumin (Varcoe *et al.*, 1975; Richards *et al.*, 1967) it has presumably been incorporated at the expense of unlabelled N. It is interesting that in Richards' study with [^{15}N]-ammonium chloride, incorporation into plasma albumin was much greater in subjects on a protein intake of 20 g than of 70 g. Indeed, it has long been known that in infants urea can substitute to some extent for non-essential amino acids in promoting growth and nitrogen retention (Snyderman *et al.*, 1962). However, *net* deposition of protein can only occur with an increased supply of non-essential N if essential amino acids are available in excess. This seems to be the case with the breast-fed baby, in which the ratio of EAA/NEAA is higher in breast milk (0.79) than in the body (0.56) (Widdowson *et al.*, 1979) and much higher than in the calculated requirement

Table 8.4. Diurnal variation in urea production (P) and salvage (T).

		mg N kg^{-1} day^{-1}	
		P	T
El-Khoury *et al.*, 1994a	Day	78	0
	Night	71	52
Meakins and Jackson, 1995	Day	90	44
	Night	94	69
El-Khoury *et al.*, 1996	Day	73	–3
	Night	84	37

(0.44) (Dewey *et al.*, 1996). Urea accounts for 10–15% of the N in human milk, and after single doses of ^{15}N urea 20–40% was retained after 48 h (Heine *et al.*, 1986; Fomon *et al.*, 1988; Donovan *et al.*, 1990). There may be other situations where NEAAs rather than EAAs are limiting.

Hydrolysis of urea may not be the only useful activity of the gut microflora. Wrong *et al.* (1985) showed that during a constant IV infusion of ^{15}N urea lasting 72 h the enrichment of free NH_3 in the colon and of total faecal N were similar, and some ten times less than the enrichment of plasma urea N. There must, therefore, be a very large flux of N within the colon, many times greater than the rate of entry from urea. This flux probably derives, at least in part, from the synthesis and breakdown of microbial protein. Many years ago, to explain how the highlanders of New Guinea were able to maintain nitrogen balance on a protein intake that was inadequate by any standards, it was suggested that they might be utilizing essential amino acids synthesized by bacteria in the colon (Oomen, 1961; Tanaka *et al.*, 1980). Several studies have been done to test this hypothesis, summarized in Chapter 10. It has been established beyond doubt that microbial syntheses of lysine and threonine, amino acids that are not transaminated, does occur in humans.

Regulation of salvage

Although urea production may initially be determined in part by nitrogen intake, it is clear that it is modulated by salvage to achieve nitrogen balance or the retention that is necessary for growth. Therefore there must be some mechanism of regulation. A number of studies have been made by adding pectin to the diet (Doherty and Jackson, 1992) or by altering the ratio of fat to carbohydrate (Jackson *et al.*, 1990) to see whether the extent of salvage can be increased by promoting bacterial activity, but they show little effect. It seems, therefore, that control must depend not on the number of bacteria but on the delivery of urea to the colon. It seems to be generally agreed that transfer of urea into the colon is a much more complicated process than simple diffusion at a rate dependent on the concentration of urea in the plasma. In the last decade there has been much interest in a family of proteins that act as urea transporters (UTs). Although UTs have been identified in many tissues (see reviews by Smith *et al.*, 1995 and Smith and Rousselet, 2001), most of the work has been done on the kidney, because urea transport from the loop of Henle and the collecting ducts into the medulla is critical for the ability of the kidney to concentrate urea to an osmolality greater than that of plasma. Analysis of rat kidney mRNA revealed two transcripts which, on translation produced two proteins, UT1 and UT2 (Smith *et al.*, 1995). In rats the expression of UT1 mRNA was responsive to changes in the protein content of the diet, whereas the expression of UT2 mRNA was sensitive to vasopressin and to the hydration state of the animal. Work is in progress on the characteristics and regulation of these transporters in the colon.

Stewart *et al.* (2005), reviewing this subject, say: 'We have identified a urea transporter in the human colon (h-UTA6) … that is likely to mediate urea transport into the colon … Interestingly, this protein is activated by cAMP and may therefore serve to regulate urea flux into the colon.' They end their review with the words: '… it therefore remains highly plausible that urea nitrogen salvage is not just simply an effect of the microbial environment but a specific regulated interaction between the gut bacteria and their host.' The key word in these quotations is 'regulated'.

8.3 References

Atkinson, D.E. and Bourke, E. (1984) The role of ureagenesis in pH homeostasis. *Trends in Biochemical Sciences* 9, 297–300.

Badaloo, A., Boyne, M., Reid, M., Persaud, C., Forrester, T., Millward, D.J. and Jackson, A.A. (1999) Dietary protein, growth and urea kinetics in severely malnourished children and during recovery. *Journal of Nutrition* 129, 969–979.

Barstow, T.J., Cooper, D.M., Sobel, E.M., Landaur, E.M. and Epstein, S. (1990) Influence of increased metabolic rate on [^{13}C] bicarbonate washout kinetics. *American Journal of Physiology* 259, R163–171.

Bowtell, J.L., Reynolds, N. and Rennie, M.J. (1994) Differential modulation of ^{13}C recovery during a ^{13}C bicarbonate infusion by dietary protein and glucose supplementation at rest and during exercise. *Clinical Science* 87, 57–58.

Bundy, R., Persaud, C. and Jackson, A.A. (1993) Measurement of urea kinetics with a single dose of [^{15}N^{15}N]-urea in free-living vegetarians on their habitual diet. *International Journal of Food Sciences and Nutrition* 44, 253–259.

Child, S.C., Soares, M.J., Reid, M., Persaud, C., Forrester, T. and Jackson, A.A. (1997) Urea kinetics varies in Jamaican women and men in relation to adiposity, lean body mass and protein intake. *European Journal of Clinical Nutrition* 51, 107–115.

Cobelli, C., Saccomeni, M.P., Tessari, P., Biolo, G., Luir, L. and Matthews, D.E. (1991) Compartmental model of leucine kinetics in humans. *American Journal of Physiology* 261, E539–550.

Coggan, A.R., Habash, D.L., Mendenhall, L.A., Swanson, S.C. and Kien, C.I. (1993) Isotopic estimation of CO_2 production during exercise before and after endurance training. *Journal of Applied Physiology* 75, 70–75.

Cohen, N.S., Kiang, F.S., Kian, S.S., Cheung, C.-W. and Raijman, L. (1985) The apparent K_m of ammonia for carbamoyl phosphate synthetase (ammonia) *in situ*. *Biochemical Journal* 229, 205–211.

Cooper, A.J.L., Nieves, E., Coleman, A.E., Filc-DeRicco, S. and Gelband, A.S. (1987) Short-term metabolic fate of [^{13}N] ammonia in rat liver *in vivo*. *Journal of Biological Chemistry* 262, 1073–1080.

Cooper, A.J.L., Nieves, E., Rosenspire, K.C., Filc-DeRicco, S., Gelband, A.S. and Brusilow, S.W. (1988) Short-term metabolic fate of ^{13}N-labeled glutamate, alanine and glutamine (amide) in rat liver. *Journal of Biological Chemistry* 263, 12268–12273.

Danielson, M. and Jackson, A.A. (1992) Limits of adaptation to a diet low in protein in normal man: urea kinetics. *Clinical Science* 83, 103–108.

Das, T.K. and Waterlow, J.C. (1974) The rate of adaptation of urea-cycle enzymes, aminotransferases and glutamate dehydrogenase to changes in protein intakes. *British Journal of Nutrition* 32, 353–373.

Dewey, K.G., Beaton, G., Fjeld, C., Lonnerdal, B. and Reeds, P. (1996) Protein requirements of infants and children. *European Journal of Clinical Nutrition* 50, Suppl. 1, S119–150.

Doherty, J. and Jackson, A.A. (1992) The effect of dietary pectin on rapid catch-up weight gain and urea kinetics in children recovering from severe malnutrition. *Acta Paediatrica* 81, 514–517.

Donovan, S.M., Lönnerdal, B. and Atkinson, S.A. (1990) Bioavailability of urea nitrogen for the low birthweight infant. *Acta Paediatrica Scandinavica* 79, 899–905.

Duda, G.D. and Handler, P. (1958) Kinetics of ammonia metabolism *in vivo*. *Journal of Biological Chemistry* 232, 303–314.

El-Khoury, A.E., Fukagawa, N.K., Sánchez, M., Tsay, R.H., Gleason, R.E., Chapman, T.E. and Young, V.R. (1994a) Validation of the tracer-balance concept with reference to leucine: 24-h intravenous tracer studies with L-[$^{1-13}$C] leucine and [^{15}N-^{15}N] urea. *American Journal of Clinical Nutrition* 59, 1000–1011.

El-Khoury, A.E., Sánchez, M., Fukagawa, N.K., Gleason, R.E. and Young, V.R. (1994b) Similar 24 h pattern and rate of carbon dioxide production by indirect calorimetry vs. stable isotope dilution, in healthy adults under standardized metabolic conditions. *Journal of Nutrition* 124, 1615–1627.

El-Khoury, A.E., Sánchez, M., Fukegawa, N.K. and Young, V.R. (1995a) Whole body protein synthesis in healthy adult humans: $^{13}CO_2$ technique vs. plasma precursor approach. *American Journal of Physiology* 268, E174–E184.

El-Khoury, A.E., Sánchez, M., Fukagawa, N.K., Gleason, R.E., Tsay, R.H. and Young, V.R. (1995b) The 24-h kinetics of leucine oxidation in healthy adults receiving a generous leucine intake via three discrete meals. *American Journal of Clinical Nutrition* 62, 579–590.

El-Khoury, A.E., Ajami, A.A., Fukagawa, N.K., Chapman, T.E. and Young, V.R. (1996) Diurnal pattern of the interrelationships among leucine oxidation, urea production and hydrolysis in humans. *American Journal of Physiology* 271, E563–573.

El-Khoury, A.E., Forslund, A., Olsson, R., Branth, S., Sjödin, A., Anderson, A., Atkinson, A., Selvaraj, A.,

Hambraeus, L. and Young, V.R. (1997) Moderate exercise at energy balance does not affect 24-h leucine oxidation or nitrogen retention in healthy men. *American Journal of Physiology* 273, E394–407.

Elia, M. (1991) Estimation of short-term energy expenditure by the labelled bicarbonate method. In: Whitehead, R.G. and Prentice, A. (eds) *New Techniques in Nutritional Research*. Academic Press, New York, pp. 207–227.

Elia, M., Fuller, N.J. and Murgatroyd, P.R. (1992) Measurement of bicarbonate turnover in humans: applicability to estimations of energy expenditure. *American Journal of Physiology* 263, E676–687.

Elia, M., Jones, M.G., Jennings, J., Poppitt, S.D., Fuller, N.J., Murgatroyd, P.R. and Jebb, S.A. (1995) Estimating energy expenditure from specific activity of urinary urea during lengthy $NaH^{14}CO_3$ infusion. *American Journal of Physiology* 269, E172–182.

Fahri, L.E. and Rahn, H. (1960) Dynamics of changes in carbon dioxide stores. *Anesthesiology* 21, 604–614.

Fomon, S.J., Bier, D.M., Matthews, D.E., Rogers, R.R., Edwards, B.B., Ziegler, E.E. and Nelson, S.E. (1988) Bioavailability of dietary urea nitrogen for the breast-fed infant. *Journal of Pediatrics* 113, 515–517.

Forrester, T., Badaloo, A.V., Persaud, C. and Jackson, A.A. (1994) Urea production and salvage during pregnancy in normal Jamaican women. *American Journal of Clinical Nutrition* 60, 341–346.

Forslund, A.H., Hambraeus, L., Olsson, R., El-Khoury, A.E., Yu, Y.-M. and Young, B.R. (1998) The 24-h whole body leucine and urea kinetics at normal and high protein intakes with exercise in healthy adults. *American Journal of Physiology* 275, E310–320.

Fowler, A.S.E., Matthews, C.M.E. and Campbell, E.J.M. (1964) The rapid distribution of ^3H_2O and $^{14}CO_2$ in the body in relation to the immediate CO_2 storage capacity. *Clinical Science* 27, 51–65.

Fuller, N.J., Murgatroyd, P.R. and Elia, M. (1990) The recovery of labelled carbon dioxide during a 36 h continuous infusion of bicarbonate in man. *Proceedings of the Nutrition Society* 49, 198A.

Gibson, J.A., Sladen, G.E. and Dawson, A.M. (1976) Protein absorption and ammonia production: the effects of dietary protein and removal of the colon. *British Journal of Nutrition* 35, 61–65.

Golden, M.H.N. and Waterlow, J.C. (1977) Total protein synthesis in elderly people: a comparison of results with [^{15}N] glycine and [^{14}C] leucine. *Clinical Science and Molecular Medicine* 53, 277–288.

Halperin, M.L., Chen, C.B., Cheema-Dadri, J., West, M. and Jungas, R.L. (1986) Is urea formation regulated primarily by acid-base balance *in vivo? American Journal of Physiology* 250, F605–612.

Hamel, N., Divertie, G., Silverberg, J., Persson, M. and Miles, J. (1993) Tracer disequilibrium in CO_2 compartments during $NaH^{14}CO_3$ infusion. *Metabolism* 42, 993–997.

Heine, W., Tiess, M. and Wutzke, K.D. (1986) ^{15}N tracer investigations of the physiological availability of urea nitrogen in mother's milk. *Acta Paediatrica Scandinavica* 75, 439–443.

Hibbert, J.M. and Jackson, A.A. (1991) Variations in measures of urea kinetics over four years in a single adult. *European Journal of Clinical Nutrition* 45, 347–352.

Hibbert, J.M., Forrester, T. and Jackson, A.A. (1992) Urea kinetics: comparison of oral and intravenous dose regimes. *European Journal of Clinical Nutrition* 46, 505–509.

Hibbert, J.M., Jackson, A.A. and Persaud, C. (1995) Urea kinetics : effect of severely restricted dietary intakes on urea hydrolysis. *Clinical Nutrition* 14, 242–248.

Hoerr, R.A., Yu, Y.-M., Wagner, D.A., Burke, J.F. and Young, V.R. (1989) Recovery of ^{13}C infused by gut and vein: effect of feeding. *American Journal of Physiology* 257, E426–438.

Irving, C.S., Wong, W.W., Shulman, R.J., Smith, E.O. and Klein, P.D. (1983) [^{13}C] bicarbonate kinetics in hormones: intra- vs. interindividual variations. *American Journal of Physiology* 245, R190–202.

Irving, C.S., Wong, W.W., Boutton, C.W., Shulman, R.J., Lifschitz, D.L., Malplus, E.W., Helge, H. and Klein, P.D. (1984) Rapid determination of whole body bicarbonate kinetics by use of a digital infusion. *American Journal of Physiology* 247, R709–716.

Jackson, A.A., Picou, D. and Landman, J. (1984) The non-invasive measurement of urea kinetics in man by a constant infusion of $^{15}N^{15}N$-urea. *Human Nutrition: Clinical Nutrition* 38C, 339–354.

Jackson, A.A., Doherty, J., de Benoist, M.-H., Hibbert, J. and Persaud, C. (1990) The effect of the level of dietary protein, carbohydrate and fat on urea kinetics in young children during rapid catch-up weight gain. *British Journal of Nutrition* 64, 371–385.

Jackson, A., Danielsen, M.S. and Boyes, S. (1993) A non-invasive method for measuring urea kinetics with a single dose of [$^{15}N^{15}N$] urea in free-living humans. *Journal of Nutrition* 123, 2129–2136.

Jackson, M.J., Beaudet, A.L. and O'Brien, W.E. (1986) Mammalian urea cycle enzymes. *Annual Review of Genetics* 20, 431–464,

Jahoor, F. and Wolfe, R.A. (1987) Reassessment of primed constant-infusion tracer method to measure urea kinetics. *American Journal of Physiology* 252, E557–564.

Krebs, H.A., Hems, R. and Lund, P. (1973) Some regulatory mechanisms in the synthesis of urea in the mammalian liver. *Advances in Enzyme Regulation* 11, 361–377.

Langran, M., Moran, B.J., Murphy, J.L. and Jackson, A.A. (1992) Adaptation to a diet low in protein: effect of complex carbohydrate upon urea kinetics in normal man. *Clinical Science* 82, 191–198.

Leese, G.P., Nicoll, A.E., Varnier, M., Thompson, J., Scrimgeour, C.M. and Rennie, M.J. (1997) Kinetics of $^{13}CO_2$ elimination after ingestion of ^{13}C bicarbonate: the effects of exercise and acid-base balance. *European Journal of Clinical Investigation* 24, 818–823.

Leijssen, D.P.C. and Elia, M. (1996) Recovery of $^{13}CO_2$ and $^{14}CO_2$ in human bicarbonate studies: a critical review with original data. *Clinical Science* 91, 665–677.

Long, L.L., Jeevandam, M. and Kinney, J.M. (1978) Metabolism and recycling of urea in man. *American Journal of Clinical Nutrition* 31, 1367–1382.

Marsolais, C., Huot, S., France, D., Garneau, M. and Brumengraber, H. (1987) Compartmentation of $^{14}CO_2$ in the perfused rat liver. *Journal of Biological Chemistry* 262, 2604–2607.

Matthews, D.E. and Downey, R.S. (1984) Measurement of urea kinetics in humans: a validation of stable isotope tracer methods. *American Journal of Physiology* 246, E519–527.

McClelland, I.-S.M., Persaud, C. and Jackson, A.A. (1997) Urea kinetics in healthy women during normal pregnancy. *British Journal of Nutrition* 77, 165–181.

McFarlane, A.S. (1963) Measurement of synthesis rates of liver-produced plasma proteins. *Biochemical Journal* 89, 277–290.

Meakins, T.S. and Jackson, A.A. (1995) Diurnal cycling in urea nitrogen hydrolysis. *Proceedings of the Nutrition Society* 54, 137A.

Meakins, T.S. and Jackson, A.A. (1996) Salvage of exogenous urea nitrogen enhances nitrogen balance in normal men consuming marginally inadequate protein diet. *Clinical Science* 90, 215–225.

Meijer, A.J., Lamers, W.M. and Chamuleau, R.A.F.M. (1990) Nitrogen metabolism and ornithine cycle function. *Physiological Reviews* 70, 701–748.

Millward, D.J. (1970) Protein turnover in skeletal muscle. 1. The measurement of rates of synthesis and catabolism of skeletal muscle proteins using [^{14}C] Na_2CO_3 to label protein. *Clinical Science* 39, 577–590.

Morris, S.M. (1992) Regulation of enzymes of urea and arginine synthesis. *Annual Reviews of Nutrition* 12, 81–101.

Oomen, H.A.P.C. (1961) The nutrition situation in Western New Guinea. *Tropical and Geographical Medicine* 13, 312–335.

Parsons, H.G., Wood, M.M. and Pencharz, P.B. (1983) Diurnal variation in urine [^{15}N] urea content, estimates of whole body protein turnover, and isotope recycling in healthy meal-fed children with cystic fibrosis. *Canadian Journal of Physiology and Pharmacology* 61, 72–80.

Picou, D. and Phillips, M. (1972) Urea metabolism in malnourished and recovered children receiving a high or low protein diet. *American Journal of Clinical Nutrition* 25, 1261–1266.

Price, G.M., Halliday, D., Percy, P.J., Quevedo, M.R. and Millward, D.J. (1994) Nitrogen homoeostasis in man: influence of protein intake on the amplitude of diurnal cycling of body protein. *Clinical Science* 86, 91–102.

Rafoth, R.J. and Onstad, G.R. (1975) Urea synthesis after oral protein loading in man. *Journal of Clinical Investigation* 56, 1170–1174.

Ram, L., Nieto, R. and Lobley, G.E. (1999) Tissue sequestration of C-labelled bicarbonate [HCO_3] in fed and fasted young sheep. *Comparative Biochemistry and Physiology A.* 122, 323–330.

Richards, P., Metcalfe-Gibson, A., Ward, E.E., Wrong, O. and Houghton, B.J. (1967) Utilization of ammonia nitrogen for protein synthesis in man, and the effect of protein restriction and uraemia. *Lancet* ii, 845–849.

Rudman, O., Difulco, T.J., Galambos, J.T., Smith, R.B., Salam, A.R. and Warren, W.D. (1973) Maximal rates of excretion and synthesis of urea in normal and cirrhotic subjects. *Journal of Clinical Investigation* 52, 2241–2249.

Rypins, E.B., Henderson, J.M., Fulenwider, J.T., Moffit, S., Galambos, J.T., Warren, W.D. and Rudman, D. (1980) A tracer method for measuring rate of urea synthesis in normal and cirrhotic subjects. *Gastroenterology* 78, 1419–1424,

Schimke, R.T. (1962a) Studies on factors affecting the levels of urea cycle enzymes in rat liver. *Journal of Biological Chemistry* 238, 1012–1018.

Schimke, R.T. (1962b) Differential effects of fasting and protein-free diets on levels of urea-cycle enzymes in rat liver. *Journal of Biological Chemistry* 237, 1921–1924.

Smith, C.P. and Rousselet, G. (2001) Urea transporters. *Journal of Membrane Biology* 181, 1–14.

Smith, C.P., Lee, W.-S., Martial, S., Knepper, M.A., You, G., Sands, J.M. and Heddiger, M.A. (1995) Cloning and regulation of expression of the rat kidney urea transporter (rUT2). *Journal of Clinical Investigation* 96, 1556–1563.

Snyderman, S.E., Holt, E., Dancis, J., Roitman, E., Boyer, A. and Balis, E.M. (1962) 'Unessential' nitrogen: a limiting factor for human growth. *Journal of Nutrition* 78, 57–72.

Speth, J. (1984) Early hominid hunting and scavaging: the role of meat as an energy source. *Journal of Human Evolution* 18, 329–343.

Steffee, W.P., Anderson, C.F. and Young, V.R. (1981) An evaluation of the diurnal rhythm of urea excretion in healthy young adults. *Journal of Enteral and Parenteral Nutrition* 5, 378–384.

Steinbrecher, H.A., Griffiths, D.M. and Jackson, A.A. (1996) Urea production in normal breast-fed infants measured with primed/intermittent oral doses of [$^{15}N^{15}N$] urea. *Acta Paediatrica* 85, 656–662.

Stewart, G.S., Fenton, R.A., Thévenot, F. and Smith, C.P. (2005) Urea movement across mouse colonic crypt plasma membranes is mediated by UT-A urea transporters. Submitted to *Gastroenterology.*

Swick, R.W. (1958) Measurement of protein turnover in rat liver. *Journal of Biological Chemistry* 231, 751–764.

Tanaka, N., Kubo, K., Shiraki, K., Koishi, H. and Yoshimura, H. (1980) A pilot study of protein metabolism in the New Guinea Highlanders. *Journal of Nutrition Science and Vitaminology* 26, 247–259.

van Hall, G. (1999) Correction factors for ^{13}C-labelled substrate oxidation at whole body and muscle level. *Proceedings of the Nutrition Society* 58, 979–986.

Varcoe, R., Halliday, D., Carson, E.R., Richards, P. and Tavill, A.S. (1975) Efficiency of utilization of urea nitrogen for albumin synthesis by chronically uraemic and normal man. *Clinical Science and Molecular Medicine* 48, 379–391.

Vilstrup, H. (1980) Synthesis of urea after stimulation with amino acids: relation to liver function. *Gut* 21, 990–995.

Vilstrup, H. (1989) On urea synthesis – regulation *in vivo. Danish Medical Bulletin* 36, 419–429.

Walser, M. (1981) Urea metabolism. In: Waterlow, J.C. and Stephen, J.M.L. (eds) *Nitrogen Metabolism in Man.* Applied Science Publishing, London, pp. 229–246.

Walser, M. and Bodenlos, L. (1959) Urea metabolism in man. *Journal of Clinical Investigation* 38, 1617–1626.

Waterlow, J.C. (1999) The mysteries of nitrogen balance. *Nutrition Research Reviews* 12, 25–54,

Wheeler, R.A., Jackson, A.A. and Griffiths, D.M. (1991) Urea production and recycling in neonates. *Journal of Pediatric Surgery* 26, 1–3.

Widdowson, E.M., Southgate, D.A.T. and Hey, E.N. (1979) Body composition of the fetus and infant. In: Visser, H.K.A. (ed.) *Nutrition and Metabolism of the Fetus and Infant.* Martinus Nijhoff, The Hague, pp. 164–178.

Wolfe, R.R., Goodenough, R.D., Wolfe, M.H., Royle, G.T. and Nader, E.R. (1982) Isotope analysis of leucine and urea metabolism in exercising humans. *Journal of Applied Physiology* 52, 458–466.

Wrong, O.M., Vince, A.J. and Waterlow, J.C. (1985) The contribution of endogenous urea to faecal ammonia in man. *Clinical Science* 68, 193–199.

Yagi, M. and Walser, M. (1990) Estimation of whole body protein synthesis from oxidation of infused [1-^{14}C] leucine. *American Journal of Physiology* 258, E151–157.

Young, V.R., El-Khoury, A.E., Raguso, C.A., Forslund, A.H. and Hambraeus, L. (2000) Rates of urea production and hydrolysis and leucine oxidation change linearly over widely varying protein intakes in healthy adults. *Journal of Nutrition* 130, 761–766.

9

The Effects of Food and Hormones on Protein Turnover in the Whole Body and Regions

Food affects protein turnover in several different ways. These include the immediate response to food; the effect of different diets in subjects adapted to them; and the effect of long-standing under-nutrition. There is a very large literature on studies in animals; the responses of individual tissues, particularly muscle and liver, are described in Chapter 15.

9.1 The Immediate Effects of Food

Two different protocols have been used: food given continuously or at frequent intervals, compared with fasting, and food given as a single meal.

9.1.1 Continuous feeding

Most studies have been made with the short protocol: an infusion of about 8–10 h, with 3–4 h postabsorptive and 4–6 h fed. Table 9.1 shows the results of 23 fasting/fed studies by the precursor method, with protein intakes in the normal range: ~0.6–1.5 g kg^{-1} day^{-1} (A). With continuous feeding at this level there was no change in whole body synthesis, but on average a fall of nearly 50% in breakdown. Oxidation, as would be expected, increased nearly twofold. The absence of any effect on synthesis fits in with the results obtained by the EP method with a single dose of tracer (Chapter 7, Table 7.3). Feeding produces a small rise (+ 20%) in plasma leucine concentration, with a very substantial increase in insulin.

When the food provided amounts of protein well below the normal range of requirements (Table 9.1B), mostly with a protein:energy ratio of only 2–3%, there was on feeding a small fall in protein synthesis and a somewhat larger fall in breakdown, but not nearly as large as that found with normal protein intakes. There was also a fall in plasma leucine, but as great an increase in insulin as with the higher protein intakes.

These figures are crude averages; they take no account of the actual amounts of protein fed nor of the possible effects of different types of protein (see section 9.2). Some particular studies need more detailed description. In the first experiments to tackle the problem, ^{14}C-leucine was infused intravenously for 24 h; for the first 12 h food was given at hourly intervals, and for the second 12 h the subjects fasted (Garlick *et al.*, 1980; Clugston and Garlick., 1982). In these experiments fasting followed feeding, and produced a 30% decrease in synthesis with no change in breakdown. When these studies were repeated with a shorter infusion of 8 h, with the first 4 h fasting and then 4 h fed, there was a slight fall in synthesis in the fed state and a very large fall in breakdown (Melville *et al.*, 1989). These divergent results were attributed to recycling during the original 24 h infusion. The plateau activity of leucine at the end of the fasting period in Clugston's study was much higher than after feeding – 1.2 *vs.* 0.8 atom % excess. This early work is well reviewed by Garlick *et al.* (1991).

The next step was fasting–feeding studies in subjects adapted to different levels of protein intake. There are three such studies, the results of which are included in Table 9.1 but are shown in more detail in Fig. 9.1 and Table 9.2. In both fed and fasted states there is a tendency for synthesis to increase with increasing intakes. If 0.6 g kg^{-1} day^{-1} is taken

Table 9.1. Response of whole-body turnover to continuous feeding in fasted–fed studies.

		Ratio of fed:fasted				
		Synthesis	Breakdown	Oxidation	Plasma leucine concentration	Plasma insulin
A. Normal protein intakes[a]		1.05	0.56	1.74	1.19	7.5
	n:	25	25	25	6	7
B. Low protein intakes[b]		0.89	0.79	0.92	0.70	7.8
	n:	8	8	8	8	5

[a]0.6 to 1.6 g protein kg^{-1} d^{-1}. n is number of subjects.
[b]< 0.6 to 1.6 g protein kg^{-1} d^{-1}.
References: Motil, 1981; Hoffer, 1985; Yang, 1986; Beaufrère, 1989; Melville, 1989, 1990; Bruce, 1990; McHardy, 1991; Pacy, 1994; Quevedo, 1994; Gibson, 1996; Tessari, 1996b; Boirie, 1997b; El-Khoury, 1997; Bouteloupe-Delange, 1998; Fereday, 1998; Forslund, 1998; MacAllan, 1998; Arnal, 2000; Millward, 2000.

as the lower limit of an adequate protein intake (see Chapter 10) the increase is greater between low and adequate (A–C) than from adequate upwards (C–D). The results for breakdown are more variable; only in C is breakdown much reduced at high protein intakes. Oxidation, as one might expect, increases: with leucine as tracer a linear relationship has been shown by Young *et al.* (1985) and by others between oxidation and plasma leucine concentration, which in turn increases with protein intake except at very low levels. The post-absorptive data are also shown in Fig. 9.1: both synthesis and breakdown increase with intake. The balances, S−B, tend to become positive at intakes of about 0.6 g kg^{-1} day^{-1}.

It may seem paradoxical that at low protein intakes the effect of food is to reduce synthesis *below* the fasting level (Table 9.1B). The reason is that at low intakes food stimulates insulin production and the consequent reduction in breakdown means that not enough amino acids are available for synthesis. One of the most interesting aspects of this work was the way in which the response to feeding changed with intake, the response being the difference between fed and fasted rates. The variation of these responses with protein intake is shown in Fig. 9.2. There is a greater response of degradation than of synthesis or oxidation. These responses are reflected by differences in nitrogen balance by day and by night. Figure 9.3 shows that the amplitude of the diurnal oscillations of balance increases with increasing protein intake. In a third study by this group protein kinetics were determined at inter-

Table 9.2. Fed:fasted ratios × 100 in subjects adapted for 7–12 days to different levels of protein intake.

Intake g protein kg^{-1} d^{-1}			Synthesis	Breakdown	Oxidation
Yang	0	A	86	69	112
Motil	0.1	A	84	78	92
Yang	0.3	B	94	69	70
Pacy	0.36	B	96	83	118
Yang	0.6	C	120	75	89
Motil	0.6	C	115	77	100
Pacy	0.8	C	104	78	149
Yang	1.5	D	122	47	220
Motil	1.5	D	100	54	261
Pacy	1.6	D	117	64	174

Measurements by IV infusion of ^{13}C-leucine. Data of Motil *et al.*, 1981; Yang *et al.*, 1986; Pacy *et al.*, 1994.

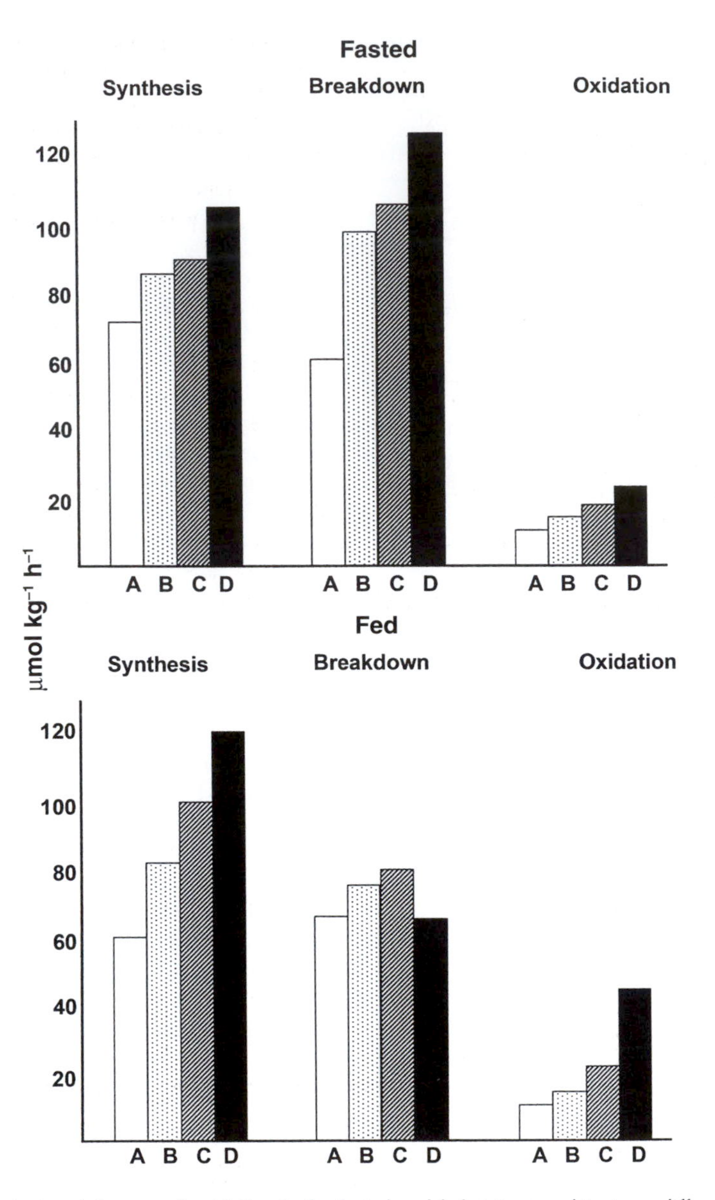

Fig. 9.1. Synthesis, breakdown and oxidation in the fasted and fed states in subjects on different levels of protein intake.
A, 0–0.1 g kg^{-1} day^{-1}; B, 0.3–0.36 g kg^{-1} day^{-1}; C, 0.6–0.8 g kg^{-1} day^{-1}; D, 1.5+ g kg^{-1} day^{-1}. Combined data of Motil *et al.* (1981); Yang *et al.* (1986); Pacy *et al.* (1994).

vals for 14 days after switching from a high to a moderate intake (1.9 to 0.8 g protein kg^{-1} day^{-1}) (Quevedo *et al.*, 1994). The most important change was again a reduction in breakdown with a small fall in oxidation.

A different approach was used by Young and his colleagues at MIT, because they were particularly interested in requirements for essential amino acids. They used test diets in which all the N was provided as crystalline amino acids, and protein kinetics were measured by intravenous infusion of labelled leucine. In their initial studies they looked at the effect of feeding varying levels of a single amino acid – leucine, lysine, threonine

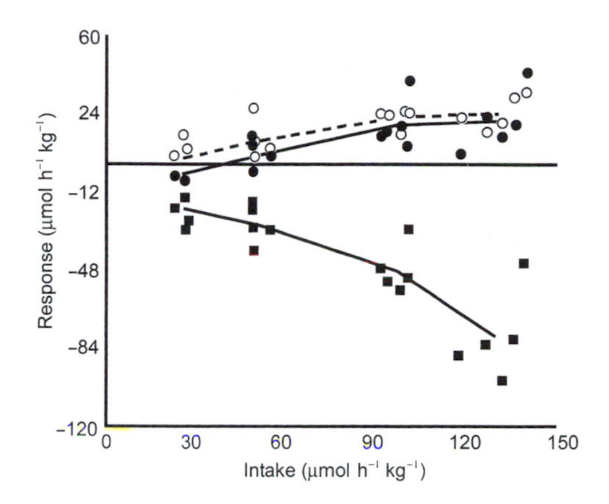

Fig. 9.2. Responses to different protein intakes after adaptation for 10 days. The response is the change between feeding and fasting.
●——●, synthesis (fed–fasted); ■——■, breakdown (fed–fasted); ○- – -○, oxidation (fed–fasted).
Reproduced from Pacy *et al.* (1994), by courtesy of *Clinical Science*.

and valine – in an otherwise complete diet after a 7-day period of adaptation (Meguid *et al.*, 1986a,b; Meredith *et al.*, 1986; Zhao *et al.*, 1986). The main finding of these experiments was that there was a sharp fall in flux when the intake of the test amino acid was clearly inadequate; however, they concluded that the amino acid flux did not provide a sensitive index of whether or not the intake met the requirement (Waterlow, 1996).

A remarkable study, which involved no less than 80 separate infusions, was undertaken by the MIT group (Marchini *et al.*, 1993). Nitrogen was provided as a mixture of amino acids equivalent to 1 g protein kg^{-1} day^{-1}, but with three different levels of the essential amino acids (EAA). The main results of this study are summarized in Table 9.3. There was little difference between the three groups in post-absorptive leucine concentration or kinetics, but in the fed state with increasing essential amino acid supply, there was a clear increase in plasma leucine concentration

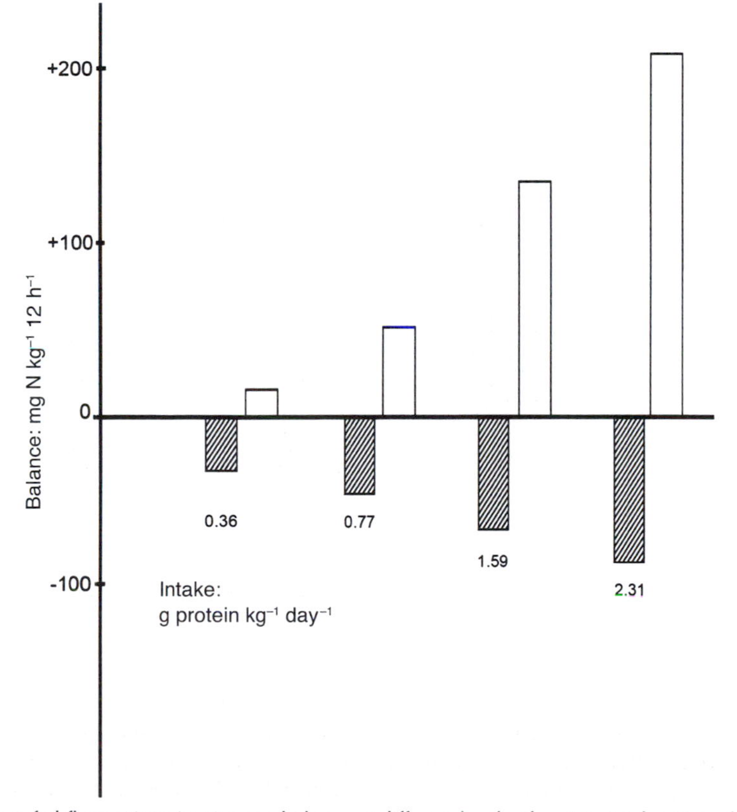

Fig. 9.3. Fasting–fed fluctuations in nitrogen balance at different levels of protein intake. Note the increasing amplitude of the day/night swings with increasing protein intake. Data of Price *et al.* (1994).
□, day; ▨, night.

and synthesis, with an accompanying fall in breakdown, resulting in a difference between the diets in net protein deposition (synthesis minus breakdown).

It may be questioned whether the effects of a diet low in protein are physiologically the same as the effects of deficiency of a single or of all essential amino acids. Absorption of crystalline amino acids will be quicker than when they are protein-bound (see below). A comparison between these different regimes is shown in Fig. 9.4. The variable shown is oxidation because it is the most sensitive kinetic parameter. There is no difference in the response of oxidation in the fed state to the supply of a protein or a limiting amino acid.

These studies of fasting and feeding have wider implications. The work of Millward's group (Pacy *et al.*, 1994; Price *et al.*, 1994; Gibson *et al.*, 1996) and of El-Khoury and colleagues (El-Khoury *et al.*, 1994a,b, 1997; Forslund *et al.*, 1998) has shown very clearly that the negative balance between protein synthesis and breakdown during fasting is cancelled out by a positive balance during feeding; this occurs at all levels of protein intake above maintenance, with bigger swings between negative and positive balance at higher protein intakes (Fig. 9.3). How this balance is achieved is the central problem of protein turnover; we return to it at the end of this book.

In summary, it seems reasonable to conclude that protein intake has only a modest influence on synthesis unless the intake is below the requirement level. Millward and his colleagues believe that the key factor determining the rate of protein deposition in response to feeding is the extent of inhibition of breakdown, as is clear from Table 9.1 and Fig. 9.2. They define the 'amino acid sensitivity of proteolysis' as the difference in protein breakdown on high and low protein diets divided by the difference in leucine intake. They showed that this quotient varied widely between individuals and was significantly related to the efficiency of protein utilization (Fereday *et al.*, 1998) (see Chapter 10).

9.1.2 Amino acid infusions

Infusions of amino acids provide some interesting contrasts with the effects of food (Table 9.4). One difference, whose importance is difficult to assess, is that amino acids are infused by vein, while food is given by mouth. Giordano *et al.* (1996) infused amino acids at 5 different levels. His results show progressive increases in synthesis and decreases in breakdown as the plasma amino acid concentrations rose. Giordano's first three levels may be regarded as not too far from physiological. Synthesis is increased more than by food, while the inhibition of breakdown is somewhat less. In the study by Louard *et al.* (1990) branched-chain amino acids alone were effective, producing even greater changes in muscle than in the whole body. The keto acids of the BCAAs had no effect (Giordano *et al.*, 2000). There is a clear relation in these studies between increases in synthesis and in plasma leucine concentration. The experiments of Goulet *et al.*

Table 9.3. Effect of feeding diets with different essential amino acid (EAA) contents on fed-state synthesis, breakdown and oxidation.

	Egg[a]	FAO[b]	MIT[b]
EAAs, per cent of total	53	10	21
Plasma [leu], mmol l^{-1}	174	86	110
[c]Synthesis, mmol kg^{-1} d^{-1}	84	68.5	75
[c]Breakdown, mmol kg^{-1} d^{-1}	56	67	62
[c]Oxidation, mmol kg^{-1} d^{-1}	31	12	18

All diets provided the equivalent of 1 g protein kg^{-1} d^{-1}.
[a]Eight separate infusions in each subject on the egg diet gave remarkably consistent results, and the averages only are given. The range for synthesis was 81–90, for breakdown 48–63.
[b]The results after 1 and 3 weeks on each diet were not consistently or significantly different, and have therefore been averaged.
[c]Fed-state results only.
Data of Marchini *et al.* (1993).

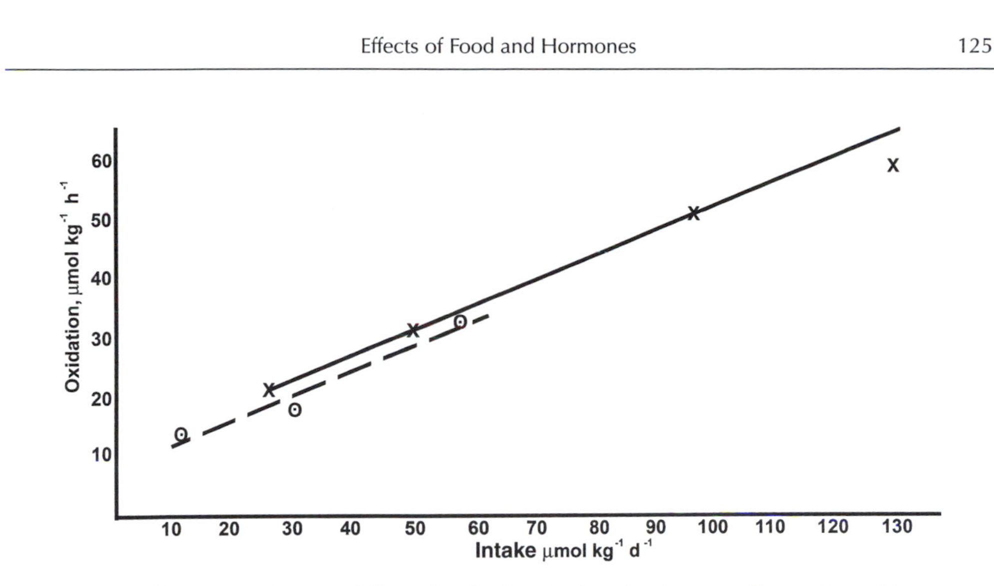

Fig. 9.4. Rate of leucine oxidation at different levels of intake from food or crystalline amino acids. ×, data of Pacy *et al.* (1994) – milk; ⊙, data of Motil *et al.* (1991) – amino acids. Reproduced by courtesy of *Proceedings of the National Academy of Sciences.*

(1993), in which adolescent boys on total parenteral nutrition (TPN) were given amino acids equivalent to 0.7, 1.5 and 2.5 g kg^{-1} day^{-1}, lead to a similar conclusion, although no baseline data were obtained. A study by Collin-Vidal *et al.* (1994) is particularly interesting. They infused intragastrically first a carbohydrate-lipid mixture, to which was then added, in randomized order, a solution of whey protein or of oligopeptides. Although the peptides produced a greater increase in synthesis than the intact protein, there was a smaller diminution in breakdown, so that the balance between synthesis and breakdown was less favourable.

In many of these studies with amino acids there was a large increase in plasma leucine concentration, much larger than with food. If the free leucine pool size is increased, the extra

Table 9.4. Response of whole-body protein turnover to intravenous amino-acid infusions.

	Ratio of infused:basal			
	Synthesis	Breakdown	Oxidation	Plasma leucine concentration
Gelfand *et al.*, 1988	1.55	0.37	3.9	2.65
Pacy *et al.*, 1988	1.40	0.85	2.0	1.87
Bennet, 1990	1.13	0.88	1.9	1.77
Tessari *et al.*, 1987	1.30	0.81	2.2	1.56
Tessari *et al.*, 1996a	1.32	–	2.15	1.79
Louard *et al.*, 1990[a]	1.73	0.88	5.4	4.0
Giordano *et al.*, 1996[b] I	1.04	0.93	1.2	1.37
II	1.07	0.87	1.3	1.58
III	1.17	0.76	2.1	2.24
IV	1.40	0.73	3.6	3.80
V	1.66	0.73	5.0	5.03
Goulet *et al.*, 1993[c] II/I	1.21	1.20	2.55	1.60
III/I	1.31	1.19	3.95	2.50

[a]Branched chain amino acids only.
[b]Amino acids infused at five different levels.
[c]No base-line data; amino acids infused in adolescent boys on total parenteral nutrition, I at 0.7, II at 1.5, III at 2.5 g kg^{-1} day^{-1}.

leucine entering the pool should be subtracted from the amount going to synthesis and added to the amount derived from breakdown. If no account is taken of this, synthesis is exaggerated and breakdown diminished. For example, in the study of Pacy *et al.* (1988) plasma leucine concentration increased by 118 μmol l^{-1}. If leucine is evenly distributed in body water, this would amount to 77 μmol kg^{-1}, or a rate of increase of 13 μmol kg^{-1} h^{-1} over the 6-h period of observation. If, as an approximate correction, this rate is subtracted from the rate of synthesis and added to that of breakdown, the relative rates compared with the control state become 122 instead of 140 for synthesis, and 100 compared with 85 for breakdown. If this correction were applied throughout, the general conclusion from Table 9.4 would be that amino acids by vein, although leading to large increases in plasma concentrations, produce a significant stimulation of synthesis and a much smaller inhibition of breakdown than food taken by mouth. A possible interpretation is that small amounts of food at frequent intervals – the usual protocol – produce only small rises in plasma amino acid concentrations, but a significant increase in insulin, whereas amino acids by vein produce larger increases in concentration but little or no increase in insulin. Insulin is, of course, a major player and we shall look at its role in more detail in the next section.

There are a few studies that enable us to compare the effects of food or of amino acids, sometimes with the addition of insulin, in forearm or leg with those in the whole body. The results in Table 9.5 suggest that synthesis in the limb responds more than in the whole body. Two studies, by Gelfand *et al.* (1988) and Cayol *et al.* (1997) on the effect of amino acids or food on splanchnic protein turnover give contradictory results. The wonderful experiments by Yu *et al.* (1990) on dogs showed that amino acids given parenterally produced a fivefold increase in splanchnic protein synthesis, compared with fasting, with a large fall in synthesis in the rest of the body. In these experiments the amount of protein given was four times that of a normal human intake.

A matter that has been much discussed is whether leucine has a specific effect on protein synthesis. In 1975 Buse and Reid found a stimulatory effect of leucine on protein synthesis by pieces of rat diaphragm incubated *in vitro* (Buse and Reid, 1975). Since then there have been many investigations of the problem, but it has not yet been completely resolved. Tischler *et al.* (1982), also working with isolated skeletal and cardiac muscle, confirmed that leucine stimulated synthesis and reduced breakdown and showed that the effect could not be attributed to increased charging of tRNA. Recently Kimball *et al.* (2002) have written of 'the unique specificity of leucine in stimulating protein synthesis in muscle'. This group has worked out the molecular mechanism in detail,

Table 9.5. Comparison of responses in whole body and limb to a meal, or to infusions of amino acids with or without insulin

	Synthesis		Breakdown	
	Whole body	Limb	Whole body	Limb
Tessari *et al.*, 1996b Meal	1.32	1.26	–	0.64
Gelfand *et al.*, 1988 Amino acids	1.55	1.82	0.37	0.58
Bennet *et al.*, 1990 Amino acids + insulin + glucose	1.11	1.13	0.84	0.70
Louard *et al.*, 1990 BCAs only	1.73	3.04	0.88	0.56
Zanetti *et al.*, 1999 Amino acids + insulin	1.25	1.99	–	1.00

Numbers are ratios of rates of synthesis and breakdown after the infusion or meal to rates before the infusion. See also Fig. 9.7.

showing that leucine injected in post-absorptive rats stimulated synthesis in muscle by enhancement of the initiation process and its signalling pathway, the effect being independent of insulin but facilitated by it (Anthony *et al.*, 2000a, 2002; see also Campbell *et al.*, 1999). In liver there was a different effect: no increase in total protein synthesis but specific stimulation of the translation of mRNAs for ribosomal proteins (Anthony *et al.*, 2001). Can these findings be reconciled with those of McNurlan *et al.* (1982) 20 years earlier, that a single dose of leucine to rats starved for 2 days had no effect on synthesis in liver, jejunal mucosa or skeletal muscle? The answer probably is that McNurlan's dose, 100 μmol leucine per 100 g rat, was much smaller than that of Anthony: ~1 mmol per 100 g.

There are few studies that describe the effect of leucine on whole body protein turnover in man. Again it is necessary to take account of the dose of leucine. In a human a normal protein intake of 1 g kg^{-1} day^{-1} would, if evenly spread over 12 h, provide leucine at a rate of ~1 μmol kg^{-1} min^{-1}. Schwenk and Haymond (1987) infused leucine at approximately this rate and found, compared with saline infusion, no effect on whole body synthesis and a 14% decrease in breakdown. Nair *et al.* (1992a,b) infused leucine at about twice the physiological rate and found no decrease in whole body breakdown but an improved balance in the limb. Frexes-Steed (1992), in dogs with a high rate of infusion (3 μmol kg^{-1} min^{-1}), found a 50% decrease in synthesis compared with saline and an even larger decrease in breakdown. Infusion of all three BCAs together in post-absorptive subjects led to a large increase in synthesis and a small fall in breakdown (Louard *et al.*, 1990). In the flooding dose method of measuring protein synthesis (see Chapter 14) a bolus dose of leucine of ~400 μmol kg^{-1} (Garlick *et al.*, 1989) stimulates muscle protein synthesis to about the same extent as a meal, but this is still a supra-physiological dose, and the same stimulation is found with other essential amino acids.

The evidence does not, in my view, support the conclusion that leucine, under physiological conditions, acts as a *regulator* of protein turnover in muscle, at least at the level of the whole body.

9.1.3 The effects of protein in a single meal

To give food by a naso-gastric infusion or in frequent small amounts for several hours in order to produce a steady state does not fit very well with what happens in real life. The alternative is a single meal, which requires non-steady state kinetics for analysis of the effects. A pioneer of this approach was Tessari, who gave a meal in which the nitrogen was provided by crystalline amino acids to normal subjects and to diabetics with and without insulin (Tessari *et al.*, 1988). The experimental design required two tracers, ^{2}H-leucine by IV infusions and ^{14}C-leucine added to the meal, which allowed 'endogenous' changes in synthesis and breakdown to be calculated separately from 'exogenous' changes reflecting the metabolism of the meal. Rates were calculated by Steele's formula (see Chapter 6, section 6.7) from the activities in plasma at intervals of 10–15 minutes over 8 h. With the meal there was a reduction of almost 30% in the endogenous breakdown. Rates of synthesis were not calculated in this paper. El-Khoury *et al.* (1997b) infused labelled leucine for 24 h, during which the subjects consumed three meals. A meal caused an increase in oxidation of leucine, but again there were no measurements of synthesis or breakdown.

The next advance was made by Beaufrère's group at the University of Clermont-Auvergne (Boirie *et al.*, 1996, 1997b). They prepared whey protein and casein labelled with ^{13}C from the milk of cows injected with ^{13}C-leucine. The whey protein was more rapidly digested and oxidized than the casein (Fig. 9.5A) and led to an increase in whole body synthesis that was maintained for 2 h (Fig. 9.5B), but with a higher rate of endogenous breakdown (Fig. 9.5C). The authors concluded that discrete meals may be more potent stimulators of whole body protein synthesis than constant feeding. They suggested that dietary leucine was mainly directed towards synthesis, whereas leucine derived from breakdown was mainly oxidized (Beaufrère *et al.* 2000).

Fouillet and her colleagues in Paris, like Beaufrère and Boirie, used protein sources (milk and soya) intrinsically labelled with ^{15}N, but added multiple sampling sites (Boirie *et al.*, 2000; see Chapter 2). They were able to estimate separately the uptake of label into visceral and peripheral proteins, and showed that the distribution was affected both by the energy supply and

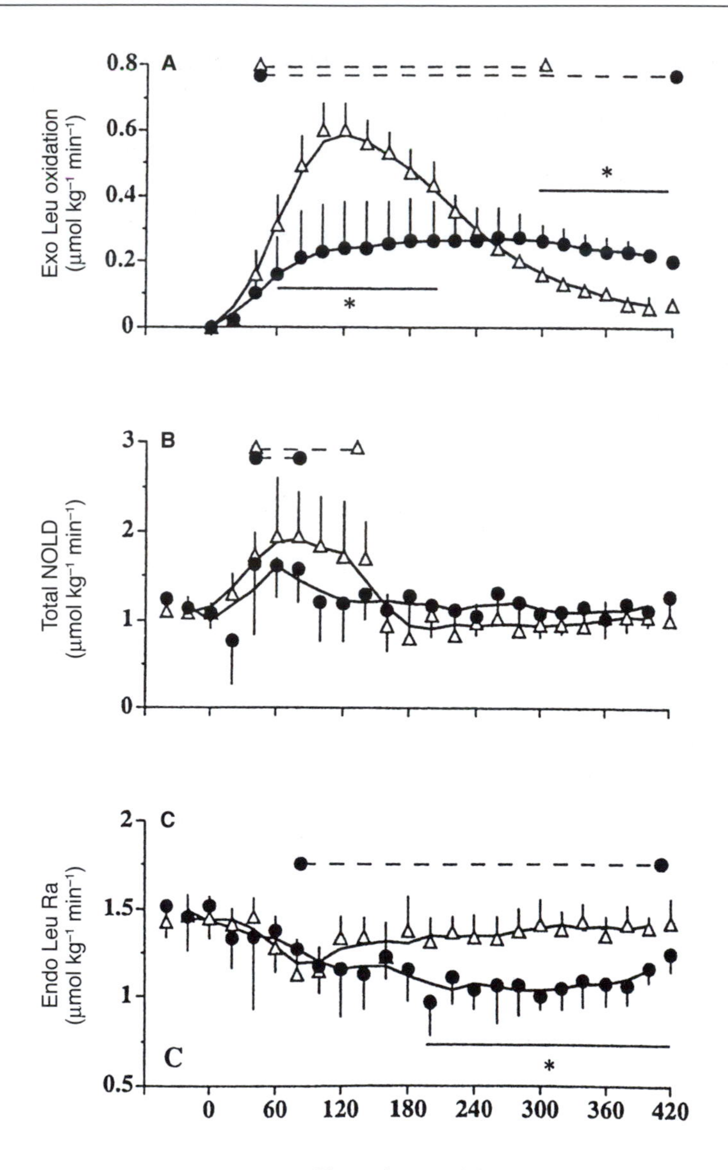

Fig. 9.5. Responses to a single meal containing whey protein or casein, labelled or unlabelled with (^{13}C)leucine. Reproduced with permission by the American Journal of Clinical Nutrition. © Am. J. Clin. Nutr. American Society for Nutrition.
A, oxidation (proteins labelled); B, whole body synthesis (proteins unlabelled); C, endogenous breakdown (proteins labelled). The dashed lines at the top of each panel indicate a significant difference from basal. ●, casein; △, whey protein. NOLD, non-oxidative leucine disposal.

by the nature of the protein (Fouillet *et al.*, 2001, 2002).

This would imply compartmentation between the precursor pools for synthesis and oxidation – a new approach to an old problem (see Chapter 5).

9.1.4 Energy intake

When the energy supply from carbohydrate and fat is inadequate to meet the needs, oxidation of amino acids increases and the body goes into negative nitrogen balance. In 1954 in a classical paper Calloway and Spector summarized the results of 460 studies on healthy young men in which nitrogen balance was related to protein and energy intake[1] (Calloway and Spector, 1954) (Fig. 9.6). The resting energy expenditure (RMR) of a healthy man is about 1750 kcal day^{-1}; it is evident that above this level the curve flattens off, so that the nitrogen sparing (reduction in negative N balance) is about 2 mg per additional kilocalorie. With intakes below the RMR the curve is much steeper, with a sparing of about 6 mg N per kilocalorie. With protein intakes below ~40 g d^{-1} N equilibrium cannot be achieved.

In starvation amino acids are oxidized as sources of energy. According to the classical equation, if the free amino acid pools are maintained in a steady state,

$$\text{protein synthesis} + \text{oxidation} = \text{protein breakdown} + \text{intake};$$

if there is a negative balance, so that oxidation is greater than intake, breakdown must be greater than synthesis. Unfortunately, the few measurements that we have found of protein turnover after short periods of fasting or low energy

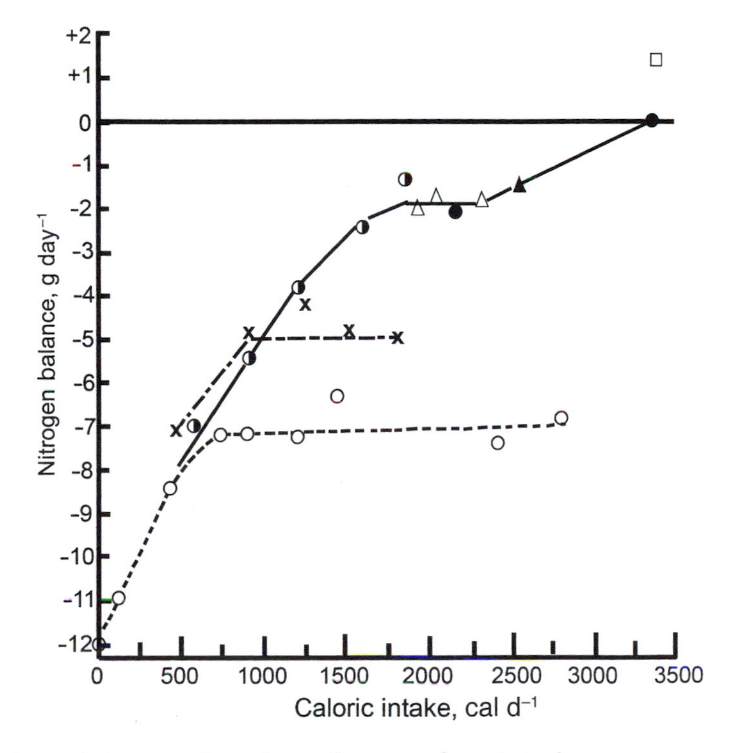

Fig. 9.6. Nitrogen balance at different levels of energy and protein intake.
Daily protein intake, g: o - - - o, 0; × - - -×, 15–31; ◑—◑ , 34–48; ●—●, 51–61; △—△, 65–74; ▲, 77; □, 96. Reproduced from Calloway and Spector (1954), by courtesy of *American Journal of Clinical Nutrition*.

[1] Since the papers cited used the kilocalorie rather than the kilojoule as the unit of energy, we shall follow their practice here.

Table 9.6. Effect of fasting or reduced energy intake on whole body protein synthesis and breakdown.

Author	Tracer	Energy intake	Duration	Ratio to basal	
		kcal d^{-1}	days	Synthesis	Breakdown
Healthy subjects					
Nair *et al.*, 1987	leucine	0	3	–	1.31
Tracey *et al.*, 1988	leucine	0	7	0.71	1.15
	[a]glycine	0	7	0.72	1.14
Obese subjects					
Winterer *et al.*, 1980	glycine	0	7	0.73	0.92
Lavivière *et al.*, 1990	leucine	0	11	0.80	0.84
Henson, 1983	glycine	0	7	–	0.75
Winterer *et al.*, 1980	glycine	~ 400	21	0.91	0.91
Bistrian *et al.*, 1981	glycine	440	21	0.70	0.72
Garlick *et al.*, 1983	glycine	500	21	1.10	1.10
Hoffer *et al.*, 1984	[b]leucine	344	21	0.85	–
	[c]leucine	344	21	0.82	–

[a] Same subjects.
[b] Protein only.
[c] Protein + CHO.

intakes (Table 9.6) give equivocal results. Synthesis nearly always fell after a week of fasting, but clear-cut increases in breakdown are only apparent in the studies of Nair *et al.* (1987) and of Tracey *et al.* (1988). These were the two studies made in normal subjects; all the others were in the obese, who were presumably maintaining their energy supply from their reserves of surplus fat. Three weeks on a low energy intake, about $^1/_4$ of the resting metabolic rate, produced with one exception a fall in synthesis, with inconsistent results on breakdown. A typical negative nitrogen balance after seven days of starvation would be about 5 g N day^{-1}, equivalent to a protein loss of some 30 g day^{-1}, which is only 10% of a protein turnover rate of ~300 g day^{-1} in an adult. It is not surprising that smaller changes in energy intake – a 20% reduction in children (Jackson *et al.*, 1983) or a 25% increase in adults (Motil *et al.*, 1981) – had no detectable effect on turnover rates. However, premature babies may be more sensitive. Duffy and Pencharz (1981) fed diets providing more or less equal amounts of nitrogen, but differing 25% in amount of energy. On the low energy diet (Vamin) both synthesis and breakdown, measured by the end-product method, were reduced by 25% – a difference that is suggestive but not significant.

No measurements of protein turnover are on record in long-term human starvation as described by Henry (1990), and it is a pity that the methods were not available when Keys made his epic study of semi-starvation during World War II (Keys *et al.*, 1950). In the rat, however, four days' starvation is equivalent to about 100 days in man, and there are good data for prolonged starvation in this small animal (see Chapter 15, section 15.2).

9.2 The Effects of Hormones on Protein Turnover in the Whole Body, Limb or Splanchnic Region

The effects of hormones on protein turnover are difficult to disentangle because they often overlap, inhibiting or supporting each other. They may also interact with food, particularly in the case of insulin, since feeding stimulates insulin secretion. Useful general reviews are those of Grizard *et al.* (1995) and Rooyackers and Nair (1997).

9.2.1 Insulin

Whole body

Much work has been done to determine the separate roles of insulin and food in regulating protein turnover. They both work in the same direction; as Munro pointed out many years ago, food depresses plasma amino acid concentrations, presumably as a result of the insulin production that is stimulated by feeding causing a decrease in protein degradation and an increase in synthesis. The outcome is a kind of self-regulating system. Table 9.7 summarizes the findings with labelled leucine or phenylalanine in 10 studies in which insulin was infused with or without amino acids. With insulin alone, plasma leucine concentrations fell and synthesis, breakdown and oxidation were all reduced. When different amounts of insulin were given the response appeared to be dose-dependent (Fukegawa *et al.*, 1985, 1988; Tessari *et al.*, 1986; Flakoll *et al.*, 1989). There was little further change when plasma insulin concentrations were increased to more than about 10 × basal, except in the study of Flakoll *et al.* (1989), in which hugely supra-physiological concentrations reduced breakdown to zero. In a study by Volpi *et al.* (1996) no exogenous insulin was given but a large rise in plasma insulin concentration was provoked by a meal of glucose + lipids; there was no significant change in either synthesis or breakdown.

Table 9.7 also shows the results when amino acids were infused as well as insulin. With the increase in plasma amino acid concentrations, of which leucine may be regarded as a marker, the suppression of synthesis is reversed. The effect is not very different from that of amino acids alone, shown in Table 9.4. The obvious conclusion is that the primary effect of insulin is to reduce protein degradation, and since all the studies in Table 9.7A were done in the post-absorptive state with no input from food, plasma amino acid concentrations fall, and this in turn results in a reduced rate of synthesis. Under natural conditions of fasting and feeding, these responses are nicely balanced, so that, as shown in Table 9.1, there is little variation in the rate of whole body synthesis.

Liu and Barrett (2002) summarize neatly the role of insulin: '... proteolysis is more sensitive than synthesis to small changes in plasma insulin, within the physiological range ... small increases in plasma insulin diminish whole body and skeletal muscle proteolysis, but higher doses are needed to stimulate synthesis.'

On the whole the results of insulin replacement in insulin-dependent (type I) diabetic patients agree with the findings in normal controls (see reviews by Charlton and Nair, 1998, and Tessari *et al.*, 1999). In the untreated diabetic plasma amino acid concentrations are high and rates of protein breakdown and oxidation are increased compared with controls; most studies have found that in the post-absorptive state synthesis was increased in diabetics. Treatment with insulin reversed these changes; both Pacy *et al.* (1991a,b) and Nair *et al.* (1995) found that not only was breakdown reduced by insulin together with oxidation and transamination of leucine to KIC, but synthesis also decreased. This rather surprising finding seems to be contrary to the supposed anabolic effect of insulin on protein metabolism; however, the negative balance characteristic of the PA state was somewhat reduced. The fall in synthesis probably results from the decreased essential amino acid concentration produced by insulin. When a meal was given as well as insulin the rate of protein breakdown was

Table 9.7. Effects on whole body turnover of infusions of insulin and amino acids in normal subjects, as ratio of infused: basal.

	Plasma [leucine]	Synthesis	Breakdown	Oxidation	
A. Insulin alone[a]	0.61	0.75	0.72	0.64	n = 10
B. Insulin + amino acids[b]	1.42	1.15	0.88	2.59	n = 6
C. Amino acids alone[c]	1.89	1.15	0.87	2.32	n = 10

n = number of studies.
[a]Studies in the post-absorptive state. Plasma insulin concentrations in the range 7–15 × basal.
[b]Excludes study of Zanetti *et al.* (1999) in which BCAs only were infused; and study IV of Castellino *et al.* (1987), in which insulin secretion was suppressed with somatostatin.
[c]From Table 9.3, excluding study by Louard *et al.* (1990), in which BCAs only were infused, and studies IV and V of Giordano *et al.* (1996) in which amino acids were infused at supraphysiological levels.
References: Fukegawa *et al.*, 1985; Tessari *et al.*, 1986, 1987, 1991; Castellino *et al.*, 1987, 1990; Flakoll *et al.*, 1989; Bennet *et al.*, 1990; Heslin *et al.*, 1992; Zanetti *et al.*, 1999.

above control levels for the first 2 h and then fell to normal (Tessari *et al.*, 1988). This is a much more physiological situation than a purely post-absorptive study.

Regional turnover

A method, applicable only in the limb, of eliminating the effect of insulin in reducing plasma amino acid concentrations, is to produce a local effect by infusing insulin into the artery supplying the limb. This elegant technique was introduced by Pozefsky *et al.* (1969) and has been particularly exploited by Barrett and his group (Gelfand and Barrett, 1987; Biolo *et al.*, 1995; Fryburg *et al.*, 1995; Hillier *et al.*, 1998). There is general agreement that in normal people insulin alone produces little change in synthesis (uptake) in the limb; only with enormous doses of insulin is synthesis significantly increased (Hillier *et al.*, 1998). On the other hand breakdown (release) is decreased, so that the negative post-absorptive balance is converted into a positive balance. Denne *et al.* (1995) concluded that the depression of breakdown in the limb (40%) could account for all the depression in the whole body. As in the whole body, when amino acids are given as well as insulin, synthesis increases and breakdown decreases even more than with insulin alone (Nygren and Nair, 2003).

A number of studies have also been made of the effect of insulin in the arm or leg of diabetic patients. Again it seems that the main effect is a decrease in degradation, which is very high in the diabetic state and is reduced by insulin (Pacy *et al.*, 1989; Nair *et al.*, 1995). There is much less information about insulin's effect on the splanchnic tissues. In the normal post-absorptive state the splanchnic bed deviates only slightly from zero balance (Tessari *et al.*, 1996b; Meek *et al.*, 1998; Nygren and Nair, 2003). Insulin alone had little effect, but addition of a high dose of amino acids produced a large increase in synthesis with a smaller reduction in breakdown in both splanchnic region and the limb (Nygren and Nair, 2003) (Fig. 9.7). Columns C in that figure probably best

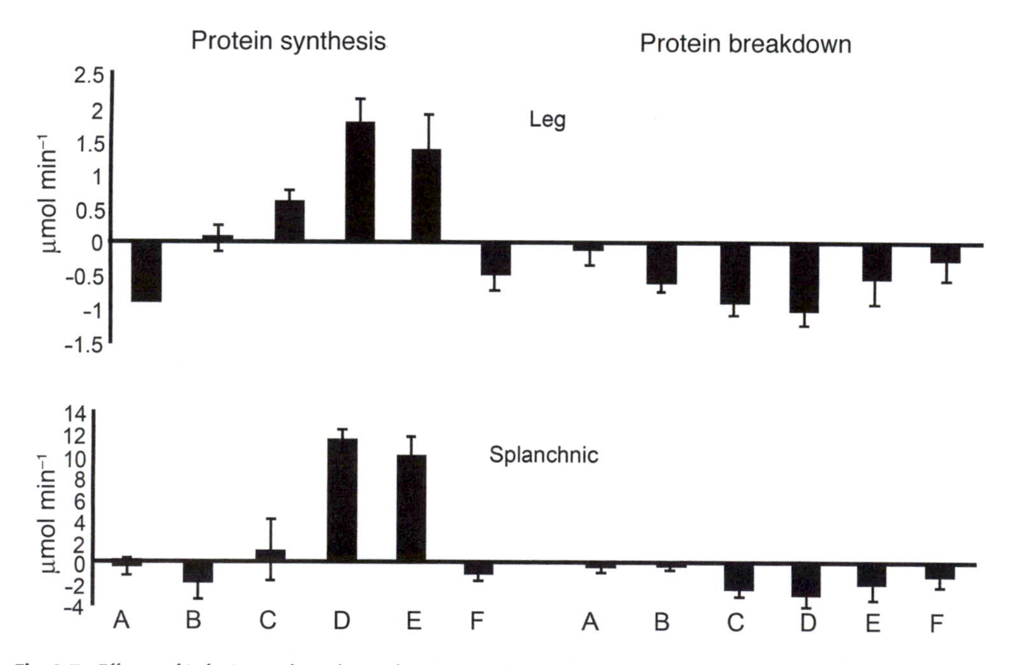

Fig. 9.7. Effects of infusions of insulin and amino acids on synthesis and breakdown in the leg and the splanchnic region.
Infusions: A, normal saline; B, insulin, 0.5 mU kg^{-1} min^{-1}; C, insulin + amino acids to maintain basal levels; D, insulin + amino acids to 2.5–3 × basal level; E, basal insulin + amino acids; F, basal insulin + normal saline. Columns represent changes from baseline. Reproduced from Nygren and Nair (2003), by courtesy of *Diabetes*.

represent the physiological response to a meal. Evidently the increase in synthesis in the limb is too small to be reflected by an increase in whole body synthesis.

It is remarkable that although in these *in vivo* studies in humans insulin had so little effect on synthesis in the limb, compared with the large effect of amino acids, yet so much attention has been given, as mentioned above, to the molecular biology of insulin action on muscle protein synthesis in the rat.

In conclusion, it has long been accepted that insulin reduces protein breakdown. It stimulates synthesis but can only do so if amino acids are available: since a lower rate of degradation reduces amino acid concentrations, the effect of insulin in any given situation must be nicely balanced. Moreover, Campbell *et al.* (1999), working with cell cultures, have shown that the stimulation by insulin of overall protein synthesis requires glucose as well as amino acids. It is strange that there is little information about the mechanism of the insulin effect on breakdown, compared with that on synthesis.

9.2.2 Insulin-like growth factor I (IGF-I)

For the human physiologist a particularly interesting property of IGF-1 is that it is very sensitive to nutritional state and to nutritional supply. Its plasma concentration is greatly reduced in children with protein-energy malnutrition and in subjects with secondary malnutrition and it is a more sensitive marker of nutritional state than the classical nutrition-related proteins albumin and transferrin (Chapter 15). In experimental fasting in adult humans the concentration of IGF-1 in plasma fell progressively over 10 days to 25% of the basal level (Fig. 9.8); restoration to normal required both protein and energy. These nutritional aspects are well reviewed by Thissen *et al.* (1994).

In the whole body in man IGF-1 increases the rate of synthesis with little or no change in breakdown (Russell-Jones *et al.*, 1994; Fryburg *et al.*, 1995; Mauras *et al.*, 1995). The effects of infusing IGF-1 in the limb are similar, although there seems to be a larger effect on protein degradation than in the whole body. Several authors have commented that infusions of IGF-1 cause an increase in blood flow through the limb; this could perhaps be the effector for the changes in

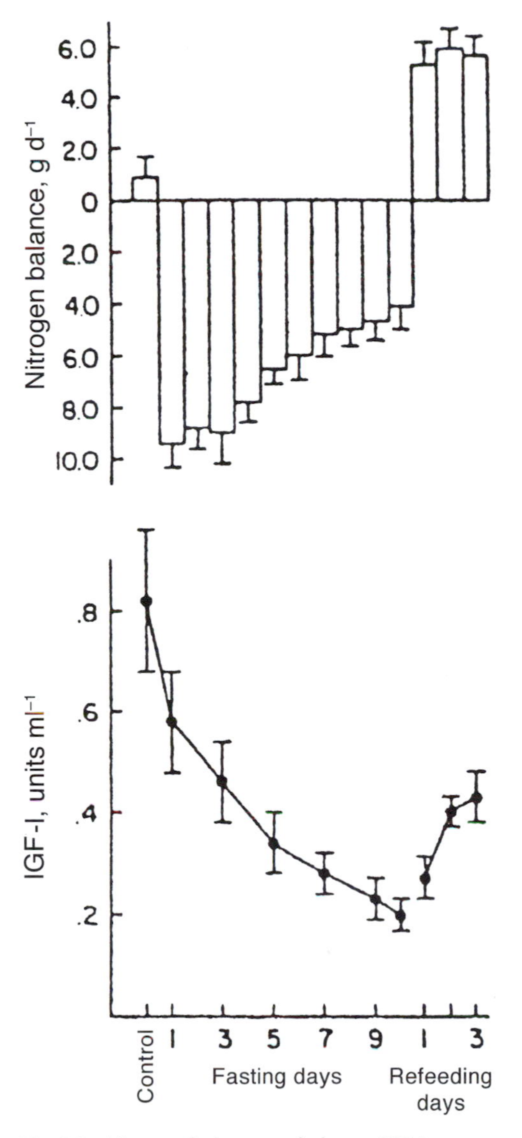

Fig. 9.8. Nitrogen balance and plasma IGF-I concentrations in subjects fasted for 9 days and then re-fed. Reproduced from Thissen *et al.* (1994), by courtesy of *Endocrine Reviews.*

protein turnover. Fryburg (1994) infused IGF-1 into one forearm, using the un-infused forearm as a control. In the un-infused arm there were considerable rises in serum IGF-1 concentration, with no effect on amino acid kinetics. He suggested that the infusion caused a high concentration locally of free IGF-1, whereas by the time it had passed through the general circulation most of it had become attached to the IGF-1 binding proteins, and so was inactivated. However, according to Rooyackers and Nair (1997), the binding can either potentiate or inhibit IGF-1 functions.

It is clear that although both IGF-1 and insulin are anabolic, their effects are different, insulin reducing protein degradation, IGF-1 promoting synthesis.

However, like insulin, IGF-1 can counter the catabolic effects of glucocorticoids (Mauras and Beaufrère, 1995; Oehri *et al.*, 1996). The latter authors claimed that IGF-1 and GH had clearly distinct effects on protein metabolism as well as on that of carbohydrate and fat.

9.2.3 Thyroid hormones

The effect of thyroid hormones on whole body protein turnover in man has not been very much studied: Garret *et al.* (1985), using [^{15}N]-glycine with ammonia as end-product, measured protein turnover in women at a series of points in the menstrual cycle. They showed a highly significant correlation between turnover rate and the ratio of T3/rT3 in plasma, with no consistent relation to the period of the cycle. We measured protein synthesis by the end-product method (Chapter 7) in refractory obese patients on a low energy diet before and after treatment with T_3 for 2 weeks. There was a barely significant increase ($P < 0.05 > 0.025$) in synthesis calculated from the end-product average, with an increase in the flux ratio Q_A:Q_U from 1.3 to 1.7, signifying an increased flux through the peripheral tissues. A detailed study of patients with thyrotoxicosis (T+) and hypothyroidism (T−) together with controls (Morrison *et al.*, 1988) showed, surprisingly, that in thyrotoxicosis synthesis and breakdown in the whole body were suppressed, synthesis rather more than breakdown. The changes were quite similar in the hypothyroid patients. On the other hand, in the leg there was a difference between the two states: in thyrotoxicosis the net effluxes of BCAs and aromatic amino acids were greatly increased, whereas in hypothyroidism they were at the normal level. All these changes were reversed by treatment. The authors interpreted these findings as showing that thyrotoxic myopathy is due primarily to depression of muscle protein synthesis.

Quite different results were obtained by Tauveron *et al.* (1995), who administered thyroid hormone for 6 weeks to normal subjects, studying them before and after treatment. The effect of the hormone was to increase both synthesis and breakdown equally, so that the negative balance characteristic of the post-absorptive state persisted unchanged. Clearly this is a subject on which further work is indicated.

9.2.4 Growth hormone

Growth hormone (GH) has long been used to promote muscle growth in farm animals (see Chapter 15, section 15.3). It is also well established that treatment with GH of hypopituitary children and adults improves nitrogen balance, and promotes an increase in muscle mass (Salomon *et al.*, 1989; Dempster *et al.*, 1990). In whole body measurements on such patients GH was found to increase protein synthesis and decrease amino acid oxidation (Beshya *et al.*, 1993; Russell-Jones *et al.*, 1993; Nørrelund *et al.*, 2001). However, the results of short-term administration of GH to normal subjects give a rather confused picture. On the one hand the findings of Horber and Haymond (1990) are quite impressive. GH for 7 days produced a 20% increase in synthesis with a substantial decrease in oxidation in both fed and fasted states. The falls in oxidation and urea excretion, promoting a positive nitrogen balance, have been generally confirmed (Russell-Jones *et al.*, 1993; Copeland and Nair, 1994; Zachwieja *et al.*, 1994; Welle *et al.*, 1996; Nørrelund *et al.*, 2003). However, Fryburg *et al.* (1991, 1992, 1993) and Welle *et al.* (1996) were unable to establish an increase in whole body synthesis. This may well be because the periods over which the hormone was administered were short compared with those used in the clinical cases.

Studies on the arm or leg have given a more consistent picture. In those of Fryburg there was an increase in synthesis with no change in breakdown, particularly when GH was infused locally into the artery. These findings were confirmed by Biolo *et al.* (2000) in trauma patients, who also showed that with GH there were increases in the inward transport of leucine and phenylalanine. A difficulty is that GH increases the plasma concentrations of insulin and IGF-1. When these rises were prevented with infusions of somatostatin and glucagon to maintain the same levels as in controls there was no stimulation of muscle protein synthesis in the leg (Copeland and Nair, 1994). Biolo *et al.* (2000) observed that GH reduced the *de novo* rate of glutamine synthesis and of net glutamine release from muscle, which might be an unfavourable effect if GH is given to patients with trauma to counteract the catabolic state (see Chapter 13, section 13.4).

9.2.5 Glucocorticoids

Glucocorticoids administered to normal human subjects in the post-absorptive state cause an increase of about 25% in whole body protein breakdown (Simmons *et al.*, 1984; Darmaun *et al.*, 1988; Beaufrère *et al.*, 1989; Brillon *et al.*, 1995), but even with pharmacological doses the increase was no greater (Beaufrère *et al.*, 1989). In Beaufrère's study a meal was given, which reduced breakdown, but it was still 30% greater with the glucocorticoid than in controls. In all these studies the glucocorticoid produced a negative nitrogen balance, breakdown being increased while synthesis was unchanged. The positive balance associated with a meal was reduced almost to zero by the hormone (Beaufrère *et al.*, 1989). The increase in breakdown produced by glucocorticoids was counteracted by growth hormone (Oehri *et al.*, 1996); likewise, the increase in synthesis produced by growth hormone was counteracted by a glucocorticoid, prednisone (Horber and Haymond, 1990). An interesting finding was that glucocorticoids increased the fluxes of glutamine and alanine through increased *de novo* synthesis (Darmaun *et al.*, 1988; Brillon *et al.*, 1995), again in opposition to the action of growth hormone: we do not know whether these extra outputs, presumably derived from muscle, are taken up in part by the visceral tissues, but presumably they contribute also to the increased excretion of nitrogen.

Interpretation of these findings is complicated by the fact that glucocorticoids stimulate insulin secretion. Brillon *et al.* (1995) clamped plasma insulin at the basal level halfway through an infusion of hydrocortisone and found that protein breakdown, as measured by the appearance rate of leucine, was greater than when insulin levels had been allowed to rise. In a forearm study Louard *et al.* (1994) showed that dexamethasone partially reversed the inhibition of proteolysis by insulin, and concluded that glucocorticoids made muscle resistant to insulin, in agreement with Brillon's findings.

Presumably in normal life there is a balance between the anabolic and catabolic hormones, the former prevailing during feeding, the latter in the post-absorptive state, amino acid supply being crucial to maintaining this balance.

9.3 References

Anthony, J.C., Gautsch, A., Kimball, S.R., Vary, T.C. and Jefferson, L.S. (2000a) Orally administered leucine stimulates protein synthesis in skeletal muscle of postabsorptive rats in association with increased IF4F formation. *Journal of Nutrition* 130, 139–145.

Anthony, J.C., Yoshizawa, F., Gautsch, A., Vary, T.C., Jefferson, L.S. and Kimball, S.R. (2000b) Leucine stimulates translation initiation in skeletal muscle of post-absorptive rats via a rapamycin-sensitive pathway. *Journal of Nutrition* 130, 2413–2419.

Anthony, T.G., Anthony, J.C., Yoshizawa, F., Kimball, S.R. and Jefferson, L.S. (2001) Oral administration of leucine stimulates ribosomal mRNA translation but not global rates of protein synthesis in the liver of rats. *Journal of Nutrition* 131, 1171–1176.

Anthony, J.C., Lang, C.H., Crozier, S.J., Anthony, T.G., MacLean, D.A., Kimball, S.R. and Jefferson, L.S. (2002) Contribution of insulin to the translational control of protein synthesis in skeletal muscle by leucine. *American Journal of Physiology* 282, E1092–1101.

Arnal, M.I., Mosoni, L., Boirie, Y., Gachon, P., Genest, M., Bayle, G., Grizard, J., Arnal, M., Antoine, M.J., Beaufrère, B. and Mirand, P.P. (2000) Protein turnover modifications induced by the protein feeding pattern still persists after the end of the diets. *American Journal of Physiology* 278, E902–909.

Beaufrère, B., Horber, F., Schweik, W.F., Marsh, H.M., Matthews, D., Gerich, J.E. and Haymond, M.W. (1989) Glucocorticosteroids increase leucine oxidation and impair leucine balance in humans. *American Journal of Physiology* 257, E712–721.

Beaufrère, B., Dangin, M. and Boirie, Y. (2000) The 'fast' and 'slow' protein concept. In: Fürst, P. and Young, V. (eds) *Proteins, Peptides and Amino Acids in Enteral Nutrition. Nestlé Nutrition Workshop Series Clinical and Performance Programme*, Vol. 3. Nestlé, Vevey and Karger, Basel, Switzerland, pp. 121–133.

Bennet, W.M., Connacher, A.A., Scrimgeour, C.M., Jung, R.T. and Rennie, M.J. (1990) Euglycemic hyperinsulinemia augments amino acid uptake by human leg tissues during hyperaminoacidemia. *American Journal of Physiology* 259, E185–194.

Beshyah, M.A., Sharp, P.S., Gelding, S.V., Halliday, P. and Johnston, D.G. (1993) Whole body leucine turnover in adults on conventional treatment for hypopituitarism. *Acta Endocrinologica* 129, 158–164.

Bistrian, B.R., Sherman, M. and Young, V.R. (1981) The mechanisms of nitrogen sparing in fasting supplemented by protein and carbohydrate. *Journal of Clinical Endocrinology and Metabolism* 53, 874–878.

Biolo, G., Inchiostro, S., Tiengo, A. and Tessari, P. (1995) Regulation of postprandial whole-body proteolysis in insulin-deprived 1DDM. *Diabetes* 44, 203–209.

Biolo, G., Iscra, F., Bosutti, A., Toigo, G., Ciocahi, B., Geatti, O., Gullo, A. and Guanieri, G. (2000) Growth hormone decreases glutamine production and stimulates protein synthesis in hypercatabolic patients. *American Journal of Physiology* 279, E323–332.

Boirie, Y., Gachon, P., Corny, S., Fauquant, J., Maubois, J.-L. and Beaufrère, B. (1996) Acute postprandial changes in leucine metabolism as assessed with an intrinsically labeled milk protein. *American Journal of Physiology* 271, E1083–1091.

Boirie, Y., Gachon, P. and Beaufrère, B. (1997a) Splanchnic and whole body leucine kinetics in young and elderly men. *American Journal of Clinical Nutrition* 65, 489–495.

Boirie, Y., Dangin, M., Gachon, P., Vasson, M.P., Manbois, J.L. and Beaufrère, B. (1997b) Slow and fast dietary proteins differently modulate post-prandial protein accretion. *Proceedings of the National Academy of Sciences, USA* 94, 14930–14935.

Bouteloupe-Demange, C., Boirie, Y., Déchelotte, P., Gachou, P. and Beaufrère, B. (1998) Gut mucosal protein synthesis in fed and fasted humans. *American Journal of Physiology* 274, E541–546.

Brillon, D.J., Zheng, B., Campbell, A.G. and Matthews, D.M. (1995) Effect of cortisol on energy expenditure and amino acid metabolism in humans. *American Journal of Physiology* 268, E501–503.

Bruce, A.C., McNurlan, M.A., McHardy, K.C., Brown, J., Buchanan, K.D., Calder, A.G., Milne, E., McGaw, B.A., Garlick, P.J. and James, W.P.T. (1990) Nutrient oxidation patterns and protein metabolism in lean and obese subjects. *International Journal of Obesity* 14, 631–646.

Buse, M.G. and Reid, S.S. (1975) Leucine: a possible regulator of protein turnover in muscle. *Journal of Clinical Investigation* 56, 1250–1261

Calloway, D.H. and Spector, H. (1954) Nitrogen balance as related to calorie and protein intake in active young men. *American Journal of Clinical Nutrition* 2, 405–412.

Campbell, L.E., Wang, X. and Proud, C.G. (1999) Nutrients differentially regulate multiple translation factors and their control by insulin. *Biochemical Journal* 344, 433–441.

Castellino, P., Luzi, L., Simonson, D.C., Haymond, M. and DeFronzo, R.A. (1987) Effect of insulin and plasma amino acid concentrations on leucine metabolism in man. Role of substrate availability on estimates of whole body synthesis. *Journal of Clinical Investigation* 80, 1784–1793.

Castellino, P., Luzi, L., Del Prato, S. and DeFronzo, R.A. (1990) Dissociation of the effects of epinephrine and insulin on glucose and protein metabolism. *American Journal of Physiology* 258, E117–125.

Cayol, M., Boirie, Y., Rambourdin, F., Pougneaud, J., Gachon, P., Beaufrère, B. and Obled, C. (1997) Influence of protein intake on whole body and splanchnic leucine kinetics in humans. *American Journal of Physiology* 272, E584–591.

Clugston, G.A. and Garlick, P.J. (1982) The response of protein and energy metabolism to food intake in lean and obese men. *Human Nutrition: Clinical Nutrition* 36C, 57–70.

Charlton, M. and Nair, K.S. (1998) Protein metabolism in insulin-dependent diabetes mellitus. *Journal of Nutrition* 128, 3235–3275.

Collin-Vidal, C., Cayol, M., Obled, C., Ziegler, F., Bommelaer, G. and Beaufrère, B. (1994) Leucine kinetics are different during feeding with whole protein or oligopeptides. *American Journal of Physiology* 267, E907–914.

Copeland, K.C. and Nair, K.S. (1994) Acute growth hormone effects on amino acid and lipid metabolism. *Journal of Clinical Endocrinology and Metabolism* 78, 1040–1047.

Dangin, M., Boirie, Y., Guillet, C. and Beaufrère, B. (2002) Influence of the protein digestion rate on protein turnover in young and elderly subjects. *Journal of Nutrition* 132, 3228S–3233S.

Darmaun, D., Matthews, D.E. and Bier, D.M. (1988) Physiological hypercortisolemia increases proteolysis, glutamine and alanine production. *American Journal of Physiology* 255, E366–373.

Dempster, D.P., Bier, D.M., Tollefsen, S.E., Rotwein, P.S., Daughaday, W.H., Jensen, M.C., Galgani, J.P., Heath-Monig, E. and Trivedi, B. (1990) Whole body nitrogen kinetics and their relationship to growth in short children treated with recombinant human growth hormone. *Pediatric Research* 28, 394–400.

Denne, S.C., Brechtel, G., Johnson, A., Liechty, E.A. and Baron, A.D. (1995) Skeletal muscle proteolysis is reduced in noninsulin-dependent diabetes mellitus and is unaltered by euglycemic hyperinsulinemia or intensive insulin therapy. *Journal of Clinical Endocrinology and Metabolism* 80, 2371–2377.

Duffy, B. and Pencharz, P. (1981) The effect of feeding route (IV or oral) on the protein metabolism of the neonate. *American Journal of Clinical Nutrition* 43, 108–111 .

El-Khoury, A.E., Fukagawa, N.K., Sánchez, M., Tsay,

R.H., Gleason, R.E., Chapman, T.E. and Young, V.R. (1994a) Validation of the tracer-balance concept with reference to leucine: 24-h intravenous tracer studies with L-[1-^{13}C] leucine and [^{15}N-^{15}N] urea. *American Journal of Clinical Nutrition* 59, 1000–1011.

El-Khoury, A.E., Sánchez, M., Fukagawa, N.K., Gleason, R.E. and Young, V.R. (1994b) Similar 24-h pattern and rate of carbon dioxide production, by indirect calorimetry vs. stable isotope dilution, in healthy adults under standardized metabolic conditions. *Journal of Nutrition* 124, 1615–1627.

El-Khoury, A.E., Sánchez, M., Fukagawa, N.K., Gleason, R.E., Tsay, R.H. and Young, V.R. (1995) The 24-h kinetics of leucine oxidation in healthy adults receiving a generous leucine intake via three discrete meals. *American Journal of Clinical Nutrition* 62, 579–590.

El-Khoury, A.E., Forslund, A., Olsson, R., Branth, S., Sjödin, A., Andersson, A., Atkinson, A., Selvaraj, A., Hambraeus, L. and Young, V.R. (1997) Moderate exercise at energy balance does not affect the 24-h leucine oxidation or nitrogen retention in healthy men. *American Journal of Physiology* 273, E394–407.

Fereday, A., Gibson, N.R., Cox, M., Pacy, P.J. and Millward, D.J. (1998) Variation in the apparent sensitivity of the insulin-mediated inhibition of proteolysis to amino acid supply determines the efficiency of protein utilization. *Clinical Science* 95, 725–733.

Flakoll, P.G., Kulaylat, M., Frexes-Steed, M., Hourani, H., Brown, L.L., Hill, J.D. and Abumrad, N.N. (1989) Amino acids augment insulin's suppression of whole body proteolysis. *American Journal of Physiology* 257, E839–847.

Forslund, A.H., Hambraeus, L., Olsson, R.M., El-Khoury, A.E., Yu, Y.-M., and Young, V.R. (1998) The 24 h whole body leucine and urea kinetics at normal and high protein intakes with exercise in healthy adults. *American Journal of Physiology* 275, E310–320.

Fouillet, H., Gaudichon, C., Mariotti, F., Mahé, S., Lescoat, P., Huneau, J.F. and Tomé, D. (2000) Compartmental modeling of postprandial dietary nitrogen distribution in humans. *American Journal of Physiology* 279, E161–175.

Fouillet, H., Gaudichon, C., Mariotti, F., Bos, C., Huneau, J.F. and Tomé, D. (2001) Energy nutrients modulate the splanchnic sequestration of dietary nitrogen in humans: a compartmental analysis. *American Journal of Physiology* 281, E248–260.

Fouillet, H., Mariotti, F., Gaudichon, C., Bos, C. and Tomé, D. (2002) Peripheral and splanchnic metabolism of dietary nitrogen are differently affected by the protein source in humans, as assessed by compartmental modelling. *Journal of Nutrition* 132, 125–133.

Frexes-Steed, M., Lacy, D.B., Collins, J. and Abumrad, N.N. (1992) Role of leucine and other amino acids in regulating protein metabolism *in vivo*. *American Journal of Physiology* 262, E925–935.

Fryburg, D.A., Gelfand, R.A. and Barrett, E.J. (1991) Growth hormone acutely stimulates forearm muscle protein synthesis in normal humans. *American Journal of Physiology* 260, E499–504.

Fryburg, D.A., Louard, R.J., Gerow, K.E., Gelfand, R.A. and Barrett, E.J. (1992) Growth hormone stimulates skeletal muscle protein synthesis and antagonizes insulin's anti-proteolytic action in humans. *Diabetes* 41, 424–429.

Fryburg, D.A. and Barrett, E.J. (1993) Growth hormone acutely stimulates skeletal muscle but not whole body protein synthesis in humans. *Metabolism* 42, 1223–1227.

Fryburg, D.A. (1994) Insulin-like growth factor I exerts growth hormone – and insulin-like actions on human muscle protein metabolism. *American Journal of Physiology* 267, E331–336.

Fryburg, D.A., Jahn, L.A., Hill, S.A., Oliveras, D.M. and Barrett, E.J. (1995) Insulin and insulin-like growth factor-1 enhance human skeletal muscle protein anabolism during hyperaminoacidemia by different mechanisms. *Journal of Clinical Investigation* 96, 1722–1729.

Fukagawa, N.K., Minaker, K.L., Rowe, J.W., Goodman, H.M., Matthews, D.E., Bier, D.M. and Young, V.R. (1985) Insulin-mediated reduction of whole body protein breakdown: dose-response effects on leucine metabolism in postabsorptive men. *Journal of Clinical Investigation* 76, 2306–2311.

Fukagawa, N.K., Minaker, K.L., Rowe, J.W., Matthews, D.E., Bier, D.M. and Young, V.R. (1988) Glucose and amino acid metabolism in aging man: differential effects of insulin. *Metabolism* 37, 371–377.

Garlick, P.J., Clugston, G.A., Swick, R.W. and Waterlow, J.C. (1980) Diurnal pattern of protein and energy metabolism in man. *American Journal of Clinical Nutrition* 33, 1983–1986.

Garlick, P.J., Wernerman, J., McNurlan, M.A., Essen, P., Lobley, G.E., Milne, E., Calder, G.A. and Vinnars, E. (1989) Measurement of the rate of protein synthesis in muscle of post-absorptive young men by injection of a 'flooding dose' of [1-^{13}C] leucine. *Clinical Science* 77, 329–336.

Garlick, P.J., McNurland, M.A. and Balmer, P.E. (1991) Influence of dietary protein intake on whole-body protein turnover in humans. *Diabetes Care* 14, 1189–1198.

Garret, D.R., Welsch, C., Arnaud, M.J. and Tourniaire, J. (1985) Relationship of the menstrual cycle and thyroid hormones to whole-body protein turnover in women. *Human Nutrition: Clinical Nutrition* 39C, 29–37.

Gelfand, R.A. and Barrett, E.J. (1987) Effect of physiologic hyperinsulinemia on skeletal muscle protein synthesis and breakdown in man. *Journal of Clinical Investigation* 80, 1–6.

Gelfand, R.A., Glickman, M.G., Castellino, P., Louard, R.J. and DeFronzo, R.A. (1988) Measurement of L-[1-^{14}C] leucine kinetics in splanchnic and leg tissues in humans. *Diabetes* 37, 1365–1372.

Gibson, N.R., Fereday, A., Cox, M., Halliday, D., Pacy, P.J. and Millward, D.J. (1996) Influences of dietary energy and protein on leucine kinetics during feeding in healthy adults. *American Journal of Physiology* 270, E282–291.

Giordano, M., Castellino, P. and DeFronzo, R.A. (1996) Differential responsiveness of protein synthesis and degradation to amino acid availability in humans. *Diabetes* 45, 393–399.

Giordano, M., Castellini, P., Ohno, A. and Defronzo, R.A. (2000) Differential effects of amino acid and ketoacid on protein metabolism in humans. *Nutrition* 16, 15–21.

Goulet, O., DePotter, S., Salas, J., Robert, J.-J., Rougier, M., Ben Hariz, M., Koziet, J., Desjeux, J.-F., Ricour, R. and Darmaun, D. (1993) Leucine metabolism at graded amino acid intake in children receiving parenteral nutrition. *American Journal of Physiology* 265, E540–546.

Grizard, J., Dardevet, D., Papet, I., Mosoni, L., Patureau, P., Attaix, M.D., Tauveron, I., Bonin, D. and Arnal, M. (1995) Nutrient regulation of skeletal muscle protein metabolism in animals. The involvement of hormones and substrates. *Nutrition Research Reviews* 8, 67–91.

Henry, C.J.K. (1990) Body mass index and the limits of human survival. *European Journal of Clinical Nutrition* 44, 329–335.

Henson, L.C. and Heber, D (1983) Whole body protein breakdown rates and hormonal adaptation in obese subjects. *Journal of Clinical Endocrinology and Metabolism* 57, 316–319.

Heslin, M.J., Newman, E., Wolf, R.F., Pisters, P.W.T. and Brennan, M.F. (1992) Effect of hyperinsulinemia on whole body and skeletal muscle leucine carbon kinetics in humans. *American Journal of Physiology* 262, E911–918.

Hillier, T.A., Fryburg, D.A., Jahn, L.A. and Barrett, E.J. (1998) Extreme hyperinsulinemia unmasks insulin's effect to stimulate protein synthesis in the human forearm. *American Journal of Physiology* 274, E1067–1074.

Hoffer, L.J., Bistrian, B.R., Young, V.R., Blackburn, G.L. and Matthews, D.E. (1984) Metabolic effects of very low calorie weight reduction diets. *Journal of Clinical Investigation* 73, 750–758.

Hoffer, L.J., Yang, R.D., Matthews, D.E., Bistrian, R.R., Bier, D.M. and Young, V.R. (1985) Effects of meal consumption on whole body leucine and alanine kinetics in young men. *British Journal of Nutrition* 53, 31–38.

Horber, F.F. and Haymond, M.W. (1990) Human growth hormone prevents the protein catabolic side effects in humans. *Journal of Clinical Investigation* 86, 265–272.

Jackson, A.A., Golden, M.H.N., Byfield, R., Jahoor, F., Royes, J. and Soutter, L. (1983) Whole body protein turnover and nitrogen balance in young children at intakes of protein and energy in the region of maintenance. *Human Nutrition: Clinical Nutrition* 37C, 433–446.

Keys, A., Brozek, J., Henschel, A., Mickleson, O. and Taylor, H.L. (1950) *The Biology of Human Starvation.* University of Minnesota Press, Minneapolis, Minnesota.

Kimball, S.R., Farrell, P.A. and Jefferson, L.S. (2002) Invited review: role of insulin in translational control of protein synthesis in skeletal muscle by amino acids or exercise. *Journal of Applied Physiology* 93, 168–180.

Lavivière, F., Wagner, D.A., Kupranyez, D. and Hoffer, L.J. (1990) Prolonged fasting as conditioned by prior protein depletion: effect on urinary nitrogen excretion and whole-body protein turnover. *Metabolism* 39, 1270–1277.

Liu, Z. and Barrett, E.J. (2002) Human protein metabolism: its measurement and regulation. *American Journal of Physiology* 283, E1105–1112.

Louard, R.J., Barrett, E.J. and Gelfand, R.A. (1990) Effect of infused branched-chain amino acids on muscle and whole-body amino acid metabolism in man. *Clinical Science* 79, 457–466.

Louard, R.J., Bhushan, R., Gelfand, R.A., Barrett, E.J. and Sherwin, R.S. (1994) Glucocorticoids antagonize insulin's antiproteolytic action on skeletal muscle in humans. *Journal of Clinical Endocrinology and Metabolism* 79, 278–284.

Macallan, D.C., McNurlan, M.A., Kurpad, A.V., de Souza, G., Shetty, P.S., Calder, A.G. and Griffin, G.E. (1998) Whole body protein metabolism in human pulmonary tuberculosis. *Clinical Science* 94, 321–331.

Marchini, J.S., Cortiella, I., Hiramatsu, T., Chapman, T.E. and Young, V.R. (1993) Requirement for indispensable amino acids in adult humans: longer-term amino acid kinetic study with support for the adequacy of the Massachusetts Institute of Technology, amino acid requirement pattern. *American Journal of Clinical Nutrition* 58, 670–683.

Mariotti, F., Mahé, S., Luengo, C., Benamouzig, R. and Tomé, D. (2000) Postprandial modulation of dietary and whole-body nitrogen utilization by carbohydrates in humans. *American Journal of Clinical Nutrition* 72, 954–962.

Mauras, N. and Beaufrère, B. (1995) Recombinant insulin-like growth factor-I enhances whole body anabolism and significantly diminishes the protein catabolic effects of prednisone in humans without a diabetogenic effect. *Journal of Clinical Endocrinology and Metabolism* 80, 864–874.

McHardy, K.C., McNurlan, M.A., Milne, E., Calder, A.G., Fearns, L.M., Brown, J. and Garlick, P.J. (1991) The effect of insulin suppression on post-prandial nutrient metabolism: studies with infusion of somatostatin and insulin. *European Journal of Clinical Nutrition* 45, 515–526.

McNurlan, M.A., Fern, E.B. and Garlick, P.J. (1982) Failure of leucine to stimulate protein synthesis *in vivo*. *Biochemical Journal* 204, 831–838.

Meek, S.E., Persson, M., Ford, C.G. and Nair, K.S. (1998) Differential regulation of amino acid exchange and protein dynamics across splanchnic and skeletal muscle beds by insulin in healthy human subjects. *Diabetes* 47, 1824–1835.

Meguid, M.M., Matthews, D.E., Bier, D.M., Meredith, C.N., Soeldner, J.S. and Young, V.R. (1986a) Leucine kinetics at graded leucine intakes in young men. *American Journal of Clinical Nutrition* 43, 770–780.

Meguid, M.M., Matthews, D.E., Bier, D.M., Meredith, C.N. and Young, V.R. (1986b) Valine kinetics at graded valine intakes in young men. *American Journal of Clinical Nutrition* 43, 781–786.

Melville, S., McNurlan, M.A., McHardy, K.C., Broom, J., Milne, E., Calder, A.G. and Garlick, P.J. (1989) The role of degradation in the acute control of protein balance in adult man: failure of feeding to stimulate protein synthesis, as assessed by L-[1-^{13}C] leucine infusion. *Metabolism* 38, 248–255.

Melville, S., McNurlan, M.A., Calder, A.G. and Garlick, P.J. (1990) Increased protein turnover despite normal energy metabolism and responses to feeding in patients with lung cancer. *Cancer Research* 50, 1125–1131.

Meredith, C.N., Wen, Z.-M., Bier, D.M., Matthews, D.E. and Young, V.R. (1986) Lysine kinetics at graded lysine intakes in young men. *American Journal of Clinical Nutrition* 43, 787–794.

Millward, D.J., Fereday, A., Gibson, N.R. and Pacy, P.J. (2000) Human adult amino acid requirements: [1-^{13}C] leucine balance evaluation of the efficiency of utilization and apparent requirements for wheat protein and lysine compared with those for milk protein in healthy adults. *American Journal of Clinical Nutrition* 72, 112–121.

Morrison, W.L., Gibson, J.N.A., Jung, R.T. and Rennie, M.J. (1988) Skeletal muscle and whole body protein turnover in thyroid disease. *European Journal of Clinical Investigation* 18, 62–68.

Motil, K.J., Matthews, D.E., Bier, D.M., Burke, J.F., Munro, H.N. and Young, V.R. (1981) Whole body leucine and lysine metabolism: response to dietary protein intake in young men. *American Journal of Physiology* 240, E712–721.

Nair, K.S., Woolf, P.D., Welle, S.L. and Matthews, D.E. (1987) Leucine, glucose and energy metabolism after 3 days of fasting in healthy human subjects. *American Journal of Clinical Nutrition* 46, 557–562.

Nair, K.S., Matthews, D.E., Welle, S.L. and Braiman, T. (1992a) Effect of leucine on amino acid and glucose metabolism in humans. *Metabolism* 41, 643–648.

Nair, S.K., Schwartz, R.G. and Welle, S. (1992b) Leucine as a regulator of whole body and skeletal muscle protein metabolism in humans. *American Journal of Physiology* 263, 928–934.

Nair, K.S., Ford, G.C., Ekberg, K., Fernqvist-Forbes, E. and Wahren, J. (1995) Protein dynamics in whole body and in splanchnic and leg tissues in type 1 diabetic patients. *Journal of Clinical Investigation* 95, 2926–2932.

Nørrelund, H., Nair, K.S., Jorgenson, J.O., Christiansen, J.S. and Møller, N. (2001) The protein-retaining effects of growth hormone during fasting involve inhibition of muscle protein breakdown. *Diabetes* 50, 96–104.

Nørrelund, H., Djurhuns, C., Jorgenson, J.O.L., Nielsen, S., Nair, K.S., Schmitz, O., Christiansen, J.S. and Møller, N. (2003) Effects of GH on urea, glucose and lipid metabolism, and insulin sensitivity in GH-deficient patients. *American Journal of Physiology* 285, E737–743.

Nygren, J. and Nair, K.S. (2003) Differential regulation of protein dynamics in splanchnic and skeletal muscle beds by insulin and amino acids in healthy subjects. *Diabetes* 52, 1377–1385.

Oehri, M., Ninnis, R., Girard, J., Frey, F.J. and Keller, U. (1996) Effects of growth hormone and IGF-I on glucocorticoid-induced protein catabolism in humans. *American Journal of Physiology* 270, E552–558.

Pacy, P.J., Garrow, J.S., Ford, G.C., Merritt, H. and Halliday, D. (1988) Influence of amino acid administration on whole body leucine kinetics and resting metabolic rate in postabsorptive normal subjects. *Clinical Science* 75, 225–231.

Pacy, P.J., Nair, K.S., Ford, G.C. and Halliday, D. (1989) Failure of insulin infusion to stimulate muscle protein synthesis in type I diabetic patients: Further evidence that anabolic effect of insulin depends upon inhibition of proteolysis. *Diabetes* 38, 618–624.

Pacy, P.J., Thompson, G.N. and Halliday, D. (1991a) Measurement of whole-body protein turnover in insulin-dependent (type 1) diabetic patients during insulin withdrawal and infusion: comparison of [^{13}C] and [$^{2}H_5$] methodologies. *Clinical Science* 80, 345–352.

Pacy, P.J., Bannister, P.A. and Halliday, D. (1991b) Influence of insulin on leucine kinetics in the whole body and across the forearm in post-absorptive insulin dependent diabetic patients. *Diabetes Research* 18, 155–162.

Pacy, J., Price, G.M., Halliday, D., Quevedo, M.R. and Millward, D.J. (1994) Nitrogen homoeostasis in man: the diurnal responses of protein synthesis and degradation and amino acid oxidation to diets with increasing protein intakes. *Clinical Science* 86, 103–118.

Pozefsky, T., Felig, P., Tobin, J.D., Soeldner, J.S. and Cahill, G.F. (1969) Amino acid balance across tissues of the forearm in postabsorptive man. Effects of insulin at two dose levels. *Journal of Clinical Investigation* 48, 2273–2282.

Price, G.M., Halliday, D., Pacy, P.J., Quevedo, M.R. and Millward, D.J. (1994) Nitrogen homoeostasis in man: influence of protein intake on the amplitude of diurnal cycling of body nitrogen. *Clinical Science* 86, 91–102.

Quevedo, M.R., Price, G.M., Halliday, D., Pacy, P.J. and Millward, D.J. (1994) Nitrogen homoeostasis in man: diurnal changes in nitrogen excretion, leucine oxidation and whole body leucine kinetics during a reduction from a high to a moderate protein intake. *Clinical Science* 86, 185–193.

Rooyackers, D.E. and Nair, K.S. (1997) Hormonal regulation of human muscle protein metabolism. *Annual Review of Nutrition* 17, 457–485.

Russell-Jones, D.L., Weissberger, A.J., Bowes, S.B., Kelly, J.M., Thomason, M., Umpleby, A.M., Jones, R.H. and Sönksen, P.H. (1993) The effects of growth hormone on protein metabolism in adult growth hormone deficient patients. *Clinical Endocrinology* 38, 427–431.

Russell-Jones, D.L., Umpleby, A.M., Hennessy, T.R., Bowes, S.B., Shojaec-Moradie, P., Hopkins, K.D., Jackson, N.C., Kelly, J.M., Jones, R.H. and Sönksen, P.H. (1994) Use of a leucine clamp to demonstrate that IGF-1 actively stimulates protein synthesis in normal humans. *American Journal of Physiology* 267, E591–598.

Salomon, F., Cuneo, R.C., Hesp, R. and Sönksen, P.H. (1989) The effects of treatment with recombinant human growth hormone on body composition and metabolism in adults with growth hormone deficiency. *New England Journal of Medicine* 321, 1797–1803.

Schwenk, F.W. and Haymond, M.W. (1987) Effects of leucine, isoleucine or threonine infusion on leucine metabolism in humans. *American Journal of Physiology* 253, E428–434.

Simmons, P.S., Miles, J.M., Gerich, J.E. and Haymond, M.W. (1984) Increased proteolysis: an effect of increases in plasma cortisol within the physiological range. *Journal of Clinical Investigation* 73, 412–420.

Tauveron, I., Charrier, S., Champedron, C., Bonnet, Y., Berry, C., Bayle, G., Pragnaud, J., Obled, C., Grizard, J. and Thiéblot, P. (1995) Response of leucine metabolism to hyperinsulinemia under amino acid replacement in experimental hyperthyroidism. *American Journal of Physiology* 269, E499–501.

Tessari, P., Trevisan, R., Inchiostro, S., Biolo, G., Nosadini, R., De Kreutzenberg, S.V., Duner, E., Tiengo, A. and Crepaldi, G. (1986) Dose-response curves of insulin on leucine kinetics in humans. *American Journal of Physiology* 251, E334–342.

Tessari, P., Inchiostro, S., Biolo, G., Trevisan, R., Fantin, G., Marescott, M.C., Iori, E., Tiengo, A. and Crepaldi, G. (1987) Differential effects of hyperinsulinemia and hyperaminoacidemia on leucine-carbon metabolism *in vivo*. Evidence for distinct mechanisms in regulation of net amino acid deposition. *Journal of Clinical Investigation* 79, 1062–1069.

Tessari, P., Pehling, G., Nissen, G.L., Gerich, J.E., Service, F.J., Rizza, R.A. and Haymond, M. (1988) Regulation of whole body leucine metabolism during mixed meal absorption in normal and diabetic humans. *Diabetes* 37, 512–519.

Tessari, P., Inchiostro, S., Biolo, G., Vincenti, E., Sabadin, L. and Vettore, M. (1991) Effects of acute systemic hyperinsulinemia on forearm muscle proteolysis in healthy man. *Journal of Clinical Investigation* 88, 27–33.

Tessari, P., Barazzoni, R., Zanetti, M., Vettore, M., Normand, S., Brutomesso, D. and Beaufrère, B. (1996a) Protein degradation and synthesis measured with multiple amino acid tracers *in vivo*. *American Journal of Physiology* 271, E733–741.

Tessari, P., Zanetti, M., Barazzoni, R., Vettore, M. and Michielan, F. (1996b) Mechanisms of postprandial protein accretion in human skeletal muscle. *Journal of Clinical Investigation* 98, 1361–1372.

Tessari, P., Barazzoni, R., Zanetti, M., Kiwanuka, E. and Tiengo, A. (1999) Protein metabolism in type 1 and type 2 diabetes in the fasted and fed states. *Diabetes, Nutrition and Metabolism* 12, 428–434.

Thissen, J.-P., Ketelslegers, J.-M. and Underwood, J.E. (1994) Nutritional regulation of the insulin-like growth factors. *Endocrine Reviews* 15, 80–101.

Tischler, M.E., Desautels, M. and Goldberg, A.L. (1982) Does leucine, leucyl-tRNA, or some metabolite of leucine regulate protein synthesis and degradation in skeletal and cardiac muscle? *Journal of Biological Chemistry* 257, 1613–1621.

Tracey, K.J., Legaspi, A., Albert, J.D., Jeevanadam, M., Matthews, D.E., Brennan, M.F. and Lowry, S.F. (1988) Protein and substrate metabolism during starvation and parenteral refeeding. *Clinical Science* 74, 123–32.

Volpi, E., Lucidi, P., Cruciani, G., Monacchia, F., Reboldi, G., Brunetti, P., Bolli, G.B. and De Feo, P. (1996) Contribution of amino acids and insulin to protein anabolism during meal absorption. *Diabetes* 45, 1245–1252.

Waterlow, J.C. (1996) The requirements of adult man for indispensable amino acids. *European Journal of Clinical Nutrition* 50, Suppl. 1, S151–S179.

Welle, S., Thornton, C., Statt, M. and McHenry, B. (1996) Growth hormone increases muscle mass and strength but does not rejuvenate myofibrillar protein synthesis in healthy subjects over 60 years old. *Journal of Clinical Endocrinology and Metabolism* 81, 3239–3243.

Winterer, J., Bistrian, B.R., Bilmazes, C.R., Blackburn, G.L. and Young, V.R. (1980) Whole body protein turnover, studied with ^{15}N-glycine, and muscle protein breakdown in mildly obese subjects during a protein-sparing diet and a brief total fast. *Metabolism* 29, 575–581.

Yang, R.D., Matthews, D.E., Bier, D.M., Wen, Z.M. and Young, V.R. (1986) Response of alanine metabolism to manipulation of dietary and protein intakes. *American Journal of Physiology* 250, E39–46.

Young, V.R., Meredith, C., Hoerr, R., Bier, D.M. and Matthews, D.E. (1985) Amino acid kinetics in relation to protein and amino acid requirements: the primary importance of amino acid oxidation. In: Garrow, J.S. and Halliday, D. (eds) *Substrate and Energy Metabolism.* Libbey, London, pp. 119–133.

Yu, Y.-M., Wagner, D.A., Tredget, A.E., Walaszewski, J.A., Burke, J.F. and Young, V.R. (1990) Quantitative role of splanchnic region in leucine metabolism: L-[$^{1-13}$C, ^{15}N] leucine and substrate balance studies. *American Journal of Physiology* 259, E36–51.

Zachwieja, J.J., Bier, D.M. and Yarakeshi, K.E. (1994) Growth hormone administration in older adults: effects on albumin synthesis. *American Journal of Physiology* 266, E840–844.

Zanetti, M., Barazzoni, R., Kiwanuka, E. and Tessari, P. (1999) Effects of branched-chain-enriched amino acids and insulin on forearm leucine kinetics. *Clinical Science* 97, 437–448.

Zhao, X.-H., Wen, Z.-N., Meredith, C.N., Matthews, D.E., Bier, D.M. and Young, V.R. (1986) Threonine kinetics at graded threonine intakes in young men. *American Journal of Clinical Nutrition.* 43, 795–802.

10

Adaptation to Different Protein Intakes: Protein and Amino Acid Requirements

10.1 Adaptation

10.1.1 General concepts

Human adaptation can be considered at three levels – genetic, physiological and behavioural (Gould and Lewontin, 1979; Waterlow, 1985). Here we are concerned mainly with physiological adaptations, although there may be genetic differences in the capacity to adapt; moreover, the differences, presumably genetic, between individuals that seem to affect most biological functions can never be left out of account. Widdowson (1962) made the point very well in her classical paper 'Nutritional Individuality'.

It is necessary to distinguish between 'adaptation' and 'response'. The difference is not only of timescale but of physiological mechanism. For example, the response to a high environmental temperature is to sweat; longer term adaptation or acclimation is to produce more sweat with a lower salt content and, according to the old literature, a fall in metabolic rate (Martin, 1930; Roberts, 1953). A distinction has already been made in the previous chapter between the immediate response to a protein meal and the longer-term adjustment to a different protein intake. It is the latter that can properly be called an adaptation.

Suggested characteristics of a physiological adaptation are shown in Table 10.1. Points number 2 and 3 in this table are difficult, perhaps controversial. There is a tendency to think 'I am normal and you are adapted', but in reality 'normal' and 'adapted' are two sides of the same coin. The physiologist von Muralt said: 'If a textbook of physiology were written by an Indian from the high Andes, he would marvel at the way in which the infant at birth adapts to the high oxygen tension at sea level – an adaptation which is not necessary for the baby born at 5000 metres.' The problem is that the third characteristic involves a value judgement. A good example of this difficulty is the stunting in height, by Western standards, of vast numbers of children in developing countries. An economist described this stunting as a favourable adaptation, because a small child needs less food and is therefore less likely to die of malnutrition. However, studies have shown that in some countries, although not in all, such children remain stunted all their lives, have reduced capacity for physical work and earning wages and impaired mental development (Grantham-McGregor, 1992). Such a state could not be regarded as beneficial or acceptable, except by the criterion of survival. There must, no doubt, be a range over which environmental stunting does not matter, but the problem remains of choosing the criteria by which to define that range. These general considerations are very relevant to the problem of adaptation to low protein intakes.

10.1.2 Adaptation to low protein intakes

This adaptation has been the subject of many experimental studies lasting a few days or weeks. The difficulties of interpretation are well illustrated by the work of Nicol and Phillips (1976) who showed that men in Nigeria were able to achieve nitrogen balance on lower intakes than those studied by Calloway *et al.* (1975) in California. N.S. Scrimshaw (personal communication) maintained that the Nigerians were

Table 10.1. Characteristics of physiological adaptation.

1. Adaptation achieves and maintains a new steady state.
2. The steady state is in a 'preferred' or 'acceptable' range of the function that is adapting.
3. The adaptation produces a 'better' state than would have been achieved without it.
4. The adaptation is usually reversible if the environment changes.
5. Most adaptations are integrations of many small changes (Barcroft, 1934).

depleted, while I maintained that they were adapted. The question could only be settled if we had information about differences in the amount and distribution of their body protein, and whether such differences had functional consequences.

According to the first criterion the subject must be in nitrogen balance. The difficulties in defining the *minimum* intake needed to secure N balance are discussed below; here, we are concerned with the transition from a higher to a lower intake or its reverse, i.e. the adaptive process. The pioneer study in humans is that of Martin and Robison (1922), who showed that when the N intake was reduced from 17 to 5 g day^{-1}, the subject came into N balance after ~7 days (Fig. 10.1). Many later studies have agreed that the time needed is 7–10 days in the adult; in the infant it is much shorter, ~2 days (Chan, 1968) and in the rat shorter still, ~30 h (Das and Waterlow, 1974).

There has been some dispute about whether 7 days is enough for full adaptation. Martin and Robison's results look like exponentials that approach an asymptote, which is always hard to define. Young and co-workers made two studies to find out whether 7 days is really long enough. They are relevant, although they involved reducing the

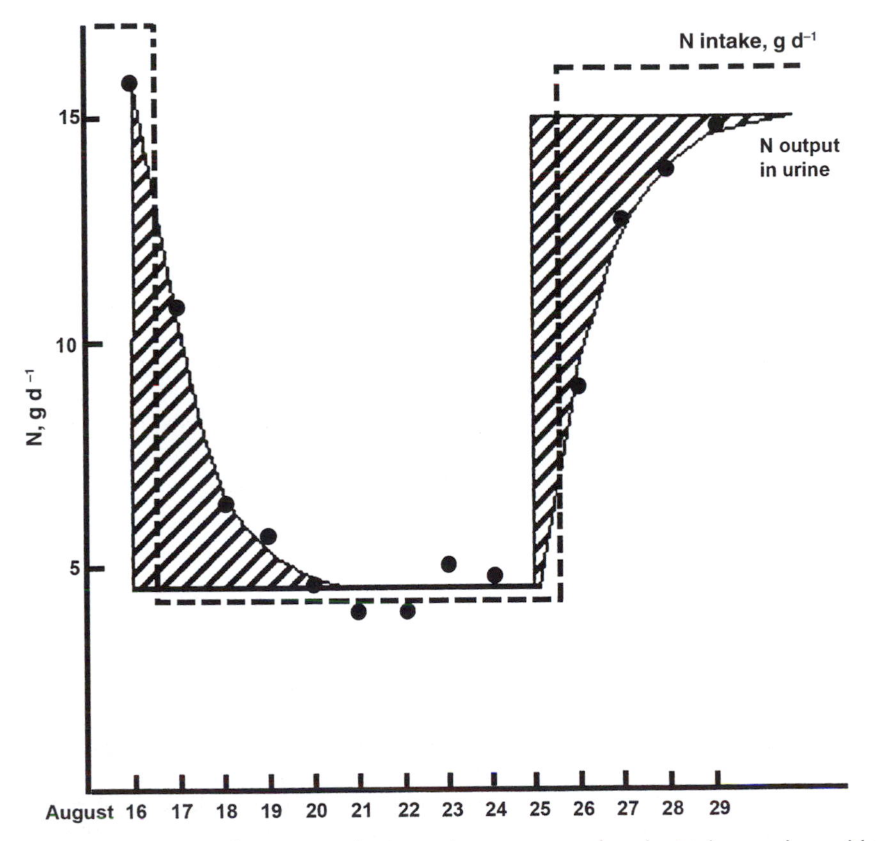

Fig. 10.1. Changes with time in total urinary nitrogen output when the intake was changed from 17 to 5 g d^{-1} and back again. Reproduced from Martin and Robison (1922), by courtesy of the *Biochemical Journal*.

intake of one essential amino acid, leucine, rather than that of total protein. In the first study (Young *et al.*, 1987) measurements were made initially on an adequate leucine intake, and again after 1 week and 3 weeks on low intakes. Leucine balance was slightly negative at 1 week and improved a little at 3 weeks. The second experiment (Marchini *et al.*, 1993) had a similar design, except that all essential amino acids, not only leucine, were reduced. Again, there were no significant differences in carbon balance after 1 and 3 weeks on the low intakes.

These results suggest that, under the conditions of the experiments, 1 week is enough for adaptation to be complete by the criterion of nitrogen balance. However, much will depend on the level to which the intake is reduced: the lower the level the longer it will take to achieve balance, and some studies have suggested that a longer time is indeed needed (section 10.1.3).

During the transition period of adaptation from a higher to a lower intake there is a net loss of N (Fig. 10.1). In rats the initial loss is from liver and gastro-intestinal tract (Addis *et al.*, 1936); but after some 2 days muscle and skin take over (Waterlow and Stephen, 1966). In the rat, in the opposite situation, when increasingly large amounts of protein were fed, the liver gained protein exponentially (Henry *et al.*, 1953). We tried, but unsuccessfully, to show that in a model with two protein pools, one fast and one slow, if the intake was reduced the system would settle down to a new steady state in which nearly all the net loss was from the slow pool. In a human switching from a generous intake of 90 g protein to a marginally adequate one of 40 g day^{-1}, the net N loss over a 7-day period of adaptation would be about 35 g, equivalent to 220 g protein or less than 2% of total body protein. If this loss is all from muscle it would be about 5% of total muscle protein. We do not know whether a loss of this degree produces any handicap. In the sarcopenia of the elderly, with clear functional deficits, the loss is much greater (see Chapter 11).

In earlier days these losses or gains of nitrogen were attributed to withdrawing or replenishing protein stores or reserves. However, no proteins have been identified that could act as stores in the way that glycogen and fat are stores for energy. It seems more likely that the losses or gains arise from a lag in the adjustment of protein turnover to the change in protein intake.

10.1.3 Mechanisms of adaptation

Kinetic changes

The fed:fasted ratio is a measure of the response to feeding. Comparisons of responses to normal and low protein intakes (Tables 9.1 and 10.2) show that a low intake decreases the responses (i.e. differences from unity) of all three variables.

Young's group at MIT in the early stages of their tracer work measured fluxes at decreasing intakes of various essential amino acids (Meguid *et al.*, 1986a,b; Meredith *et al.*, 1986; Zhao *et al.*, 1986 – summarized in Waterlow, 1996). However, the only study we know of in which successive kinetic measurements with [^{13}C] leucine were made during the transition from a high to a lower protein intake is that of Quevedo *et al.* (1994) (Table 10.3). In this study the transition was from a generous protein intake (1.82 g kg^{-1} day^{-1}) to a normal one (0.77 g), rather than from normal to low; nevertheless, the results are relevant. In the fasting state there was a small fall in oxidation, which was complete in 7 days, and a decrease in degradation by day 3 with little further change. In the fed state there was a substantial fall in oxidation, with no further change after 7 days. Garlick has said: '… changes during the fasting state are additional evidence that protein metabolism not only responds to the immediate intake of food but also undergoes adaptation to varying intakes of protein over a longer period.'

Table 10.2. Responses to feeding: comparison between subjects on normal and low protein intakes.

Previous protein intake	Percentage change on feeding over levels during fasting		
	Synthesis	Breakdown	Oxidation
Normal	+5	−44	+74
Low	−11	−21	−8

From the data in Table 9.1.

Table 10.3. Responses of protein synthesis and breakdown to feeding on changing from a high to a low protein intake.

	Synthesis	Breakdown
Days	mmol kg^{-1} day^{-1}	
0	+16	−80
3	+2	−39
7	+5	−31
14	+6	−30

The response is the difference between feeding and fasting rate. Data of Quevedo *et al.*, 1994.

The responses to feeding, i.e. the differences between feeding and fasting rates, are discussed in Chapter 9 and summarized in Table 9.1. What they imply is that the characteristic of the adaptation is a change in the amplitude of diurnal cycling. This is well shown by changes in the size of the urea pool, which could be measured over 12-h periods (Fig. 10.2). The conclusion from this experiment is that in this adaptation to a different protein intake there is a hierarchy in the kinetic changes: oxidation > breakdown > synthesis. Moreover, the adaptation appears to be complete after 7 days and nearly complete by 3 days. It would be interesting if this experiment were repeated with a change from a normal to a low protein intake.

Adaptive changes in enzymes

In Quevedo's experiment changes in urea excretion paralleled changes in leucine oxidation, as might be expected. It was a major contribution to our understanding of the mechanism of adaptation to a change in protein intake when Schimke (1962) showed in rats that it involved coordinate changes in the activity of the urea cycle enzymes. Das and Waterlow (1974) repeated this work with special attention to the time-scale of the adaptation, showing coincidence between the enzyme changes and the change in urinary N excretion (Fig. 10.3). In the case of two of the enzymes, arginase and ornithine transcarbamylase, they showed that the change in activity represented a change in the

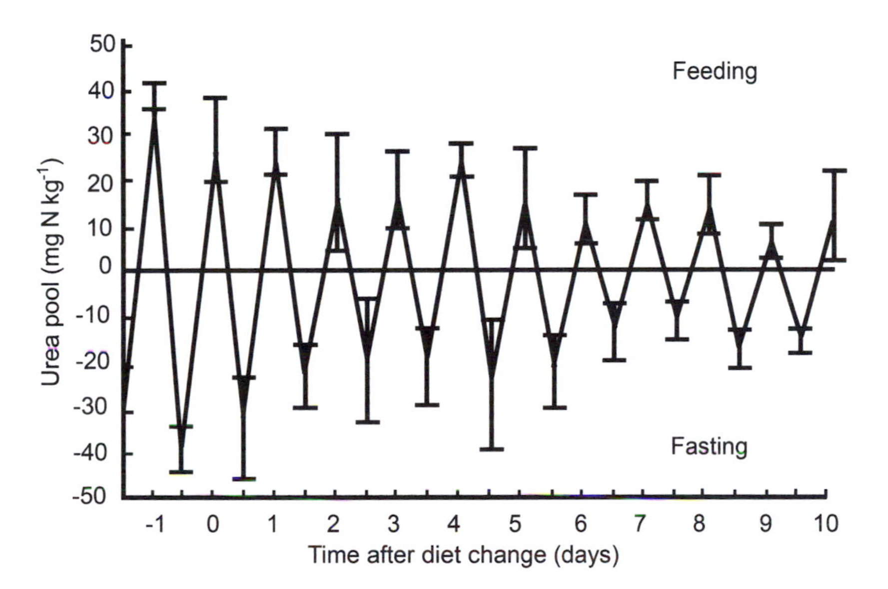

Fig. 10.2. Diurnal changes in size of the urea pool in subjects changing from a high protein to a moderate protein intake. The values were measured at the end of consecutive 12 h dietary periods of either fasting or feeding. Reproduced from Quevedo *et al.* (1994), by courtesy of *Clinical Science*.

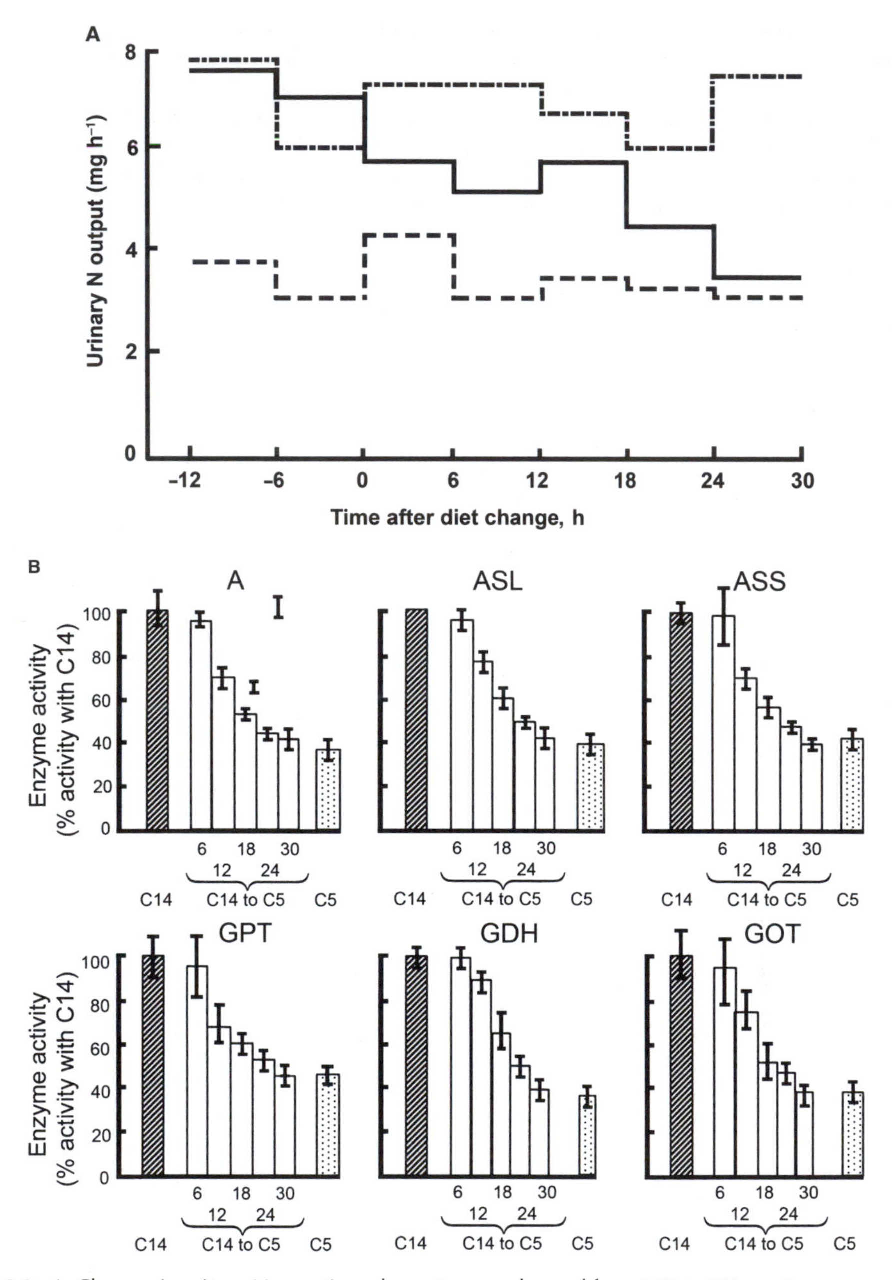

Fig. 10.3. A. Changes in urinary N excretion when rats were changed from 14% to 5% casein. — – — – — 14% casein throughout, —— 14%→5%, – – – – 5% throughout. B. Changes in rat liver enzymes when the diet was changed from 14% to 5% casein. A, arginase; ASL, argininosuccinate lyase; ASS, argininosuccinate synthetase; GPT, alanine aminotransferase; GDH, glutamate dehydrogenase; GOT, aspartate aminotransferase. Clear bars: rats changed from 14% to 5% casein. Shaded bars: rats maintained throughout on either: 14% ▨ or 5% ⬚ casein Reproduced from Das and Waterlow (1974), by courtesy of the *British Journal of Nutrition.*

amount of enzyme (Fig. 10.4; Schimke, 1964). The only evidence of whether this kind of adaptation occurs in man is that in malnourished children the activity of argininosuccinase in the liver was only 72% of that found after recovery (Stephen and Waterlow, 1968).[1]

Changes in activity have been produced in many enzymes other than those of the urea cycle and by many inducers or repressors such as amino acids, glucocorticoids or glucagon. They therefore provide a particularly good opportunity for studying protein kinetics in a non-steady state. A vast amount of work was done on this subject in the 1960s and '70s – see reviews by Schimke (1970) and Kenney (1970) and our earlier book (Waterlow *et al.*, 1978). The basic assumptions are, firstly: that synthesis is a zero order and breakdown a first order process; secondly, that a semilog plot of the specific activity of the enzyme enables calculation of the fractional breakdown rate; and of total activity in a tissue, calculation of the synthesis rate (Koch, 1962). Schimke (1970) gives an equation for the new steady-state level of an enzyme, E′, when the synthesis rate changes from S to S′ and the fractional degradation rate changes from k to k′:

$$E'/E_o = S'/k'E_o - (S'/k'E_o - 1)e^{-k't}$$

Figure 10.4 is a good example of results obtained in this way, showing that the decrease in arginase on adapting from a high to a low protein intake involved a decrease in synthesis with an initially large increase in degradation until a new steady state was achieved. The time course of the adaptation is determined only by the fractional rate of degradation, whereas the level of the new steady state is determined by the ratio of the synthesis rate to the fractional degradation rate, S/k_d.

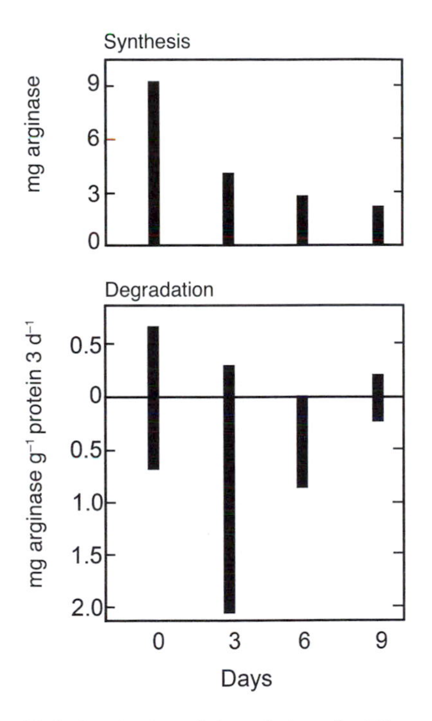

Fig. 10.4. Synthesis and degradation of rat liver arginase on changing from a high to a low protein diet (70% casein to 8% casein). Reproduced from Schimke (1964), by courtesy of the *Journal of Biological Chemistry*.

10.1.4 Longer-term 'adaptation'

The discussion so far has been concerned with adaptation within a range of protein intakes, but above the level of minimum requirement. Suppose the intake is reduced to less than the minimum requirement, which is currently estimated to be, on average, 0.6 g protein kg^{-1} day^{-1} provided that the protein is of good quality (see next section). A man weighing 70 kg would need an intake of 42 g protein day^{-1} to meet this level. Suppose that his intake was reduced to 30 g day^{-1}: his weight would fall until, at 48 kg, the intake of 30 g day^{-1} would provide 0.625 g protein kg^{-1} day^{-1} and so be meeting the requirement. The position is shown in Fig. 10.5. By the criterion of nitrogen balance the subject would be regarded as adapted, but this adapted state cannot be regarded as satisfactory; he would have lost 30% of his initial body weight, mainly at the expense of muscle (Barac-Nieto *et al.*, 1978; Soares *et al.*, 1991). This is a 'thought' experiment. A natural experiment is provided by the poor Indian labourers studied by Soares *et al.* (1991, 1994). These men had low body weights compared with controls (students). Their body mass indices were in the region of 16, which is about 1 SD below the lower limit of normal,

[1] We commented: 'These changes may reasonably be regarded as part of a compensating mechanism which may be an important factor for human survival'. At that time it was regarded as ethical to perform liver biopsies in children in whom the mortality rate even in hospital was some 20%.

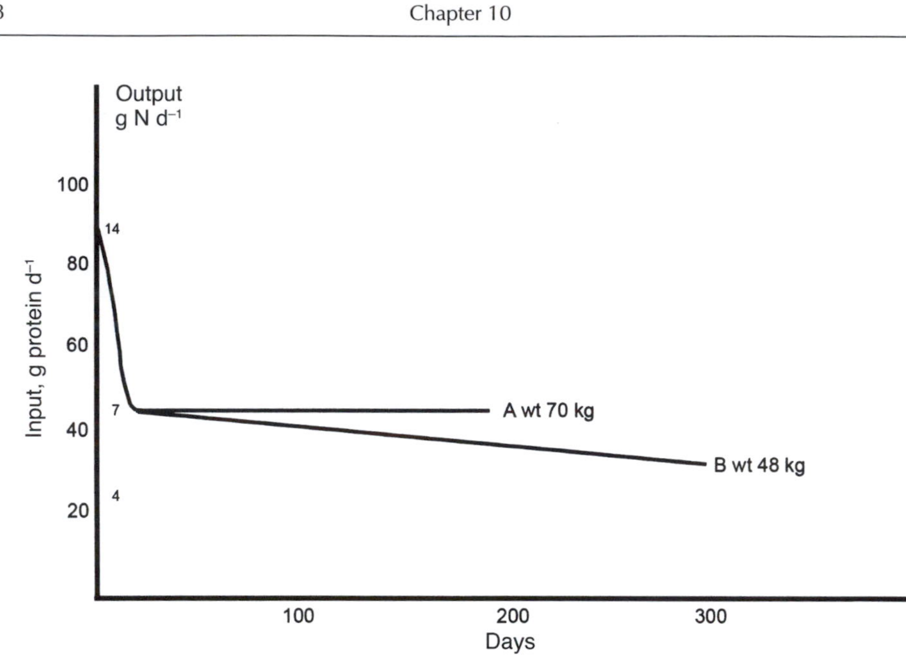

Fig. 10.5. Hypothetical scheme showing that if a subject has an intake of protein substantially below his requirement at a normal body weight, he will eventually come into N balance at the expense of a large loss of body N and body weight.

according to our provisional definition (James *et al.*, 1988).

The characteristics of these labourers, compared with student controls, are shown in Table 10.4. We have no information about the previous protein intake of these men, but it may well have been as low as 30 g day^{-1}, after correction for digestibility and quality. On this intake, to provide for a safe level of protein at 0.75 g kg^{-1} day^{-1} (FAO/WHO/UNU, 1985) the body weight would have to fall to 40 kg. These men have therefore 'adapted' to their low protein intake and survive by losing weight; they lost, or failed to gain, 10 kg of muscle but only 1 kg of non-muscle tissue. Whole body protein turnover per kg fat-free mass was unimpaired, but the flux ratio showed a lower flux through peripheral tissues. These labourers were also 9 cm shorter than the student controls, suggesting a life-long

Table 10.4. Characteristics of poor Indian labourers.

	Students (controls)	Labourers
Weight (kg)	60	42
Height (m)	1.70	1.61
BMI (kg/m^2)	20.8	16.7
Body fat (kg)	10.8	4.6
Fat-free mass FFM (kg)	49.4	38.6
Muscle mass (kg)	26.0	16.2
Non-muscle mass (kg)	23.4	22.4
Protein synthesis (g 12 h^{-1} kg^{-1} FFM)	3.05	3.07
Q_A/Q_U	0.89	0.61

Data of Soares *et al.*, 1994.

[1] The values obtained in different studies are somewhat variable (Fig. 10.5) but 50 mg kg^{21} d^{21} is that accepted by the UN Organizations (FAO/WHO/UNU, 1985: Table 3). The obligatory loss also has an allometric relation to body weight, with an exponent of 0.75 (Henry and Collingwood, 1998).

deprivation. The adaptation enabled them to work, but with what efficiency is not known. The long time over which the adaptation developed must be important, because these Indian men were very different from Keys' volunteers; they were on a low energy diet for 6 months, and reached the same BMI of about 16, but were in a far worse physiological and psychological state (Keys *et al.*, 1950).

There is a large literature showing that low body weight and in particular stunting in height are associated with reduced physical capacity and earning power and impaired mental development (Satyanarayana *et al.*, 1977, 1980; Spurr, 1983; Grantham-McGregor, 1992).

To conclude, adaptation is a necessary concept, but because it involves subjective value judgements, impossible to define in any simple way.

10.2 Requirements for Protein and Amino Acids

10.2.1 Protein requirements

We are interested here not so much in the quantitative estimates that have been obtained for human protein and amino acid requirements but in the contributions of kinetic studies to this very difficult problem. Traditionally the protein requirement of humans has been determined by the method of nitrogen balance as the minimum amount needed to balance obligatory losses in urine, faeces and by the skin. These losses, which have to be replaced, amount, in adults on a protein-free diet, to ~50 mg N kg^{-1} day^{-1}.[1] There may be small ethnic differences in the obligatory loss, as suggested by Fig. 10.6, and a good deal of between-subject variation (Fig. 10.7). Determination of the requirement

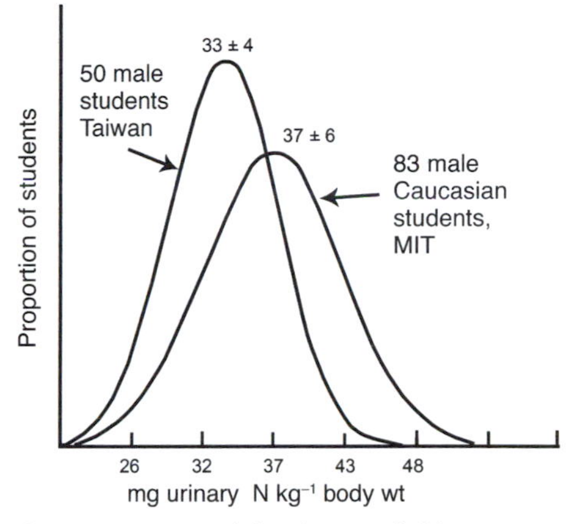

Fig. 10.6. Mean and distribution of obligatory urinary N losses in two populations of young adult men. Reproduced from Young and Scrimshaw (1979), by courtesy of the *American Journal of Clinical Nutrition.* © Am. J. Clin. Nutr. American Society for Nutrition.

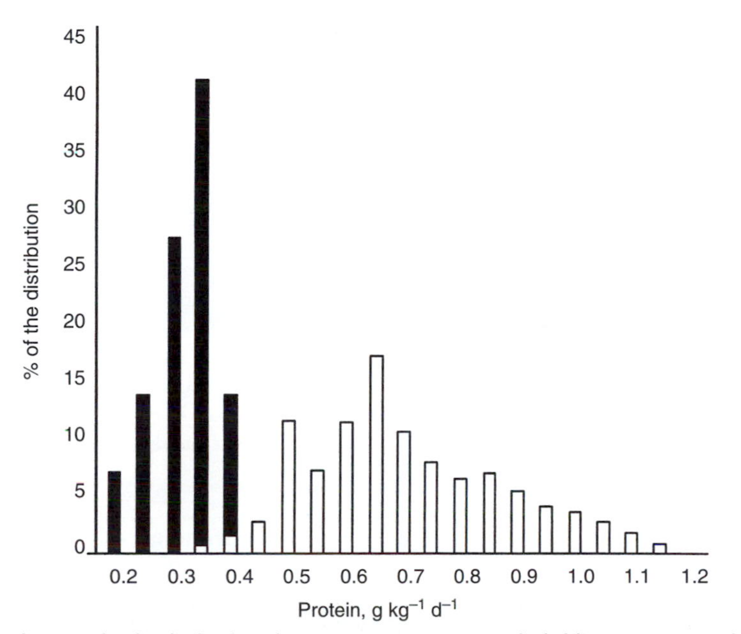

Fig. 10.7. Distribution of individual values for protein requirement and of obligatory nitrogen loss, reported by Rand *et al.* (2003), by courtesy of the *British Journal of Nutrition.* Reproduced from Millward (2003). ■, nitrogen loss; □, protein requirement.

involves a series of studies at different levels of protein intake, each after a period of adaptation to that intake, in order to find the minimum intake that will produce nitrogen balance. Young *et al.* (1989) have set out the drawbacks and uncertainties of the method, and the very large literature has been summarized in a meta-analysis by Rand *et al.* (2003). The conclusion is that the minimum requirement of adult humans for good quality protein such as meat or milk is ~0.60 g protein or ~100 mg N kg^{-1} day^{-1}. Thus the efficiency of utilization, i.e. the loss to be replaced divided by the intake needed to replace it, comes to 50/100 or 0.5. It has long been a puzzle why the efficiency even of good quality highly digestible protein should be so low.

The work of Millward and co-workers (Pacy *et al.*, 1994; Price *et al.*, 1994), described briefly in Chapter 9, has thrown new light on this problem. They showed in fasting/feeding studies of protein turnover that in the post-absorptive phase oxidation occurs at a rate that is determined by the habitual protein intake. This loss is termed the 'regulatory loss'. Millward developed the concept of 'metabolic demand', which is the sum of the obligatory loss and the regulatory loss. Both are assumed to continue at the same rate during the post-absorptive and post-prandial periods; the post-absorptive loss occurs at the expense of protein breakdown and is restored by synthesis on feeding at a rate that is determined by the habitual protein intake (Fig. 10.8). In addition there is a meal-related loss resulting from the oxidation of excess substrate. It is tempting to suppose that the regulatory loss, related to the habitual diet, results from re-setting of the urea cycle enzymes, as described above. This concept has implications for the calculation of efficiency, since in the equation: efficiency = losses restored/intake, the numerator becomes obligatory loss + regulatory loss instead of obligatory loss alone, thus raising the net protein utilization of a good quality protein from ~0.5 to ~0.8. This is an important result of turnover studies; the calculations are explained in detail in Millward and Pacy (1995), Millward *et al.* (2002) and Millward (2003).

An approach that has been tried in rats, but not in man, is to give a carbon-labelled amino acid and after it has become uniformly labelled in tissue proteins – about 20 days in the rat – the rate of loss of label provides an estimate of the amount to be replaced, and hence of the requirement (Neale and Waterlow, 1974, 1977).

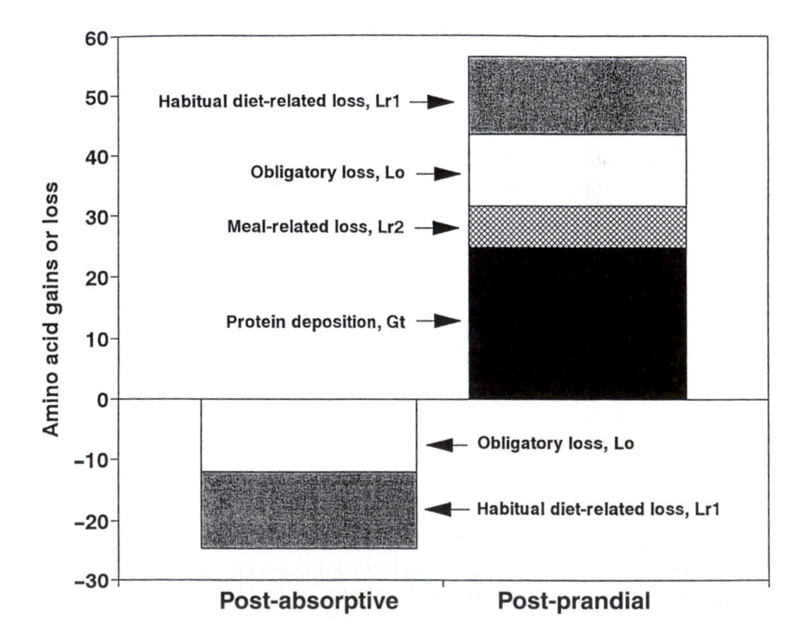

Fig. 10.8. Millward's concept of diurnal amino acid gains or losses. The losses in the post-absorptive phase are balanced by protein synthesis in the post-prandial phase. The continuing losses during food intake are covered by the food. Reproduced, by permission, from Millward (personal communication).

10.2.2 Requirements for essential amino acids

The principles by which the estimates of requirements have been adjusted for age-groups other than adults and for physiological states such as pregnancy are set out in the 1985 UN report (FAO/WHO/UNU, 1985). For infants and young children the so-called factorial approach has been used, in which the requirement for growth, calculated from the protein content of the body, is added to the requirement for maintenance, determined by N balance. An excellent discussion is that of Dewey *et al.* (1996).

Young and his colleagues at the Massachusetts Institute of Technology approached the problem by exploring the requirements for individual essential amino acids, which need to be known for assessing the biological value of the protein in a diet. Previous attempts to do this had used nitrogen balance as a marker of adequacy and had succeeded in identifying certain amino acids, such as lysine and methionine, as likely to be limiting in many natural diets. As soon as [^{13}C]-labelled amino acids became available Young realized that the way was open for measuring the requirements for individual amino acids by carbon balance instead of nitrogen balance. There followed a remarkable series of studies, still continuing, over a period of some 15 years. Initially the amino acid used was [1-^{13}C]-L leucine; the protocol was to feed a crystalline amino acid mixture corresponding to a fully adequate protein intake of 0.8–1.0 g kg^{-1} day^{-1}, with only the leucine content being altered. This diet was fed with varying levels of leucine, each after an adaptation period of about 7 days. The protocol was progressively refined; [^{13}C]-leucine was infused intravenously, initially in the fed state only (Motil *et al.*, 1981a,b), then for 3 h fasting, 5 h feeding (e.g. Cortiella *et al.*, 1988; Marchini *et al.*, 1993), and finally in heroic experiments in which the infusion was continued for 24 h, with 12 h fasted and 12 h fed (El-Khoury *et al.*, 1994a,b). These studies set the leucine requirement at about 40 mg kg^{-1} day^{-1}. Almost identical results were obtained by Kurpad *et al.* in both healthy and undernourished Indian men (Kurpad *et al.*, 2001, 2003).

Young proposed that the pattern of amino acids making up the obligatory loss and which therefore had to be replaced, must be the same as the pattern of amino acids in body proteins (Young *et al.*, 1989; Young and El-Khoury, 1995). In our view, and that of Millward, this hypothesis is not well grounded, since the obligatory loss must represent the summed outcome of the extent to which each amino acid is either conserved or irreversibly lost through various pathways, summarized in Table 10.5, which have nothing to do with their concentration in body protein. On Young's hypothesis once the requirements for one essential amino acid, i.e. leucine, was known the requirements for the others could be deduced from the composition of body protein. On this basis Young proposed the MIT pattern of essential amino acid requirements (Table 10.5). This pattern was compared with the original pattern based on nitrogen balances (FAO/WHO/UNU, 1985) and with the pattern in whole egg in the experiment of Marchini *et al.* (1993) alluded to above.

The estimates of essential amino acid require-

Table 10.5. Metabolites other than proteins, derived from essential and semi-essential amino acids.

Amino acid	Metabolite
Histidine	Histamine, methyl histidine
Lysine	Carnitine
Methionine	Polyamines, CH_3-compounds
Cysteine	Taurine
Phenylalanine	Catecholamines
Tyrosine	Catecholamines, thyroid hormones
Tryptophan	Serotonin, nicotinic acid
Glycine	Creatine, purines
Arginine	Creatine

There seems to be no product from leucine and the other BCAAs except oxidation. Modified from Young and Marchini, 1990.

ments in the MIT pattern were 2–3 times those in the original FAO pattern. This discrepancy represents a serious public health problem. A detailed examination of the carbon-balance method was therefore made in an attempt to identify any major source of error, but none was found (Waterlow, 1996). A possibility is that oxidation of leucine when it is at a low level in a complete amino acid mixture may be greater than when the total protein intake is low. Figure 9.4 suggests that this is not the case. Raguso *et al.* (1999) found that leucine oxidation on a protein-free diet was 25 mg kg^{-1} day^{-1}, compared with 7 mg kg^{-1} day^{-1} on a leucine-free but otherwise adequate diet. This is just the opposite of a difference that would explain the discrepancy. There are not many data to take this comparison further, but Fig. 9.4 shows that the relation of oxidation to leucine intake in the fed state is much the same, whether the intake is altered by varying amounts of protein or varying amounts of leucine in a complete amino acid mixture. Raguso's finding remains unconfirmed and unexplained. It is to be noted also that all these studies were done with tracer leucine infused intravenously and calculations based on KIC enrichment. However, as set out in Chapter 5, plasma KIC enrichments are very much the same, whether the tracer is given orally or intravenously.

The discrepancy between the MIT and FAO estimates of requirements is particularly serious for lysine, which is the limiting amino acid in cereal-based diets in developing countries. For lysine a metabolite that might play the same role as KIC for leucine, as a proxy for the precursor, is 2-amino-adipic acid (Arends and Bier, 1991). In one study [^{13}C]-lysine was infused for 24 h and fluxes and oxidation rates compared when based on the enrichment of lysine and of 2-amino-adipic acid (El-Khoury *et al.*, 1998). For reasons set out in detail in that paper, it was concluded that estimates of lysine oxidation should be based on plasma ^{13}C-lysine enrichment with 24 h oral infusion of labelled lysine.

A difficulty with lysine, and perhaps also with threonine, is the large size and flexibility of the free pool, so that movement of lysine into the free pool during the feeding phase could save a substantial proportion of it from being oxidized; this conservation of lysine would reduce the requirement for it. If no such expansion of the pool is possible, as for example with leucine, which has a very small free pool, it would mean that the relative requirements for the two amino acids would not be the same as their proportions in body protein. This point is well discussed by El-Khoury *et al.* (1998). For amino acids other than the BCAs the precursor problem remains a difficulty.

In 1993 an alternative method, known as the indicator amino acid oxidation method (IAAO), was introduced by Pencharz and his group in Toronto, based on earlier studies on rats and pigs (Ball and Bayley, 1984; Harper and Benjamin, 1984). This approach avoids the precursor problem; the essence of it is that the test amino acid, unlabelled, is fed at various levels while an 'indicator' amino acid, which is labelled, is fed at a constant level above the requirement. In theory, at low levels of the test amino acid synthesis will be limited so that all other amino acids, including the indicator, are in relative excess and have to be oxidized. Each addition of the limiting test amino acid will increase the uptake of dietary amino acids for protein synthesis and reduce their oxidation, including that of the indicator amino acid. Once the intake of the test amino acid reaches the requirement there will be no further increase in synthesis and oxidation of the indicator will become constant. Thus, when oxidation of the indicator is plotted against test amino acid level, the result is two lines, meeting at a 'breakpoint', which represents the requirement of the test amino acid. The results for lysine obtained by Zello *et al.* (1993) by this method are shown in Fig. 10.9. In a later study by the same group, with the indicator phenylalanine given by both oral and IV routes, the breakpoints gave the same estimate of lysine requirement (~38 mg kg^{-1} day^{-1}) but were less clear cut (Kriengsinyos *et al.*, 2002).

A modification of the IAAO method was to use leucine as indicator, with 2 h infusions at various levels of lysine, so that the lysine requirement could be determined not only by the breakpoint but also by the lysine intake at which leucine balance was achieved. This is known as the 'indicator amino acid balance' method (IAAB).

Three studies by this method in healthy Indian men gave remarkably consistent results – a lysine requirement of 28–31 mg kg^{-1} day^{-1} (Kurpad *et al.*, 1998, 2001b, 2002), two or more times greater than the traditional estimate of 12 mg kg^{-1} day^{-1} obtained by nitrogen balance and

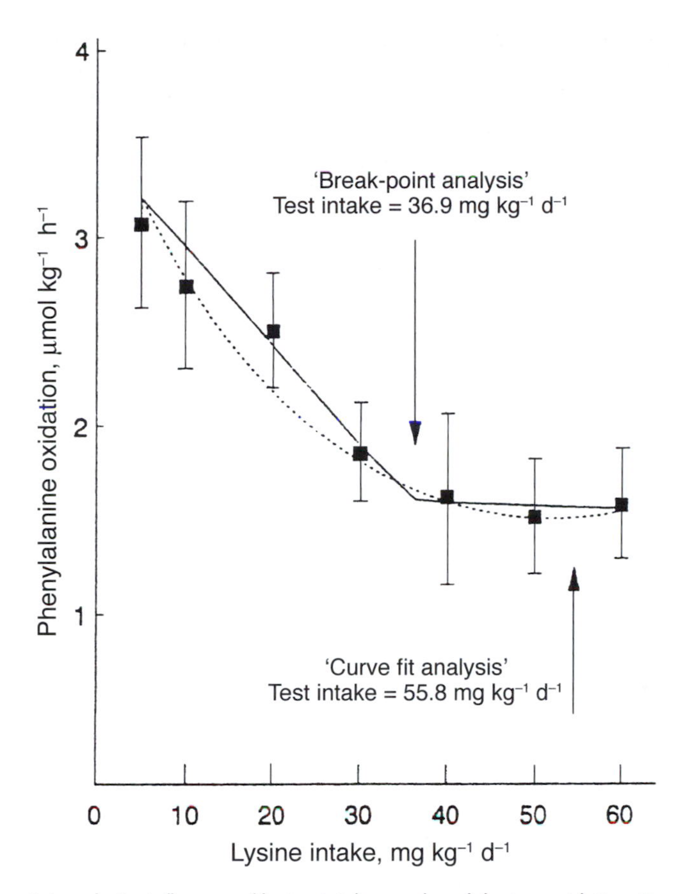

Fig. 10.9. Break-point analysis: influence of lysine intake on phenylalanine oxidation. Reproduced from Zello *et al.* (1993), by courtesy of the *American Journal of Physiology.*

50% higher than Millward's estimate of 18–20 mg kg^{-1} day^{-1} based on the utilization of milk and wheat protein (Millward *et al.*, 2002). It was also shown that extending the period of adaptation from 7 to 21 days made no difference to the results (Kurpad *et al.*, 2002), in agreement with earlier findings of the MIT group (Marchini *et al.*, 1993). A fourth study, this time in undernourished Indians, showed a significantly higher requirement – 44 mg kg^{-1} day^{-1} – than in their well-nourished compatriots (Kurpad *et al.*, 2003). Perhaps this could be explained by their having a relatively smaller ratio of muscle mass to visceral mass (Soares *et al.*, 1991).

A further point, which seems not to have excited comment, is that examination of the flux data in these studies shows that in general synthesis *decreases* as the lysine intake increases, which is contrary to the theory of the indicator method; the better balance seems to be achieved by a fall in breakdown, which allows a decrease in oxidation.

Both the direct carbon balance method and the indicator method have been used to determine the requirements for other indispensable amino acids and the extent to which, for the aromatic amino acids, tyrosine can spare phenylalanine and for the sulphur amino acids cystine can spare methionine. The large amount of work on these two subjects by the MIT and Toronto groups has been discussed by Young (2001) and by Bos *et al.* (2002); the latter paper gives a valuable discussion of the pros and cons of the different methods.

Jackson reminds us that the so-called non-essential amino acids must not be forgotten in considering estimates of protein requirements. In a recent paper subjects were changed from a habitual protein intake of 1.1 g kg^{-1} day^{-1} to the safe level, according to FAO/WHO/UNU (1985), of 0.75 g kg^{-1} day^{-1}. This involved a 40% reduction in their glycine intake. Uptake of glycine into protein was well maintained, but there was an increase in *de novo* glycine synthesis, perhaps made possible by greater salvage of urea-N (Gibson *et al.*, 2002) (see Chapter 8).

10.2.3 Microbial amino acid synthesis

There is one other aspect of amino acid requirements, touched on earlier (Chapter 8), that has involved turnover studies – the possibility of microbial synthesis of essential amino acids in the gut. Some 50 years ago Australian nutritionists observed that the highlanders of Papua New Guinea were able to maintain activity and health on a diet which did not by any standard fulfil the requirement for protein. It was suggested that perhaps they were able to obtain essential amino acids from synthesis by microbes in the large intestine. Tanaka *et al.* (1980) found in these people some labelling of lysine-N in plasma proteins after feeding [^{15}N]-urea, but the level was very low. Later work confirmed and extended these results (Table 10.6). Since lysine and threonine are not transaminated, the appearance of label in them is good evidence of *de novo* synthesis. Most authors have given [^{15}N]-urea or ammonium chloride; Jackson *et al.* (1999) used lactose [^{15}N]-ureide, a compound that is not absorbed in the small intestine but is split in the colon.

The important question is: how much is synthesized and absorbed, and does it make a significant contribution to the requirement? As a first approximation the amount synthesized could be calculated as: (enrichment of amino acid in plasma × amino acid flux)/enrichment of amino acid in microbial protein at the site of synthesis and absorption. It is the denominator that is the problem. Most of the results in the table were calculated on the assumption that the precursor enrichment is that of the amino acid of the microbial protein in the faeces, but the lysine enrichment was shown to be 5 × higher in the faeces than in the ileal digesta of patients with ileostomies (Metges *et al.*, 1999a; Metges, 2000). If the ileal figures are used they lead to absurdly high estimates of the amount of lysine synthesized. It has been suggested that perhaps the synthesis of lysine takes place specifically in microbes adhering to the mucosa of the lower ileum, which would also be the site of absorption; hence the lysine synthesized and absorbed might have a much higher enrichment than that of the microbial flora as a whole.

10.2.4 The practical problem

What is the outcome of all the work on protein requirements? A practical approach has been developed by Millward and Jackson (2003). Their starting point is that since food is eaten to fulfil the requirement for energy, what matters is the protein:energy ratio (P/E) of the diet. Beaton, in a contribution to the 1985 FAO/WHO/UNU report on protein and energy requirements, had pointed out the statistical difficulties of arriving at a 'safe' P/E ratio,[1] which seems to have discouraged further use of that ratio. However, Whitehead *et al.* (1977) had shown in studies in Africa a close relation between the P/E of the diets of children and the development, or not, of severe clinical malnutrition. Millward and Jackson (2003) suggest, based on current estimates of protein requirements, that the diet may supply inadequate amounts of protein in some groups, such as the elderly, whose intake of energy is low, even though the P/E is satisfactory. The same applies to populations in less developed countries whose diet, based on cereals or starchy roots, has a low P/E ratio. Admittedly we do not know how far functional failures in such populations, such as low birth weight, reduced working capacity, perhaps reduced immune responses, are caused by deficiency of protein or indeed of any nutrient, rather than by infections or other concomitants of poverty – nor would anyone wish to repeat the mistake of the 1970s, described by Maclaren (1975) as 'The Protein Fiasco', when immense resources were devoted to the development of high protein supplements. Supplements provided from outside are not a good way of countering life-long deficiencies in whole populations. There is perhaps greater promise in genetic modification to produce cereals and tubers with higher yield and quality of protein, as has been done for maize and is now being developed for starchy roots and tubers; a

[1] It is assumed that the P/E ratio is corrected both for biological value and digestibility of the protein.

Table 10.6. Studies on the contribution of microbial synthesis in the gut to the production of essential amino acids.

Authors:	Torrallardona *et al.* (1996)	Torrallardona *et al.* (1994)	Metges *et al.* (1999a)	Metges *et al.* (1999b)	Jackson *et al.* (1999)
Species	Rat	Pig	Human	Human	Human
Source of ^{15}N	NH_4Cl	NH_4Cl	(a) NH_4Cl (b) Urea	(a) NH_4Cl (b) Urea	Lactose-ureide
Amino acid measured	Lysine	Lysine	Threonine	Lysine	Lysine
Site of measurement	Whole body protein	Whole body protein	Plasma free AA	Plasma free AA	Urine free AA
Microbial precursor	Faeces	Ileum	Faeces	Faeces	Faeces
Estimated amount of amino acid synthesized mg kg^{-1} day^{-1}	62	43	(a) 24 (b) 12	(a) 44 (b) 29	32

Full references are given at the end of the chapter.

valuable review of the current state is that of Christou and Twyman (2004). The contribution of the work on nitrogen and carbon balance has been to clarify the specific nutritional goals to which these new studies should be directed.

10.3 References

Addis, T., Poo, L.J. and Lew, W. (1936) Protein loss from liver during 2-day fast. *Journal of Biological Chemistry* 115, 117–118.

Arends, J. and Bier, D.M. (1991) Labelled amino acid infusion studies of *in vivo* protein synthesis with stable isotope tracers and gas chromatography – mass spectrometry. *Analytica Chimica Acta* 247, 255–263.

Ball, R.O. and Bayley, H.S. (1984) Tryptophan requirement of the 2.5 kg piglet determined by the oxidation of an indicator amino acid. *Journal of Nutrition* 114, 1741–1746.

Barac-Nieto, M., Spurr, G.B., Maksud, M.G. and Lotero, H. (1978) Aerobic work capacity in chronically undernourished adult males. *Journal of Applied Physiology* 44, 209–215.

Barcroft, J. (1934) *Features in the architecture of physiological function.* Cambridge University Press.

Bos, C., Gaudichon, C. and Tome, D. (2002) Isotopic studies of protein and amino acid requirements. *Current Opinion in Clinical Nutrition and Metabolic Care* 5, 55–61.

Calloway, D.H. (1975) Nitrogen balance of men with marginal intakes of protein and energy. *Journal of Nutrition* 105, 914–923.

Chan, H. (1968) Adaptation of urinary nitrogen excretion in infants to changes in dietary protein intakes. *British Journal of Nutrition* 22, 315–323.

Christou, P. and Twyman, R.M. (2004) The potential of genetically enhanced plants to address food insecurity. *Nutrition Research Reviews* 17, 23–42.

Cortiella, J., Matthews, D.E., Hoerr, R.A., Bier, D.M. and Young, V.R. (1988) Leucine kinetics at graded intakes in young men: quantitative fate of dietary leucine. *American Journal of Clinical Nutrition* 48, 998–1009.

Das, T.K. and Waterlow, J.C. (1974) The rate of adaptation of urea cycle enzymes, aminotransferases and glutamic dehydrogenase to changes in dietary protein intake. *British Journal of Nutrition* 32, 353–373.

Dewey, K.G., Beaton, G., Fjeld, C., Lonnerdal, B. and Reeds, P. (1996) Protein requirements of infants and children. *European Journal of Clinical Nutrition* 50, S119–150.

El-Khoury, A.E., Fukegawa, N.K., Sanchez, M., Tsay, R.H., Gleason, R.E., Chapman, A.E. and Young, V.R. (1994a) Validation of the tracer-balance concept with reference to leucine: 24-h intravenous tracer studies with L-[1-^{13}C] leucine and [^{15}N^{15}N] urea. *American Journal of Clinical Nutrition* 59, 1000–1011.

El-Khoury, A.E., Fukegawa, N.K., Sanchez, M., Tsay, R.H., Gleason, R.E., Chapman, R.E. and Young, V.R. (1994b) The 24-h pattern and rate of leucine oxidation, with particular reference to tracer estimates of leucine requirements in healthy adults. *American Journal of Clinical Nutrition* 59, 1012–1020.

El-Khoury, A.E., Basile, A., Beaumier, L., Wang, S.Y., Al-Amiri, H.A., Selvaraj, A., Wong, S., Atkinson, A., Ajami, A.M. and Young, V.R. (1998) Twenty-four hour intravenous and oral tracer studies with L-2-aminoadipic [1-^{13}C] acid and L-lysine [1-^{13}C] as tracers at generous nitrogen and lysine intakes in healthy adults. *American Journal of Clinical Nutrition* 68, 827–839.

FAO/WHO/UNU (1985) Energy and protein requirements: report of a joint expert consultation. *WHO Technical Report Series* no. 724, WHO, Geneva.

Gibson, N.R., Jahoor, F., Ware, L. and Jackson, A.A. (2002) Endogenous glycine and tyrosine production is maintained in adults consuming a marginal protein diet. *American Journal of Clinical Nutrition* 75, 511–518.

Gould, S.J. and Lewontin, R.C. (1979) The spandrels of San Marco and the Panglossian paradigm: a critique of the adaptionist programme. In: Maynard Smith, J. and Holliday, R. (eds) *The Evolution of Adaptation by Natural Selection.* The Royal Society, London, pp. 147–164.

Grantham-McGregor, S. (1992) The effect of malnutrition on mental development. In: Waterlow, J.C. *Protein-energy Malnutrition.* (ed.) Edward Arnold, London, pp. 344–360.

Harper, A.E. and Benjamin, E. (1984) Relationship between intake and rate of oxidation of leucine and α-ketoisocaproate *in vivo* in the rat. *Journal of Nutrition* 114, 431–440.

Henry, C.J.K. and Collingwood, A. (1998) Allometry of fasting urinary nitrogen loss and basal metabolic rate in homeotherms. *Nutrition Research* 18, 49–60.

Henry, K.M., Kosterlitz, H.W. and Quenouille, M.H. (1953) A method for determining the nutritive value of a protein by its effect on liver protein. *British Journal of Nutrition* 7, 51–67.

Jackson, A.A., Bundy, R., Hounslow, A., Murphy, J.L. and Wootton, J.A. (1999) Metabolism of lactose-[^{13}C] ureide and lactose [^{15}N,^{15}N] ureide in normal adults consuming a diet marginally adequate in protein. *Clinical Science* 97, 547–555.

James, W.P.T., Ferro-Luzzi, A. and Waterlow, J.C. (1988) Definition of chronic energy deficiency in adults. *European Journal of Clinical Nutrition* 42, 469–481.

Kenney, F.T. (1970) Hormonal regulation of synthesis of liver enzymes. In: Munro, H.N. (ed.) *Mammalian Protein Metabolism*, vol. IV. Academic Press, New York, pp. 131–177.

Keys, A., Brozek, J., Henschel, A., Mickelson, O. and Taylor, H.L. (1950) *The Biology of Human Starvation.* University of Minnesota Press, Minneapolis, Minnesota.

Koch, A.L. (1962) The evaluation of the rates of biological processes from tracer kinetic data. 1. The influence of labile metabolic pools. *Journal of Theoretical Biology* 3, 283–303.

Kriengsinyos, W., Wykes, L.J., Ball, R.O. and Pencharz, P.B. (2002) Oral and intravenous tracer protocols of the indicator amino acid oxidation method provide the same estimate of the lysine requirement in healthy men. *Journal of Nutrition* 132, 2251–2257.

Kurpad, A.V., El-Khoury, A.E., Beaumer, S., Srivatsa, A., Kuriyan, R., Raj, T., Borgonha, S., Ajami, A.A. and Young, V.R. (1998) An initial assessment, using 24-h [^{13}C] leucine kinetics, of the lysine requirement of healthy adult Indian subjects. *American Journal of Clinical Nutrition* 67, 58–66.

Kurpad, A.V., Raj, T., El-Khoury, A.E., Kuriyan, R., Maruthi, K., Borgonha, S., Chandukudlu, D., Regan, M.M. and Young, V.R. (2001a) Daily requirement for and splanchnic uptake of leucine in healthy adult Indians. *American Journal of Clinical Nutrition* 74, 747–755.

Kurpad, A.V., Raj, T., El-Khoury, A.E., Beaumier, L., Kuriyan, R., Srivatsa, A., Borgonha, S., Selvaraj, A., Regan, M.M. and Young, V.R. (2001b) Lysine requirements of healthy adult Indian subjects, measured by an indicator amino acid balance technique. *American Journal of Clinical Nutrition* 73, 900–907.

Kurpad, A.V., Regan, M.M., Raj, T., El-Khoury, A.E., Kuriyan, R., Vaz, M., Chandukudlu, D., Venkataswamy, V.G., Borgonha, S. and Young, V.R. (2002) Lysine requirements of healthy adult Indian subjects receiving long-term feeding, measured with a 24-h indicator amino acid oxidation and balance technique. *American Journal of Clinical Nutrition* 76, 404–412.

Kurpad, A.V., Regan, M.M., Raj, T., Vasudevan, J., Kuriyan, R., Gnanou, J. and Young, V.R. (2003) Lysine requirements of chronically undernourished Indian men, measured by a 24-h indicator amino acid oxidation and balance technique. *American Journal of Clinical Nutrition* 77, 101–108.

Marchini, J.S., Cortiella, J., Hiramatsu, T., Chapman, T.E. and Young, V.R. (1993) Requirements for indispensable amino acids in adult humans; longer-term amino acid kinetic study with support for the adequacy of the Massachusetts Institute of Technology amino acid requirement pattern. *American Journal of Clinical Nutrition* 58, 670–683.

Martin, C.J. (1930) Thermal adjustment of man and animals to external conditions. *Lancet* 2, 617–620.

Martin, C.J. and Robison, R. (1922) The minimum nitrogen expenditure of man and the biological value of various proteins for human nutrition. *Biochemical Journal* 16, 407–422.

McLaren, D.S. (1975) The protein fiasco. *Lancet* 2, 93–96.

Meguid, M.M., Matthews, D.E., Bier, D.M., Meredith, C.N., Soeldner, J.S. and Young, V.R. (1986a) Leucine kinetics at graded leucine intakes in young men. *American Journal of Clinical Nutrition* 43, 770–780.

Meguid, M.M., Matthews, D.E., Bier, D.M., Meredith, C.N., Soeldner, J.S. and Young, V.R. (1986b) Valine kinetics at graded valine intakes in young men. *American Journal of Clinical Nutrition* 43, 781–786.

Meredith, C.N., Wen, Z.-M., Bier, D.M., Matthews, D.E. and Young, V.R. (1986) Lysine kinetics at graded lysine intakes in young men. *American Journal of Clinical Nutrition* 43, 787–794.

Metges, C.C. (2000) Contribution of microbial amino acids to amino acid homeostasis of the host. *Journal of Nutrition* 130, 1857–1864S.

Metges, C.C., Petzke, K.J., Henneman, L., Grant, I., Bedri, S., Regan, M.M., Fuller, M.F. and Young, V.R. (1999a) Incorporation of urea and ammonia nitrogen into ileal and fecal microbial proteins and plasma free amino acids in normal men and ileostomates. *American Journal of Clinical Nutrition* 70, 1046–1058.

Metges, C.C., El-Khoury, A.E., Henneman, L., Petzke, K., Grant, I., Bedri, S., Pereira, P.P., Ajami, A.M., Fuller, M.F. and Young, V.R. (1999b) Availability of intestinal microbial lysine for whole body homeostasis in human subjects. *American Journal of Physiology* 277, E597–607.

Millward, D.J. (2003) An adaptive metabolic demand model for protein and amino acid requirements. *British Journal of Nutrition* 90, 249–260.

Millward, D.J. and Jackson, A.A. (2003) Protein/energy ratios of current diets in developed and developing countries compared with a safe protein/energy ratio: implications for recommended protein and amino acid intakes. *Public Health Nutrition* 7, 387–405.

Millward, D.J. and Pacy, P.J. (1995) Postprandial protein utilization and protein quality assessment in man. *Clinical Science* 88, 597–606.

Millward, D.J., Fereday, A., Gibson, N.R., Cox, M.C. and Pacy, P.J. (2002) Efficiency of utilization of wheat and milk protein in healthy adults and apparent lysine requirements determined by a single meal

[1-^{13}C] leucine balance protocol. *American Journal of Clinical Nutrition* 76, 1326–1334.

Motil, K.J., Matthews, D.E., Bier, D.M., Burke, J.F., Munro, H.N. and Young, V.R. (1981a) Whole body leucine and lysine metabolism: response to dietary protein intake in young men. *American Journal of Physiology* 240, E712–721.

Motil, K.J., Bier, D.M., Matthews, D.E., Burke, J.F. and Young, V.R. (1981b) Whole body leucine and lysine metabolism studied with [1-^{13}C] leucine and [α-^{15}N] lysine: response in healthy young men given excess energy intake. *Metabolism* 30, 783–791.

Neale, R.J. and Waterlow, J.C. (1974) Critical evaluation of a method for estimating amino acid requirements for maintenance in the rat by measurement of the rate of ^{14}C-labelled amino acid oxidation *in vivo*. *British Journal of Nutrition* 32, 257–272.

Neale, R.J. and Waterlow, J.C. (1977) Endogenous loss of leucine and methionine in adult male rats. *British Journal of Nutrition* 37, 259–268.

Nicol, B.M. and Phillips, P.G. (1976) The utilization of dietary protein by Nigerian men. *British Journal of Nutrition* 36, 337–351.

Pacy, P.J., Price, G.M., Halliday, D., Quevedo, M.R. and Millward, D.J. (1994) Nitrogen homeostasis in man: the diurnal responses of protein synthesis and degradation and amino acid oxidation to diets with increasing protein intakes. *Clinical Science* 86, 103–118.

Price, G.M., Halliday, D., Pacy, P.J., Quevedo, M.R. and Millward, D.J. (1994) Nitrogen homoeostasis in man: influence of protein intake on the amplitude of diurnal cycling of body nitrogen. *Clinical Science* 86, 91–102.

Quevedo, M.R., Price, G.M., Halliday, D., Pacy, P.J. and Millward, D.J. (1994). Nitrogen homoeostasis in man: diurnal changes in nitrogen excretion, leucine oxidation and whole body leucine kinetics during a reduction from a high to a moderate protein intake. *Clinical Science* 86, 185–193,

Raguso, C.A., Pereira, P. and Young, V.R. (1999) A tracer investigation of oxidative amino acid losses in healthy young adults. *American Journal of Clinical Nutrition* 70, 474–483.

Rand, W.M., Pellett, P.L. and Young, V.R. (2003) Meta-analysis of nitrogen balance studies for estimating protein requirements in healthy adults. *American Journal of Clinical Nutrition* 77, 109–127.

Roberts, D.F. (1953) Basal metabolism, race and climates. *American Journal of Physical Anthropology* 11, 533–556.

Satyanarayana, K., Naidu, A.N., Chatterjee, B. and Rao, N.S. (1977) Body size and work output. *American Journal of Clinical Nutrition* 30, 322–325.

Satyanarayana, K., Naidu, A.N. and Rao, B.S.N. (1980) Agricultural employment, wage earnings and nutritional status of teenage rural Hyderabad boys. *Indian Journal of Nutrition and Dietetics* 17, 281–285.

Schimke, R.T. (1962) Adaptive characteristics of urea cycle enzymes in the rat. *Journal of Biological Chemistry* 237, 459–468.

Schimke, R.T. (1964) Importance of both synthesis and degradation in the control of arginase levels in rat liver. *Journal of Biological Chemistry* 239, 3808–3817.

Schimke, R.T. (1970) Regulation of protein degradation in mammalian tissue. In: Munro, H.N. (ed.) *Mammalian Protein Metabolism*. Academic Press, New York, pp. Vol. IV 178–228.

Soares, M.J., Piers, L.S., Shetty, P.S., Robinson, S., Jackson, A.A. and Waterlow, J.C. (1991) Basal metabolic rate, body composition and whole-body protein turnover in Indian men with differing nutritional status. *Clinical Science* 81, 419–425.

Soares, M.J., Piers, L.S., Shetty, P.S., Jackson, A.A. and Waterlow, J.C. (1994) Whole body protein turnover in chronically undernourished individuals. *Clinical Science* 86, 441–446.

Spurr, G.B. (1983) Nutritional status and physical work capacity. *Yearbook of Physical Anthropology* 26, 1–35.

Stephen, J.M.L. and Waterlow, J.C. (1968) Effect of malnutrition on two enzymes concerned with amino acid metabolism in human liver. *Lancet* i, 118–119.

Tanaka, N., Kubo, K., Shiraki, K., Koishi, H. and Yoshimura, H. (1980) A pilot study of protein metabolism in New Guinea highlanders. *Journal of Nutritional Science and Vitaminology* 26, 247–259.

Torrallardona, D.C., Harris, C.L., Milne, E. and Fuller, M.F. (1994) The contribution of intestinal microflora to amino acid requirements in pigs. In: Souffrant, W.-B. and Hagemeister, H. (eds) *Proceedings of 16th International Symposium of Digestive Physiology in Pigs*, pp. 245–248.

Torrallardona, D.C., Harris, I.C., Coates, M.E. and Fuller, M.F. (1996) Microbial amino acid synthesis and utilization in rats: incorporation of ^{15}N from $^{15}NH_4$ into lysine in the tissues of germ-free and conventional rats. *British Journal of Nutrition* 76, 689–700.

Waterlow, J.C. (1985) What do we mean by adaptation? In: Blaxter, K. and Waterlow, J.C. (eds) *Nutritional Adaptation in Man*. John Libbey, London, pp. 1–11.

Waterlow, J.C. (1996) The requirement of adult man for indispensable amino acids. *European Journal of Clinical Nutrition* 50, Suppl. 1, S151–179.

Waterlow, J.C. and Stephen, J.M.L. (1966) Adaptation of the rat to be a low protein diet: the effect of a reduced protein intake on the pattern of incorporation of L-[^{14}C] lysine. *British Journal of Nutrition* 20, 461–484.

Waterlow, J.C., Garlick, P.J. and Millward, D.J. (1978) *Protein Turnover in Mammalian Tissues and in the Whole Body*. North Holland/Elsevier, Amsterdam.

Whitehead, R.G., Coward, W.A., Lunn, P.G. and Rutishauser, I. (1977) A comparison of the pathogenesis of protein-energy malnutrition in Uganda and The Gambia. *Transactions of the Royal Society of Tropical Medicine and Hygiene* 71, 189–195.

Widdowson, E.M. (1962) Nutritional individuality. *Proceedings of the Nutrition Society* 21, 121–128.

Young, V.R. (1987) 1987 McCollum Award Lecture: kinetics of human amino acid metabolism: nutritional implications and some lessons. *American Journal of Clinical Nutrition* 46, 709–725.

Young, V.R. (1994) Adult amino acid requirements: the case for a major revision of current recommendations. *Journal of Nutrition* 124, 1517S–1523S.

Young, V.R. (2001) Got some amino acids to spare? *American Journal of Clinical Nutrition* 74, 709–711.

Young, V.R. and El-Khoury, A.E. (1995) Can amino acid requirements for nutritional maintenance in adult humans be approximated from the amino acid composition of body mixed proteins? *Proceedings of the National Academy of Science, USA* 92, 300–304.

Young, V.R. and Marchini, J.S. (1990) Mechanisms and nutritional significance of metabolic responses to altered intakes of protein and amino acids, with reference to nutritional adaptation in humans. *American Journal of Clinical Nutrition* 51, 270–289.

Young, V.R. and Scrimshaw, N.S. (1979) Genetic and biological variability in human nutrient requirements. *American Journal of Clinical Nutrition* 32, 486–500.

Young, V.R., Gucalp, C., Rand, W.M., Matthews, D.E. and Bier, D.M. (1987) Leucine kinetics during three weeks at submaintenance-to-maintenance intakes of leucine in men: adaptation and accommodation. *Human Nutrition: Clinical Nutrition* 41C, 1–18.

Young, V.R., Bier, D.M. and Pellett, P.L. (1989) A theoretical basis for increasing current estimates of the amino acid requirements in adult man, with experimental support. *American Journal of Clinical Nutrition* 50, 80–92.

Zello, G.A., Pencharz, P.B. and Ball, R.O. (1993) Dietary lysine requirement of young adult males determined by oxidation of L-[1-^{13}C] phenylalanine. *American Journal of Physiology* 264, E677–685.

Zhao, X., Wen, Z.-M., Meredith, C.N., Matthews, D.E., Bier, D.M. and Young, V.R. (1986) Threonine kinetics at graded threonine intakes in young men. *American Journal of Clinical Nutrition* 43, 795–802.

11

Physiological Determinants of Protein Turnover

11.1 Body Size – the Contribution of Allometry

Allometry has for long been a subject of interest to biologists and physiologists because it enables comparisons between different animals. The reason for introducing a note on it here is that it might contribute to a more general view on the physiology of protein turnover. For example, the proteins of a rat turn over five or six times faster than those of a human (Chapter 15, section 15.2); we cannot explain why this should be so, but the fact is worth looking at in a wider context. Much of the discussion centres on oxygen turnover rather than protein turnover, but it is relevant because the two seem to be closely linked. Heusner (1987), in a review written 15 years ago, said that 'to date about 750 power functions or allometric equations have been reported', and his references go back nearly 150 years.

The essential feature of allometry is the description of some feature as a power function of body mass, of the form:

$F = aM^b$

or: $\log F = \log a + b \log M$ where F is some body characteristic such as metabolic rate, M is body mass and a and b are constants; a is a coefficient determined by the units in which F is expressed and b is an exponent or slope.

For mass-specific rates, e.g. metabolic rate per kg, the equation is:

$$F/M = \frac{aM^b}{M} = aM^{(b-1)}.$$

The slopes are obtained by linear regression of log F on log M, and it is not always realized that although the correlations may be excellent, the slope has a standard deviation and confidence intervals which may be quite large (Prothero, 1982), so that there may not be any real difference between slopes that differ by less than, say, 0.1 log unit. Good discussions of these statistical matters are given by Smith (1984), Peters (1988) and Diamond (1993). Table 11.1 lists a few of the properties and functions that have been studied, with their allometric scaling factors.

11.1.1 Interspecies comparisons

The relationship of metabolic rate (MR) to body mass (M), developed independently by Brody and Kleiber in 1932 for adult animals ranging from mouse to elephant gave an exponent of 0.73 (Brody and Procter, 1932) and 0.75 (Kleiber, 1932), and $M^{0.75}$ has come to be known as the 'metabolic mass' (Blaxter, 1989: 125). The exponent per unit body mass would thus be –0.25. The original analyses of MR in mammals were extended by Hemmingsen (1960) to cover poikilotherms and single-celled organisms. He showed that the exponents b were very similar in the three groups, and the relations differed only in their elevation, a. Of particular interest is the work of Else and Hulbert (1981), who compared various aspects of energy production in a 'mammal machine' and a 'reptile machine'. The two 'machines' were a mouse and a lizard, with equal body weights and body temperatures but a mouse has a standard MR per kg about four times that of a lizard. This finding illustrates an essential difference, not related to environment, between homoeotherms and poikilotherms. Sadly, their protein turnover rates were not measured.

Table 11.1. Some allometric relationships of energy and protein metabolism to body weight in mature animals of different species.

Function	Unit	Constant (a)	Exponent (b)	Reference
Resting metabolism	kJ d^{-1}	293	0.75	Kleiber (1975)
Protein synthesis				
whole body	g d^{-1}	15.8	0.72	Waterlow (1984)
Obligatory urinary				Henry and Collingwood
nitrogen loss	mg d^{-1}	272	0.75	(1998)
Liver weight	g	0.09	0.87	Prothero (1982)
Muscle weight	g	0.42	1.01	Else and Hulbert (1985)
Albumin breakdown	mg d^{-1}	5.83	0.68	Wetterfors (1985)
Total liver RNA	mg liver	?	0.755	Munro and Downie (1964)
rRNA turnover	nmol d^{-1}	~15	0.69	Schöch and Topp (1994)
tRNA turnover	nmol d^{-1}	~2000	0.78	Schöch and Topp (1994)
mRNA turnover	μmol d^{-1}	~2.53	0.75	Schöch and Topp (1994)
Mitochondrial membrane				
surface area	m^2 per whole organ			
Liver		0.98	0.64	Else and Hulbert (1985)
Muscle		5.10	0.78	
Oxygen consumption	mg min^{-1} mg^{-1}			
of hepatocytes	dry weight	6.83	–0.18	Porter and Brand (1995)

From the equation; $F = aW^b$, where F is the function and M = body mass

The pattern of body composition also changes with increasing body weight, and this makes an important contribution to whole body scaling. The liver, for example, accounts for 7.6% of body mass in the shrew, weighing ~1 g, but only 0.9% in the whale, weighing ~10^8 g (Prothero, 1982). The rapidly metabolizing organs scale against body mass with slopes between 0.69 and 0.85 (Prothero, 1982; Else and Hulbert, 1985a), although Stahl (1965) gives somewhat higher values; the exponent for muscle mass is ~1.0. It is perhaps not surprising that the exponents for the whole body should lie between these limits.

Probing more deeply into the allometry of energy production has produced some fascinating results. Apart from body components, there is also a scaling of cellular metabolic rate. The oxygen consumption of isolated hepatocytes, per mg dry weight, is 9 × higher in the mouse than in the horse (Porter and Brand, 1995). Else and Hulbert (1985b) showed that in mammals, over a 100-fold range of body mass, the total mitochondrial membrane surface area, summated over all the major tissues, scaled as body mass$^{0.75}$. Taylor (1987) pointed out that in skeletal muscle the work per unit time is proportional to the speed of shortening, which gets slower with increasing body size. The key factor in energy transduction in the shortening cycle is the enzyme actomyosin ATPase, and the activity of this enzyme per unit weight of muscle was shown to scale as $M^{-0.25}$.

Thirty years ago we suggested that there might be a relationship between protein turnover and energy turnover (Waterlow, 1968). When information became available from measurements in different animals it was found that protein turnover also scales with a slope of 0.72, not significantly different from 0.75. The total RNA content of the liver, representing the capacity for protein synthesis, scales as $M^{0.75}$ (Munro and Downie, 1964) and Schöch and co-workers have found by measurement of the urinary excretion of specific methylated nucleosides, that the turnover rates of all three RNA species also scale as $M^{\sim 0.75}$ (Schöch *et al.*, 1990) (Fig. 11.1).

Many have felt that there must be a physiological principle underlying what seems to be a universal relationship. Hemmingsen (1960) wrote: 'We fail to see how it is possible to underestimate the theoretical significance of the uniform regression of metabolic rate with size common to the whole organismic world.' Kleiber (1947) said,

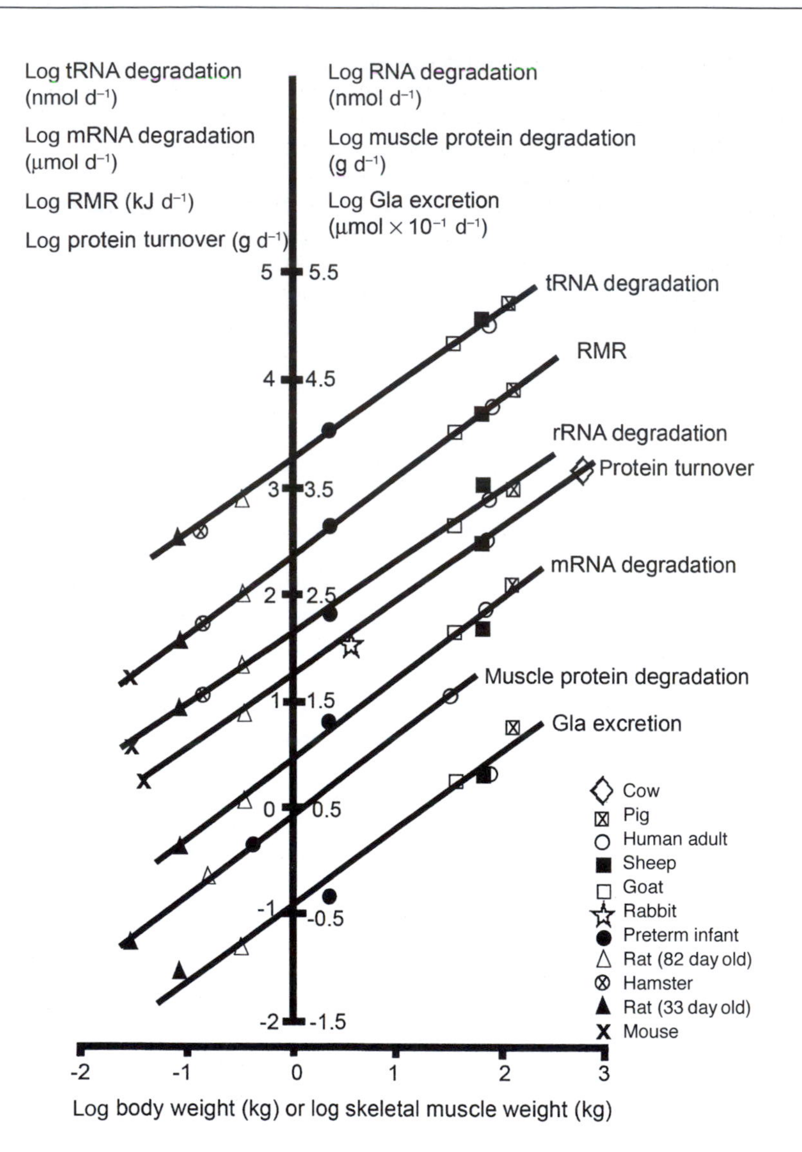

Fig. 11.1. Log-log plots of protein turnover, resting metabolism rate and whole body RNA degradation rates against body weight. Gla = γ-carboxyglutamic acid.
Reproduced from Schöch and Topp (1994), by courtesy of Raven Press (New York) and Nestec (Lausanne).

'Such equations should be a start, rather than end, to a physiologist's further reasoning'; and in 1975, 'Whether or not such a relationship has physiological significance depends on the physiologist, just as it depends on the listener whether he hears a symphony or a series of noises.'

As far as metabolic rate is concerned, two hypotheses have been proposed in the past. The first, going back to Rubner, is that it depends on surface area, which would give an exponent of 0.67, statistically different from 0.75. Kleiber (1947) gave six reasons for refuting the surface area relationship. Moreover, although surface area is obviously an important determinant of the rate of heat loss, to suggest that it drives heat production seems to be putting the cart before the horse.

Another hypothesis is that of McMahon (1973, 1975) who suggested that, at least in mammals, the design of the animal must take account of mechanical considerations, and that similarity of design might be based on elastic stress, i.e. the capacity for load-bearing without collapse. This, combined with dimensional analysis leads to the observed relationship of metabolic rate with body mass. However, the hypothesis has been criticized in detail by Heusner (1987). Moreover, similarity of elastic stress would hardly apply to aquatic mammals and fish.

Modern opinion seems to be moving against any physiological 'explanation' of allometric relationships. Some, e.g. Peters (1988), have argued that they are purely statistical. Certainly the slope and elevation of the log–log line are determined statistically, sometimes uncritically, with an unstated margin of error.

Heusner (1987) concluded: '… to date there is no biologically satisfactory theoretical explanation of the 0.75 power of mass.'

However, it is difficult not to agree with Hemmingsen's remarks quoted above; it is hard to suppose that it is simply a matter of chance that so many different functions and properties share the same or closely similar exponent. One proposed explanation of the 0.75 exponent is that it derives from the fractal geometry of the organism, since exchanges with the environment depend on the surface area of branching structures such as the capillaries and the pulmonary alveoli (West *et al.*, 1999). Recently a more powerful hypothesis has been proposed by Andersen *et al.* (2002). This is based on the concept, derived from physics, that maximum efficiency of a process, in the sense of minimal overall entropy production, is achieved when the rate of entropy production is constant over time; however, 'time' here refers not to ordinary clock time but to a physiological time scale, such as the time between heartbeats or the life-span of the animal (Schmidt-Nielsen, 1984). The allometric equation for these functions is:

$$\tau = pM^{0.25}$$

where τ is physiological time and p is a constant.

The rate of entropy production (E) is given by the metabolic rate, so

$$E = qM^{0.75}$$

when q is a constant.

Thus: the rate of entropy production per unit body mass $= E.\tau / M = \dfrac{(pM^{0.25} \times qM^{0.75})}{M} = pq$, which is a constant.

This implies that the total entropy production over a life-span per unit body mass is the same for all animals. This is a necessary condition for minimum entropy production, and shows that over the whole range covered by the Kleiber–Brody equation the animals are operating at maximal efficiency.

The idea is attractive because a large, though unknown, proportion of energy expenditure is devoted not to physical work, such as that of the diaphragm or heart, but to the negative entropy of maintaining what Larsen and Larsen (1995) have called dissipative functions, such as protein synthesis and ion gradients. It seems to cover both energy and protein turnover in a way that the other hypotheses do not. Whether one drives the other we do not know.

11.1.2. Intra-species relationships

Kleiber (1975: 211) concluded that when functions are scaled against body mass in the same species '… it is reasonable to adopt for intra-specific prediction … the metabolic unit of body size found most suitable for interspecific prediction, namely the $^3/_4$ power of body weight.' However, this may not really be true. A baby is not just a small adult. In addition to what may be called 'the size effect', body composition changes with increasing maturity, but not necessarily in the same way as it changes with increasing body mass in interspecies comparisons. Next, the young animal is growing while the adult is not, and growth has effects on both energy and protein metabolism which will be discussed in the next section. Moreover, it is possible that the tissues of immature animals have an intrinsically faster metabolism than those of mature ones, but to our knowledge this has not been tested. In view of the number of factors affecting metabolism during development, one would not expect to find it related to body weight by a single scaling factor over the whole course of life and development. The within-species scaling factors for a number of animals have been summarized by Altman and Dittmer (1968) and by Albritton (1954). In farm animals, at least in dairy cattle and pigs, the exponent decreases as the animals grow older and heavier.

In man there is plenty of information on resting metabolic rate at different ages and body masses. Karlberg (1952) made a detailed study of the older literature and found that in reports on infants from four countries, France, Germany, UK and USA, over a range of 1–10 kg the value of the scaling factor was between 0.91 and 1.11, and never as low as 0.75. A similar slope was recorded by Hill and Rahimtulla (1965). Figure 11.2 shows a log-log plot of MR against body mass in man, derived from the data compiled by Schofield (1985). Up to 15 kg in weight, corresponding to about 3 years in age, the scaling factor is ~1. Thereafter, from infancy to maturity, it decreases to ~0.5. It seems that, between birth and maturity, the scaling factor changes, and that these different rates are probably segments of a curve. We conclude that within species there is no single scaling factor relating MR to body weight, but that the factor changes as animals mature.

One would expect the same to apply to whole body protein turnover, but we need more information. There are good data on protein turnover in the rat at different ages from the work of Millward *et al.* (1975) and Goldspink and Kelly (1984). Goldspink studied rats from ~6 days before birth to senility; over most of this age-range there was a single linear slope of 0.93 between log synthesis rate and log body weight until the rats became senile at the age of 2 years, when the rates fell off (Fig. 11.3).

Data are available for man at only four ages, apart from senescence: prematures, neonates, infants and adults. Except for adults and the elderly they are scanty and variable (see Chapter 12). A log-log plot of protein synthesis against age is shown in Fig. 11.4. In prematures the metabolic rate is about the same as in neonates – ~250 kJ kg^{-1} day^{-1}, but the rate of protein synthesis is much higher, nor, from the three points available does there seem to be any change in slope after infancy.

A difficulty with intra-specific allometry is that slopes are seldom derived from more than about six points and therefore there is a good deal of uncertainty about them. For comparing functions at different ages a less sophisticated but adequate approach may simply be to give ratios between young and old, as in the next section.

The purpose of this excursion into allometry is to explore in more detail, however inadequately, the relations between energy metabolism and protein metabolism. As Webster (1988) has said, 'The power of these scaling laws lies not in the fact that they make all animals appear the same, but that they can reveal the genuine anatomical and physiological basis for differences in energy exchange within and between species'. This thought applies equally well to protein turnover.

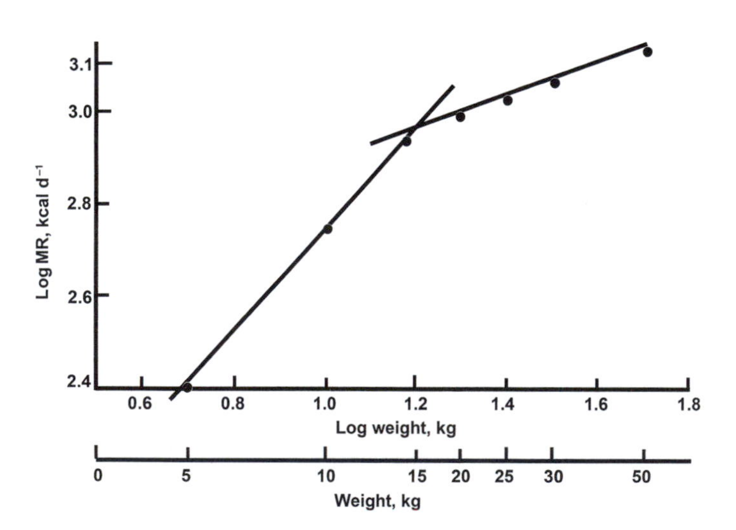

Fig. 11.2. Log-log plots of resting metabolic rate against body weight in humans at different ages. Data of Schofield (FAO/WHO/UNU, 1985). Mean body weight of males at different ages.

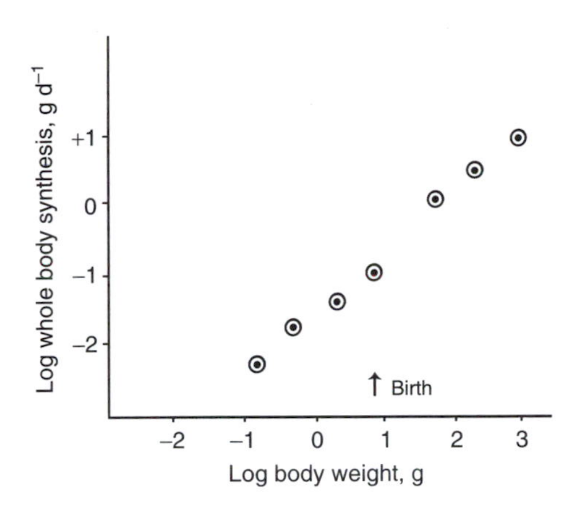

Fig. 11.3. Log-log plot of whole body protein synthesis against body weight in the rat. Data of Goldspink *et al.* (1984).

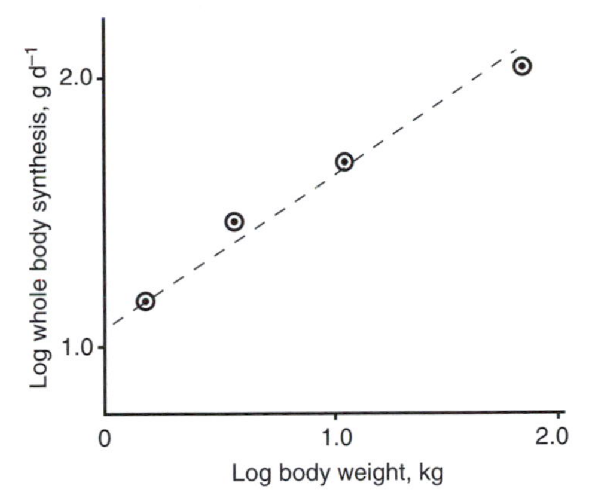

Fig. 11.4. Log-log plot of whole body protein synthesis against body weight in humans. Data from Chapters 7 and 11.

11.2 Growth and its Cost

11.2.1 Changes in tissue composition during growth

In the early stages of development there is rapid multiplication of cells – hyperplasia. It has long been established that the DNA content per cell is the same in all tissues except the liver, where a number of cells are polyploid, having failed to divide after mitosis and therefore contain twice or even four times the normal amount of DNA per nucleus (Jacob *et al.*, 1954). In muscle, where the cells are multinucleate, the concept of the 'DNA unit' was pioneered by Cheek (1971). In the later stages of growth hyperplasia decreases or stops, at varying times in different tissues and species, and growth continues by hypertrophy – increase of cytoplasmic constituents, mainly protein.

Since it is a central dogma of molecular biology that DNA makes RNA and RNA makes protein, DNA may be regarded as the final controller of what happens in the cytoplasmic territory that it controls. It was therefore of some importance, from the point of view of nutrition, when Kosterlitz (1947) and Thomson *et al.* (1953) showed that in adult rats the DNA content of the liver was unaffected by dietary changes; on protein-deficient diets the activity of the genome may be altered but its size is not, whereas the cytoplasm, with its content of protein and phospholipid, shrinks round the nucleus. Thus DNA provides a standard of reference; for example, we showed early on that in malnourished children the ratio of protein:DNA in the liver was substantially reduced compared with after recovery (Waterlow and Weisz, 1956). Since the concentration of protein per g of tissue does not change, the fall in protein/DNA is the only way of assessing the extent of protein depletion or accretion.

It has long been known that RNA and its associated proteins constitute the machinery of protein synthesis. Therefore in any discussion of growth the RNA concentration [RNA] in a tissue is very important, since it represents the *capacity* for synthesis. Since most of the RNA (~80%) is ribosomal, the synthesis rate per unit RNA (k_{RNA}) is a measure of ribosomal *activity*. In the description of growth that follows, the capacity for synthesis is for the most part expressed as RNA/protein; it might be more logical to relate RNA to DNA.

Table 11.2 shows a change with growth in the composition of liver and muscle of two very different species – rat and salmon trout. The latter was chosen because the work of Peragón *et al.* (1998, 2001) provides very complete information and it represents a non-mammalian species and hence a wider view.

Hyperplasia was going on throughout the period of growth, as shown by the increases in total DNA in both tissues and both species. The 12-fold increase in DNA in trout muscle is remarkable, considering that the starting point was later than in the rat, the growth period shorter and the temperature was only 15°C. Hypertrophy occurring at the same time is evident from the increases in total protein and protein per cell (protein/DNA), which are greater in muscle than liver. The increase in RNA goes hand in hand with that of protein, but at only half the rate, so that the values of the RNA/protein ratio do not vary very much.

11.2.2 Changes in protein turnover during growth

The data in Table 11.3 again come from two very different species and two very different tissues. The results in rat and fish are similar: is it reasonable to conclude that the machinery of protein synthesis – the ratio of synthesis to DNA and RNA and hence to the processes of transcription and translation – is, under normal conditions remarkably constant and stable?

Protein deposition during growth obviously requires an increased rate of synthesis above the maintenance level. This is typically accompanied by an increased rate of breakdown (Bates and Millward, 1981), a phenomenon also found in some pathological states (Chapter 16). A good illustration is the work of Millward's group on rat muscle, summarized in Tables 11.4 and 11.5. There were two genetically different strains of rats, and although exact comparison is difficult because of differences in age, it is clear that the CFY rats grew faster, had higher efficiencies of deposition and reached a larger final body weight than the hooded rats, in spite of synthesis and

Table 11.2. Changes in body and tissue composition during growth in the rat and the rainbow trout. The numbers are the ratio of old:young values.

	Whole body	Liver		Muscle			
	A	B	C	D	E	F	G
Interval, weeks	3–42	3–42	14–40	14–40	3–44	3–44	3–47
Whole body or tissue weight, g	12	12	15	33	–	–	14
Total DNA, mg	5	7	8	12	6	3	6
Total protein, g	14	9	10	30	21	10	22
Protein/DNA, g mg^{-1}	2.8	1.3	1.25	2.5	3.5	2.7	3.7
Total RNA, mg	6	6.5	5	11	9	6	9
RNA/DNA, mg mg^{-1}	1.2	0.9	0.6	0.95	1.5	2.0	1.5
RNA/protein, mg g^{-1}	0.43	0.72	0.50	0.37	0.42	0.60	0.71

A. Goldspink *et al.*, 1984, rat; B. Goldspink *et al.*, 1984, rat; C. Peragón *et al.*, 1998, rainbow trout at 15°C; D. Peragón *et al.*, 2001, rainbow trout at 15°C; E. Lewis *et al.*, 1984, rat: soleus muscle (slow twitch oxidative); F. Lewis *et al.*, 1984, rat: tibialis muscle (fast-twitch glycolytic); G. Millward *et al.*, 1975, rat: gastrocnemius + quadriceps (mixed).

Table 11.3. Changes in protein turnover during growth in the rat and the rainbow trout. The numbers are ratios old:young.

	Whole body	Liver		Muscle			
	A	B	C	D	E	F	G
k_S	0.39	0.80	0.32	0.77	0.50	0.38	0.29
k_d	0.50	1.00	0.23	0.91	0.71	0.71	0.45
S/RNA (k_{RNA})	0.91	1.11	0.62	2.0	1.25	1.67	0.83
S/DNA	1.05	1.05	0.47	0.91	2.0	1.25	0.77

k_d = synthesis minus growth rate. A. Goldspink *et al.*, 1984, rat; B. Goldspink *et al.*, 1984, rat; C. Peragón *et al.*, 1998, rainbow trout at 15°C; D. Peragón *et al.*, 2001, rainbow trout at 15°C; E. Lewis *et al.*, 1984, rat: soleus muscle (slow twitch oxidative); F. Lewis *et al.*, 1984, rat: tibialis muscle (fast-twitch glycolytic); G. Millward *et al.*, 1975, rat: gastrocnemius + quadriceps (mixed).

breakdown rates that were lower, at least initially. Bates and Millward (1981) make the point that the fast-growing rats had a life span of only a year, compared with 2 years for the hooded rats. Breakdown rates tended to be higher in the slower-growing hooded rats. Data on DNA and RNA are only available for the CFY rats. Both capacity (S/DNA) and efficiency (S/RNA) of synthesis settle down after the first month to a level that remains constant while body weight increases sixfold.

The same pattern was seen when hypertrophy of muscle was induced by hanging a weight on the anterior latissimus dorsi muscle of a fowl (Laurent *et al.*, 1978; Laurent and Millward, 1980; see next section). There was a large increase in synthesis rate, with a parallel but smaller increase in degradation. Because increased synthesis during growth is accompanied by an increase in breakdown, the net deposition of protein, synthesis minus breakdown, is always much less than the total synthesis (see below).

A more efficient way of promoting growth would be simply to reduce the rate of breakdown. Thus Condé and Scornik (1977) in studies on mouse liver found that in the neonatal mouse the rate of synthesis in the liver was the same as in the adult, but the rate of breakdown was only about half. From studies on regenerating liver, Scornik and Botbol (1976) concluded that 'changes in the rate of protein degradation are the single most important factor determining the increase in protein content of the liver during compensatory growth' after partial hepatectomy. Scornik measured breakdown rate both by the usual method of subtracting net deposition from synthesis and by following the rate of loss of label after injecting ^{14}C-carbonate.

Table 11.4. Changes with age in the turnover rates of skeletal muscle protein in two strains of rat.

	Age days	Body weight g	Synthesis % d^{-1}	Breakdown % d^{-1}	Deposition: Net synthesis % d^{-1}	Deposition: Efficiency 100 × [net/ total synthesis]
CFY rats	25	75	15.6	9.8	5.8	37
	32	129	15.2	9.5	5.7	37.5
	52	289	7.3	4.4	2.9	40
	101	546	5.2	4.1	1.1	21
	320	716	4.5	4.5	0	0
Hooded rats	23	37	28.6	22.4	6.1	21
	46	70	16.1	13.1	3.0	18.5
	65	116	11.5	9.8	1.7	15
	130	233	5.3	4.6	0.7	13
	330	511	4.9	4.9	0	0

Recalculated from data of Bates and Millward, 1981. Breakdown rates were calculated by subtracting growth rates from synthesis rates. The hooded rats deposited protein with lesser efficiency than the CFY rats; they grew more slowly but lived twice as long.

Table 11.5. Changes with age in nucleic acid concentrations and efficiency of synthesis in skeletal muscle of CFY rats.

Age days	Protein/DNA mg mg^{-1}	RNA/protein mg g^{-1}	RNA/DNA mg mg^{-1}	S/RNA mg mg^{-1}	S/DNA mg mg^{-1}
25	167	12.4	2.1	12.4	26
32	180	9.6	1.7	15.9	27
52	290	7.3	2.1	10	21
101	400	4.5	1.8	11.7	21
320	452	4.1	1.9	11.6	22

S = mg protein synthesized d^{-1} g^{-1} protein. Data from Bates and Millward, 1981.

Two invertebrate species show a similar pattern of growth. In the mussel, *Mytilus edulis*, with increased protein absorption, which allows better growth, there was very little increase in synthesis but breakdown fell almost to zero (Hawkins, 1985). It was argued that in this species higher growth stemmed from lower intensity of synthesis, so that the energy required for maintenance was less and more was available for growth (Hawkins, 1985, 1991). In *Octopus vulgaris* synthesis increased with increasing growth rate, but at high growth rates breakdown was again almost completely abolished (Houlihan *et al.*, 1990). These are examples of the most efficient way of producing growth – to allow synthesis to continue at a minimal rate, with degradation greatly reduced.

It would be very important to find out how widely these two patterns are distributed in the animal kingdom. Two explanations have been advanced for the inefficiency of deposition; the first is that higher rates of synthesis will allow greater temporary production of enzymes, and so more flexibility in an organism which is developing as well as growing in size. The second is that an increased synthesis rate will produce more coding errors and more abnormal proteins which require to be rapidly destroyed, and for this an increased rate of degradation is needed.

11.2.3 The energy cost of protein synthesis and deposition

Theoretical calculations

The calculation must include not only the energy cost of synthesizing peptide bonds but also the cost of associated processes: the synthesis of messenger RNA, the cost of amino acid transport into cells, the cost of protein degradation and perhaps, as suggested by Reeds and Garlick (1984), the cost of regulatory mechanisms.

Peptide bond synthesis

The stoichiometry of the different steps in peptide bond synthesis has been well worked out. Amino acid activation and coupling to t-RNA require two ATP per amino acid, and chain elongation two GTP per peptide bond. This amounts to four mole ATP + GTP per mole peptide bond. If the average molecular weight of amino acids is taken as 100, there would be 0.01 mol peptide bonds per gram of protein synthesized. The energy cost of ATP or GTP formation varies slightly with the substrate, but in round figures $-\Delta H$ could be taken as 80 kJ mol^{-1} (Blaxter, 1989). The cost of peptide bond synthesis would therefore work out at 3.2 kJ per gram of protein. As Blaxter pointed out, many proteins, particularly secreted ones, are synthesized as pre-pro-proteins which are much larger than the final product. In the case of insulin, for example, the final molecule has 49 peptide bonds, compared with 119 in pre-pro-insulin. How far this will cause error in estimating synthesis rate depends on the time taken for the measurement of synthesis compared with the rate of formation of the final product from its precursor. In liver, for example, measurements by the flooding dose method (Chapter 14), which only needs about 1 h, will include, at least in part, the synthesis of pre-albumin, while by the constant infusion method, which takes longer, most of the precursor would have been broken down to albumin and secreted into the plasma. To the extent that this kind of process occurs, more peptide bonds may be synthesized than are taken into account, so the figure of 3.2 kJ per gram of protein will be a minimum estimate.

RNA synthesis

As mentioned earlier, Schöch and his co-workers have developed a method of measuring the turnover rates of tRNA, rRNA and mRNA from the rate of excretion of methylated nucleotides that are derived from the breakdown of RNAs and are not re-utilized. These rates are shown in Table 11.6. The mature RNAs are derived from precursors from which many nucleotides are removed after transcription. mRNAs and their precursors will be of many different sizes, so only an approximate estimate can be given of their nucleotide number. In Table 11.6 it is assumed that it is the precursor that has to be synthesized.

A tentative estimate of the energy cost of synthesising precursor mRNA, before it is processed, is shown in Table 11.6. This cost amounts to about 0.5 kJ kg^{-1} day^{-1}. The cost of peptide bond synthesis at 4 g protein kg^{-1} day^{-1} would be 12.8 kJ kg^{-1} day^{-1}, so that mRNA synthesis would add less than 5%. From the turnover rates of RNA species given by Sander *et al.* (1986) the cost of synthesizing rRNA and tRNA would be small. The energy cost of total RNA synthesis probably adds about 5% to the overall cost of protein synthesis.

Amino acid transport

For determining the contribution of amino acid transport to the total cost of protein synthesis we need to know two things: the amount in unit time of amino acid transported from the blood into the tissues, and the amount taken up per gram of protein synthesized. The ratio of these two we have called the 'transport index' (Chapter 4). Baños *et al.* (1973) examined the rate of uptake of 20 amino acids into rat muscle and found that the rates for individual amino acids varied by about two orders of magnitude, from 16 for threonine to 0.2 nmol min^{-1} g^{-1} for aspartate. Moreover, entry rates varied with the plasma concentration. Also to be considered are differences between tissues in the permeability of their capillaries, and the inhibiting effect of one amino acid on the transport of another that shares the same carrier (Pratt, 1981). In humans Hovorka *et al.* (1999), by a very simple analysis, arrived at a figure of 6.7 μmol min^{-1} kg^{-1} for the inward transport of leucine in the whole body. Release of leucine by breakdown amounted to 1.93 μmol min^{-1} kg^{-1}, if subjects were in a steady state, so that if S = B, the transport index would be 6.7/1.93 = ~3.5.

Further information comes from studies on forearm and leg, summarized in Chapter 6. In adult humans in the post-absorptive state the mean transport indices for the three amino acids that have been used were:

leucine	1.6	(8 studies)
phenylalanine	2.5	(6 studies)
lysine	1.0	(4 studies)

In one study (Tessari *et al.*, 1996a) the index for leucine in splanchnic tissue was 2.0. If the same ratio applied to all amino acids in all tissues, one might conclude that very roughly 2 × as many amino acids are transported as are synthesized to protein.

The proportion of total transport that is sodium-dependent seems not to be known, but is probably large – say $^2/_3$. In different tissues either 1 or 2 Na^+ ions are used for the transport of one amino acid (Eddy, 1987). The stoichiometry usually supposed for the sodium pump is 3 Na^+ to 1 ATP. Putting these figures together gives an average ATP cost of 0.5 mol ATP per mol amino acid transported. If as suggested twice as many moles amino acid are transported as are synthesized, transport would add 1 ATP to the 4 ATP used for peptide bond synthesis.

Table 11.6. Estimated energy cost of RNA synthesis in adults.

	tRNA	rRNA	mRNA
A. Turnover rate, μmol kg^{-1} d^{-1}	0.63	0.037	0.99[a]
B. Nucleotides per molecule	100	13,500	8000[b]
C. Inter-nucleotide bonds synthesized, per day A × B	63	500	7900
D. Cost of this synthesis, kJ kg^{-1} d^{-1}	40×10^{-3}	31×10^{-3}	500×10^{-3}
E. Per cent of cost of protein synthesis	0.31	0.24	3.9

A. Sander *et al.*, 1986. B. H. Topp, personal communication. D. Blaxter, 1989; 2 ATP per bond, and 31 kJ per mol ATP. E. Protein turnover in adult taken as 4 g kg^{-1} d^{-1} at cost of 3.2 kJ g^{-1}.

[a] H. Topp, personal communication.

[b] This is the number of nucleotides in the precursor or cap-mRNA. The number in the final mRNA is much smaller – about 1200 (Sander *et al.*, 1986).

If whole body prtotein synthesis is taken as 4gkg^{-1} d^{-1}, and if an average protein has MW 40,000 and contains 40 amino acids, then $4 \times {}^{400}/_{40,000}$ = 40 mmol of peptide bonds are synthesized daily. If each amino acid is represented by three bases, then 120 mmol nucleotides are turned over per kg per day. If the coding length of an average mRNA needed for the synthesis of an average protein is taken to be 1200, then 120 mmol/1200 = 100 μmol RNA kg^{-1} d^{-1} are involved in protein synthesis. This is 100 times higher than the turnover rate of mRNA, which means that one RNA molecule on average is translated 100 times. I am indebted to H. Topp for this calculation.

With these approximations the cost of synthesizing 1 g protein may be estimated as:

Peptide bond synthesis:	3.2 kJ
RNA synthesis:	0.3
Transport:	1.0
Total	~4.5 kJ g^{-1} protein.

In theory some allowance should also be made for the cost of degradation by the ATP-dependent ubiquitin process (Chapter 17), but as only a few ubiquitin molecules are coupled to one protein molecule, this extra cost will be negligible. However, clearly more research is needed on this subject. An unknown amount of ATP will also be needed to maintain the internal acidity of the lysosomes at a pH of ~5.

Indirect estimates

An approach widely used in the animal production industry has been to relate, in growing animals, the metabolizable energy intake less the maintenance requirement, to the amount and composition, measured at slaughter, of the new tissue laid down. Deducting the heat content (enthalpy) of the tissue deposited made it possible to calculate, per g, the cost of depositing it. This method in pigs gave very variable estimates of the cost of protein synthesis, averaging about 20 kJ g^{-1}, much larger than the theoretical estimate (Kielanowski, 1976; Fuller *et al.*, 1987).

This might be called the 'operational' method of determining the cost of protein since it answers the question: how much extra food must be given to produce 1 kg protein gain?

A modification of this approach was used by Spady *et al.* (1976) in infants recovering from malnutrition and gaining weight very rapidly. The energy cost of growth was determined from the slope of the linear relationship between weight gain and intake. Energy expenditure was measured by the calibrated heart rate method. Deducting expenditure from intake gave the energy balance, or energy stored. It was assumed that the new tissue contained 15% protein (Fomon, 1967). Deducting again the enthalpy of this protein (23.6 kJ g^{-1}) gave a figure for the energy cost of depositing it of 32 kJ g^{-1}. A similar value (32 kJ g^{-1}) was found by Roberts and Young (1988) working with premature infants. A further study, but one that is difficult to quantify, is the finding by Brooke and Ashworth (1972) in children recovering from malnutrition of a close linear relation between the rate of growth and the extra energy expenditure after a meal. We attributed this to protein synthesis, providing a new explanation of the old concept of specific dynamic action.

There followed the very important study of Welle and Nair (1990) in which resting metabolic rate and leucine flux were measured at the same time on nearly 50 adult subjects. When leucine flux was converted to g protein synthesized per day, and both rates were related to total body potassium, it emerged that the metabolic rate increased by 4.35 kJ for each increase of 1 g protein synthesized – an exact correspondence with the theoretical estimate. Note that since these were adult subjects there was no deposition of protein and no growth. This relationship showed that protein synthesis accounted for ~20% of resting energy expenditure.

The main reason for the large discrepancy between some of the indirect estimates and the theoretical cost of ~5 kJ g^{-1} protein synthesized is that deposition is not the same as synthesis. As mentioned above, increased synthesis is usually accompanied by increased breakdown. When direct measurements of the rate of protein synthesis in man became available it was shown by Golden *et al.* (1977) that in children recovering from malnutrition 1.4 g protein were synthesized per g deposited. Reeds and Harris (1981), working with pigs, found a ratio of 2 g synthesized per g deposited (Fig. 11.5). Lobley (1998) has collated results on pigs, sheep and cattle which show similar differences between synthesis and deposition. In the liver of rainbow trout the ratio of synthesis to deposition fell from 6 at a body weight of 5 g to 2 at 400 g (Peragón *et al.*, 1998). The inverse of this ratio may properly be termed the 'efficiency' of synthesis.

In order to get a realistic estimate of the energy cost of protein synthesis by the indirect method it is necessary to have 4 independent measurements: the energy gained; the composition of the tissue gained, particularly the fat content; the protein gained and the rate of protein synthesis. The study that, as far as we know, is closest to meeting these criteria is that of Jéquier's group on premature infants in the Gambia (Cauderay *et al.*, 1988). In this study the energy available for growth was determined as

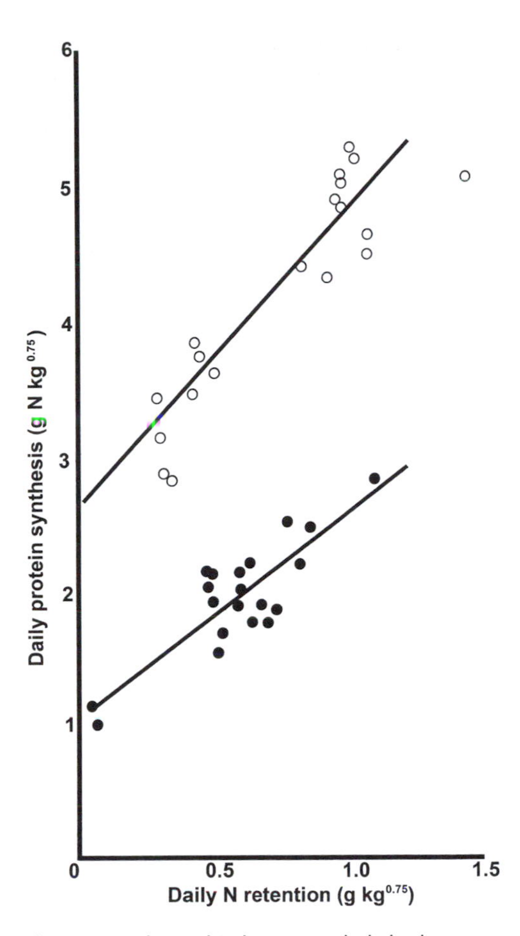

Fig. 11.5. Relationship between whole body protein synthesis and nitrogen retention in children recovering from malnutrition (●-----●) and in growing pigs (○-----○).
Reproduced from Reeds and Harris (1981) by courtesy of Applied Science Publishers.

the difference between intake and expenditure, measured by calorimetry. Nitrogen balance was measured over 3 days, and during this time protein synthesis and breakdown were measured by the [^{15}N] glycine-end-product method. An important point, confirming the validity of the methods, was that protein gain as synthesis minus breakdown was identical with gain from N balance. The calculations are summarized in Table 11.7. The table shows that calculation on the basis of difference in enthalpy leaves no margin for the variable we are interested in, the energy cost of depositing protein, as opposed to its enthalpy. What is needed is an independent measure of the fat deposited. If the fat gained were 3 g instead of 3.3, the energy cost of synthesising 2.1 g protein would be 12 kJ or 5.7 kJ g^{-1}. Unfortunately, in the human it is impossible with present techniques to measure the fat gained with sufficient accuracy. In animals it can be done and was done at slaughter by Kielanowski and others, and if the synthesis rates had been measured and adjusted for breakdown, the energy cost of protein synthesis would have come much closer to the theoretical figure. The reason for the high figures for the energy cost of protein sythesis is that a large part of the sythesis is 'wasted' by balancing the increased breakdown.

11.3 The Effect of Muscular Activity and Immobility on Protein Turnover

The metabolic changes that occur in exercise have been studied by physiologists in great detail. A recent review is that of Rennie and Tipton (2000). In earlier times one focus of interest was the possible influence of physical exertion on human protein requirements. More recently it has shifted to the effect of exercise in counteracting the atrophy of muscle that occurs with immobilization and in old age, as discussed in the next chapter. It is difficult to treat the subject coherently because there are so many different kinds of exercise and routines: short bouts, long endurance, prolonged training; resistance, eccentric, dynamic exercise, all of which may have different effects.

11.3.1 Nitrogen balance and amino acid oxidation

Liebig, the great German chemist of the 19th century, believed that muscle protein was the only source of fuel for muscle contraction. This view was conclusively refuted by a classical experiment of Fick and Wislicenus (1866) in which they collected urine before, during and after a 2000 m climb of the Faulhorn mountain in the Bernese Oberland. Their results are shown in Table 11.8. They made a rough calculation of the work done and concluded that it was impossible for it to have been fuelled by the combustion of

Table 11.7. Cauderay's study on protein and energy metabolism of premature infants appropriate for gestational age.

Variable	Units	
A. Energy available for growth (intake – expenditure)	kJ kg^{-1} body wt	176
B. Protein synthesis	g kg^{-1} d^{-1}	9.7
Protein breakdown	g kg^{-1} d^{-1}	7.6
Protein gain	g kg^{-1} d^{-1}	2.1
C. Protein gain by N balance, N × 6.25		2.1
D. Enthalpy of protein gained at 23.4 kJ g^{-1}		49
E. Enthalpy of fat gained, calculated as A–D		127
F. Fat gain, calculated as E/38.3 g^{-1} of weight gain		3.3
G. Body weight gain,	g kg^{-1}	18.2
H. Fat gain as % weight gain	%	18.2%

Data of Cauderay *et al.* (1988). If F were 3.0 g instead of 3.3, enthalpy of fat gain would be 115 kJ, leaving 12 kJ, or 5.7 kJ g^{-1} as the cost of synthesising 2.1g protein.

muscle protein. Since then it was widely held that physical work had no effect on nitrogen balance and protein requirements, a view that was maintained in a United Nations report of 1973 on protein requirements (FAO/WHO, 1973). Gontzea *et al.* (1975) found that increased physical activity resulted in an initial increase in nitrogen excretion, followed by adaptation over 3 weeks, resulting in restoration of N balance. There is only scanty and somewhat contradictory information about nitrogen excretion while exercise is actually going on. Rennie *et al.* (1981) found that urea excretion fell during nearly 4 h of exercise at 50% of $VO_{2\ max}$. On the other hand Refsum *et al.* (1979) recorded that in man in a cross-country ski race of 70 km lasting $4^1/_2$–6 h, urinary urea excretion increased twofold, and in a study by Calles-Escandon *et al.* (1984) at a similar level of intensity but of shorter duration the loss of urea doubled; however, much of this loss was in sweat, which was not measured in the other studies. There seems to be general agreement that at least after exercise has stopped there is an increase in urea excretion which lasts for many hours (Rennie *et al.*, 1981; Dohm *et al.*, 1982), but over a longer time it returns to normal. Some authors have reported an increase in 3-methylhistidine excretion, a measure of muscle protein breakdown, e.g. Dohm *et al.* (1982), Fielding *et al.* (1991), although again there are contradictory results in the literature.

It may be argued, although it is debatable (see Chapter 8), that urea production is a better measure of overall amino acid oxidation than urea excretion. In several studies urea production has been measured with ^{15}N urea during short periods of exercise but in none of them did it change (Wolfe *et al.*, 1982; Carraro *et al.*, 1993; El-Khoury *et al.*, 1997; Forslund *et al.*, 1998). Carraro's study is particularly interesting because he showed that in exercise a larger fraction of urea production was hydrolysed in the gut and recycled to protein.

Table 11.8. The experiment of Fick and Wislicenus: nitrogen excretion during a climb of 2000 m.

		Urinary nitrogen excretion, g N h^{-1}	
		Fick	Wislicenus
Overnight, before climb	11 h	0.63	0.61
During climb	8 h	0.41	0.39
Rest after climb	6 h	0.40	0.40
Following night	10.5 h	0.45	0.51

Data of Fick and Wislicenus, 1866.

11.3.2 Kinetic studies with [^{13}C] leucine

The oxidation of infused [1^{13}C] leucine, as a measure of total amino acid oxidation, was convincingly validated by the 24-h studies of Young's group at MIT (El-Khoury *et al.*, 1994). All work with this approach has shown an increase in leucine oxidation during exercise with return to usual rates at rest, so that over 24 h leucine balance is achieved (Rennie *et al.*, 1981; Wolfe *et al.*, 1982; El-Khoury *et al.*, 1997; Bowtell *et al.*, 1998; Forslund *et al.*, 1998). It was found also that the recovery of $^{13}CO_2$ was greatly increased at the beginning of exercise (see Chapter 8) but afterwards fell rapidly to below the usual level. When account is taken of this, the estimate that leucine oxidation is increased is still valid (El-Khoury *et al.*, 1997). Bowtell *et al.* (1998) discussed the mechanism of this increase, and showed that in exercise a greater proportion of the leucine flux is transaminated to KIC. They suggested that the consequent increase in KIC concentration inhibits the phosphorylation of the BCA-dehydrogenase, thus activating the complex (see Chapter 4) and increasing the oxidation of leucine. Against this is the finding of Kasparek (1989), that a rise in BCA oxoacid concentration did not correspond in time with increase in activity of the dehydrogenase.

These kinetic studies also provided estimates of whole body protein synthesis and breakdown, and showed almost without exception that during exercise synthesis is decreased while breakdown tends to be increased. The changes interact with the diurnal cycling of fasting and feeding, so that exercise during fasting makes the balance more negative and during feeding makes it more positive (Table 11.9). These results relate to short bouts of exercise only. A 16-week programme of aerobic exercise had no effect on whole body protein turnover (Short *et al.*, 2004). We do not know whether there are any differences in protein turnover and amino acid catabolism in people habitually engaged in heavy physical work.

11.3.3 Muscle protein turnover

Athletes know very well that continuous or repeated use of muscles leads to increases in their size and strength. It is well established experimentally that continuous stretching of a muscle causes it to hypertrophy, with increased rates of both synthesis and breakdown of protein (Goldspink, 1977a; Laurent *et al.*, 1978). Laurent produced this result by hanging a weight on a chicken's wing, Goldspink by enclosing the hind-limb of rats in plaster so that either the extensors or the flexors were stretched. Later Goldspink *et al.* (1995) showed in the extensor digitorum muscle of the rabbit that while electrical stimulation had no effect, when combined with stretch there was a more than additive increase in synthesis (Table 11.10), which resulted mainly from an increase in RNA concentration. There were also increases in some of the lysosomal proteinases, which suggested a concomitant rise in degradation.

The last decade has seen a spate of studies in man with muscle biopsies and sometimes measurements of arterio-venous differences in a limb. Biopsies obviously cannot be taken while exercise is actually going on, but within about 3 h from the end of exercise the rate of muscle protein synthesis (MPS) is significantly increased,

Table 11.9. The effect of 90 minutes of resistance exercise on the balance between synthesis and breakdown in the whole body in fasted and fed states.

	S – B, μmol leucine kg^{-1} h^{-1}		
	Before	During	After
Fasting			
El-Khoury *et al.*, 1997	–20	–36	–19
Forslund *et al.*, 1998	–20	–37	–20
Fed			
El-Khoury *et al.*, 1997	+29	–14	+12
Forslund *et al.*, 1998[a]	+23	+8	+6

[a]Subjects receiving 1 g protein kg^{-1} d^{-1}.
Turnover measured by constant infusion of [1^{-13}C] leucine.

Table 11.10. Response of rabbit muscle to electrical (10 Hz) and mechanical stimulation.

	Stimulation		
	Electrical	Mechanical	Both
	% increase above control		
RNA/protein	+11	+94	+231
Fractional synthesis, % d^{-1}	0	+138	+345
Synthesis, g g^{-1} RNA	−8	+27	+45
IGF-I mRNA, fold increase	×5	×12	+40

Data of D. Goldspink *et al.*, 1995.

sometimes doubled and sometimes with an increase in breakdown as well (Chesley *et al.*, 1992; Biolo *et al.*, 1995, 1997; MacDougall *et al.*, 1995). Biolo's studies show convincingly that the increase in synthesis is mediated by increased amino acid inflow, perhaps at the expense of protein breakdown in the gastro-intestinal tract (Williams *et al.*, 1996). These changes are transient and disappear 36 h after the exercise.

A somewhat different protocol has been used, particularly in the elderly, and we must admire their fortitude in submitting to repeated biopsies. In these studies MPS has been measured at an unstated interval (e.g. after an overnight fast) after an exercise programme that may have lasted from 2 weeks to 3 months. Here the results are less consistent – either an increase in synthesis (Yarasheki *et al.*, 1992, 1993, 1995, 1999; Short *et al.*, 2004) or no change (Welle *et al.*, 1995). In North's study mild aerobic exercise – bicycling for 20 minutes 3–4 times a day at 70% of maximum heart rate – caused a 20% increase in muscle protein synthesis.

The conclusion is that exercise causes an increase in muscle protein turnover but we do not know how long it lasts after the exercise programme ends. There is also evidence that the post-exercise increase in synthesis is enhanced by amino acids (Biolo *et al.*, 1997; Børsheim, 2002) and breakdown reduced by insulin. The responses also vary with the muscle; in the rat an increase in synthesis was much more pronounced in the fast-twitch brachialis, whereas in the slow-twitch soleus there was no increase in synthesis but a decrease in breakdown. The enhancement of synthesis is greater in the myofibrillar than in the sarcoplasmic fraction (Balangopal, 1997a) and involves particularly the myosin heavy chain (Welle *et al.*, 1995; Balagopal, 1997b), with shifts between the different isoforms (Gregory *et al.*, 1990; Caiozza *et al.*, 1996; Sharman *et al.*, 2001). Chesley *et al.* (1992) recorded no change in total RNA concentration after a single bout of resistance exercise; since the synthesis rate was doubled, there must have been an increase in RNA activity.

An excellent review of the sarcopenia of ageing is that of Short and Nair (1999).

11.3.4 Immobilization

Immobilization is the other side of the coin from exercise. Fifty years ago Schønheyder *et al.* (1954) performed a classical experiment in which they encased three subjects in a plaster cast up to the hips. Urine was collected for 5 days after a single dose of ^{15}N glycine and compartmental analysis of the curve of tracer excretion showed that the negative nitrogen balance that ensued resulted from a fall in synthesis rather than an increase in breakdown. Gibson *et al.* (1987) studied patients in whom one leg had been immobilized in a long-leg plaster for 5 weeks. At the end of this time, after infusion of [^{13}C]-leucine biopsies were taken from the quadriceps muscle in the injured and normal legs. The results are shown in Table 11.11. There was no change in the distribution of fibre types, but significant atrophy of the slow fibres. There was a 10% loss of fibre volume, but RNA concentration increased, so that although the fractional rate of synthesis was reduced by only 25%, there was a 50% fall in RNA activity. This is a very remarkable finding, which to my knowledge has not been further investigated.

Confinement to bed for as little as 2 weeks resulted in a fall in whole body synthesis with no significant change in breakdown (Ferrando *et al.*,

Table 11.11. Muscle protein synthesis in a leg immobilized after fracture compared with the control leg.

	Control	Immobilized
Fractional synthesis rate, % h^{-1}	0.046	0.034
RNA concentration, μg mg $protein^{-1}$	1.8	2.5
Efficiency of synthesis, mg protein h^{-1} mg^{-1} RNA	0.27	0.14

From Gibson *et al.*, 1987.

1996). In these studies arterio-venous measurements on the hind-limb together with muscle biopsies showed a 50% decrease in muscle protein synthesis. Even in as short a time as 9 h in bed whole body synthesis and breakdown were both reduced by 25% (Bettany *et al.*, 1996). It seems likely that the negative nitrogen balance after trauma is caused as much by the consequent immobility as by the trauma itself and the same may apply in the elderly (see Chapter 12, section 12.5).

In rats experimental weightlessness in the hind-limbs was produced by suspending them in a harness with the fore-limbs supported (Goldspink, 1986); this resulted in atrophy and a greatly decreased rate of protein synthesis of the normally weight-bearing hind-limb muscles. The changes were reversed if the muscles were at the same time passively stretched.

The muscles of astronauts also atrophy in the microgravity of space flights (see review by Fitts *et al.*, 2000). The extensors are most severely affected; contractile proteins are lost more than sarcoplasmic and actin more than myosin, and there is a fall in protein synthesis.

In the sarcopenia both of ageing and weightlessness there appears to be a shift in the distribution of myosin heavy chain isoforms from IIa to IIb which can be reversed by exercise (Sharman *et al.*, 2001) and is reflected also by changes in mRNA abundance (Caiozzo *et al.*, 1996; Welle *et al.*, 1999), although there is some disagreement here (Marx *et al.*, 2002). These changes in isoform distribution have effects on muscle function: the unloaded shortening velocity was found to be highly dependent on the myosin heavy chain isoform composition (Larsson *et al.*, 1997). In exercised human subjects a strong correlation has been observed between the time to peak level of delayed force increase and the isoform composition of myosin heavy chains (Hilber *et al.*, 1999). In elegant studies on single fibres the atrophy of ageing reduced the sliding speed of actin filaments in both rodents and humans (Hook *et al.*, 2001). These are striking demonstrations of the dependence of function on molecular structure.

Mechanisms

It seems from the work of Goldspink (1977a,b), Gibson *et al.* (1987) and Chesley *et al.* (1992) that stretch and activity increase synthesis rate by an increase in translational efficiency. Recent studies have indeed concentrated on changes in activation of the initiation complex and of its signalling pathway (Farrell *et al.*, 2000; Rennie and Tipton, 2000; Kimball *et al.*, 2002). Jefferson's group have now proposed that hypertrophy of muscle involves two processes, one acute, manifested within minutes to hours, speeding up translation by involvement of the initiation complex, and one slower, increasing the transcription of ribosomal proteins (Bolster *et al.*, 2004). This second process seems to be the one that is operating after a time-lag in the experiment illustrated in Table 11.11. In the adaptive response to 6 weeks aerobic exercise a large number of genes were identified whose expression, up or down, changed by a factor of more than 1.5. Clusters of gene transcripts were generated that went in the opposite direction to changes observed in animal models of muscle atrophy (Lecker *et al.*, 2004; Timmons and Sundberg, 2004). So the pathway from muscle activity to gene to muscle protein turnover rate is now beginning to be identified.

For muscular activity to affect the synthesis rates of muscle proteins there must be mechanoreceptors converting mechanical stimuli into biochemical signals. According to Hornberger and Esser (2004) there are two possibilities for mechanoreceptors: folding and deformation of the lipid bilayer of the plasma membrane, or through integrins, transmembrane gluco-proteins which, in the words of Kjaer (2004), 'establish a mechanical continuum along which forces can be transmitted from the outside to the inside of the cell'. Within the cell the message is transmitted by a chain of protein kinases. The complex signalling pathways have been discussed in detail by Thompson and Palmer (1998); they recall an observation that may be important, but seems to be neglected in recent work: that when rabbit muscle is stretched *in vitro* there is an increase in protein synthesis that is stimulated by the prostaglandin PF2 or its precursor, arachidonic acid, and prevented by the cyclooxygenase inhibitor indomethacin (Smith *et al.*, 1983).

11.4 Conclusion

Exercise, if intense enough, in the short term increases protein breakdown and amino acid oxidation. From a teleological point of view it has been suggested that the 'purpose' is to provide additional substrate for energy production, but only a small proportion – 5–10% – of the extra energy consumed is covered by oxidation of protein. On this Fisk and Wislicenus were clearly correct.

Whole body protein synthesis is decreased during and immediately after exercise, particularly in the fasting state, but is rapidly restored with feeding; synthesis of muscle proteins is increased, primarily that of the myofibrillar proteins. In the longer term exercise produces hypertrophy, and immobility the reverse. Current work aims at identifying the pathways linking muscular contraction to protein turnover.

11.5 References

Albritton, E.D. (ed.) (1954) *Standard Values in Nutrition and Metabolism*. W.B. Saunders, Philadelphia, Pennsylvania.

Altman, P.L. and Dittmer, D.S. (eds) (1968) *Metabolism*. Federation of American Societies for Experimental Biology, Bethesda, Maryland.

Andersen, B., Shiner, J.S. and Uehlinger, D.E. (2002) Allometric scaling and maximum efficiency in physiological eigen-time. *Proceedings of the National Academy of Sciences* 99, 5822–5824.

Balagopal, P., Ljungqvist, O. and Nair, K.S. (1997a) Skeletal muscle myosin heavy-chain synthesis rate in healthy humans. *American Journal of Physiology* 272, E45–50.

Balagopal, P., Rooyackers, O.E., Adey, D.B., Ades, P.A. and Nair, K.S. (1997b) Effects of aging on *in vivo* synthesis of skeletal muscle myosin heavy chain and sarcoplasmic protein in humans. *American Journal of Physiology* 273, E790–800.

Baños, G., Daniel, P.M., Moorhouse, S.R. and Pratt, O.E. (1973) The movement of amino acids between blood and skeletal muscle in the rat. *Journal of Physiology* 235, 459–475.

Bates, P.C. and Millward, D.J. (1981) Characteristics of skeletal muscle growth and protein turnover in a fast-growing rat strain. *British Journal of Nutrition* 46, 7–13.

Bettany, G.E.A., Ang, B.C., Georgicanos, S.N., Halliday, D. and Powell-Tuck, J. (1996) Bed rest decreases whole-body protein turnover in post-absorptive man. *Clinical Science* 90, 73–75.

Biolo, G., Maggi, S.P., Williams, B.D., Tipton, K.D. and Wolfe, R.R. (1995) Increased rates of muscle protein turnover and amino acid transport after resistance exercise in humans. *American Journal of Physiology* 268, E514–520.

Biolo, G., Tipton, K.D., Klein, S. and Wolfe, R.R. (1997) An abundant supply of amino acids enhances the metabolic effect of exercise on muscle protein. *American Journal of Physiology* 273, E122–129.

Blaxter, K.L. (1989) *Energy Metabolism in Animals and Man*. Cambridge University Press, Cambridge, UK.

Bolster, D.R., Kubica, N., Crozier, S.J., Williamson, D.L., Farrell, P.A., Kimball, S.R. and Jefferson, L.R. (2004) Understanding skeletal muscle hypertrophy: integration of cell signalling. *Physiology News* 55, 18–19.

Børsheim, E., Tipton, K.D., Wolf, S.E. and Wolfe, R.R. (2002) Essential amino acids and muscle protein recovery from resistance exercise. *American Journal of Physiology* 283, E648–657.

Bowtell, J.A., Leese, J.P., Smith, K., Watt, P.W., Nevill, A., Rooyackers, O., Wagenmakers, A.J.M. and Rennie, M.J. (1998) Modulation of whole body protein metabolism, during and after exercise, by variation of dietary protein. *Journal of Applied Physiology* 85, 1744–1752.

Brody, S. and Procter, R.C. (1932) Growth and development with special reference to domestic animals: further investigations of surface area in energy metabolism. *University of Missouri Agricultural Experiment Station, Research Bulletin no. 116.*

Brooke, O.G. and Ashworth, A. (1972) The influence of malnutrition on postprandial metabolic rate and respiratory quotient. *British Journal of Nutrition* 27, 407–415.

Caiozzo, V.J., Haddad, F., Baker, M.J. and Baldwin, K.M. (1996) Influence of mechanical loading on myosin heavy-chain protein and mRNA expression. *Journal of Applied Physiology* 80, 1503–1512.

Calles-Escandon, J., Cunningham, J.J., Snyder, P., Jacob, R., Huszar, G., Loke, J. and Felig, P. (1984) Influence of exercise on urea, creatinine and 3-methylhistidine excretion in normal human subjects. *American Journal of Physiology* 246, E334–338.

Carraro, F., Kimbrough, T.D. and Wolfe, R.R. (1993) Urea kinetics in humans at two levels of exercise intensity. *Journal of Applied Physiology* 75, 1180–1185.

Cauderay, M., Schulz, Y., Micheli, J.-L., Calame, A. and Jéquier, E. (1988) Energy-nitrogen balances and protein turnover in small and appropriate for gestational age low birthweight infants. *European Journal of Clinical Nutrition* 42, 125–136.

Cheek, D.B., Holt, A.B., Hill, D.E. and Tulbert, J.L. (1971) Skeletal muscle mass and growth: the concept of the deoxyribonucleic acid unit. *Pediatric Research* 5, 312–328.

Chesley, A., MacDougall, J.D., Tarnopolsky, M.A., Atkinson, S.A. and Smith, K. (1992) Changes in human muscle protein synthesis after resistance exercise.

Condé, R.D. and Scornik, O.A. (1977) Faster synthesis and slower degradation of liver protein during developmental growth. *Biochemical Journal* 166, 115–121.

Diamond, J.M. (1993) Quantitative design of life. *Nature* 366, 405–406.

Dohm, G.L., Williams, R.T., Kasperek, G.J. and Van Rij, A.M. (1982) Increased excretion of urea and N^{τ}-methylhistidine by rats and humans after a bout of exercise. *Journal of Applied Physiology* 52, 27–33.

Eddy, A.A. (1987) The sodium gradient hypothesis of organic solute transport with special reference to amino acids. In: Yudilevich, D.L. and Boyd, C.A.R. (eds) *Amino Acid Transport in Animal Cells.* Manchester University Press, Manchester, UK, pp. 47–86.

El-Khoury, A.E., Sánchez, M., Fukagawa, N.E., Gleason, R.E. and Young, V.R. (1994) Similar 24-h pattern and rate of carbon dioxide production, by indirect calorimetry vs. stable isotope dilution, in healthy adults under standardized metabolic conditions. *Journal of Nutrition* 124, 1615–1627.

El-Khoury, A.E., Forslund, A., Olsson, R., Branth, S., Sjödin, A., Andersson, A., Atkinson, A., Selvaraj, A., Hambraeus, L. and Young, V.R. (1997) Moderate exercise at energy balance does not affect 24-h leucine oxidation and nitrogen retention in healthy men. *American Journal of Physiology* 273, E394–407.

Else, P.L. and Hulbert, A.J. (1981) Comparison of the 'mammal machine' and the 'reptile machine': energy production. *American Journal of Physiology* 240, R3–9.

Else, P.L. and Hulbert, A.J. (1985a) An allometric comparison of the mitochondria of mammalian and reptilian tissues: the implications for the evolution of endothermy. *Journal of Comparative Physiology* B 156, 3–11.

Else, P.L. and Hulbert, A.J. (1985b) Mammals: an allometric study of metabolism at tissue and mitochondrial level. *American Journal of Physiology* 248, R415–421.

FAO/WHO (1973) Energy and protein requirements. *World Health Organization Technical Report Series,* no. 522, Geneva.

Farrell, P.A., Hernandez, J.M., Fedele, M.J., Vary, T.C., Kimbell, S.R. and Jefferson, L.S. (2000) Eukaryotic initiation factors and protein synthesis after resistance exercise in rats. *Journal of Applied Physiology* 88, 1036–1042.

Ferrando, A.A., Lane, H.W., Stuart, C.A., Davis-Street, J. and Wolfe, R.R. (1996) Prolonged bed rest decreases skeletal muscle and whole body protein synthesis. *American Journal of Physiology* 270, E627–633.

Fick, A. and Wislicenus, J. (1866) On the origin of muscular power. *The London, Edinburgh and Dublin Philosophical Magazine and Journal of Science* 41, Suppl. 31, 485–503.

Fielding, R.A., Meredith, C.N., O'Reilly, K.P., Frontera, W.R., Cannon, J.G. and Evans, W.J. (1991) Enhanced protein breakdown after eccentric exercise in young and older men. *Journal of Applied Physiology* 71, 674–679.

Fitts, R.H., Riley, D.R. and Widnick, J.J. (2000) Microgravity and skeletal muscle. *Journal of Applied Physiology* 89, 823–839.

Fomon, S.J. (1967) Body composition of the male reference infant during the first year of life. *Pediatrics* 40, 863–870.

Forslund, A.H., Hambraeus, L., Olsson, R.M., El-Khoury, A.E., Yu, Y.-M. and Young, V.R. (1998) The 24-h whole body leucine and urea kinetics at normal and high protein intake with exercise in healthy adults. *American Journal of Physiology* 275, E310–320.

Fuller, M.F., Cadenhead, A., Mollison, G. and Seve, B. (1987) Effects of the amount and quality of dietary protein on nitrogen metabolism and heat production in growing pigs. *British Journal of Nutrition* 58, 277–285.

Garby, L. and Larsen, P.S. (1995) *Bioenergetics.* Cambridge University Press, Cambridge, UK.

Gibson, J.N.A., Halliday, D., Morrison, W.L., Stoward, P.J., Hornsby, G.A., Watt, P.W. and Murdoch, G. (1987) Decrease in human quadriceps muscle protein turnover consequent upon leg immobilization. *Clinical Science* 72, 503–509.

Golden, M.H.N., Waterlow, J.C. and Picou, D. (1977) The relationship between dietary intake, weight change, nitrogen balance and protein turnover in man. *American Journal of Clinical Nutrition* 30, 1345–1348.

Goldspink, D.F. (1977a) The influence of immobilization and stretch on protein turnover of skeletal muscle. *Journal of Physiology* 264, 267–282.

Goldspink, D.F. (1977b) The influence of activity on muscle size and protein turnover. *Journal of Physiology* 264, 283–296.

Goldspink, D.F. and Kelly, F.J. (1984) Protein turnover and growth in the whole body, liver and kidney of the rat from the foetus to senility. *Biochemical Journal* 217, 507–516.

Goldspink, D., Morton, A.J., Loughna, P. and Goldspink, G. (1986) The effect of hypokinesia and hypodynaemia on protein turnover and the growth of four skeletal muscles of the rat. *Pflügers Archiv* 407, 333–340.

Goldspink, D.F., Cox, V.M., Smith, S.K., Eaves, L.A., Osbaldeston, M.F., Lee, D.M. and Mantle, D. (1995) Muscle growth in response to mechanical stimuli. *American Journal of Physiology* 268, E288–297.

Gontzea, I., Sutsesan, R. and Dimitrache, S. (1975) The influence of adaptation to physical effort on nitrogen balance in man. *Nutrition Reports International* 11, 231–236.

Gregory, P., Gagnon, J., Essig, D.A., Reid, S.K., Prior, G. and Zak, R. (1990) Differential regulation of actin and myosin isoenzyme synthesis in functionally overloaded skeletal muscle. *Biochemical Journal* 265, 525–532.

Hawkins, A.J.S. (1985) Relationships between the synthesis and breakdown of protein, dietary absorption and turnovers of nitrogen and carbon in the blue mussel, *Mytilus edulis* L. *Oecologia* 66, 42–49.

Hawkins, A.J.S. (1991) Protein turnover: a functional appraisal. *Functional Ecology* 5, 222–233.

Hemmingsen, A.M. (1960) Energy metabolism as related to body size and respiratory surfaces and its evolution. *Reports of the Steno Memorial Hospital and the Nordisk Insulin Laboratorium* 9.1.

Henry, C.J.K. and Collingwood, A. (1998) Allometry of fasting urinary nitrogen loss and basal metabolic rate in homeotherms. *Nutrition Research* 18, 49–60.

Heusner, A.A. (1987) What does the power function reveal about structure and function in animals of different sizes. *Annual Review of Physiology* 49, 121–133.

Hilber, K., Galler, S., Gohlsch, B. and Pette, D. (1999) Kinetic properties of myosin heavy chain isoforms in single fibres from human skeletal muscle. *FEBS Letters* 455, 267–270.

Hill, J.R. and Rahimtulla, K.A. (1965) Heat balance and the metabolic rate of new-born babies in relation to environmental temperature; and the effect of age and of weight on basal metabolic rate. *Journal of Physiology* 180, 239–265.

Hook, P., Sriramoju, V. and Larsson, L. (2001) Effects of aging on actin sliding speed on myosin from single skeletal muscle cells of mice, rats and humans. *American Journal of Physiology* 280, C782–788.

Hornberger, T.A. and Esser, K.A. (2004) Mechanotransduction and the regulation of protein synthesis in skeletal muscle. *Proceedings of the Nutrition Society* 63, 331–335.

Houlihan, D.F., McMillan, D.N., Agnisola, C., Genoina, I.T. and Foti, L. (1990) Protein synthesis and growth in *Octopus vulgaris*. *Marine Biology* 106, 251–259.

Hovorka, R., Carrol, P.V., Gowrie, I.J., Jackson, N.C., Russell-Jones, D.L. and Umpleby, M. (1999) Surrogate measure of whole body leucine transport across the cell membrane. *American Journal of Physiology* 276, E573–579.

Jacob, M., Mandel, L. and Mandel, P. (1954) Étude de la consummation d'oxygène et le teneur en acide désoxyribonucléique du foie à divers âges chez le rat. *Experientia* 10, 218–222.

Karlberg, P. (1952) Determination of standard energy metabolism (basal metabolism) in normal infants. *Acta Paediatrica Scandinavica* 41, Suppl. 89, 1–151.

Kasparek, G.T. (1989) Regulation of branched-chain 2-oxo acid dehydrogenase activity during exercise. *American Journal of Physiology* 256, E186–190.

Kielanowski, J. (1976) In: Cole, D.J.A., Borrman, K.N., Buttery, P.J., Lewis, D., Neale, R.J. and Swan, H. (eds) *Protein Metabolism and Nutrition. European Association for Animal Production,* publication 16. Butterworth, London.

Kimball, F.C., Farrell, P.A. and Jefferson, L.S. (2002) Exercise effects on muscle insulin signalling and action: role of insulin in translational control of protein synthesis in skeletal muscle by amino acids or exercise. *Journal of Applied Physiology* 93, 1168–1180.

Kjaer, M. (2004) Role of the extracellular matrix in adaptation of tendon and skeletal muscle to mechanical loading. *Physiological Reviews* 84, 649–698.

Kleiber, M. (1932) Body size and metabolism. *Hilgardia* 6, 315–353.

Kleiber, M. (1947) Body size and metabolic rate. *Physiological Reviews* 27, 511–541.

Kleiber, M. (1975) *The Fire of Life: an introduction to animal energetics.* Robert E. Krieger, New York.

Kosterlitz, H.W. (1947) The effects of changes in dietary protein on the composition and structure of the liver cells. *Journal of Physiology* 106, 194–210.

Larrson, L., Li, X. and Frontera, W.R. (1997) Effects of aging on shortening velocity and myosin isoform composition in single human skeletal muscle cells. *American Journal of Physiology* 272, C638–649.

Laurent, G.J. and Millward, D.J. (1980) Protein turnover during skeletal muscle hypertrophy. *Federation Proceedings* 39, 42–47.

Laurent, G.J., Sparrow, M.P. and Millward, D.J. (1978)

Turnover of muscle protein in the fowl: changes in rates of protein synthesis and breakdown during hypertrophy of the anterior and posterior latissimus dorsi muscles. *Biochemical Journal* 176, 407–417.

Lecker, S.H., Jagoe, R.T., Gilbert, A., Gomes, M., Baracos, V., Bailey, J., Price, S.R., Mitch, W.E. and Goldberg, A.L. (2004) Multiple types of skeletal muscle atrophy involve a common program of changes in gene expression. *FASEB Journal* 18, 39–51.

Lewis, S.E.M., Kelly, S.J. and Goldspink, D.F. (1984) Pre- and post-natal growth and protein turnover in smooth muscle, heart and slow- and fast-twitch skeletal muscles of the rat. *Biochemical Journal* 217, 517–526.

Lobley, G.E. (1998) Nutritional and hormonal control of muscle and peripheral tissue metabolism in farm species. *Livestock Production Science* 56, 91–114.

MacDougall, J.D., Gibala, M.J., Tarnopolsky, M.A., MacDonald, J.R., Iterisano, S.A. and Yarasheki, H.E. (1995) The time-course for elevated muscle protein synthesis following heavy resistance exercise. *Canadian Journal of Applied Physiology* 20, 480–486.

Marx, J.O., Kraemer, W.J., Nindl, B.C. and Larsson, L. (2002) Effects of aging on human skeletal muscle myosin heavy-chain mRNA content and protein isoform expression. *Journal of Gerontology* A 57, B232–238.

McMahon, T.A. (1973) Size and shape in biology. *Science* 179, 1201–1204.

McMahon, T.A. (1975) Using body size to understand the structural design of animals: quadruped locomotion. *Journal of Applied Physiology* 39, 619–627.

Millward, D.J., Garlick, P.J., Stewart, R.J.C., Nnanyelugo, D.O. and Waterlow, J.C. (1975) Skeletal-muscle growth and protein turnover. *Biochemical Journal* 150, 235–243.

Munro, H.N. and Downie, E.D. (1964) Relationship of liver composition to intensity of protein metabolism in different animals. *Nature* 203, 603–604.

Peragón, J., Barroso, J.B., de la Higuera, M. and Lupiáñez, J.A. (1998) Relationship between growth and protein turnover rates and nucleic acids in the liver of rainbow trout (*Onchorhynchus mykiss*) during development *Canadian Journal of Fish and Aquatic Science* 55, 649–657.

Peragón, J., Barroso, J.B., García-Salguero, L., de la Higuera, M. and Lupiáñez, J.A. (2001) Growth, protein-turnover rates and nucleic-acid concentrations in the white muscle of rainbow trout during development. *International Journal of Cell Biology and Biochemistry* 33, 1227–1238.

Peters, R.H. (1988) The relevance of allometric comparisons to growth, reproduction and nutrition in primate and man. In: Blaxter, K. and Macdonald, I. (eds) *Comparative Nutrition*, John Libbey, London, pp. 1–20.

Porter, R.K. and Brand, M.D. (1995) Cellular oxygen consumption depends on body mass. *American Journal of Physiology* 269, R226–228.

Pratt, O.E. (1981) The entry of amino acids into cells. In: Waterlow, J.C. and Stephen, J.M.L. (eds) *Nitrogen Metabolism in Man.* Applied Science Publishers, London, pp. 17–37.

Prothero, J.W. (1982) Organ scaling in mammals: the liver. *Comparative Biochemistry and Physiology A,* 71, 567–577.

Reeds, P.J. and Harris, C.I. (1981) Protein turnover in animals: man in his context. In: Waterlow, J.C. and Stephen, J.M.L. (eds) *Nitrogen Metabolism in Man.* Applied Science Publishers, London, pp. 391–408.

Reeds, P.J. and Garlick, P.J. (1984) Nutrition and protein turnover in man. *Advances in Nutrition Research* 6, 93–138.

Refsum, H.E., Gjessing, L.R. and Strømme, S.P. (1979) Changes in plasma amino acid distribution and urine amino acids excretion during prolonged heavy exercise. *Scandinavian Journal of Clinical and Laboratory Investigation* 39, 407–413.

Rennie, M.J. and Tipton, K.D. (2000) Protein and amino acid metabolism during and after exercise and the effects on nutrition. *Annual Reviews of Nutrition* 20, 457–483.

Rennie, M.J., Edwards, R.H.T., Krywawych, S., Davies, C.T.M., Halliday, D., Waterlow, J.C. and Millward, D.J. (1981) Effect of exercise on protein turnover in man. *Clinical Science* 61, 627–639.

Roberts, S.B. and Young, V.R. (1988) Energy costs of fat and protein deposition in the human infant. *American Journal of Clinical Nutrition* 48, 951–955.

Sander, G., Topp, H., Heller-Schöch, G., Wieland, J. and Schöch, G. (1986) Ribonucleic acid turnover in man: RNA catabolites in urine as measure for the metabolism of each of the three major species of RNA. *Clinical Science* 71, 367–374.

Schmidt-Nielsen, K. (1984) *Scaling: Why Animal Size is so Important.* Cambridge University Press, Cambridge, UK.

Schöch, G. and Topp, H. (1994) Interrelations between the degradation rates of RNA and protein and the energy turnover rates. In: Räihä, C.C. (ed.) *Protein Metabolism in Infancy.* Raven Press, New York.

Schöch, G., Topp, H., Held, A., Heller-Schöch, G., Ballau, F.A., Manz, F., and Sander, G. (1990) Interrelation between whole body turnover rates of RNA and protein. *European Journal of Clinical Nutrition* 44, 647–658.

Schofield, W.N. (1985) Predicting basal metabolic rate, new standards and review of previous work. *Human Nutrition: Clinical Nutrition* 39C, Suppl. 1, 5–41.

Schønheyder, F., Heilshov, N.C.S. and Olesen, K. (1954) Isotopic studies on the mechanism of negative nitrogen balance produced by immobilization.

Scandinavian Journal of Clinical and Laboratory Investigation 6, 178–188.

Scornik, O.A. and Botbol, V. (1976) Role of changes in protein degredation in the growth of regenerating liver. *Journal of Biological Chemistry* 251, 2891–2897.

Sharman, M.J., Newton, R.U., Triplett-McBride, T., McGuigan, M.R., McBride, J.M., Hakkinen, A., Hakkinen, K. and Kraemer, W.J. (2001) Changes in myosin heavy chain composition with heavy resistance training in 60- to 75-year old men and women. *European Journal of Applied Physiology* 84, 127–132.

Short, K.R. and Nair, K.S. (1999) Mechanisms of sarcopenia of aging. *Journal of Endocrinological Investigation* 22, 95–105.

Short, K.R., Vittore, J.L., Bigelow, M.L., Procter, D.N. and Nair, K.S. (2004) Age and aerobic exercise training effects on whole body and muscle protein metabolism. *American Journal of Physiology* 286, E92–101.

Smith, R.H., Palmer, R.M. and Reeds, P.J. (1983) Protein synthesis in isolated rabbit forelimb muscles: the possible role of metabolites of arachidonic acid in the response to intermittent stretching. *Biochemical Journal* 214, 153–161.

Smith, R.J. (1984) Allometric scaling in comparative biology: problems of concept and method. *American Journal of Physiology* F246, R152–160.

Spady, D.W., Payne, P.R., Picou, D. and Waterlow, J.C. (1976) Energy balance during recovery from malnutrition. *American Journal of Clinical Nutrition* 29, 1076–1088.

Stahl, W.R. (1963) Organ weights in primates and other mammals. *Science* 150, 1038–1041.

Taylor, C.R. (1987) Structural and functional limits to oxidate metabolism: insights from scaling. *Annual Review of Physiology* 49, 135–146.

Tessari, P., Garibotto, G., Inchiostro, S., Robando, C., Saffioti, S., Vettore, M., Zanetti, M., Russo, R. and Deferrari, G. (1996) Kidney, splanchnic, and leg protein turnover in humans. *Journal of Clinical Investigation* 98, 1481–1492.

Thompson, M.G. and Palmer, R.M. (1998) Signalling pathways regulating protein turnover in muscle. *Cellular Signalling* 10, 1–11.

Thomson, R.Y., Heagy, F.C., Hutchison, W.C. and Davidson, J.N. (1953) The deoxyribonucleic acid content of the rat cell nucleus and its use in expressing the results of tissue analysis, with particular reference to the composition of liver tissue. *Biochemical Journal* 53, 460–474.

Timmons, J.A., and Sundberg, C.J. (2004) Determinants of human exercise performance. *Physiology News* 55, 41–42.

Waterlow, J.C. (1968) Observations on the mechanism of adaptation to low protein intakes. *Lancet* ii, 1091–1092.

Waterlow, J.C. (1984) Protein turnover with special reference to man. *Quarterly Journal of Experimental Physiology* 69, 409–438.

Waterlow, J.C. and Weisz, T. (1956) The fat, protein and nucleic acid content of the liver in malnourished human infants. *Journal of Clinical Investigation* 35, 346–354.

Webster, A.J.F. (1988) Comparative aspect of the energy exchange. In: Blaxter, K.L. and Macdonald, I. (eds) *Comparative Nutrition*. John Libbey, London, pp. 37–51.

Welle, S. and Nair, K.S. (1990) Relationship of resting metabolic rate to body composition and protein turnover. *American Journal of Physiology* 258, E990–998.

Welle, S., Thornton, C. and Statt, M. (1995) Myofibrillar protein synthesis in young and old human subjects after three months of resistance training. *American Journal of Physiology* 268, E422–427.

Welle, S., Bhatt, K. and Thornton, C.A. (1999) Stimulation of myofibrillar synthesis by exercise is mediated by more efficient translation of mRNA. *Journal of Applied Physiology* 86, 1220–1225.

West, G.B., Brown, J.H. and Enquist, B.J. (1999) The fourth dimension of life: fractal geometry and allometric scaling of organisms. *Science* 284, 1677–1679.

Wetterfors, J. (1965) General aspects on the metabolism of serum albumin. In: Koblet, M., Vesin, P., Diggelman, H. and Barundun, S. (eds) *Physiology and Pathophysiology of Protein Metabolism.* Hans Huber, Berne, Switzerland, pp. 83–88.

Williams, B.D., Wolfe, R.R., Bracy, D.P. and Wasserman, D.H. (1996) Gut proteolysis contributes essential amino acids during exercise. *American Journal of Physiology* 270, E85–90.

Wolfe, R.R., Goodenough, R.D., Wolfe, M.H., Royle, G.T. and Nadel, E.R. (1982) Isotopic analysis of leucine and urea metabolism in exercising humans. *Journal of Applied Physiology* 52, 458–466.

Yarasheki, K.E., Campbell, J.A., Smith, K., Rennie, M.J., Holloszy, J.O. and Bier, D.M. (1992) Effect of growth hormone and resistance exercise on muscle growth in young men. *American Journal of Physiology* 262, E261–267.

Yarasheki, K.E., Zachwieja, J.J. and Bier, D.M. (1993) Acute effects of resistance exercise on muscle protein synthesis rate in young and elderly men and women. *American Journal of Physiology* 265, E210–214.

Yarasheki, K.E., Zachwieja, J.J., Campbell, J.A. and Bier, D.M. (1995) Effect of growth hormone and resistance exercise on muscle growth and strength in older men. *American Journal of Physiology* 268, E268–276.

Yarasheki, K.E., Pak-Loduca, J., Hasten, D.L., Obert, K.A., Brown, M.B. and Sinacore, D. (1999) Resistance exercise training increases mixed muscle protein synthesis in frail women and men ≥ 76 yr old. *American Journal of Physiology* 277, E118–125.

12

Whole Body Protein Turnover at Different Ages and in Pregnancy and Lactation

12.1 Premature Infants

The first study in this age-group was made by Nicholson by the EP method with ammonia as end-product as long ago as 1970. Eleven years later the subject was taken up by Jackson *et al.* (1981), who studied three babies, fed expressed breast milk (EBM) with multiple oral dosage of ^{15}N-glycine for 72 h. They found that even after this long period no tracer at all appeared in urea. Jackson's interpretation was that glycine is a semi-essential amino acid, so that the rapidly growing infant cannot meet the demand for it by *de novo* synthesis; all available glycine is needed for incorporation into tissue protein and none is available for conversion to urea. This finding has been confirmed by several others (e.g. Stack *et al.*, 1989; Van Langen *et al.*, 1992; Van Goudoever *et al.*, 1995a). A factor contributing to the low labelling of urea may be that human milk, fed in many of the studies, contains substantial amounts of preformed urea, which could dilute urea synthesized from glycine. This point was investigated by Darling *et al.* (1993), who showed that the enrichment of urea decreased as the amount of preformed urea in the feed increased. However, the failure of urea to become labelled has not always been found. A number of studies, summarized at C in Table 12.1, give synthesis rates that agree with those obtained with NH_3 (A and B) (Cauderay *et al.*, 1988).

Table 12.1. Whole body protein synthesis in premature infants in the fed state.

	Tracer and route	Synthesis mg protein kg^{-1} h^{-1}	n	N	References
A	Glycine, single dose, oral, EP NH_3	428	4	39	1–4
B	Glycine, multiple dose, oral, EP NH_3	401	5	95	5–9
C	Glycine, multiple dose, oral, EP urea	423	6	84	10
D	Leucine, IV infusion, oral feeding[a]	450	4	50	11–14
E	Leucine or phenylalanine, IV infusion, TPN, healthy	457	2	11	15–17
F	Leucine, IV infusion, TPN, sick	370	4	32	18–21

A–E: healthy infants; F: respiratory distress syndrome or diarrhoea. n, number of studies; N, number of subjects; TPN, total parenteral nutrition; EP, end-product.

[a]Results based on leucine enrichment have been corrected × 1.25 to bring them into line with those calculated from KIC (see Chapter 6).

References: 1. Nicholson, 1970; 2. Stack *et al.*, 1989; 3. Wutzke *et al.*, 1992; 4. Plath *et al.*, 1996; 5. Jackson *et al.*, 1981; 6. Catzeflis *et al.*, 1985; 7. Pencharz *et al.*, 1989a; 8. Darling *et al.*, 1993; 9. van Goudoevor *et al.*, 1995a; 10. Cauderay *et al.*, 1988; 11. de Benoist *et al.*, 1984; 12. Beaufrère *et al.*, 1990; 13. Beaufrère *et al.*, 1992; 14. van Goudoever *et al.*, 1995a; 15. Battista *et al.*, 1996; 16. Poindexter *et al.*, 2001; 17. Duffy *et al.*, 1981; 18. Mitton, 1991; 19. Mitton and Garlick, 1992; 20. van Goudoever *et al.*, 1995b; 21. Rivera *et al.*, 1993.

The data in Table 12.1, A and B, relate to healthy babies born at 28–32 weeks of gestation, mostly with weights appropriate to gestational age (AGA), and studied at varying times after birth. Where information is available, they were gaining weight at a rate of about 15 g kg^{-1} day^{-1}. Protein intakes were in the region of 1–3 g kg^{-1} day^{-1}, and within this range seemed to have no influence on the synthesis rate. There is no evidence that human milk promotes higher rates of synthesis than formulae. Energy intakes were about 330–420 kJ kg^{-1} day^{-1}, which allows an adequate margin for the energy cost of growth, since the resting metabolic rate of prematures has been estimated as about 200–250 kJ kg^{-1} day^{-1} as soon as the infants have begun to grow rapidly (Beaufrère *et al.*, 1990; Bauer *et al.*, 1997; Kingdon *et al.*, 2000; Leitch and Denne, 2000). In their pioneer study, Nissim *et al.* (1983) found linear correlations between net protein gain and protein and energy intakes; perhaps more interesting is that there was also a correlation between the proportion of total protein breakdown derived from muscle, measured by methylhistidine excretion and conceptual age between 26 and 37 weeks, illustrating the rapid development of muscle in the fetus at this stage (Fig. 12.1).

Comparisons of synthesis rates of babies appropriate (AGA) or small (SGA) for gestational age have been inconclusive:

		Method
SGA > AGA:	Pencharz *et al.*, 1981	^{15}N, urea EP
	Van Goudoever, 1995a	^{15}N, NH_3 EP
SGA = AG:	Van Goudoever, 1995a	Leucine
SGA < AGA:	Cauderay, 1988	^{15}N, urea EP

Turnover rates of healthy prematures have also been measured by the precursor method, with results shown in Table 12.1, D–E. De Benoist *et al.* (1984) introduced the idea of measuring leucine enrichment in urine rather than in plasma to make the method less invasive. In pilot studies they found that the ratio of the enrichment of leucine in plasma to that in urine was on average 0.93, while Wykes *et al.* (1990) found the enrichments to be identical. Measurement of enrichment in urine might well be more widely used. The rates in these four studies agree well with those obtained by the end-product method. Van Goudoever *et al.* (1995a) used [^{15}N]-glycine and [^{13}C]-leucine together and obtained identical results. It is interesting that in the study of Beaufrère *et al.* (1992) with protein intakes ranging from 1.0 to 4.5 g kg^{-1} day^{-1}, there was no relation between intake and synthesis, but a strong negative correlation with breakdown and hence a positive correlation with leucine balance.

The parenterally fed infants in E were said to be clinically stable and not requiring artificial ventilation, whereas those in F were suffering

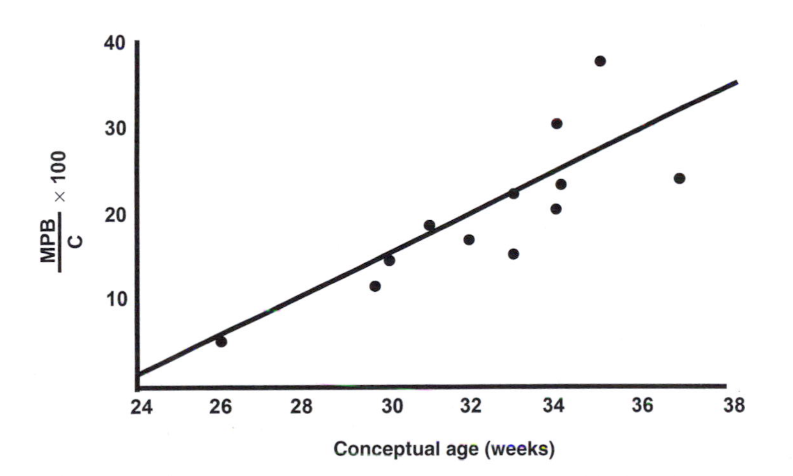

Fig. 12.1. Percentage of total body protein breakdown (C) derived from muscle protein breakdown (MPB) in premature infants at different conceptional ages. Whole body breakdown was measured as the end-product average flux after a single intravenous dose of [^{15}N] glycine. Muscle protein breakdown was determined from excretion of 3-methylhistidine. The regression equation is: $y = -54 + 2.3x$, $r = 0.865$. Reproduced from Nissim *et al.* (1983), by courtesy of the *Journal of Pediatric Gastroenterology and Nutrition.*

from the respiratory distress syndrome or had diarrhoea, which evidently depressed the synthesis rate. In a study by Denne *et al.* (1996) extremely premature infants (26 weeks gestation, ~0.8 kg) failed to respond to parenteral feeding in the way that normal neonates did, with a decrease in protein degradation.

The overall conclusion is that protein synthesis in healthy premature infants is about 2.5 times that in adults, consistent with the difference in their metabolic rates.

12.2 Neonates

It is arguable whether infants as studied by Chien *et al.* (1993) *in utero* just before Caesarian delivery can legitimately be regarded as neonates. Be that as it may, the calculated fetal rate of protein synthesis (uptake–oxidation) was remarkably low, at 183 mg protein $kg^{-1} h^{-1}$.

The results of other studies are summarized in Table 12.2. In spite of differences in clinical state, method of feeding (oral or TPN) and tracer, they are reasonably consistent. The weighted average is, as might be expected, somewhat lower than in prematures.

12.3 Infants 6 months–2 years

The only information available in this age-group comes from work on infants who have recovered from malnutrition. These children are of normal weight for height, but still stunted in height for their age.

In the 1960s, when we were working in Jamaica, severe malnutrition in children was common and had a high death rate. It seemed possible that treatment failed because of an irreversible breakdown of the machinery of protein synthesis. This hypothesis stimulated development of the method of measuring whole body protein turnover by constant infusion of ^{15}N-glycine with urea as end-product (Chapter 7). In the first study (Picou and Taylor-Roberts, 1969) a variety of problems were tackled: comparison between intravenous and intragastric infusion of tracer and the effect of level of protein intake, neither of which showed an important difference; and a comparison of glycine with ^{15}N-labelled egg protein.[1] The question here was whether the metabolism of glycine could be regarded as representative of that of total amino-N (see Chapter 7), but as there was only enough material for one trial with egg protein the results showed only that they were of the same order of magnitude:

Table 12.2. Whole body protein synthesis in neonates in the fed state.

	Author	Feeding	Tracer	S mg $kg^{-1} h^{-1}$	N
Healthy	Denne *et al.*, 1991	Oral	Leucine IV[c]	355	11
	Denne *et al.*, 1996	TPN	Phenylalanine[d]	364	10
After surgery[a]	Pencharz *et al.*, 1989b	TPN	Glycine, EP av.	337	10
	Duffy and Pencharz, 1986	Oral	Glycine, EP urea	525	12
		IV	Glycine, EP urea	362	12
	Jones *et al.*, 1995	TPN	Leucine, IV	401	18
Sick[b]	Bresson *et al.*, 1991	TPN	Leucine, IV,	354	8
			Weighted mean:	390	

[a]At least 5 days after operation for gastrointestinal problems.
[b]With respiratory distress syndrome or diarrhoea.
[c]Corrected × 1.25 for KIC.
[d]Corrected × 2.5 for molar concentration relative to leucine, and × 1.25 for precursor activity.
The study of Denne and Kalhan (1987) is omitted from this table because the babies were in the post-absorptive state. They were infused with leucine. After correcting for KIC mean S was 266 mg protein $h^{-1} kg^{-1}$.

[1] The labelled egg protein was prepared in two steps by the late Dr N.W. Pirie and Dr Marie Coates, to whom we are grateful. First, labelled leaf protein was produced from plants in which $^{15}NH_3$ was the sole source of N. This labelled leaf protein was then fed to a hen.

^{15}N-glycine 210; ^{15}N egg-protein, white 163, yolk 268 mg protein kg^{-1} h^{-1}.

Several years passed before the work on protein turnover in infants was taken up again. A modification of the original method was introduced: instead of tracer being infused intravenously or by naso-gastric tube it was given in multiple doses added to each feed over a period of 48 h (Golden *et al.*, 1977a). It was found that increasing the protein and energy intake in the days before the test had no effect on synthesis but caused a marked decrease in breakdown. The next step was the introduction of ammonia as a second end-product, so that end-product averages and ratios could be calculated (Golden *et al.*, 1977b; Waterlow *et al.*, 1978) (see Chapter 7). Comparison of S_{av} by single dose and multiple dose showed good agreement, so Table 12.3 shows the mean of the results by the two methods.

Golden *et al.* (1977b), reviewing the work in children who had recovered from malnutrition, found that synthesis rate was related to food intake but as protein and energy intakes were not varied independently no causal conclusions could be drawn. The relationship of breakdown to food intake was much less tight (Fig. 12.2).

Later Jackson *et al.* (1983) explored in more detail the relations of protein turnover to protein and energy intake. They gave two levels of protein, 1.7 and 0.66 g kg^{-1} day^{-1}, and three levels of energy – 80, 90 and 100 kcal kg^{-1} day^{-1}. The measurements were made in the fed state. Over this rather narrow range of energy intakes the results show no important differences with energy intake, and so have been combined in Table 12.3. The authors noted that there was more variation within each dietary group than between the groups. The synthesis rates were very close with high and low protein intakes when measured with urea as end-product; with ammonia as end-product both synthesis and breakdown rates on the low protein intake were only about half the rates on high protein. This finding emphasizes again the value of using both end-products. The contrast with the lack of sensitivity to protein intake found by Picou and Taylor Roberts (1969) is probably because Jackson's low protein intake (0.66 g kg^{-1} day^{-1}) was below the children's requirement and below that of Picou (1.2 g kg^{-1} day^{-1}).

A final study with the EP method by the Jamaican group (Golden and Golden, 1992) explored the effect during recovery of different levels of zinc supplementation. Three studies were done in each child: first, when they had lost oedema and were entering the stage of rapid growth; again when they had regained about half their weight deficit; and finally when they had regained normal weight for height. Eleven children were divided into three groups who received throughout different amounts of supplementary zinc. We give here (Table 12.3) only the results after full recovery (see Chapter 14). It appears that zinc in sufficient dosage not only increased the rate of synthesis, but, from the ratio of

Table 12.3. Whole body synthesis rates in previously malnourished recovered children, measured with [^{15}N]-glycine by the end-product method.

Author	S mg protein kg^{-1} h^{-1}	S_A/S_U	N
Golden *et al.*, 1977a	262[a]	–	5
Waterlow *et al.*, 1978	212[b]	0.72	8
Golden and Golden, 1992			
Low zinc	273	0.72	4
Medium zinc	252	0.95	4
High zinc	329	1.36	3
Jackson *et al.*, 1983			
High protein (1.7 g kg^{-1})	230	0.76	5
Low protein (0.7 g kg^{-1})	167	0.42	2
Weighted average	250[c]		

S_A, S_U, synthesis rate with ammonia or urea as end-product. N, number of subjects.
[a]Results based on urea only as end-product. All others are the arithmetic average of S_A and S_U.
[b]Results with single and multiple dosage did not differ significantly and so have been combined.
[c]Excluding Jackson, low protein.

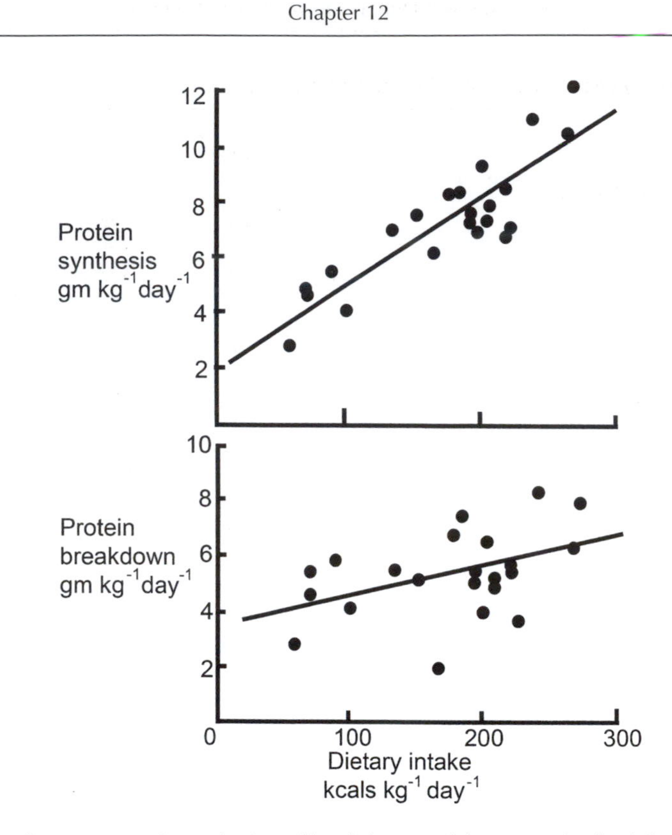

Fig. 12.2. Relation between protein synthesis and breakdown and dietary intake (kcal day^{-1}) in children recovering from malnutrition. Since the composition of the diet was constant, the protein intake varied with the energy intake. Reproduced from Golden *et al.* (1977b), by courtesy of the *American Journal of Clinical Nutrition.* © Am. J. Clin. Nutr. American Society for Nutrition.

S_A/S_U, promoted it particularly in peripheral tissues. The authors concluded that zinc promoted the deposition of lean tissue rather than fat, and we may add the possibility that it does so particularly in muscle.

12.4 Older Children

Data on protein turnover in older children are extremely scanty. The only results that we have found are shown in Table 12.4. They are too variable for useful conclusions to be drawn, except that in these pubertal or prepubertal children the whole body synthesis rate was roughly halfway between the adult and the premature infant.

12.5 Pregnancy

In pregnancy the mother has to adapt her protein metabolism to provide for the increasing needs of the fetus, the uterus, the placenta and the mammary glands. It has been calculated that a normal pregnancy requires the deposition of ~900 g of new protein, ~1 g day^{-1} in the first trimester, 3 g day^{-1} in the second and 6 g day^{-1} in the third. There has only been a handful of studies of protein turnover in pregnancy to show how these needs are met and, as always, there are methodological differences.

These differences should, in theory, disappear if the results during pregnancy are related to those in the non-pregnant state (Table 12.5). Unfortunately, in most of the studies the pregnant and non-pregnant women were not the same, but at least the measurements were made by the same method. The pioneer longitudinal study of de Benoist *et al.* (1985) in Jamaica gave results out of line with the others. Jackson has attempted to make a correction, but the original numbers are included here. It seems reasonable to conclude that in pregnancy there is a modest increase in whole body synthesis – about 15% – which is

Table 12.4. Protein synthesis in older children and adolescents.

Author	Average age years		Synthesis mg protein kg^{-1} h^{-1}
A. Measurement with ^{15}N-glycine			
Kien and Camitta, 1983, EP urea	10.8	Fed	175
Kien *et al.*, 1978, EP urea	12	Fed	158
Vaisman *et al.*, 1992, NH_3 and urea	17.3	Fed	143
B. Measurement with ^{13}C leucine			
Kien *et al.*, 1996, leucine/KIC	8.7	PA	223
		Fed	211
Avslanian, 1996, leucine/KIC	10.5	PA	288
Salman *et al.*, 1996, leucine/KIC?	11.2	PA	177
Avslanian and Kalhan, 1996, leucine/KIC	13.6	PA	265

PA, post-absorptive.
All these values are from control children in studies of various pathological states – cystic fibrosis, sickle cell disease, burns.

Table 12.5. Whole body protein synthesis in pregnancy.

Author		Per cent of non-pregnant		
		Trimester		
		1	2	3
[a]Fitch and King, 1987	Glycine SD oral, EP NH_3	–	–	1.10
[b]de Benoist *et al.*, 1985	Glycine, primed intermittent, EP av	1.53	1.53	1.12
[c]Willomet *et al.*, 1992	Glycine, SD oral EP av	0.97	1.28	1.25
[d]Jackson *et al.*, 2000; Duggleby and Jackson, 2001, 2002	Glycine, SD oral EP av	–	0.94	1.00
[e]Thompson and Halliday, 1992	Leucine Fasting	1.06	1.22	1.41
[f]Kalhan *et al.*, 1998	Leucine			
	Fasting	0.99	1.04	1.06
	Fed	0.91	1.22	1.09
[g]Whittaker *et al.*, 2000	Leucine, fasting	–	–	1.16
	+ AA/insulin	–	–	1.18

[a]P and non-P were different women.
[b]Sequential studies on the same women.
[c]P and non-P different women. P studied sequentially through pregnancy.
[d]P and non-P different women. Same women at two stages of pregnancy. The value for NPN is given in Jackson and Grove (2000).
[e]Non-P controls a different group. Sequential measurements on P women.
[f]P and non-P different groups. Some P studied sequentially.
[g]Non-P measured on same women 12 weeks after delivery.

established by mid-pregnancy. This is important for the outcome, since Duggleby and Jackson (2001, 2002) have shown a relationship between increase in synthesis and weight and length at birth. It is in the third trimester that the growth in fetal protein mass is most rapid, and Naismith and Morgan (1976) and Naismith *et al.* (1982), from experiments in rats, made the interesting suggestion that in the second trimester the mother stores protein in muscle which is drawn upon in the third trimester to support the growth of the fetus. The rate of intravascular albumin synthesis is also substantially increased in late pregnancy (Olufemi *et al.*, 1991).

The information about breakdown in pregnancy is somewhat conflicting; in the fasted state it was either higher (Whittaker *et al.*, 2000) or did not change (Thompson and Halliday, 1992; Kalhan *et al.*, 1998), while in the fed state it increased more or less in parallel with synthesis (Willomett *et al.*, 1992; Kalhan *et al.*, 1998). According to Whittaker *et al.* (2000) pregnancy did not alter the sensitivity of protein breakdown to insulin. Therefore the adaptive response evidently does not involve a conservation of protein by reduction in breakdown.

The second adaptation is a decrease in amino acid oxidation and urea synthesis, particularly in the first and second trimesters. Duggleby and Jackson (2002) showed a correlation between decrease in urea production and the baby's birth weight. Some data from Kalhan *et al.* (1998) are shown in Table 12.6. The fall in urea production closely parallelled the fall in transamination of leucine to KIC. Moreover, the proportion of urea production hydrolysed in the colon also increased in the second trimester (McClelland *et al.*, 1997), thus further promoting the economy of nitrogen.

All the studies mentioned above were done in healthy apparently adequately nourished women with body weights of ~60 kg, in the USA, the UK, the Gambia (Willomett *et al.*, 1992) and in Jamaica (McClelland *et al.*, 1997). In the Jamaican women there was an increased urinary output of α-oxoproline compared with pregnant women in the UK, the discrepancy growing larger as pregnancy progressed (Jackson *et al.*, 1997). This finding suggests a marginal deficiency of glycine (see Chapter 4), an amino acid for which there is a particularly high demand in pregnancy (Widdowson *et al.*, 1979).

So far we have focused on protein turnover in the pregnant mother, but what about her baby? Here there is a dearth of information. In a seminal study by Chien *et al.* (1993), protein turnover was measured by infusions of labelled leucine and phenylalanine in mothers undergoing Caesarian section. Some results with leucine, calculated from their data, are shown in Table 12.7. The inflow of leucine to the human fetus is only ~10% of the total maternal flux and the uptake for synthesis and oxidation is only ~3% of the maternal flux. Synthesis by the fetus appears to be rather inefficient, since the ratio of synthesis:deposition is nearly 7, compared with 5 in prematures and about 2 in older infants (Chapter 11, section 11.2). Nevertheless, the rate of deposition, when translated into gain of tissue containing 12% protein, amounts to a weight gain of 16 g kg^{-1} day^{-1}, agreeing with the results found in premature infants.

The amino acid exchanges between mother

Table 12.6. Urea production and rate of leucine transamination in pregnant women.

		Transamination μmol kg^{-1} d^{-1}	Urea production μmol kg^{-1} h^{-1}
Fasting:	non-pregnant	80	282
	T1[a]	56	204
	T2	62	174
	T3	44	156
Fed:	non-pregnant	89	252
	T1	70	216
	T2	80	216
	T3	57	150

[a]T, trimester.
Data of Kalhan *et al.*, 1998.

Table 12.7. Leucine turnover in the human and the ovine fetus.

	A. Human	B. Sheep
Weight of mother, kg	80	35
Weight of fetus, kg	3.4	3.3
Total maternal + fetoplacental leucine Flux, μmol h^{-1}	7620	7350
Total leucine inflow by umbilical vein μmol h^{-1}	740	1400
Fetal synthesis μmol h^{-1}	1150	810
Fetal breakdown μmol h^{-1}	980	380
Fetal accretion μmol h^{-1}	170	430
Fetal oxidation μmol h^{-1}	70	420
Inflow, % of maternal flux	10	19
Oxidation + accretion, % of inflow	32	61

A. Recalculated from Chien *et al.*, 1993.
B. Recalculated from Ross *et al.*, 1996.

and fetus have been studied in great detail in sheep by Battaglia and his colleagues. Their results, from a study with leucine (Ross *et al.*, 1996) are also shown in Table 12.7. There are some interesting differences between the two species: the fetal weights and the total maternal + feto-placental fluxes are the same, although the woman's body weight is twice that of the sheep. The rate of accretion of fetal protein is more than twice as high in the sheep, and this is achieved with greater economy, since synthesis/accretion is close to 2, compared with 7 in the woman; on the other hand the fetal lamb oxidizes a much higher proportion of the inflow than the human fetus. Finally, the fetal lamb disposes, by accretion and oxidation, of a larger proportion of its amino acid supply than the human. If one can generalize from this single study, it would seem that in the woman the umbilical amino acid supply should provide a comfortable margin of safety for the fetus. We do not know how far this would apply in undernourished populations where the proportion of low birthweight babies is very high.

The data in the table apply only to leucine. Battaglia's group has also explored the transport of other essential amino acids from placenta to fetus (Paolini *et al.*, 2001). One factor affecting the flux across the placenta is the concentration of amino acid in the maternal plasma. There are also significant differences in transplacental clearance, which the authors attribute to competition for shared transporters and the use of separate transport systems. An additional point, not made in Paolini's paper, is a highly significant correlation between the umbilical steady-state uptake of the different essential amino acids and their molar concentrations in fetal protein (Fig. 12.3), assuming that Widdowson's data for humans can be applied to sheep.

It is not known what proportion of the amino acids used for synthesis are derived from protein breakdown, but it is interesting that the ovine fetus, like the premature human infant, is insensitive to the action of insulin in reducing breakdown that occurs in adults (Liecht and Denne, 1998).

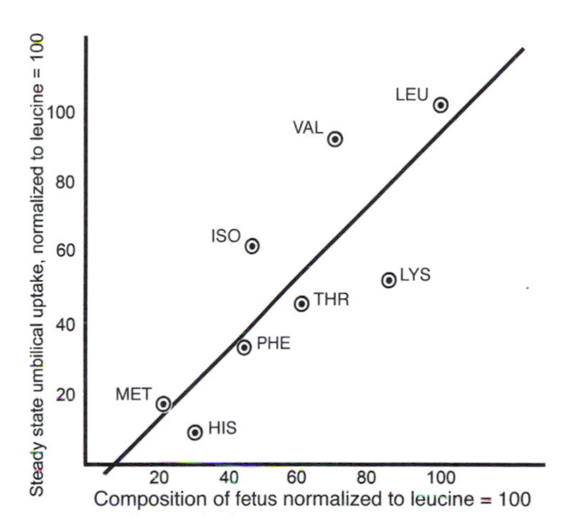

Fig. 12.3. Uptake of essential amino acids by the fetus related to the amino acid composition of the fetus. Data on uptake from Paolini *et al.*, 2001; on composition of fetus from Widdowson *et al.*, 1979.

Low birthweight is extremely common in less developed countries and many epidemiological studies have shown that it may have serious consequences for the future health and development of the baby: yet we know almost nothing about its metabolic causes and whether they include preventable deviations in protein and amino acid metabolism.

12.6 Lactation

Very little has been done on protein turnover in lactating women. In contrast, there have been many studies on farm animals, because of the economic importance of the relation between food supply and milk intake. For an excellent review of this subject see Bequette (1998). He states that at the beginning of lactation in the cow protein synthesis in the mammary gland increases 1000-fold, compared with +20% or so in the gastro-intestinal tract and liver. This huge increase in synthesis needs to be supported by an inflow of amino acids from other tissues, mainly from muscle, as in pregnancy (Naismith *et al.*, 1982; Pine *et al.*, 1994) but also from skin (Baracos *et al.*, 1991).

According to Bequette, there are genetic differences between cows in their ability to mobilize these reserves. We have no information about whether the same applies to humans. In the cow the ratio of total protein synthesis in the mammary gland to synthesis of milk protein is ~2.5:1, similar to the ratio of whole body synthesis to protein deposition found in young growing animals (see Chapter 11, section 11.2). About $^{1}/_{3}$ of the milk protein synthesized is degraded without being secreted. Pacheco *et al.* (2003) have measured in the sheep the partition to the mammary gland of the whole body fluxes of each essential amino acid. On average 28% (range 17–32%) of the whole body flux goes to the mammary gland. A relationship can be calculated between the supply of each amino acid and its concentration in milk protein (Fig. 12.4). It appears that the branched chain AAs are provided in excess; for the other EAAs the supply is approximately equal to the demand. The non-essentials (to the right of the line of identity in Fig. 12.4) are presumably synthesized *de novo* in the mammary gland to make up for the deficit in their supply.

Compared with the cow, lactation is a much smaller burden on protein synthesis in the human mother. A cow weighing 500 kg might produce

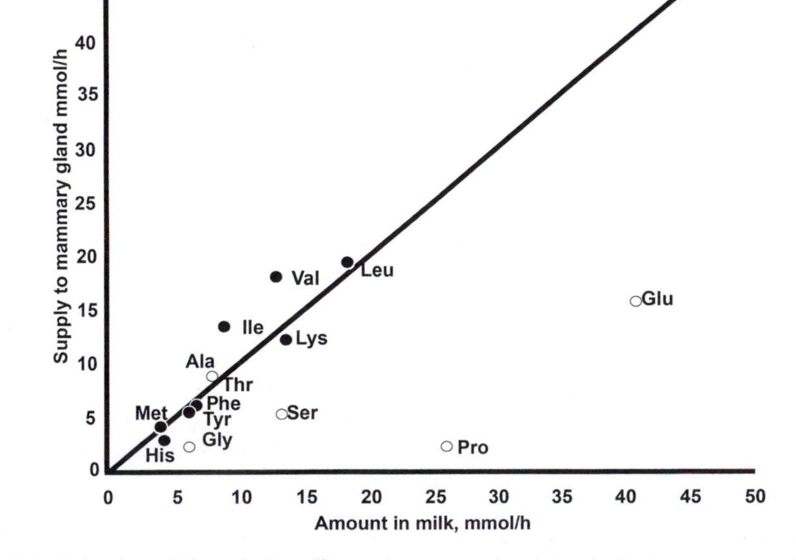

Fig. 12.4. Essential amino acid output in milk protein compared with uptake by the mammary gland. Data by courtesy of Dr Pacheco.

700 g of milk protein in a day, or 1.4 g kg^{-1} body weight (Pecheco *et al.*, 2003), and a goat produces a similar amount (Baracos *et al.*, 1991), whereas a nursing mother weighing 60 kg produces only about 10 g milk protein, or 0.17 g kg^{-1} day^{-1} – barely more than $^1/_{10}$ of the cow on the basis of body weight. Moreover, this amount of milk protein corresponds to only ~5% of the whole body flux compared with ~30% in the cow, so it is not surprising that studies of protein turnover in lactating women have not shown any drastic changes. Such studies as there have been have all come from the group at the Children's Nutrition Research Center in Houston, Texas, and they are rather contradictory; Motil *et al.* (1989a,b) found no difference in leucine flux or synthesis in the fasted state between lactating and non-lactating women, but those who were lactating had a significantly more negative nitrogen balance in spite of a larger energy intake. Within this group there was a positive correlation between whole body synthesis and milk production. However, in a later study on women in the fed state (1 g protein kg^{-1} day^{-1}) those who were lactating had significantly lower rates of both synthesis and breakdown than normal controls, but with higher rates of N retention (Motil *et al.*, 1996) because the reduction of breakdown was greater than that of synthesis; this downregulation of protein metabolism was considered to be an adaptive change. It was suggested that insulin and cortisol modulate the mobilization of endogenous stores of protein, while thyroid hormones modulate the diversion of these stores to milk protein synthesis (Motil *et al.*, 1994).

To our knowledge no studies have been made in lactating women on protein kinetics in the breast analogous to those in the human arm or leg; however, one study by Thomas *et al.* (1991), from the same laboratory as Motil, is of interest, if only from the point of view of methodology. They gave single doses of [^{13}C]-lysine IV and [^{15}N]-lysine by mouth to lactating women and controls and analysed the results with an 11-compartment model containing fast and slow protein pools (see Chapter 2). The standard deviations of the various rate-constants and fluxes were very high, and the only significant difference between lactating women and controls was that in the former the uptake of lysine into the slow protein pool was significantly lower. There were no independent measures of protein breakdown in the slow pool, but as far as it goes this is the only real evidence, in lactating women, of diversion of protein synthesis away from muscle. There was no significant difference in synthesis in the fast pool.

12.7 The Elderly

As the number of old people in most parts of the world increases, much research is being devoted to finding out in what ways their metabolism differs from that of the young. The elderly are, of course, a heterogeneous group, varying in age, in immobility and so on. A very obvious question is whether or not they have lower rates of whole body protein synthesis than healthy young people, in keeping with their apparently lower basal metabolic rates per kg body weight (FAO/WHO/UNU, 1985).

Our analysis of a number of studies (Table 12.8) shows a slightly higher rate of whole body synthesis in the elderly of both sexes when related to lean body mass but a lower rate related to body mass. Short *et al.* (2004) summarized 20 studies comparing protein turnover per kg body mass and per kg fat-free mass (FFM) in young and old subjects. In the majority there was no difference between the two ages; in a few synthesis was lower in the elderly, in two it was higher. In addition to their analysis of the literature Short's own most recent study seems to give a definitive answer to the question, partly because it was very large – they had 78 subjects of both sexes, ranging in age from 20 to 80 years – and partly because of their method of analysis. They showed that after covariate adjustment for fat-free mass whole body synthesis declined with age in both sexes at 4% per decade. Synthesis was also closely correlated with metabolic rate. The argument that it is erroneous to express such functions per unit body mass or FFM without adjustment has a long history (see Toth *et al.*, 1993).

Another important finding is that there was no difference between the young and the old in the efficiency of protein utilization (Millward *et al.*, 1997; Fereday *et al.*, 1997).

It is well recognized that fat-free mass or lean body mass as a proportion of body weight decreases with age, at least in Western societies, and that the decrease is largely at the expense of muscle, the visceral tissues being better pre-

Table 12.8. Ratio of whole body synthesis rate, old/young.

	Synthesis ratio	N[a]	n[b]
A. Per kg body weight			
Men	0.895	78	10
Women	0.875	42	7
Both sexes[c]	0.87	144	20
B. Per kg lean body mass			
Men	1.06	54	6
Women	1.06	42	7
Both sexes[c]	1.035	111	15

[a]N, number of subjects.
[b]n, number of groups studied.
[c]Includes groups in which results for men and women were not reported separately.
Sources: Winterer *et al.*, 1976; Uauy *et al.*, 1978; Fukegawa *et al.*, 1988, 1989; Lehmann *et al.*, 1989; Welle *et al.*, 1994; Benedik *et al.*, 1995; Balagopal *et al.*, 1997a; Boirie *et al.*, 1997; Morais *et al.*, 1997; Arnal, *et al.*, 2000.

served. Muscle mass has traditionally been estimated from creatinine excretion and more recently by dual-photon absorptiometry (Heymsfield *et al.*, 1990; Balagopal *et al.*, 1997b) or by magnetic resonance imaging (e.g. Morais *et al.*, 2000). By these various methods muscle mass has been found to account for ~45% of LBM in old people, compared with ~60% in the young.

A decreased proportion of muscle would be expected to increase the rate of whole body synthesis per unit LBM, as was observed in the undernourished Indian labourers studied by Soares *et al.* (1991) (Chapter 10). The relatively high rates of whole body synthesis per kg found by Bos *et al.* (2000) may also be explained in this way, because their elderly subjects were stated to be undernourished and therefore probably had a particularly low muscle mass.

The loss of muscle in the elderly, termed sarcopenia, seems to be caused by an overall reduction in the number of fibres and a specific atrophy of the type IIA fast-twitch fibres (Lexell *et al.*, 1988; Lexell and Downham, 1992; Singh *et al.*, 1999). The turnover rate of the depleted muscle protein has been investigated in several ways. Excretion of 3-methyl histidine (3 MH) has been proposed as a measure of the breakdown of myofibrillar proteins (see Chapter 13), so that the ratio of 3 MH:creatinine excretion should provide a normalized estimate of the fractional breakdown rate. This ratio was not found to differ between the old and the young (Uauy *et al.*, 1978; Fielding *et al.*, 1991; Morais *et al.*, 1997, 2000). However, doubt has been cast on the reliability of the method, since 3 MH is produced in tissues other than muscle.

Volpi *et al.* (1998, 1999, 2000) measured amino acid kinetics in the leg, using Biolo's model (Biolo *et al.*, 1992). They found no difference between old and young in amino acid uptake and release, or in delivery or transport of amino acids into muscle. In muscle biopsies the findings are conflicting: in Volpi's study (1999) there was no age-related difference in the fractional synthesis rate of mixed muscle proteins in the basal state, whereas others record a clear decrease with age (Table 12.9) (Yarasheki *et al.*, 1993; Short *et al.*, 2004).

A further question is whether age affects the response of muscle proteins to amino acids and insulin. Here again there is some difference in the findings of different authors: according to Volpi *et al.* (1999) in the leg there was no difference between young and old in the stimulation of uptake produced by infusions of amino acids; this was in spite of a greater first-pass splanchnic uptake in the elderly, which presumably would divert amino acids from muscle. Similarly, in muscle biopsies age made no difference to the stimulation of synthesis by amino acids. On the other hand Balagopal *et al.* (1997a) found a small, barely significant, decrease in response in

Table 12.9. The effect of age on mixed muscle protein synthesis (MPS).

	Age	N	MPS % h^{-1}
Young	19–38	20	0.046
Middle-aged	40–55	11	0.040
Elderly	60–74	20	0.037

MPS in young and elderly significantly different ($p < 0.025$); in middle-aged not significantly different from either young or old. Data of Short *et al.* (2004).

the elderly, depending on the choice of precursor, and Cuthbertson *et al.* (2004) observed that both the sensitivity of muscle protein synthesis to essential amino acids and the maximal response are reduced in the old.

As regards insulin, Volpi *et al.* (2000) observed that when glucose was added to the infusate to promote insulin secretion the response to amino acids was blunted. This recalls the old observation of Baillie and Garlick (1991) that in rats muscle protein synthesis was increased in the young but not in the old in response to feeding, with its consequent secretion of insulin. It is indeed recognized that insulin-resistance is commonly found in the elderly.

There seems, therefore, to be good evidence that muscle protein synthesis decreases with age: this would over the years lead to the sarcopenia that is observed even if the fractional breakdown rate remained unchanged. In the muscle-wasting of starvation, infection and weightlessness there is stimulation of the ubiquitin-proteasome system of protein degradation (Medina *et al.*, 1995; Taillandier *et al.*, 1996; Voisin *et al.*, 1996). However, Attaix and Taillandier, in a review in 1998, stated that there was no evidence yet of this process being concerned in sarcopenia of the elderly in man. Moreover, in the elderly there are often latent, chronic infections causing increased production of interleukins with their well-known effects on muscle protein breakdown (Morley *et al.*, 2001).

The picture becomes more complicated when the different muscle proteins are considered separately. Welle *et al.* (1993, 1994, 1995) showed that in both the fasting and fed states the FSR of myofibrillar protein was ~30% lower in the old than in the young. Extending these observations, Balagopal *et al.* (1997b) found a decline with ageing of almost 50% in the FSR of myosin heavy chains with no change in FSR of the sarcoplasmic proteins. The FSR of mitochondrial proteins in muscle decreased by nearly 50% by middle age, with no further change as age advanced (Rooyackers *et al.*, 1996). There were also decreases with age in the mitochondrial enzymes cytochrome oxidase and citrate synthase, indicating a decline in oxidative capacity and mitochondrial function. Cuthbertson *et al.* (2004) find in the elderly reduced concentrations of proteins concerned with signal transduction. All these changes are being explored in more detail at the molecular level (Barazzoni *et al.*, 2000; Dardevet *et al.*, 2003).

The loss of muscle mass is accompanied by decreased force production and endurance and increased fatigability (Fig. 12.5). What is the causal relationship? Do these functional effects result simply from loss of muscle mass, which in turn results from reduced synthesis of contractile proteins? There have been many studies of the relationship between functional changes and the degree of muscle atrophy as measured by the cross-sectional area (CSA) of a group of muscles, e.g. of the thigh. Jubrias *et al.* (1997) have reviewed this subject and find that atrophy accounts for only about half the loss of force; there was no relation to the isoform content of the muscles (see Chapter 11, section 11.3). They discuss the physiological changes that might explain the decline in specific force (F/CSA).

The question remains whether the immobility characteristic of old age is an effect or a cause, since, irrespective of age, immobility leads to muscle atrophy and decreased turnover rate of muscle proteins (Chapter 11). This could indeed be a vicious cycle. It is not surprising that in recent years a great deal of work has been done on the extent to which resistance exercise can reverse the changes of ageing and increase the rate of muscle protein synthesis, including that of myosin heavy chains. Welle *et al.* (1995) found that a progressive exercise programme lasting for 3 months produced no increase in MPS, so they

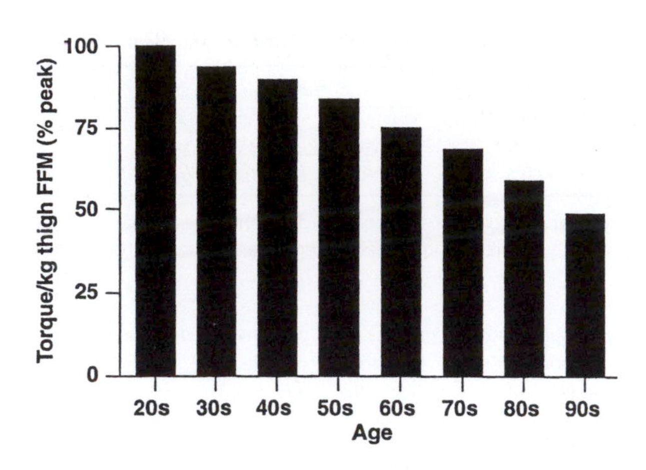

Fig. 12.5. Decline in muscle quality with age. The peak torque is normalized to the fat-free mass of the thigh. Reproduced from Short and Nair (1999), by courtesy of the *Journal of Endocrinological Investigation.*

concluded that the low rate was an irreversible effect of old age, whereas Yarasheki *et al.* (1993) found that only 2 weeks of resistance exercise more than doubled the rate of MPS. Results have been summarized in an excellent review by Parise and Yarasheki (2000).

In addition to immobility hormones may be involved. Insulin resistance develops with ageing and there are low plasma concentrations of T4, IGF-1 and the sex hormones. Testosterone production in men declines with age, and Urban *et al.* (1995) claimed that administration of testosterone to elderly men increased muscle protein synthesis and strength. The possible influence of growth hormone and IGF-1 is discussed by Proctor *et al.* (1998), and a useful general review is that of Roth *et al.* (2000).

The literature on sarcopenia in the elderly is already enormous and rapidly expanding.

12.8 References

Arnal, M.A., Mosoni, L., Boirie, Y., Gachon, P., Genest, M., Bayle, S., Grizard, J., Arnal, M., Antoine, J.M., Beaufrère, B. and Mirand, P.P. (2000) Protein turnover modification induced by the protein feeding pattern still persist after the end of the diets. *American Journal of Physiology* 278, E902–909.

Attaix, D. and Taillandier, D. (1998) The critical role of the ubiquitin-proteasome pathway in muscle wasting in comparison to lysosomal and Ca^{2+}-dependent systems. *Advances in Molecular and Cell Biology* 27, 235–266.

Avslanian, S.A. and Kalhan, S.C. (1996) Protein turnover during puberty in normal children. *American Journal of Physiology* 270, E79–84.

Baillie, A.G.S. and Garlick, P.J. (1991) Attenuated responses of muscle protein synthesis to fasting and insulin in adult female rats. *American Journal of Physiology* 262, E1–5.

Balagopal, P., Ljungqvist, O. and Nair, K.S. (1997a) Skeletal muscle myosin heavy-chain synthesis rate in healthy humans. *American Journal of Physiology* 272, E45–50.

Balagopal, P., Rooyackers, O., Adey, D.B., Ades, P.A. and Nair, K.S. (1997b) Effects of aging on in vivo synthesis of skeletal muscle myosin heavy-chain and sarcoplasmic protein in humans. *American Journal of Physiology* 273, 790–800.

Baracos, V.E., Brun-Bellut, J. and Marie, M. (1991) Tissue protein synthesis in lactating and dry goats. *British Journal of Nutrition* 66, 461–465.

Barrazoni, R., Short, K.R. and Nair, K.S. (2000) Effects of aging on mitochondrial DNA copy number and cytochrome C oxidase expression in rat, skeletal muscle, liver and heart. *Journal of Biological Chemistry* 275, 3343–3347.

Battista, M., Price, P.T. and Kalhan, S.C. (1996) Effect of parenteral amino acids on leucine and urea kinetics in preterm infants. *Journal of Pediatrics* 128, 130–134.

Bauer, K., Pasel, K., Uhrig, C., Sperling, P. and Versmold, H. (1997) Comparison of face mask, head hood and canopy for breath sampling in flow-through indirect calorimetry to measure oxygen consumption and carbon dioxide production of preterm infants $<$ 1500 grams. *Pediatric Research* 41, 139–144.

Beaufrère, B., Putet, G., Pachiandi, C. and Salle, B.

(1990) Whole body protein turnover measured with ^{13}C-leucine and energy expenditure in preterm infants. *Pediatric Research* 28, 147–152.

Beaufrère, B., Fournier, V., Salle, B. and Putet, G. (1992) Leucine kinetics in fed low-birth-weight infants: importance of splanchnic tissues. *American Journal of Physiology* 263, E214–220.

Benedik, C., Berclaz, P.-Y., Jéquier, E. and Schutz, Y. (1995) Resting metabolic rate and protein turnover in apparently healthy elderly Gambian men. *American Journal of Physiology* 268, 1083–1088.

Bequette, B.J., Backwell, F.R.C. and Crompton, L.A. (1998) Current concepts of amino acid and protein metabolism in the mammary gland of the lactating ruminant. *Journal of Dairy Science* 81, 2540–2559.

Biolo, G., Chinkes, D., Zhang, X.-J. and Wolfe, R.R. (1992) A new model to determine in vivo the relationship between amino acid transmembrane transport and protein kinetics in muscle. *Journal of Enteral and Parenteral Nutrition* 16, 305–315

Boirie, Y., Gachon, P. and Beaufrère, B. (1997) Splanchnic and whole-body leucine kinetics in young and elderly men. *American Journal of Clinical Nutrition* 65, 489–495.

Bos, C., Benemouzig, R., Bruhat, A., Roux, C., Mahé, S., Valenci, P., Gaudichn, C., Ferrière, F., Rautureau, J. and Tomé, D. (2000) Short-term protein and energy supplementation activates nitrogen kinetics and accretion in poorly nourished elderly subjects. *American Journal of Clinical Nutrition* 71, 1129–1137.

Bresson, J.L., Bader, B., Rocchiccioli, F., Mariotti, A., Ricour, C., Sachs, C. and Rey, J. (1991) Protein-metabolism kinetics and energy-substrate utilization in infants fed parenteral solutions with different glucose-fat ratios. *American Journal of Clinical Nutrition* 54, 370–376.

Catzeflis, C., Schutz, Y., Micheli, J.-L., Welsch, C., Arnaud, M.J. and Jéquier, E. (1985) Whole body protein synthesis and energy expenditure in very low birth weight infants. *Pediatric Research* 19, 679–687.

Cauderay, M., Schutz, Y., Micheli, J.-L., Calame, A. and Jéquier, E. (1988) Energy-nitrogen balances and protein turnover in small and appropriate for gestational age low birthweight infants. *European Journal of Clinical Nutrition* 42, 125–136.

Chien, P.F.W., Smith, K., Watt, P.W., Scrimgeour, C.M., Taylor, D.J. and Rennie, M.J. (1993) Protein turnover in the human fetus studied at term using stable isotope tracer amino acids. *American Journal of Physiology* 265, E31–35.

Cuthbertson, D.J.R., Smith, K., Babraj, J., Leese, G.P., Waddell, T., Atherton, P.J., Wackerhage, H. and Rennie, M.J. (2004) Anabolic signalling deficits underlie amino acid resistance of wasting, ageing muscle. *FASEB Journal* 19, 422–424.

Dardevet, D., Rieu, I., Fafournoux, P., Sornet, C., Combaret, L., Bruhat, A., Mordier, S., Mosoni, L. and Grizard, J. (2003) Leucine: a key amino acid in ageing-associated sarcopenia? *Nutrition Research Reviews* 16, 61–70.

Darling, P., Wykes, L.J., Clarke, R., Papageorgion, A. and Pencharz, P.B. (1993) Utilization of non-protein nitrogen in whey-dominant formulae by low-birth-weight infants. *Clinical Science* 84, 543–548.

Davis, T.A., Fiorotto, M.L. and Reeds, P.J. (1993) Amino acid compositions of body and milk protein change during the suckling period in rats. *Journal of Nutrition* 123, 947–956.

de Benoist, B., Abdulrazzak, Y., Brooke, O.G., Halliday, D. and Millward, D.J. (1984) The measurement of whole body protein turnover in the preterm infant with intragastric infusion of L-[1-^{13}C] leucine and sampling of the urinary leucine pool. *Clinical Science* 66, 155–164.

de Benoist, B., Jackson, A.A., Hall, J.St E. and Persaud, C. (1985) Whole body protein turnover in Jamaican women during normal pregnancy. *Human Nutrition: Clinical Nutrition* 39C, 167–179.

Denne, S.C. and Kalhan, S.C. (1987) Leucine metabolism in human newborns. *American Journal of Physiology* 253, E605–615.

Denne, S.C., Rossi, E.M. and Kalhan, S.C. (1991) Leucine kinetics during feeding in normal newborns. *Pediatric Research* 30, 23–27.

Denne, S.C., Karn, C.A., Ahrlriches, J.A., Dorotheo, A.R., Wang, J. and Liechty, E.A. (1996) Proteolysis and phenylalanine hydroxylation in response to parenteral nutrition in extremely premature and normal newborns. *Journal of Clinical Investigation* 97, 746–754.

Duffy, B. and Pencharz, P. (1986) The effect of feeding route (IV or oral) on the protein metabolism of the neonate. *American Journal of Clinical Nutrition* 43, 108–111.

Duffy, B., Gunn, T., Collinge, J. and Pencharz, P. (1981) The effect of varying protein quality and energy intake on the nitrogen metabolism of parenterally fed very low birthweight (< 1600 g) infants. *Pediatric Research* 15, 1040–1044.

Duggleby, S.L. and Jackson, A.A. (2001) Relationship of maternal protein turnover and lean body mass during pregnancy and birth length. *Clinical Science* 101, 65–72.

Duggleby, S.L. and Jackson, A.A. (2002) Higher weight at birth is related to decreased maternal amino acid oxidation during pregnancy. *American Journal of Clinical Nutrition* 76, 852–7.

FAO/WHO/UNU (1985) Energy and protein requirements: Report of an expert consultation. *WHO Technical Report Series* no. 724, WHO, Geneva.

Fereday, A., Gibson, N.R., Cox, M., Pacy, P.J. and Millward, D.J. (1997) Protein requirements and ageing: metabolic demand and efficiency of utilization. *British Journal of Nutrition* 77, 685–702.

Fielding, R.A., Meredith, C.N., O'Reilly, K.P., Frontera, W.R., Cannon, J.G. and Evans, W.J. (1991) Enhanced protein breakdown after essential exercise in young and older men. *Journal of Applied Physiology* 71, 674–679.

Fitch, W.L. and King, J.C. (1987) Protein turnover and 3-methylhistidine excretion in non-pregnant, pregnant and gestationally diabetic women. *Human Nutrition: Clinical Nutrition*, 327–340.

Fukegawa, N.K., Minaker, K.L., Rowe, J.W., Matthews, D.E., Bier, D.M. and Young, V.R. (1988) Glucose and amino acid metabolism in aging man: differential effects of insulin. *Metabolism* 37, 371–377.

Fukegawa, N.K., Minaker, K.L., Young, V.R., Matthews, D.E., Bier, D.M. and Rowe, J.W. (1989) Leucine metabolism in aging humans: effect of insulin and substrate availability. *American Journal of Physiology* 256, 288–294.

Golden, B.E. and Golden, M.H.N. (1992) Effect of zinc on lean tissue synthesis during recovery from malnutrition. *European Journal of Clinical Nutrition* 46, 697–706.

Golden, M.H.N., Waterlow, J.C. and Picou, D. (1977a) Protein turnover, synthesis and breakdown before and after recovery from protein-energy malnutrition. *Clinical Science and Molecular Medicine* 53, 473–477.

Golden, M.H.N., Waterlow, J.C. and Picou, D. (1977b) The relation between dietary intake, weight change, nitrogen balance, and protein turnover in men. *American Journal of Clinical Nutrition* 30, 1345–1348.

Heymsfield, S.B., Smith, R., Aulet, M., Bensen, B., Lichtman, S., Wang, J. and Pierson, R.N. (1990) Appendicular skeletal muscle mass: measurement by dual proton absorptiometry. *American Journal of Clinical Nutrition* 52, 214–218.

Jackson, A.A., Shaw, J.C.L., Barber, A. and Golden, M.H.N. (1981) Nitrogen metabolism in preterm infants fed human donor breast milk: the possible essentiality of glycine. *Pediatric Research* 15, 1454–1461.

Jackson, A.A., Golden, M.H.N., Byfield, R., Jahoor, F., Royes, J. and Souter, L. (1983) Whole body protein turnover and nitrogen balance in young children at intakes of protein and energy in the region of maintenance. *Human Nutrition: Clinical Nutrition* 37C, 433–446.

Jackson, A.A., Persaud, C., Werkmeister, G., McClelland, I.F., Badeloo, A. and Forrester, T. (1997) Comparison of urinary 5-L-oxoproline [L-pyroglutamate) during normal pregnancy in women in England and Jamaica. *British Journal of Nutrition* 77, 183–196.

Jackson, A.A., Duggleby, S.L. and Grove, G. (2000) Whole body protein turnover can be measured non-invasively in women using the end product method with [^{15}N] glycine to show changes with the menstrual cycle and pregnancy. *European Journal of Clinical Nutrition* 54, 1–8.

Jones, M.O., Pierro, A., Garlick, P.J., McNurlan, M.A., Donnell, S.C. and Lloyd, D.A. (1995) Protein metabolism kinetics in neonates: effect of intravenous carbohydrate and fat. *Journal of Pediatric Surgery* 30, 458–462.

Jubrias, S.A., Odderson, I.B., Esselman, P.C. and Cowley, K.E. (1997) Decline in isokinetic force with age: muscle cross-sectional area and specific force. *Pflügers Archiv – European Journal of Physiology* 434, 246–253.

Kalhan, S.C., Rossi, K.Q., Gruca, L.L., Super, D.M. and Savin, S.M. (1998) Relation between transamination of branched-chain amino acids and urea synthesis: evidence from human pregnancy. *American Journal of Physiology* 275, E423–431.

Kien, C.L. and Camitta, B.M. (1983) Increased whole body protein turnover in sick children with newly diagnosed leukemia or lymphoma. *Cancer Research* 43, 5586–5592.

Kien, C.L., Rohrbaugh, D.K., Burke, J.F. and Young, V.R. (1978) Whole body protein synthesis in relation to basal energy expenditure in healthy children and in children recovering from bone injury. *Pediatric Research* 12, 211–216.

Kien, C.L., Zipf, W.B., Horswill, C.A., Denne, S.C., McCoy, K.S. and O'Dorisio, T.M. (1996) Effects of feeding on protein turnover in healthy children and in children with cystic fibrosis. *American Journal of Clinical Nutrition* 64, 608–614.

Kingdon, C.C., Mitchell, F., Bodamer, O.F. and Williams, A.F. (2000) Measurement of carbon dioxide production in very low birth weight babies. *Archives of Diseases in Childhood: Fetal and Neonatal Edition* 83, F50–55.

Lehmann, A.B., Johnston, D. and James, O.R.W. (1989) The effects of old age and immobility on protein turnover in human subjects with some observations on the possible role of hormones. *Age and Ageing* 18, 148–157.

Leitch, C.A. and Denne, S.C. (2000) Energy expenditure in the extremely low-birth weight infant. *Clinics in Perinatology* 27, 181–195.

Lexell, J. and Downham, D. (1992) What is the effect of ageing on type 2 muscle fibres? *Journal of Neurological Science* 107, 250–251.

Lexell, J., Taylor, C.C. and Sjostrom, M. (1988) What is the cause of the ageing atrophy? Total number, size and proportion of different fiber types studies in whole vastus lateralis muscle in 15- to 83-year-old men. *Journal of Neurological Science* 84, 275–295.

Liechty, E.A. and Denne, S.C. (1998) Regulation of

fetal amino acid metabolism: substrate or hormonal regulation? *Journal of Nutrition* 128, 342–465.

McClelland, I.S.M., Persaud, C. and Jackson, A.A. (1997) Urea kinetics in healthy women during normal pregnancy. *British Journal of Nutrition* 77, 165–181.

Medina, R., Wing, S.S. and Goldberg, A.L. (1995) Increase in levels of polyubiquitin and proteasome mRNA in skeletal muscle during starvation and denervation atrophy. *Biochemical Journal* 301, 631–637.

Millward, D.J., Fereday, A., Gibson, N. and Pacy, P.J. (1997) Aging, protein requirements and protein turnover. *American Journal of Clinical Nutrition* 66, 774–786.

Mitton, S.G. and Garlick, P.J. (1992) Changes in protein turnover after the introduction of parenteral nutrition in premature infants: comparison of breast milk and egg protein-based amino acid solutions. *Pediatric Research* 32, 447–454.

Mitton, S.G., Calder, A.G. and Garlick, P.J. (1991) Protein turnover rates in sick, premature neonates during the first few days of life. *Pediatric Research* 30, 418–422.

Morais, J.A., Gougeon, R., Pencharz, P.B., Jones, P.J.H., Ross, R. and Marliss, E.B. (1997) Whole body protein turnover in the healthy elderly. *American Journal of Clinical Nutrition* 66, 880–889.

Morais, J.A., Ross, R., Gougeon, R., Pencharz, P.B., Jones, P.J.H. and Marliss, E.B. (2000) Distribution of protein turnover changes with age in humans as assessed by whole-body magnetic resonance image analysis to quantity tissues' volumes. *Journal of Nutrition* 130, 784–791.

Morley, J.E., Baumgartner, R.N., Roubenoff, R., Mayor, J. and Nair, K.S. (2001) Sarcopenia. *Journal of Laboratory and Clinical Medicine* 137, 231–243.

Motil, K.J., Montandon, C.M., Hachey, D.L., Boulton, T.W., Klein, P.D. and Garza, C. (1989a) Relationships among lactation performance, maternal diet and baby protein metabolism in humans. *European Journal of Clinical Nutrition* 43, 681–691.

Motil, K.J., Montandon, C.M., Hachey, D.L., Boutton, T.W., Klein, P.D. and Garza, C. (1989b) Whole body protein metabolism in lactating and non-lactating women. *Journal of Applied Physiology* 66, 370–376.

Motil, K.J., Thotatchucery, M., Montandon, C.M., Hachey, D.L., Boutton, T.W., Klein, P.D. and Garza, C. (1994) Insulin, cortisol and thyroid hormones modulate maternal protein status and milk production and composition in humans. *Journal of Nutrition* 124,

Motil, K.J., Davis, T.A., Montandon, C.M., Wong, W.W., Klein, P.D. and Reeds, P.J. (1996) Whole body protein turnover in the fed state is reduced in response to dietary protein restriction in lactating women. *American Journal of Clinical Nutrition* 64, 32–39.

Naismith, P.J. and Morgan, B.L.G. (1976) The biphasic nature of protein metabolism during pregnancy in the rat. *British Journal of Nutrition* 36, 563–566.

Naismith, D.J., Richardson, D.P. and Pritchard, A.E. (1982) The utilization of protein and energy in the rat, with particular regard to the use of fat accumulated in pregnancy. *British Journal of Nutrition* 48, 433

Nicholson, J.F. (1970) Rate of protein synthesis in premature infants. *Pediatric Research* 4, 384–397.

Nissim, I., Yudkoff, M., Pereira, G. and Segal, S. (1983) Effects of conceptual age and dietary intake on protein metabolism in premature infants. *Journal of Pediatric Gastroenterology and Nutrition* 2, 507–516.

Olufemi, O.S., Whittaker, P.G., Halliday, D. and Lind, T. (1991) Albumin metabolism in fasted subjects during late pregnancy. *Clinical Science* 81, 161–168.

Pacheco, D., Tavendale, M.H., Reynolds, G.W., Barry, J.N., Lee, J. and McNabb, W.C. (2003) Whole body fluxes and partitioning of amino acids to the mammary gland of cows for fresh pasture at two levels of intake during early lactation. *British Journal of Nutrition* 90, 271–281.

Paolini, C.L., Meschia, G., Fennessey, P.V., Pike, A.W., Teng, C., Battaglia, F.C. and Wilkening, R.B. (2001) An *in vivo* study of ovine transport of essential amino acids. *American Journal of Physiology* 280, E31–39.

Parise, G. and Yarasheki, K.E. (2000) The utility of resistance exercise training and amino acid supplementation for reversing age-associated decrements in muscle protein mass and function. *Current Opinion in Clinical Nutrition and Metabolic Care* 3, 489–495.

Pencharz, P.B., Masson, M., Desgranges, F. and Papageorgiou, A. (1981) Total-body protein turnover in human premature neonates: effects of birth weight, intra-uterine nutritional status and diet. *Clinical Science and Molecular Medicine* 52, 485–498.

Pencharz, P.B., Clarke, R., Archibald, E.M. and Vaisman, N. (1988) The effect of a weight-reducing diet on the nitrogen metabolism of obese adolescents. *Canadian Journal of Physiology and Pharmacology* 66, 1469–1474.

Pencharz, P., Beesley, J., Sauer, P., van Aerde, J., Canagarayar, U., Renner, J., McVey, M., Wesson, D. and Swyer, P. (1989a) Total body protein turnover in parenterally fed neonates: effect of energy source studied by using [^{15}N] glycine and [1-^{13}C] leucine. *American Journal of Clinical Nutrition* 50, 1395–1400.

Pencharz, P., Beesley, J., Sauer, P., van Aerde, J., Canagarayar, U., Renner, J., McVey, M., Wesson, D. and Swyer, P. (1989b) A comparison of estimates of whole-body protein turnover in parenterally fed neonates obtained using three different end-products. *Canadian Journal of Physiology and Pharmacology* 67, 624–628.

Picou, D. and Taylor-Roberts, T. (1969) The measurement of total protein synthesis and catabolism and nitrogen turnover in infants in different nutritional states and receiving different amounts of dietary protein. *Clinical Science* 36, 286–296.

Pine, A.P., Jessop, N.S., Allan, G.F. and Oldham, J.N. (1994) Maternal protein reserves and their influence on lactational performance in rats. 4. Tissue protein synthesis and turnover associated with mobilization of maternal protein. *British Journal of Nutrition* 72, 831–844.

Plath, C., Heine, W., Wutzke, K.D. and Uhlemannn, M. (1996) ^{15}N tracer studies in formula-fed protein infants: the role of glycine supply in protein turnover. *Journal of Pediatric Gastroenterology and Nutrition* 23, 287–297.

Poindexter, B.B., Karn, C.A., Leitch, C.A., Liechty, E.A. and Denne, S.C. (2001) Amino acids do not suppress proteolysis in premature neonates. *American Journal of Physiology* 281, E472–478.

Proctor, D.N., Balagopal, P. and Nair, K.S. (1998) Age-related sarcopenia in humans is associated with reduced synthetic rates of specific muscle proteins. *Journal of Nutrition* 128, 351–355S.

Rivera, A., Bell, E.F. and Bier, D.M. (1993) Effect of intravenous amino acids on protein metabolism of pre-term infants during the first three days of life. *Pediatric Research* 33, 106–111.

Rooyackers, O.A., Adly, D.B., Ades, P.A. and Nair, K.S. (1996) Effect of age on *in vivo* rates of mitochondrial protein synthesis in human skeletal muscle. *Proceedings of the National Academy of Science, USA* 93, 15364–15369.

Ross, J.C., Fennessy, P.V., Wilkening, R.B., Battaglia, F.C. and Meschia, G. (1996) Placental transport and fetal utilization of leucine in a model of fetal growth retardation. *American Journal of Physiology* 270, E491–503.

Roth, S.M., Ferrell, R.A. and Hurley, B.F. (2000) Strength training for the prevention and treatment of sarcopenia. *Journal of Nutrition, Health and Aging* 4, 143–155.

Salman, E.K., Haymond, M.W., Bayne, E., Sager, B.K., Wiisann, H., Pitel, P. and Darmaun, D. (1996) Protein and energy metabolism in pre-pubertal children with sickle cell anemia. *Pediatric Research* 40, 34–40.

Short, K.R. and Nair, K.S. (1999). Mechanisms of sarcopenia of ageing. *Journal of Endocrinological Investigation* 22, 95–105.

Short, K.R., Vittone, J.L., Bigelow, M.L., Proctor, D.N. and Nair, K.S. (2004) Age and aerobic exercise training effects on whole body and muscle protein metabolism. *American Journal of Physiology* 286, E92–101.

Singh, M.A., Ding, W., Manfredi, T.J., Solares, G.S., O'Neill, E.F., Clements, K.M., Ryan, N.D., Kehayias, J.J., Fielding, R.A. and Evans, W.J. (1999) Insulin-like growth factor I in skeletal muscle after weight-lifting exercise in frail elders. *American Journal of Physiology* 277, E135–143.

Soares, M.J., Piers, L.S., Shetty, P.S., Robinson, S., Jackson, A.A. and Waterlow, J.C. (1991) Basal metabolic rate, body composition and whole-body protein turnover in Indian men with differing nutritional status. *Clinical Science* 81, 419–425.

Stack, T., Reeds, P.J., Preston, T., Hay, S., Lloyd, D.J. and Agett, P. (1989) ^{15}N tracer studies of protein metabolism in low birth weight preterm infants: a comparison of ^{15}N glycine and ^{15}N yeast protein hydrolysate and of human milk and formula-fed babies. *Pediatric Research* 25, 167–172.

Taillandier, D., Aurousseau, E., Meynial-Denis, D., Bechet, D., Ferrara, M., Cottins, P., Ducastaing, A., Bigard, X., Guizennen, C.Y., Schmid, H.-P. and Attaix, D. (1996) Coordinate activation of lysosomal, Ca^{2+}-activated and ATP-ubiquitin-dependent proteases in the unweighted rat soleus muscle. *Biochemical Journal* 316, 65–72.

Thomas, M.R., Irving, C.S., Reeds, P.J., Malphus, E.W., Wong, W.W., Boutton, T.W. and Klein, P.D. (1991) Lysine and protein metabolism in the young lactating women. *European Journal of Clinical Nutrition* 45, 227–242.

Thompson, G.N. and Halliday, D. (1992) Protein turnover in pregnancy. *European Journal of Clinical Nutrition* 46, 411–417.

Toth, M.J., Goran, M.J., Ades, P.A., Howard, D.B. and Poehlman, E.T. (1993) Examination of data normalization procedures for expressing peak VO_2 data. *Journal of Applied Physiology* 75, 2288–2292.

Uauy, R., Winterer, J.C., Bilmazes, C., Haverberg, L.N., Scrimshaw, N.S., Munro, H.N. and Young, V.R. (1978) The changing pattern of whole body protein metabolism in aging humans. *Journal of Gerontology* 33, 663–671.

Urban, R.J., Bodenburg, Y.H., Gilkinson, C., Foxworth, J., Coggan, A.R., Wolfe, R.A. and Ferrando, A. (1995) Testosterone administration to elderly men increases skeletal muscle strength and protein synthesis. *American Journal of Physiology* 269, E820–826.

Vaisman, N., Clarke, R., Rossi, M., Goldberg, E., Zello, G.A. and Pencharz, P.B. (1992) Protein turnover and resting energy expenditure in patients with undernutrition and chronic lung disease. *American Journal of Clinical Nutrition* 55, 63–69.

van Goudoever, J.B., Sulkers, E.J., Halliday, D., Degenhart, H.J., Carnielli, V.P., Wattimena, J.L.D.

and Sauer, P.J.J. (1995a) Whole-body protein turnover in preterm appropriate for gestational age and small for gestational age infants: comparison of [^{15}N] glycine and [1-^{13}C] leucine administered simultaneously. *Pediatric Research* 37, 381–388.

van Goudoever, J.B., Cohen, T., Wattimena, J.L.D., Huijmans, J.G.M., Carnielli, V.P. and Sauer, P.J.J. (1995b) Immediate commencement of amino acid supplementation in preterm infants: effect on serum amino acid concentrations on the first day of life. *Journal of Pediatrics* 127, 458–465.

van Lingen, R.A., van Goudoever, J.B., Luijendijk, I.H.T., Wattimena, J.L.D. and Sauer, P.J.J. (1992) Effects of early amino acid administration during total parenteral nutrition on protein metabolism in pre-term infants. *Clinical Science* 82, 199–203.

Voison, L., Breuillé, D., Combaret, L., Pouyet, C., Taillandier, D., Aurousseau, E., Obled, C. and Attaix, D. (1996) Muscle wasting in a rat model of long-lasting sepsis results from activation of lysosomal, Ca^{2+}-activated and ubiquitin-proteasome proteolytic pathways. *Journal of Clinical Investigation* 97, 1610–1617.

Volpi, E., Ferrando, A.A., Yecker, C.W., Tipton, K.D. and Wolfe, R.R. (1998) Exogenous amino acids stimulate net muscle protein synthesis in the elderly. *Journal of Clinical Investigation* 101, 2000–2007.

Volpi, E., Mittendorfer, B., Wolf, S.E. and Wolfe, R.R. (1999) Oral amino acids stimulate muscle protein metabolism in the elderly despite higher first pass extraction. *American Journal of Physiology* 277, E5132–520.

Volpi, E., Mittendorfer, B., Rasmussen, B.B. and Wolfe, R.R. (2000) The response of muscle protein anabolism to combined hyperaminoacidemia and glucose-induced hyperinsulinaemia is impaired in the elderly. *Journal of Clinical Endocrinology and Metabolism* 85, 4481–4490.

Waterlow, J.C., Golden, M.H.N. and Garlick, P.J. (1978) Protein turnover in man measured with ^{15}N: comparison of end products and dose regimes. *American Journal of Physiology* 235, E165–174.

Welle, S., Thornton, C., Jozefowicz, R. and Statt, M. (1993) Myofibrillar protein synthesis in young and old men. *American Journal of Physiology* 264, E693–698.

Welle, S., Thornton, C., Statt, M. and McHenry, B. (1994) Post prandial myofibrillar and whole body synthesis in young and old human subjects. *American Journal of Physiology* 267, E599–604.

Welle, S., Thornton, C. and Statt, M. (1995) Myofibrillar protein synthesis in young and old human subjects after three months of resistance training. *American Journal of Physiology* 268, 422–427.

Whittaker, P.G., Lee, C.H. and Taylor, R. (2000) Whole body protein kinetics in women: effect of pregnancy and IDDM during anabolic stimulation. *American Journal of Physiology* 279, E978–988.

Widdowson, E.M., Southgate, D.A.T. and Hey, E.N. (1979) Composition of the fetus and infant. In: Visscher, H.E.A. (ed.) *Nutrition and Metabolism of the Fetus and Infant.* Martinus Nijhoff, The Hague, Netherlands, pp. 169–178.

Willommet, L., Schutz, Y., Whitehead, R., Jéquier, E. and Fern, E.B. (1992) Whole body protein metabolism and resting energy expenditure in pregnant Gambian women. *American Journal of Physiology* 263, E624–631.

Winterer, J.C., Steffee, W.P., Davy, W., Perera, A., Uauy, R., Scrimshaw, N.S. and Young, V.R. (1976) Whole body protein turnover in aging man. *Experimental Gerontology* 11, 79–87.

Wutzke, K.D., Heine, W., Plath, C., Müller, M. and Uhlemann, M. (1992) Whole body protein parameters in premature infants: a comparison of different ^{15}N tracer substances and different methods. *Pediatric Research* 31, 95–101.

Wykes, L.J., Ball, R.O., Menendez, C.E. and Pencharz, P.B. (1990) Urine collection as an alternative to blood sampling: a non-invasive means of determining isotopic enrichment to study amino acid flux in neonates. *European Journal of Clinical Nutrition* 44, 605–608.

Yarasheki, K.E., Zachwieja, J.J. and Bier, D.M. (1993) Acute effects of resistance exercise on muscle protein synthesis rate in young and elderly men and women. *American Journal of Physiology* 265, E210–214.

13

Protein Turnover in Some Pathological States: Malnutrition and Trauma

13.1 Malnutrition

13.1.1 Infants

The effects of malnutrition on protein turnover have been studied in young children more than in adults. The findings after recovery have been described in the previous chapter. In their original study Picou and Taylor-Roberts (1969) reported results on five infants who were measured twice, once when malnourished and again after recovery. Mean synthesis rates with intragastric infusions of [^{15}N]-glycine in the fed state and urea as end-product gave synthesis rates that were twice as high in the malnourished as in the recovered – 467 *vs.* 254 mg protein kg^{-1} h^{-1}. However, these children were not studied immediately on admission to hospital but after 7–10 days, when infections had been treated and they were clinically stable, on the verge of taking off into rapid growth. Later work in Jamaica showed that when protein turnover was measured in severely malnourished infants within 1–2 days of admission synthesis was very low; within about 2 weeks, when the child was still very undernourished in terms of their weight, but beginning to eat well and gain weight rapidly, synthesis rose to very high levels, to fall again when normal weight for length had been regained (Table 13.1). These changes are paralleled by changes in the metabolic rate – initially low, followed by an overshoot (Montgomery, 1962; Brooke and Cocks, 1974).

An interesting study was made by Golden and Golden (1992) to test the hypothesis that on diets based on milk protein deposition might be limited by inadequate zinc. Supplements of zinc had little effect on synthesis calculated from urea as end-product; with ammonia as end-product the estimate of synthesis was increased, so that S_A/S_U was doubled, rising from 0.7 to 1.4. This is the highest ratio that has ever been found in our experience. According to the Fern hypothesis (Chapter 7), it implies a large increase of synthesis in peripheral tissues.

Most severely malnourished children in developing countries are also suffering from infec-

Table 13.1. Synthesis rates in malnourished, recovering and recovered infants, measured with ^{15}N-glycine and either urea or urea + ammonia as end-products.

		Synthesis in mg protein kg^{-1} h^{-1}		
		Malnourished	Recovering	Recovered
Golden *et al.*, 1977	n	5	2	5
	S_U	167	422	262
Waterlow *et al.*, 1978[a]	n	5	5	8
	S_U	173	306	247
	S_A	92	185	174
	S_{av}	132	244	212
	S_A/S_U	0.54	0.61	0.72

[a]Mean of single dose and multiple dose measurements.

tions. A study was carried out in Malumfashi, northern Nigeria on 22 children divided into four groups, as shown in Table 13.2 (Tomkins *et al*., 1983).

In acute infection (Groups I and II) synthesis was very much increased, while in severe malnutrition (Group IV) it was well below normal. The changes in breakdown were even greater. The children with acute infections were in negative nitrogen balance (S–B), while the severely malnourished were just able to maintain balance. Similar increases in synthesis and breakdown were found in children with malaria in The Gambia, falling to near normal levels on treatment (Berclaz *et al*., 1996). These increases with infection seem to be another example of synthesis and breakdown changing together, as in rapid growth (see Chapter 11); however, we are unable to tell which is the driver and which the driven: whether in infection synthesis is driven by breakdown and in growth it is the other way round.

In recent years turnover rates in malnourished children have been measured with [^{13}C]-leucine by Manary *et al.* (1997, 1998) in Malawi and Reid *et al.* (2002) in Jamaica. Clinically the children are of two types: those with oedema, commonly termed 'kwashiorkor', and those without oedema, referred to as 'marasmus'. Only Manary studied both types. Reid made three sets of measurements: (i) within about 2 days of admission to hospital; (ii) some 8 days later, when infections had been treated; and (iii) after full recovery. The results in Table 13.3 show remarkable agreement between studies in the oedematous cases, in spite of differences in method: synthesis on admission is depressed; it is unchanged with treatment of infections (Reid, 2001) and is restored to normal on full recovery. It looks as if, in this type of malnutrition, the body is unable to mount the increase in turnover that one might expect with infection. There remains Manary's remarkable observation that in the marasmic children synthesis and breakdown rates were more than twice as high as in those with oedema and higher than in normal infants. We do not know whether this was because their infections were more severe. Marasmus is a state of starvation, of balanced deprivation; this finding recalls the high synthesis rate of Stroud and his colleague, at the end of their crossing of the Antarctic, when they had lost about 20 kg and were near to death (Stroud *et al.*, 1996). There is a long-standing controversy, still unresolved, of whether the oedema of severe childhood malnutrition is caused by a diet inade-

Table 13.2. Protein turnover in children with varying degrees of infection and malnutrition in Malumfashi, northern Nigeria.

	Group[a]				
	I	I(T)	II	III	IV
No. of subjects	5	5	6	6	5
Weight for height % of standard	87	87	75	65	63
Plasma albumin g l^{-1}	38	34	38	19	17
Leucocytes, 10^{-3} per mm^3	13.5	8.0	7.1	9.1	5.8
Protein intake during test g kg^{-1} d^{-1}	0.52	–	0.80	1.12	1.17
Protein synthesis mg kg^{-1} h^{-1}	433	372	448	261	147
Protein breakdown mg kg^{-1} h^{-1}	577	418	500	252	131
Methyl histidine excretion, μmol kg^{-1} d^{-1}	15.7	9.1	5.2	4.2	2.6

Measurements in fed state with single dose of [^{15}N]-glycine, end-product ammonia.
[a]Group I: mildly undernourished, acute measles; I(T): the same children after treatment for 6 days; II: moderately undernourished, acute infection (pneumonia, septicaemia); III: severely malnourished with post-acute infection (pneumonia, septicaemia), five were oedematous; IV: severely malnourished, no evidence of infection, four were oedematous.
Data, with acknowledgement, from Tomkins *et al.*, 1983.

Table 13.3. Protein turnover in severely malnourished children – comparison of different studies.

Study	Time of study[a]	Infection	Oedema	Synthesis mg protein	Breakdown kg^{-1} h^{-1}
Reid *et al.*, 2002[b]	2–3	+	+	175	125
	11[f]	–	0	160	120
	~70	–	0	223	164
Tomkins *et al.*, 1983[c]	initial	–	+	147	131
Manary *et al.*, 1997, 1998[d]	<1	+	+	159	169
	<1	+	0	354	371
Normal[e]	–	–	0	200–250	–

All children were also infected except those Tomkins *et al.* (1983) (see Table 13.2) and normals.
[a]Days after admission to hospital, when known.
[b]Intravenous leucine infusion in the fed state. Calculations based on KIC.
[c]Group IV of Table 13.2. Oral glycine, end-product ammonia.
[d]Intravenous leucine infusion, post-absorptive.
[e]See Table 12.1.
[f]After treatment of infections.

quate in protein but relatively high in energy, or by free radical damage resulting from infection (see Waterlow, 1992).

Jackson in 1986 discovered that children with oedema had significantly lower red cell glutathione concentrations than those without oedema (Jackson, 1986). Glutathione, a tripeptide of glycine, glutamic acid and cysteine, in its reduced form GSH, has important functions in antioxidant defence and maintaining redox potentials. The group in Jamaica went on to show that in malnourished infected children with oedema the synthesis rate of glutathione was reduced, as well as the concentrations of glutathione and of free cysteine (Reid *et al.*, 2000; Badaloo *et al.*, 2002). They suggested that these changes resulted from an inadequate supply of sulphur amino acids in the diet (see also Roediger, 1995), compounded by a reduced inflow from protein breakdown, as demonstrated by Manary *et al.* (1998). These findings go far towards reconciling the two theories about the cause of oedema, if an inadequate protein intake leads to failure of the antioxidant defences.

13.1.2 Adults

In industrialized countries malnutrition in adults is usually associated with systemic disease, such as cancer. Perhaps the best example of primary malnutrition is anorexia nervosa but, surprisingly, we have not found any papers describing whole body turnover in this condition.

In developing countries the position is different The body mass index [wt (kg)/ht^2 (m)] (BMI) is widely used as a measure of nutritional state. A BMI of 18.5 may be taken as the lower limit of 'normal' (James *et al.*, 1988). In developing countries there are many people, apparently fit, with BMIs in the region of 16. Two studies of such subjects were made by Soares *et al.* (1991, 1994) and have already been mentioned in Chapter 10 and Table 10.3. Fat was determined from skin-folds, muscle mass from creatinine excretion and non-muscle lean body mass by difference. Protein turnover was calculated from the end-product average with [^{15}N] glycine. The results in Table 13.4 show that BMR kg^{-1} FFM was higher in the undernourished group, as might be expected from their larger proportions of non-muscle mass with a great reduction in muscle mass. The whole-body synthesis rate was the same in the two groups, but there was a striking difference in the flux ratio. Q_A was correlated with the muscle mass, Q_U with the non-muscle mass, in agreement with the hypothesis developed in Chapter 7. It is also possible to calculate the fractional synthesis rate of the non-muscle component; on the assumption that the FSR of muscle protein is 2% day^{-1}, the FSR of non-muscle protein comes out as 15% lower in the undernourished group, compared with the well nourished. This might be regarded as an adaptation to a low protein intake over a lifetime. On the other hand, such a calculation relies on many approximations and assumptions regarded as an adaptation to a low protein intake over a lifetime.

Table 13.4. Body composition and whole body protein turnover in undernourished Indian labourers (UN) compared with normal-weight controls (NW).

	NW	UN
Body wt, kg	67.5	43.6
BMI	22.1	16.6
Fat-free mass, kg	53.8	38.8
Muscle, kg	31.7	18.9
Non-muscle, kg	22.2	19.9
Resting metabolic rate, kJ kg^{-1} FFM day^{-1}	119	131
Protein synthesis, g kg^{-1} FFM day^{-1}	5.4	5.7
Q_A/Q_U	0.77	0.59

Data, with acknowledgement, from Soares *et al.*, 1991. The data from a subsequent paper (Soares *et al.*, 1994) are very similar.

13.2 Trauma

The word 'trauma' is used to describe a multitude of pathological states – burns, soft-tissue injury, fractures, multiple organ failure, generalized sepsis. The body's response seems to be much the same in all these conditions, so they will not be considered separately. The response to trauma has two components, hormonal and metabolic; both involve a catabolic state and the question arises, but cannot be answered here, whether the hormonal changes are 'drivers' or whether they are simply part of a general reaction to injury.

13.2.1 Metabolic changes

Here Cuthbertson, who was essentially a physiologist working with both animals and human subjects, was the pioneer (Cuthbertson, 1930, 1942, 1954). He introduced but did not define, the terms 'ebb' and 'flow' to describe the 'dwindling and rising tides of heat loss'. The ebb phase lasts 1–3 days; the changes are mainly circulatory and the condition is often referred to as 'shock'. Later work has concentrated almost entirely on the 'flow' phase, in which there is increased oxygen uptake and a massive negative nitrogen balance (Fig. 13.1). Although these two changes appear to be related, they are generally not statistically correlated.

Early isotopic studies, e.g. on children with severe burns, showed that the negative balance resulted from a massive increase in whole body protein breakdown (Kien *et al.*, 1978); in fed studies there is also sometimes an increase in synthesis which, however, is not large enough to balance the breakdown (Table 13.5). Smaller changes have been found after elective surgery (e.g. Crane *et al.*, 1977), quite large ones after vaccination even though it causes only a small rise in body temperature and white cell count (Garlick *et al.*, 1980; Cayol *et al.*, 1995). In one study vaccination produced an increase in hepatic protein synthesis as well as in whole body turnover (Grizard and Obled, 1995).

In this connection it is worth pointing out that many of the earlier tracer studies, such as those of Jeevanandam *et al.* (1993), were done with [^{15}N]-glycine by the end-product method. They are in general agreement with results by the precursor method (infusions of leucine or phenylalanine) but are more informative because they cover a longer period (infusions of at least 24 h *vs.* 4 or 8 h) and include feeding, so they are less influenced by diurnal fluctuations.

The nutritional state is also important; Macallan *et al.* (1998) found no increase in synthesis in undernourished patients with pulmonary tuberculosis and suggested that amino acids were diverted from synthesis to oxidation. In children who developed an infection while malnourished there was little increase in either synthesis or breakdown (Table 13.2, group III). An early observation by Cuthbertson's group (Munro and Chalmers, 1945) was that the nitrogen loss in urine was proportional to the pre-injury level, and that in the rat it could be eliminated entirely if the animals had been on a low protein diet. It was therefore proposed that the catabolic loss of N in the flow phase is at the expense of 'labile protein reserves'. Nowadays, however, the concept of protein reserves is not generally accepted, and 'labile' protein is thought to be simply protein

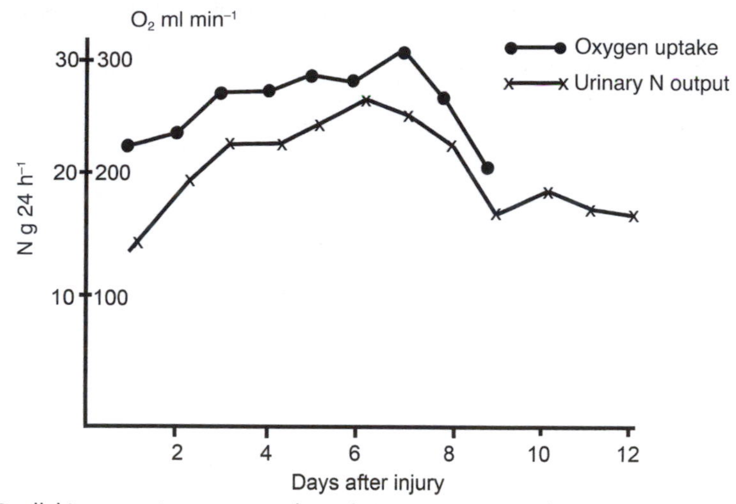

Fig. 13.1. Parallel increases in oxygen uptake and nitrogen excretion after injury (fracture of tibia). After Cuthbertson (1976), Figure 2, by courtesy of Pitmans Medical.

that is turning over rapidly. The behaviour of individual tissues in traumatized subjects is therefore relevant. Cuthbertson suggested that the greater part of the catabolic nitrogen was derived from muscle, because the losses of N were accompanied by losses of sulphate, phosphate, potassium, magnesium and zinc more or less in the proportions in which they occur in muscle. There was also an output of creatine. The increased breakdown of muscle protein was confirmed by Long *et al.* (1981), who found larger increases in the outputs of 3-methyl-histidine in patients with trauma.

Ideally the response of muscle protein turnover

Table 13.5. Examples of the effect of trauma on whole body protein breakdown and synthesis.

	mg protein kg^{-1} h^{-1}	
	Control	Trauma
A. Birkhahn *et al.*, 1980		
Breakdown	139	249
Synthesis	121	181
Urinary N, g day^{-1}	5	25
B. Jeevanandam *et al.*, 1993		
Breakdown	101	150
Synthesis	77	90
N balance mg kg^{-1} day^{-1}	−99	−222
C. Arnold *et al.*, 1993		
Breakdown	108	246
Synthesis	136	200

A, skeletal trauma, infused for 11 h with 1[^{14}C] leucine; IV glucose but no food: Calculations from ε leucine: corrected × 1.25 for KIC. B, multiple fractures or gunshot wounds: infused for 24 h with [^{15}N] glycine, with calculation from EP average; fasted throughout. Elderly patients were also studied but their data are not included here. C, multiple organ failure, infused for 4 h with [^{13}C] leucine; post-absorptive; calculations from KIC.
Note the large negative balance (difference between synthesis and breakdown) in all the trauma cases.

should be confirmed by direct measurements on muscle biopsies. A study by Essen *et al.* (1998) showed, surprisingly, that in critically ill patients the fractional rate of muscle protein synthesis was normal, although there was a large scatter. The fractional synthesis rates of lymphocyte protein and of plasma albumin were slightly higher than normal. Obviously measurements of tissue protein turnover are easier in experimental animals than in man. In rats with long-lasting sepsis the FSR of protein in muscle and skin was decreased, while in the liver and gut it was increased (Breuillé *et al.*, 1998; O'Leary *et al.*, 2001).

Interpretation of these findings, as so often, is made difficult by lack of direct information on tissue protein breakdown. This deficiency was made good by Biolo *et al.* (2002), who studied amino acid kinetics in the leg in severely burnt patients and showed a large increase in protein breakdown; with some increase in synthesis, which they attributed to the greater amino acid availability. A convincing argument was put forward by Reeds *et al.* (1994) that 'the mobilization of muscle protein confers an adaptive advantage to the organism'. They showed that the acute-phase proteins (APP) are richer in aromatic amino acids than muscle protein, and to produce 1 g of acute-phase proteins would require the breakdown of about 2 g of muscle protein. From an approximate estimate of the amount of acute-phase proteins produced in response to a typical injury (Colley *et al.*, 1983) they calculated that the nitrogen 'wasted' in the APP response might amount to 130 mg N per kg body weight, a figure very close to the typical catabolic loss of N following trauma. In addition to the APP, there is, of course, an increased production of other polypeptides, notably glutathione (Chapter 4), although this has not been quantified for the whole body.

A condition that has been described as 'catabolic' and sharing some of the features of trauma is chronic kidney disease and the treatment of it by haemodialysis. Dialysis itself has been said to be catabolic, but there is evidence that this is not so as regards the whole body (Lim *et al.*, 1993; Raj *et al.*, 2004b). Two studies in which patients were compared with controls gave somewhat contradictory results: when patients were not dialysed whole body protein breakdown was less than in controls (Lim *et al.*, 1998); when they were dialysed there was no difference. The picture becomes clearer when it comes to studies on arm or leg in patients during dialysis: breakdown is greatly increased, with a modest increase in synthesis, leading to a net negative balance across the limb (Ikizler *et al.*, 2002; Raj *et al.*, 2004a, 2005). Acidosis, which often accompanies chronic kidney damage, is a complicating factor. In both man and rat acidosis greatly increased whole body proteolysis (May *et al.*, 1987; Reaich *et al.*, 1992; Ballmer *et al.*, 1995). Experimental acidosis causes a 25% decrease in albumin synthesis (Ballmer *et al.*, 1995). Mitch *et al.* (1994) found that in isolated muscle of acidotic rats the mRNAs of ubiquitin and some of the proteasome subunits were increased up to fourfold. There is increased transamination and decarboxylation of the leucine released through stimulation of the activity of the BCA dehydrogenase.

In patients receiving haemodialysis the fractional synthesis rates of albumin and fibrinogen were increased (Raj *et al.*, 2004b).

Glutamine is the amino acid that has been most widely studied in severely ill or traumatized patients. The results are somewhat contradictory, but on the whole there seems to be a consensus that plasma and muscle glutamine concentrations are reduced (Jackson *et al.*, 1999; Mittendorfer *et al.*, 1999; Biolo *et al.*, 2000; Gore and Jahoor, 2000), as is *de novo* glutamine synthesis (Gore and Jahoor, 1994). Opinion is divided about whether glutamine efflux from muscle is increased or decreased. Gore and Wolfe (2002) found no change with glutamine supplementation in muscle protein kinetics, but they infused glutamine for only 8 h. The conclusion is that in injury or disease the requirement for glutamine is increased, above the level that can be met from glutamine stores in muscle or *de novo* synthesis, but the details of this requirement remain obscure. The effects of glutamine supplementation have been summarized by Wilmore (2001); they appear to be mainly improvements in immune response and in gut function.

13.2.2 Hormones

In trauma plasma concentrations of the catabolic hormones, glucocorticoids and glucagon, are increased and that of growth hormone decreased. This has been a consistent finding; Cuthbertson (1941) was well aware of the possible role of the anterior pituitary. It was well established that the catabolic loss of N in trauma could not be completely abolished by giving extra protein, but it was found that infusions of dextrose, amino acids and insulin did decrease the rate of breakdown and increase that of synthesis, with a reduction in N loss (O'Keefe *et al.*, 1981; Jeevanandam *et al.*, 1991). It was therefore natural that many clinicians, concerned to reduce the loss of muscle, should consider whether the same effect could not be produced by replacement of growth hormone during the flow phase. A very thorough study by Jeevanandam *et al.* (1996), in which protein turnover was measured with 24-h infusions of [^{15}N]-glycine, gave the results shown in Fig. 13.2. Growth hormone maintained the modest increase in synthesis that followed the trauma but had little effect on the increase in whole body breakdown that occurred in the week after the trauma. On the other hand, muscle protein breakdown was greatly reduced, as was the urinary loss of N. Similarly, Jiang *et al.* (1989) had shown that growth hormone given after elective surgery decreased N excretion and improved N balance but did not decrease whole body protein breakdown. Ferrando *et al.* (1999) found that insulin given to burns patients increased protein synthesis in the limb, with no change in breakdown. The evidence, therefore, is difficult to interpret: growth hormone reduces muscle protein breakdown, but probably has little effect on breakdown in the rest of the body. Moreover, Biolo *et al.* (2000) have shown that GH decreases muscle glutamine production and release. Therefore trying to reduce the effects of injury may do more harm than good. It is thus still an open question whether the body's response to injury is a series of reactions which individually are harmful and need to be counteracted; or whether the response is a coordinated one in

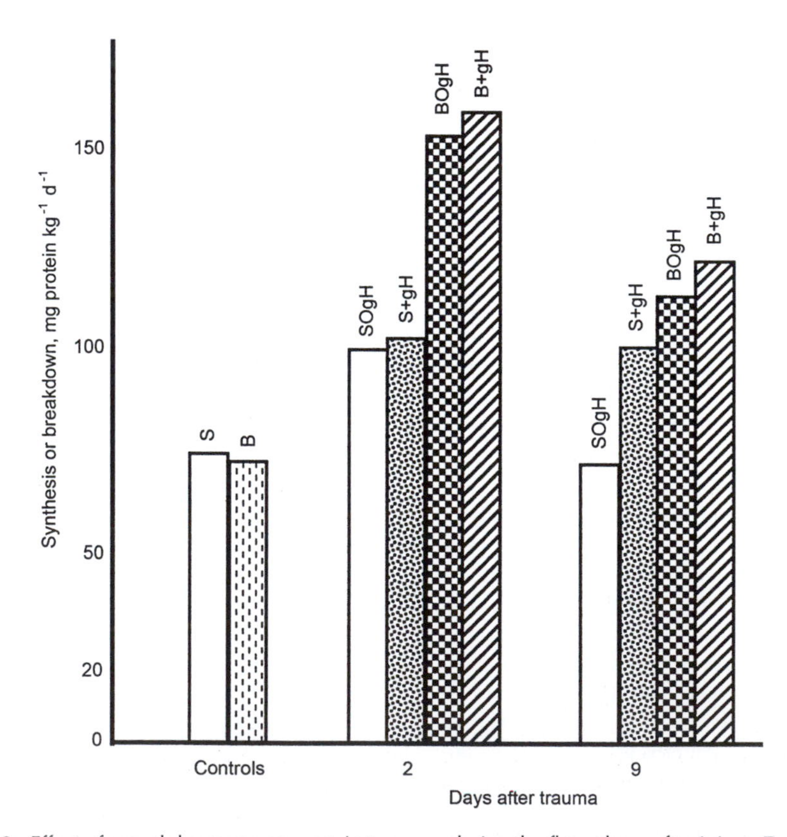

Fig. 13.2. Effect of growth hormone on protein turnover during the flow phase after injury. Drawn from the data of Jeevanandam *et al.* (1996). Synthesis, no growth hormone SOgH. Breakdown, no growth hormone BOgH. Synthesis plus growth hormone S+gH. Breakdown plus growth hormone B+gH.

which, at the expense of muscle, amino acids are made available for repair in other parts of the body. This would fit in with the view expressed more than 200 years ago by John Hunter (Hunter, 1794), based on clinical observation, that 'the injury has a tendency to produce both the disposition and the means of cure'.

11.3 References

Arnold, J., Campbell, I.T., Samuels, T.A., Devlin, J.C., Green, C.J., Hipkin, L.J., Macdonald, I.A., Scrimgeour, C.M., Smith, K. and Rennie, J.M. (1993) Increased whole body breakdown predominates over increased whole body synthesis in multiple organ failure. *Clinical Science* 84, 655–661.

Badaloo, A., Reid, M., Forrester, T., Heird, C. and Jahoor, F. (2002) Cysteine supplementation improves the erythrocyte glutathione synthesis rate in children with severe edematous malnutrition. *American Journal of Clinical Nutrition* 76, 646–652.

Ballmer, P.E., McNurlan, M.A., Hulter, H.N., Andersen, S.E., Garlick, P.J. and Krapf, R. (1995) Chronic metabolic acidosis decreases albumin synthesis and induces negative nitrogen balance in humans. *Journal of Clinical Investigation* 95, 39–45.

Berclaz, P.-Y., Benedek, C., Jequier, Y. and Schutz, Y. (1996) Changes in protein turnover and resting energy expenditure after treatment of malaria in Gambian children. *Pediatric Research* 39, 401–409.

Biolo, G., Isera, F., Rosutti, A., Toigo, G., Cocchi, B., Gealti, O., Gullo, A. and Guarnieri, G. (2000) Growth hormone decreases muscle glutamine production and stimulates protein synthesis in hypercatabolic patients. *American Journal of Physiology* 279, E323–332.

Biolo, G., Fleming, R.Y.D., Maggi, S.P., Nguyen, T.T., Herndon, D.N. and Wolfe, R.R. (2002) Inverse regulation of protein turnover and amino acid transport in skeletal muscle of hypercatabolic patients. *Journal of Clinical Endocrinology and Metabolism* 87, 3378–3384.

Birkhahn, R.H., Long, C.L., Fitkin, D., Geiger, J.W. and Blakemore, W.S. (1980) Effects of major skeletal trauma on whole body protein turnover in man measured with L [1-^{14}C] leucine. *Surgery* 88, 294–300.

Breuillé, D., Arnal, M., Rambourdin, F., Bayle, G., Levieux, D. and Obled, C. (1998) Sustained modifications of protein metabolism in various tissues in a rat model of long-lasting sepsis. *Clinical Science* 94, 413–423.

Brooke, O.G. and Cocks, T. (1974) Resting metabolic rate in malnourished babies in relation to total body potassium. *Acta Paediatric Scandinavica* 63, 817–825.

Cayol, M., Tauveron, I., Rambourdin, F., Prugnaud, J., Gachon, P., Thieblot, P., Grizard, J. and Obled, C. (1995) Whole body protein turnover and hepatic protein synthesis are increased by vaccination in man. *Clinical Science* 89, 389–396.

Colley, C.M., Fleck, A., Goode, A.W., Muller, B.R. and Myers, M.A. (1983) Early time course of acute phase protein response in man. *Journal of Clinical Pathology* 36, 203–207.

Crane, C.W., Picou, D., Smith, R. and Waterlow, J.C. (1977) Protein turnover in patients before and after elective orthopedic operations. *British Journal of Surgery* 64, 129–133.

Cuthbertson, D.P. (1930) The disturbance of metabolism produced by bony and non-bony injury, with notes on certain abnormal conditions of bone. *Biochemical Journal* 24, 1244–1263.

Cuthbertson, D.P. (1942) Post-shock metabolic response. *Lancet* I, 433–437.

Cuthbertson, D.P. (1954) Inter-relationship of metabolic changes consequent to injury. *British Medical Bulletin* 10, 33–37.

Cuthbertson, D.P. (1976) Surgical metabolism: historical and evolutionary aspects. In: Wilkinson, A.W. and Cuthbertson, D.P. (eds) *Metabolism and the Response to Injury.* Pitmans Medical, Tunbridge Wells, UK, pp. 1–34.

Cuthbertson, D.P., Shaw, G.P. and Young, F.G. (1941) The anterior pituitary gland and protein metabolism. II. The influence of anterior pituitary extract on the metabolic response of the rat to injury. *Journal of Endocrinology* 2, 465.

Essen, P., McNurlan, M.A., Gamrin, L., Hunter, K., Calder, G., Garlick, P.J. and Wernerman, J. (1998) Tissue protein synthesis rates in critically ill patients. *Critical Care in Medicine* 26, 92–100.

Ferrando, A.A., Chinkes, D.L., Wolf, S.E., Matin, S., Herndon, D.N. and Wolfe, R.R. (1999) A submaximal dose of insulin promotes net skeletal muscle synthesis in patients with severe burns. *Annals of Surgery* 224, 11–18.

Garlick, P.J., McNurlan, M.A., Fern, E.B., Tomkins, A.M. and Waterlow, J.C. (1980) Stimulation of protein synthesis and breakdown by vaccination. *British Medical Journal* 26 (July), 263–265.

Golden, B.E. and Golden, M.H.N. (1992) Effect of zinc on lean tissue synthesis during recovery from malnutrition. *European Journal of Clinical Nutrition* 46, 697–706.

Golden, M.H.N., Waterlow, J.C. and Picou, D. (1977) Protein turnover, synthesis and breakdown before and after recovery from protein-energy malnutrition. *Clinical Science and Molecular Medicine* 53, 473–477.

Gore, D.C. and Jahoor, F. (1994) Glutamine kinetics in burn patients. *Archives of Surgery* 129, 1318–1323.

Gore, D.C. and Wolfe, R.R. (2002) Glutamine supplementation fails to affect muscle protein kinetics in critically ill patients. *Journal of Parenteral and Enteral Nutrition* 26, 342–349.

Grizard, J. and Obled, C. (1995) Whole body protein turnover and hepatic protein synthesis are increased by vaccination in man. *Clinical Science* 89, 389–396.

Hunter, J. (1794) Quoted in Cuthbertson, 1976.

Ikizler, T.A., Pupim, L.B., Bronillelte, J.R., Levenhagen, D.K., Farmer, K., Hakim, R.M. and Flakoll, P.J. (2002) Haemodialysis stimulates muscle and whole body protein loss and alters substrate oxidation. *American Journal of Physiology* 282, E107–116.

Jackson, A.A. (1986) Blood glutathione in severe malnutrition in childhood. *Transactions of the Royal Society of Tropical Medicine and Hygiene* 80, 911–913.

Jackson, N.C., Carroll, P.V., Russell-Jones, D.L., Sönksen, P.H., Treacher, D.F. and Umpleby, A.M. (1999) The metabolic consequences of critical illness: acute effects on glutamine and protein metabolism. *American Journal of Physiology* 276, E163–170.

James, W.P.T., Ferro-Luzzi, A. and Waterlow, J.C. (1988) Definition of chronic energy deficiency in adults. *European Journal of Clinical Nutrition* 42, 969–981.

Jeevanandam, M., Leland, D., Shamos, R.F., Casano, S.F. and Schiller, W.R. (1991) Glucose infusion improves endogenous protein synthesis efficiency in multiple trauma victims. *Metabolism* 40, 1199–1206.

Jeevanandam, M., Peterson, S.R. and Shamos, R.T. (1993) Protein and glucose fuel kinetics and hormonal changes in elderly trauma patients. *Metabolism* 42, 1255–1262.

Jeevanandam, S., Holaday, N.J. and Petersen, S.R. (1996) Integrated nutritional, hormonal, and metabolic effects of recombinant human growth hormone (rhGH) supplementation in trauma patients. *Nutrition* 12, 777–787.

Jiang, Z.-M., He, G.-Z., Zhang, S.-Y., Wang, X.-R., Yang, N.-F., Zhu, Y. and Wilmore, D. (1989) Low-dose hormone and hypocaloric nutrition attenuate the protein-catabolic response after major operation. *Annals of Surgery* 210, 513–525.

Kien, C.L., Rohrbaugh, K., Burke, J.F. and Young, V.R. (1978) Whole body protein synthesis in relation to basal energy expenditure in healthy children and in children recovering from burn injury. *Pediatric Research* 12, 211–216.

Lim, V.S., Bier, D.M., Flanigan, M.J. and Sum-Ping, S.T. (1993) The effect of haemodialysis on protein metabolism. A leucine kinetic study. *Journal of Clinical Investigation* 91, 2429–2436.

Lim, V.S., Yarasheki, K.E. and Flanigan, M.J. (1998) The effect of uraemia, acidosis and dialysis treatment on protein metabolism: a longitudinal leucine kinetic study. *Nephrology, Dialysis, Transplantation* 13, 1723–1730.

Long, C.S., Birkhahn, R.H., Geiger, J.W. and Blakemore, W.S. (1981) Contribution of skeletal muscle protein in elevated rates of whole body catabolism in trauma patients. *American Journal of Clinical Nutrition* 34, 1087–1093.

Macallan, D.C., McNurlan, M.A., Kurpad, A.V., De Souza, G., Shelty, P.S., Calder, A.G. and Griftin, G.E. (1998) Whole body protein metabolism in human pulmonary tuberculosis and undernutrition: evidence for anabolic block in tuberculosis. *Clinical Science* 94, 321–331.

Manary, M.J., Brewster, D.R., Broadhead, D.L., Crowley, J.R., Fjeld, C.R. and Yarasheki, K.E. (1997) Protein metabolism in children with edematous malnutrition and acute lower respiratory infection. *American Journal of Clinical Nutrition* 65, 1005–1010.

Manary, M.J., Broadhead, R.L., and Yarasheki, K.E. (1998) Whole body protein kinetics in marasmus and kwashiorkor during acute infection. *American Journal of Clinical Nutrition* 67, 1205–1209.

May, R.C., Hara, Y., Kelly, R.A., Block, K.P., Buse, M.G. and Mitch, W.E. (1987) Branched chain amino acid metabolism in rat muscle: abnormal regulation in acidosis. *American Journal of Physiology* 252, E712–718.

Mitch, W.E., Medina, R., Grieber, S., May, R.C., England, B.K., Price, S.R., Bailey, J.L. and Goldberg, A.L. (1994) Metabolic acidosis stimulates muscle protein degradation by activating the adenosine triphosphate-dependent pathway involving ubiquitin and proteasomes. *Journal of Clinical Investigation* 93, 2127–2133.

Mittendorfer, B., Gore, D., Herndon, D. and Wolfe, R. (1999) Accelerated glutamine synthesis in critically ill patients cannot maintain normal intramuscular free glutamine concentration. *Journal of Parenteral and Enteral Nutrition* 235, 243–252.

Montgomery, R.D. (1962) Changes in the basal metabolic rate of the malnourished infant and their relation to body composition. *Journal of Clinical Investigation* 41, 1653–1663.

Munro, H.N. and Chalmers, M.I. (1945) Fracture metabolism at different levels of protein intake. *British Journal of Experimental Pathology* 29, 396–404.

O'Keefe, S.J.D., Moldawer, L.L., Young, V.R. and Blackburn, G.L. (1981) The influence of intravenous nutrition on protein dynamics following surgery. *Metabolism* 40, 1199–1206.

O'Leary, M.J., Ferguson, C.M., Rennie, M.J., Hinds, C.J., Coakley, J.H. and Preedy, V.R. (2001) Sequential changes in *in vivo* muscle and liver protein synthesis and plasma and tissue glutamine levels in sepsis in the rat. *Clinical Science* 101, 295–304.

Picou, D. and Taylor-Roberts, T. (1969) The measurement of total protein synthesis and catabolism and nitrogen turnover in infants in different nutritional states and receiving different amounts of dietary protein. *Clinical Science* 36, 283–296.

Raj, D.S., Zager, P., Shah, V.O., Dominic, E.A., Ardeniyi, O., Blandon, P., Wolfe, R. and Ferrando, A. (2004a) Protein turnover and amino acid transport kinetics in end-stage renal disease. *American Journal of Physiology* 286, E136–143.

Raj, D.S., Dominic, E.A., Wolfe, R., Shah, V.O., Bankhurst, A., Zager, P.G. and Ferrando, A. (2004b) Coordinated increase in albumin, fibrinogen and muscle protein synthesis during haemodialysis: role of cytokines. *American Journal of Physiology* 286, E658–664.

Raj, D.S., Welbourne, T., Dominic, E.A., Waters, D., Wolfe, R. and Ferrando, A. (2005) Glutamine kinetics and protein turnover in end-stage renal disease. *American Journal of Physiology* 288, E37–46.

Reaich, D., Channon, S.M., Scrimgeour, C.M. and Goodship, T.H.J. (1992) Ammonium chloride-induced acidosis increases protein breakdown and amino acid oxidation in humans. *American Journal of Physiology* 263, E735–739.

Reeds, P.J., Fjeld, C.R. and Jahoor, F. (1994) Do the differences between the amino acid compositions of acute-phase and muscle proteins have a bearing on nitrogen loss in traumatic states? *Journal of Nutrition* 124, 906–910.

Reid, M., Badaloo, A., Forrester, T., Morlese, J.F., Frazer, M., Heird, W.C. and Jahoor, F. (2000) *In vivo* rates of erythrocyte glutathione synthesis in children with severe protein-energy malnutrition. *American Journal of Physiology* 278, E405–412.

Reid, M., Badaloo, A., Forrester, T., Heird, W.C. and Jahoor, F. (2002) The response of splanchnic and whole body leucine kinetics to treatment of children with edematous protein-energy malnutrition. *American Journal of Clinical Nutrition* 76, 633–640.

Roediger, W.E. (1995) New views on the pathogenesis of kwashiorkor: methionine and other amino acids. *Journal of Pediatric Gastroenterology and Nutrition* 21, 130–136.

Soares, M.J., Piers, L.S., Shetty, P.S., Robinson, S., Jackson, A.A. and Waterlow, J.C. (1991) Basal metabolic rate, body composition and whole-body protein turnover in Indian men with differing nutritional status. *Clinical Science* 81, 419–425.

Soares, M.J., Piers, L.S., Shetty, P.S., Jackson, A.A. and Waterlow, J.C. (1994) Whole-body protein turnover in chronically undernourished individuals. *Clinical Science* 86, 441–446.

Stroud, M.A., Jackson, A.A and Waterlow, J.C. (1996) Protein turnover rates of two human subjects during an unassisted crossing of Antarctica. *British Journal of Nutrition* 76, 165–174.

Tomkins, A.M., Garlick, P.J., Schofield, W.N. and Waterlow, J.C. (1983) The combined effects of infection and malnutrition on protein metabolism in children. *Clinical Science* 65, 318–324.

Waterlow, J.C. (1992) *Protein-Energy Malnutrition.* Edward Arnold, London.

Waterlow, J.C., Golden, M.H.N. and Garlick, P.J. (1978) Protein turnover in man measured with ^{15}N: comparison of end-products and dose regimes. *American Journal of Physiology* 235, E165–E174.

Wilmore, D. (2001) The effect of glutamine supplementation in patients following elective surgery and accidental injury. *Journal of Nutrition* 131, 2543S–2549S

14

Protein Turnover in Individual Tissues: Methods of Measurement and Relations to RNA

The first indication that the proteins of different tissues turn over at different rates came from the experiments of Schoenheimer *et al.* in 1939 and Ratner *et al.* in 1940. Their results are summarized in Table 14.1 and do not differ much in relative rates from those found today. The first systematic study of the turnover rates of proteins from different tissues was that of Shemin and Rittenberg (1944). They fed ^{15}N-glycine for 3 days, killed the animals thereafter at intervals for a week and traced the decay curves. Their conclusion that liver proteins had an average half-life of 7 days is clearly too long because of recycling of tracer, but the paper raises three important points: first, that carcass proteins incorporated glycine from other organs faster than they transferred glycine-N to other amino acids; secondly, that the non-protein nitrogen (NPN) fractions of different tissues had different isotope concentrations, and ‘even though the components of the NPN are freely diffusible, each organ has its own characteristic composition’. This was the first indication that precursor activities could differ, even in a steady state. Thirdly, they showed that the concentration of ^{15}N in urinary urea was very nearly the same as in the guanidine group of liver protein arginine – a finding that was very important for later work.

Table 14.1. ^{15}N content of protein nitrogen obtained from different organs after feeding [l-^{15}N] leucine (25 mg N per day) to rats for 3 days. Calculated for 100 atom per cent in compound administered.

	Atom %
Serum proteins	1.67
Intestinal wall	1.49
Kidney	1.38
Spleen	1.10
Liver	0.94
Heart	0.89
Testes	0.77
Muscle	0.31
Haemoglobin	0.29
Skin	0.18

From Schoenheimer, 1939.

14.1 Measurement of Breakdown

14.1.1 Indirect estimates *in vivo*

In the steady state breakdown is assumed to be equal to synthesis and does not need to be measured separately. However, the steady state is an ideal: even within a single day there are fluctuations produced by meals (see Chapters 6 and 15). In growth there is no steady state: as we have seen in Chapter 12, section 12.2 breakdown can be determined as synthesis minus protein deposited, but even in the young rapidly growing rat it takes several days to get a reasonably accurate estimate of deposition. The work of Millward *et al.* (1975, 1976) illustrates the successes and limitations of this method (see Chapter 15).

14.1.2 Tracer decay after dosage of labelled amino acids

In the early days of the isotope era in a remarkable experiment Thompson and Ballou (1956) gave tritiated water to rats for 124 days. During this time they were mated, the pups given the same labelled water and killed at intervals of up to 300 days. The curves of radioactivity in the various tissues could be resolved into two exponential components, which suggested that each tissue contained two groups of proteins, with half-lives of roughly 22 and 130 days, the slow components accounting for nearly half the total tissue. It had been shown that collagen is virtually inert in the mature animal (Neuberger *et al.*, 1951); this paper further emphasized the heterogeneity of tissues, and the likelihood that the turnover of substantial proportions of body protein may be missed in short-term experiments.

Another important experiment was that of Buchanan (1961) who fed rats for 50 days a diet in which all the carbon was labelled. Sucrose was labelled biosynthetically with ^{14}C and yeast was grown on a medium in which ^{14}C-sucrose was the sole source of carbon; the labelled sucrose and yeast provided all the carbon in the diet. The results showed differences in the replacement rates of the proteins and of other components of the tissues. Replacement rate was calculated as the 100 × SA (specific radioactivity) of the product divided by the SA of the food. The results are summarized in Table 14.2.

These experiments demonstrated the heterogeneity of turnover within tissues as well as between them; this remains a problem whenever measurements are made on whole tissue (see Garlick *et al.*, 1976). Another difficulty was that these long-term experiments could not provide accurate quantitative estimates of protein turnover because of reutilization of labelled amino acid (see Chapter 6). Swick (1958) tackled this problem by feeding $Ca^{14}CO_3$ to rats for 8 days. He found, as had Shemin and Rittenberg (1944), that the specific activity of the guanidine group of arginine in liver protein was virtually the same as that of urea, which had reached a plateau in about 0.5 days, and therefore could be used as a measure of the precursor activity during the rest of the experiment. He argued that arginine entering the free pool could be regarded as a non-reutilizable label, because its uptake into urea is many times greater than into protein. Swick also measured the rate of formation of arginine carboxyl carbon and calculated that the probability of its being reutilized was about 50%. This type of experiment, comparing the 'true' half-life obtained from the guanidine-C of arginine with the 'apparent' half-life from the carboxyl-C was repeated by Stephen and Waterlow (1966), who were particularly interested in the effects of a low protein diet. They found replacement rates almost identical to those of Swick (1958) and obtained some evidence that on the low protein diet the breakdown rate of both liver and serum proteins was decreased and the recycling rate increased, suggesting an economy of amino acids.

The use of the C6 group of arginine as an effective way of eliminating reutilization is probably confined to the liver as the site of urea formation. In muscle this route of disposal does not exist. Millward (1970) found that in muscle, after labelling with $Na^{14}HCO_3$ there was quite significant reutilization of guanidine-labelled arginine. The carboxyl groups of all amino acids became labelled, but judging by their more rapid decay rates, the carboxyl groups of aspartate and

Table 14.2. Replacement rates of tissue proteins after feeding rats on a diet in which all the protein was labelled.

	Per cent replaced	
Days of feeding:	20	50
Serum proteins	100	100
Liver	90	95
Gastrointestinal tract	80	80
Kidney	70	90
Heart	60	80
Spleen	75	80
Brain	40	60
Muscle	50	70

The numbers are approximate, taken from the published curves. From Buchanan, 1961.

glutamate were the least recycled, presumably because of their very rapid exchange with the body's extremely large bicarbonate pool, which, like urea, acts as a 'sink' from which there is no return. These are also the two amino acids which show particularly extensive fixation of ^{14}C in their carboxyl groups after dosage with ^{14}C bicarbonate (Swick and Handa, 1955).

An ingenious method for measuring the turnover rates of liver fractions and enzymes synthesized in the liver was developed by Arias *et al.* (1969). In this double isotope method one tracer, e.g. [^{14}C]-leucine, is injected at zero time and another, e.g. [^{3}H]-leucine, 3–10 days later. The animal is sacrificed shortly after the second injection, the protein or fraction of interest isolated, and the ratio [^{14}C]:[^{3}H] determined. [^{3}H] represents the initial labelling and [^{14}C] the labelling after several days, so that the lower the ratio the more rapid the turnover rate. It was assumed that degradation followed first order kinetics. Glass and Doyle (1972) got round the problem of reutilization by constructing a calibration curve relating the [^{14}C]:[^{3}H] ratio to degradation rates measured with guanidino labelled arginine. With this calibration curve (Fig. 14.1) the isotope ratio could be used to determine the degradation rates of proteins that were not present in sufficient amounts for the labelled arginine to be measurable by the techniques then available. Results were obtained for a number of enzymes. However, Zak *et al.* (1977) found it impossible to obtain accurate results on different muscle fractions, because the half-lives were extremely sensitive to the timing of the two tracers.

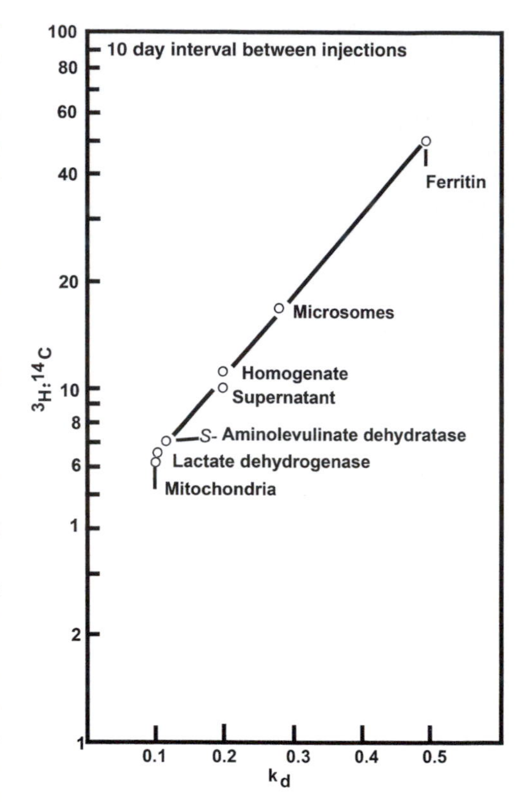

Fig. 14.1. Logarithm of ^{3}H:^{14}C ratio plotted against fractional degradation rate determined with [guanidino -^{14}C] arginine. From Glass and Doyle, 1972. Reproduced by courtesy of the *Journal of Biological Chemistry*.

14.1.3 Urinary excretion of 3-methyl histidine

Another approach that avoids reutilization is the excretion of this methylated amino acid (MeH). Since it is widely used, particularly in clinical situations, it is necessary to consider in a little detail the controversy that has surrounded it.

Young and co-workers in 1973 proposed that the urinary excretion of MeH might be used as a measure of the rate of breakdown of skeletal muscle protein (Young *et al.*, 1973). The basis for the idea was that MeH is formed by immediate post-translational methylation of histidine. When the protein breaks down the MeH liberated cannot be reutilized for synthesis and is excreted. It was known that MeH is found only in actin and myosin, and Haverberg *et al.* (1975) had claimed that in the rat 98% of the MeH was present in skeletal muscle. Young and Munro (1978) showed that MeH does not charge tRNA and so cannot be synthesized into protein, and that when labelled MeH was injected it was quantitatively excreted in the urine, with no labelled CO_2 being formed.

In 1980 Millward *et al.*, from studies on the rate of synthesis of MeH in muscle, gut and skin, claimed that the turnover of MeH in skeletal muscle was not enough to account for all the MeH excreted in the urine, so that there must be a significant contribution, up to 50%, from smooth muscle of the gut and other tissues in which MeH turns over rapidly. This conclusion was contested by Harris (1981), although later work by

Millward's group showed that MeH in muscle turned over at approximately the same rate as myosin and actin (Bates *et al.*, 1983; Millward and Bates, 1983). The root of the difficulty seems to be one of numbers: a difference between Haverberg *et al.* (1975) and Millward *et al.* (1980) in their estimates of the amounts of MeH in tissues of the rat other than skeletal muscle. The subject is of some importance because MeH excretion is widely used as a measure of skeletal muscle breakdown. The problem may be resolved, at least for humans, by the remarkable study of Afling *et al.* (1981) on a patient with neuromuscular dystrophy in whom, at post-mortem, no skeletal muscle at all could be found. This patient's excretion of MeH was 30% of that of controls, but the ratio of MeH to creatinine was 1.7 times as high. A similar situation occurred in patients with muscular dystrophy who, on the basis of a high urinary MeH:creatinine ratio, were diagnosed as having increased rates of muscle protein breakdown (Ballard *et al.*, 1979) but later, from direct measurements, were shown to have decreased rates of muscle protein synthesis (Halliday *et al.*, 1988).

The conclusion seems to be that tissues other than muscle do make a substantial contribution to MeH excretion; and that a high urinary ratio of MeH:creatinine may be misleading when loss of muscle mass leads to a low creatinine excretion.

14.1.4 Measurements on limbs and organs

Protein breakdown can be determined in a limb or the splanchnic region by measurement of blood flow and of arterial and venous amino acid concentrations and enrichments. The key measurement is dilution of the tracer between entering and leaving the tissues by unlabelled amino acid derived from proteolysis (see Chapter 6).

A variant of the method, applied to muscle, is to infuse tracer until a plateau has been reached in the free pool; then to stop the infusion and measure the rate of decay of enrichment in arterial blood and the muscle free pool over the next 60 minutes (Zhang *et al.*, 1996). This allows calculation of the rate of dilution of intracellular tracer by unlabelled amino acid from protein breakdown.

Perfusion of the liver and heart has been used very successfully to measure both synthesis (Peavy and Hansen, 1976) and degradation in these organs (Mortimore and Pösö, 1984). Perfusions were done in the single-pass mode: degradation was determined from the release of tyrosine or other amino acids into the medium, with synthesis blocked by cycloheximide.

14.1.5 Measurements on isolated tissues, suspensions of hepatocytes and cell cultures

This is an approach that has been widely used, particularly by Goldberg and his group at Harvard, with concentration on muscle because muscle wasting is a serious complication in so many nutritional and pathological states. With muscle it is important to maintain adequate oxygenation and normal tension (Baracos and Goldberg, 1986) and for this reason the diaphragm, being very thin, has been a popular choice, but even so there may be discrepancies with *in vivo* measurements. In one study synthesis *in vitro* was only one half that *in vivo*, even though the incubated muscles were supplied with generous amounts of amino acids (Preedy *et al.*, 1986). Breakdown is often measured by tyrosine release, since tyrosine is not oxidized in muscle.

Total degradation = net release + uptake by synthesis, which may either be determined with a separate amino acid or abolished by cycloheximide. There are some reports that cycloheximide has an inhibitory effect on breakdown (Amenta *et al.*, 1978; Bacchino, 1982), but in general the two methods agree well. Alternatively, with modern instrumentation it is possible to measure the output of 3-methylhistidine by tissues or cells *in vitro* (Thompson *et al.*, 1996). Incubated muscle has the great advantage that inhibitors of the three major proteolytic systems can be used to sort out the contributions of each of them. Scornik developed a method in which the accumulation of polypeptides was measured in liver and muscle after bestatin had been given, which inhibits the hydrolysis of peptides to amino acids (Botbol and Scornik, 1997).

In many recent studies degradation has been assessed by measuring the activities of the degradative enzymes and/or the concentrations of their mRNAs. With this approach there is always the possibility that the amount of mRNA may reflect the stability of the messenger rather than the extent of its gene expression. It has been claimed that there is not a close correspondence between the amount of a protein and the amount of its messenger (Anderson and Anderson, 1998).

To conclude, nowadays synthesis is measured much more often than breakdown, because, except for *in vitro* studies, the methods are quicker and probably more reliable, and all the results that follow have been obtained by measuring synthesis. However, an important advantage of measurements of degradation, which is often forgotten, is that they can be applied in the non-steady state, for example in tissues of a growing animal or in studies of enzyme turnover after induction or repression. In such situations the rate of decay of total activity or enrichment gives the rate of breakdown and the decay of specific activity the rate of synthesis (Koch, 1962; Schimke *et al.*, 1965). This approach was put to good use by Millward (1970) in early studies of muscle protein turnover in growing rats.

14.2 Measurements of Synthesis

14.2.1 Constant infusion of tracer

Early studies on animals were done with unprimed constant infusions, in which, in principle, the synthesis rates of tissue proteins were calculated from the basic equation relating the activity in protein to that of the precursor amino acid at plateau:

$$S_B/S_{Amax} = 1 - e^{-ks.t}$$

where S_B is the activity in protein, S_{Amax} is the plateau activity of the precursor, k_s is the fractional rate of synthesis and *t* is the time of measurement. This simple formula assumes that the plateau activity of the precursor is reached instantaneously. With lysine with its large free pool this is very far from being the case. In rats it was found that in liver the activity of free lysine took about 3 h to reach plateau, and in muscle more than 6 h (Waterlow and Stephen, 1968). Therefore equations were developed to take account of the rate of rise of the precursor activity to plateau, this rate being different in liver and muscle (Swick, 1958; Waterlow and Stephen, 1968; Garlick *et al.*, 1973). These equations are seldom used nowadays and the interested reader is referred to the original papers. A study by Halliday and McKeran (1975) – probably the first to measure human muscle protein synthesis – got round the problem by infusing [^{15}N] lysine for 20–30 h; plateau enrichment of free lysine in plasma was reached in 14 h, after which two biopsies of muscle were taken. This avoided the rising phase of precursor activity. Another rather sophisticated way of defining the curve to plateau on only one animal, proposed by Dudley *et al.* (1998), is to infuse six different tracers at intervals of an hour or so before sacrifice, so that the last one infused will give the earliest point on the time curve.

After lysine, the next amino acid to be used routinely was tyrosine (James *et al.*, 1976) and finally leucine, which have much smaller free pools than lysine, so that the rise to plateau enrichment is much faster and several measurements can be made at plateau within a reasonable time.

14.2.2 The flooding dose method

The flooding dose method has been used extensively by Garlick and his co-workers for measurements on human tissues. It has the great advantage of taking only about 1–1$^1/_2$ h, compared with a minimum of 3–4 h with a constant infusion, even with priming. Therefore it is important to establish its validity.

The first mention of the method is in the work of Loftfield and Harris (1956), Loftfield and Eigner (1958), Kean (1959) and Henshaw *et al.* (1971). It was extensively used by Scornik in experiments on mice (Scornik, 1974; Conde and Scornik, 1976). The aim of the flood is to reduce differences in labelling of the precursor in plasma and tissues and between the tissue-free amino acid and tRNA. The labelling of the intracellular pool depends on the relative inflows of labelled amino acids from plasma and unlabelled amino acids from protein breakdown. It is assumed that a flooding dose will increase the entry from plasma to a point at which the relative contribution from protein breakdown becomes negligible. In agreement with this assumption it was early shown that with a labelled flood the larger the dose the greater the incorporation of label into protein (Kean, 1959; Scornik, 1974). Incorporation of tracer is, of course, not the same as synthesis.

Garlick's initial study in rats with a flooding

dose of phenylalanine accompanied by constant infusions of lysine and threonine showed that the fractional synthesis rates given by these two amino acids were not affected by the phenylalanine flood (Garlick *et al.*, 1980). In subsequent studies on rats, sheep, dogs and man, synthesis rates with flood and constant infusion were compared for a range of different tissues (see reviews by Garlick *et al.*, 1994, and Rennie *et al.*, 1994). Since different amino acids were used, with different choices of precursor, the results were rather variable and contradictory. However, Davis *et al.* (1999), with flooding doses of phenylalanine, obtained excellent agreement between the enrichments in tissue fluid and tRNA in both liver and muscle. In man particular attention was concentrated on muscle because this tissue can be easily sampled by biopsy. There is no disagreement that in post-absorptive subjects the apparent rate of muscle protein synthesis is almost doubled by a flooding dose of either leucine or phenylalanine, the amino acids most commonly used. The question is whether this increase is an artefact or real.

The general protocol of a flooding measurement is as follows: a commonly used dose is 50 mg kg^{-1}, which with leucine or phenylalanine would increase the total body pool four or five times. McNurlan *et al.* (1994) explored the use of a smaller dose – about 30 mg kg^{-1} – and found that it made no difference. After an initial peak the plasma activity of the flooding amino acid falls in man at a rate of about 25% per h (Fig. 14.2), although much faster in the rat (Table 14.3). A muscle biopsy is taken initially and again after about 90 min. In three heroic studies 5–6 biopsies were made at intervals up to 90–120 minutes (Garlick *et al.*, 1989; McNurlan *et al.*, 1994; Smith *et al.*, 1998); incorporation of tracer into protein was linear with time. The rate of muscle protein synthesis was calculated as:

protein enrichment $(t_2 - t_1)$/(area under curve of precursor enrichment $\times$ time)

There is the perennial problem of defining the precursor. It has been shown in rats (Obled *et al.*, 1989), sheep (Attaix *et al.*, 1986; Schaeffer *et al.*, 1986; Lobley *et al.*, 1992) and man (Garlick *et al.*, 1989; McNurlan *et al.*, 1991; Essén *et al.*, 1992) that after a flooding dose the activity of the free amino acid in muscle becomes progressively closer to that in plasma until by 40–60 minutes they are virtually the same, fulfilling the aim of equalizing the enrichment of possible precursor

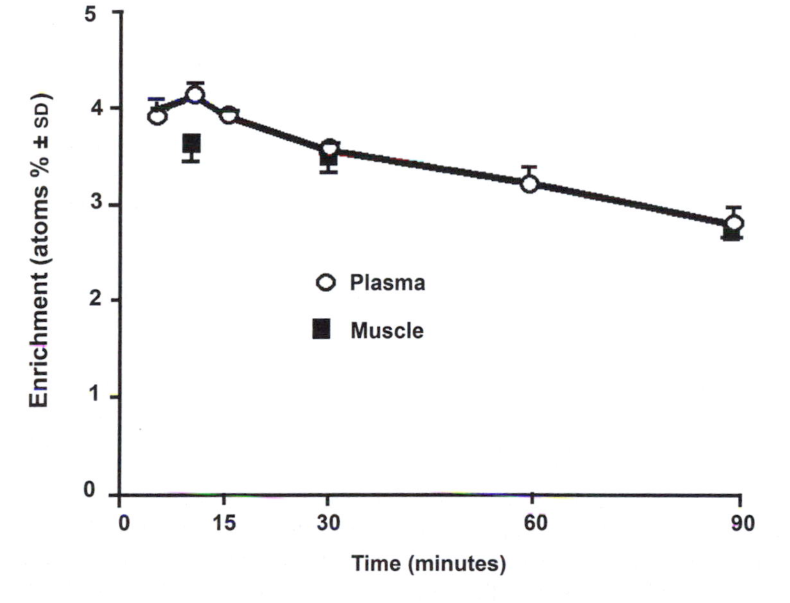

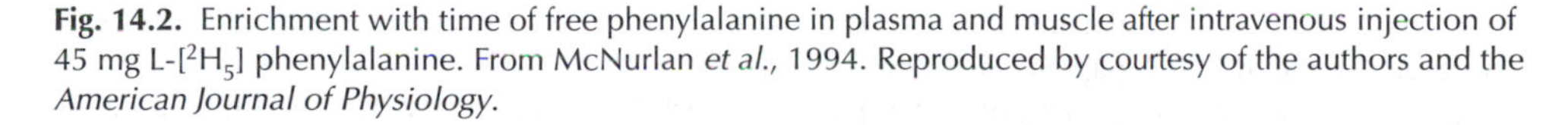
Fig. 14.2. Enrichment with time of free phenylalanine in plasma and muscle after intravenous injection of 45 mg L-[$^2H_5$] phenylalanine. From McNurlan *et al.*, 1994. Reproduced by courtesy of the authors and the *American Journal of Physiology*.

Table 14.3. Rate of decay of activity in plasma after a flooding dose of a labelled amino acid.

	Rate of decay % h^{-1}
Rats	
Garlick *et al.*, 1980	60
Obled *et al.*, 1989	76
Sheep	
Attaix *et al.*, 1986	6
Lobley *et al.*, 1992 high dose	0
low dose	27
Southorn *et al.*, 1992	11
Man	
Garlick *et al.*, 1989	25
McNurlan *et al.*, 1991	27

pools. Thus there will be little error in taking the enrichment of the precursor as that of the amino acid in plasma, or in the case of leucine, that of KIC. Admittedly there is still an error in supposing that the intracellular amino acid enrichment accurately reflects that of tRNA, but the error is not likely to be large (see Chapter 5).

There are two ways of testing the effect of the flood. In one the measurements are made from the flood alone and compared with earlier values obtained by constant infusion; in the other a labelled amino acid is infused and a different amino acid used for flooding. The results by both variants are shown in Table 14.4.

The flooding dose apparently increases by 50–100% the estimate of muscle protein synthesis (MPS) obtained by constant infusion. Garlick *et al.* (1989) and Smith *et al.* (1998) found two- to three-fold increases in intracellular concentration after flooding doses of leucine, phenylalanine or threonine. The question then is whether such increases directly stimulate protein synthesis. Chinkes *et al.* (1993) apparently postulated a mass action effect of the higher amino acid concentration; they proposed a method of correcting the result to what it would have been without the flooding, which seems to contradict their belief that the stimulation was real. Attaix *et al.* (1986), working on sheep, gave massive doses of valine, which increased the total body pool 10-fold, but found a negligible effect on protein synthesis in liver and muscle. On the other hand, the effect of amino acid infusions (Chapter 9) and the work on the limb showing that higher plasma concentrations lead to higher inward transport into tissues (Chapter 4) are consistent with a cause and effect relation between amino acid concentrations and synthesis rate. When increasing amounts of amino acids were infused, there was a stimulation of muscle protein synthesis which seemed to be related more closely to plasma- than intracellular amino acid concentration (Bohé *et al.*, 2003). This would fit in with the earlier suggestion (Chapter 5) that the enrichment of tRNA is closer to that of the extracellular than the intracellular amino acid. Finally, the idea that the increased IC concentration might somehow distort the precursor–product relationship is a non-starter; it would require the IC or tRNA enrichment to be higher than the enrichment in plasma, and this is impossible.

If the increase in synthesis produced by the flood is real, it must involve increased incorporation of all amino acids, and not just that of the flooding dose, the other amino acids being presumably drawn from the muscle free pool. This is clearly shown by the experiments of Smith *et al.* (1992 and 1998 – C and E in Table 14.4) in which the amino acid used for measuring synthesis was not the same as that of the flood. The enrichment of the constantly infused amino acid in muscle was not changed by the flood, but the estimate of synthesis was increased.

In the first six studies in Table 14.4 the initial measurement of muscle protein synthesis (MPS) was in the fasted state. Garlick's study (F) showed that in the fed state the pre-flood rate of MPS was 60% higher than in the fasted state. In study G (Tjader *et al.*, 1996), in which an amino acid mixture was infused, there was no difference between pre- and post-flood rates of MPS. This suggests that the rate was already maximal and could not be increased by the flood. From the evidence available so far, the flood has to be an essential amino acid: leucine, phenylalanine, threonine and valine are all effective in stimulating MPS, while no effect was found with arginine, glycine or serine (Smith *et al.*, 1998). It is interesting, therefore, to compare the effect of a complete protein given in a single dose with that of a flooding dose of an amino acid. Very illuminating in this context is a study of Boirie *et al.* (1996) on the effects of a single meal of whey protein, which is rapidly absorbed. The meal provided 380 μmol kg^{-1} of leucine, which is exactly the same as the usual flooding dose (50 mg kg^{-1}), and produced an approximately two-fold increase in whole body protein synthesis, peaking

Table 14.4. Synthesis rate of mixed muscle protein with or without a flooding dose of labelled amino acid.

	Amino acid		Muscle protein synthesis, % h^{-1}		
	For CI	For flood	From CI before or without flood	From CI after flood	From flood
A Garlick *et al.*, 1989	–	Leucine	–	–	0.081
B McNurlan *et al.*, 1991	Leucine	–	0.043	–	–
	–	Phenylalanine	–	–	0.072
	–	Phe + valine	–	–	0.067
	–	Phe + leucine[a]	–	– From phe	0.069
	–	Phe + leucine[a]	–	– From leu	0.075
C Smith *et al.*, 1992	Valine	Leucine	0.043	0.065	0.060
D McNurlan *et al.*, 1994	–	Phenylalanine	–	–	0.069
E Smith *et al.*, 1998	Leucine	Phenylalanine	0.036	0.067	–
	Leucine	Threonine	0.037	0.070	–
F Garlick *et al.*, 1994	Leucine	Leucine	0.040		
		Post-absorptive	0.046	–	0.087
		Fed	0.075	–	0.097
G Tjader *et al.*, 1996	Valine	Leucine, fed	0.091	0.095	0.093
	Leucine	Valine, fed	0.083	0.087	0.107

CI, constant infusion. For constant infusions of leucine or valine, muscle protein synthesis rates were calculated from enrichment of plasma KIC (α-ketoisocaproic acid) or KIV (isoketovaleric acid).
[a]Leucine flood was enriched as well as that of phenylalanine.
[b]All studies were made in the post-absorptive state except that of Garlick *et al.* (F) and Tjader *et al.* (G), where the subjects were receiving total parenteral nutrition, and that of Garlick *et al.* (1994), where the subjects were few.

at about 90 min and lasting 2–3 h. In this study the plasma leucine concentration increased from ~100 μmol l^{-1} to a peak of ~350 μmol l^{-1}, returning to baseline after 3–4 h. This is not very different, either in pattern or amount, from the curve of leucine concentration after a flood, except that with the meal the peak is later. Our conclusion is that a flooding dose of an essential amino acid mimics the effect of a meal – but probably not all workers would agree with this.

14.3 RNA Content and Activity

Brachet (1941) and Caspersson (1941) identified RNA in tissues by selective staining or ultraviolet absorption and observed that RNA was particularly abundant in tissues in which protein synthesis was believed to be particularly rapid. Campbell and Kosterlitz (1947) showed that when rats were fed a protein-free diet the DNA content of the liver was preserved while the concentration of RNA was reduced by 35%. Henshaw *et al.* (1971) and Millward *et al.* (1973) proposed that, since 80% of RNA is ribosomal, the amount of protein synthesized in unit time per g RNA could be used as a measure of ribosomal activity, denoted k_{RNA}.[1] Figure 14.3 is an example of the early use of this relationship. The data of Henshaw *et al.* in Table 14.5A on liver and muscle show that the very large difference in k_s between liver and muscle results partly from a lower [RNA] and partly from a lower RNA activity. The results of Millward *et al.* (1973) in Table

[1] The term 'activity' is more appropriate than 'efficiency' because the latter has been used in connection with animals that are growing to denote the ratio of protein deposited to protein synthesized (see Chapter 12).
k_{RNA} is used to express RNA activity and [RNA] to express the concentration of RNA relative to that of protein (mg g^{-1}). The usual expression of fractional synthesis rate, FSR, is amount synthesized per 100 mg protein. Therefore k_{RNA} = FSR × 10/[RNA].

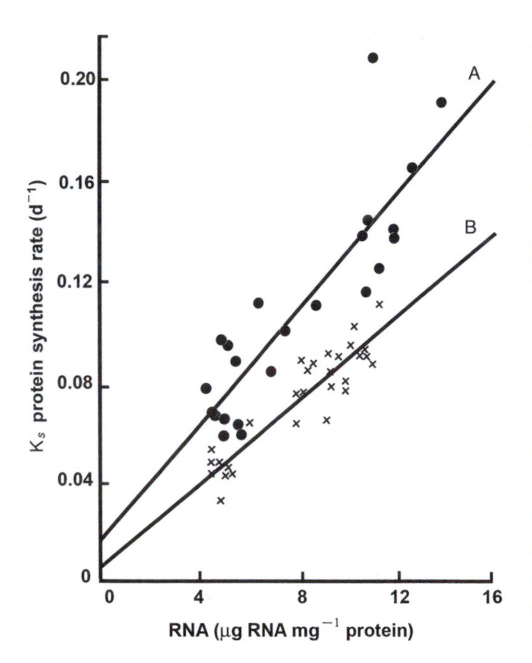

Fig. 14.3. Relationship between rates of protein synthesis and RNA concentrations in rat muscle. From Millward *et al.*, 1973. Reproduced by courtesy of the authors and *Nature* (*London*).
A, ●—●, rats on normal diet; B, x—x, rats on low protein diet.

14.5B show that after only one day on a protein-free diet there was a 30% fall in synthesis, denoted k_s, mainly due to a decrease in k_{RNA}; but after 21 days the loss of RNA had become much greater and was the main determinant of the low k_s. Henshaw *et al.* in their classical paper of 1971 concluded that: '... changes in ribosome content, in chain initiation and in polyribosome activity are of roughly equal importance in determining the depression of protein synthesis in fasting.' Whether or not they are of equal importance will depend on the situation, but it is clear that in each case account must be taken of two *separate* processes, loss of ribosomes ([RNA]) and decreased ribosomal activity [k_{RNA}]. These two studies laid the foundation for much later work, which will be described in the next chapter.

Millward's data in Table 14.5B suggest that the loss of ribosomes occurs more slowly than the fall in their activity. This is in keeping with the slower turnover rate of ribosomal than of messenger RNA that has been demonstrated by Schöch and co-workers (Table 11.6). A reduction in k_{RNA} probably results from a block in initiation, as suggested by Henshaw *et al.*; this has been the subject of intensive investigation in recent years, particularly in muscle (e.g. Kimball *et al.*, 2002). The other possibility is a reduction in mRNA, which would not show up in the crude measurement of total RNA, since mRNA is such a small proportion of it. Current work is concentrating on identifying and quantifying specific mRNAs that are reduced or increased in particular situations.

All the results discussed above were from the rat: what of other species?

Data on RNA activity are sparse; such as we have found are quoted in the next chapter. A remarkable finding emerges from figures given by Munro (1969) for the total RNA concentrations in liver and muscle of a variety of animals (Table 14.6). It appears that the relative concen-

Table 14.5. Early data on ribosomal activity.

	k_s % d^{-1}	[RNA] mg g^{-1} protein	k_{RNA} mg synthesized mg^{-1} RNA
A. Henshaw *et al.*, 1971 (recalculated)			
Liver	73	49	15
Muscle	8	13	6
B. Millward *et al.*, 1973			
Muscle.			
Normal diet	15.3	11.5	13.1
Protein-free diet			
1 day	9.8	10.5	9.3
21 days	4.3	4.8	9.1
Protein-free/normal			
1 day	0.64	0.91	0.71
21 days	0.28	0.42	0.69

Table 14.6. Relative total RNA concentrations per kg tissue in liver and muscle of mammals of different sizes.

	Weight kg	Concentration in liver/ concentration in muscle
Mouse	0.03	8.1
Rat	0.16	8.1
Rabbit	1.5	7.3
Dog	25.5	7.5
Bullock	450	8.1
Horse	690	7.95

Calculated from data of Munro (1969).

trations in liver and muscle are virtually constant, at about 8, over a range of mammals varying in weight over more than four orders of magnitude. What does this mean? In a fish – rainbow trout – from the data of Peragon *et al.* (1998, 2001) the ratio is about 5, at least until the age of 40 weeks.

14.4. References

Afling, E.-G., Bernhardt, W., Janzen, R.W.C. and Röthig, H.-J. (1981) Quantitative importance of non-skeletal-muscle N^{τ}-methylhistidine and creatine in human urine. *Biochemical Journal* 200, 449–452.

Amenta, J.S., Sargus, M.J. and Baccino, F.M. (1978) Inhibition of basal protein degradation in rat embryo fibroblasts by cycloheximide: correlation with activities of lysosomal proteases. *Journal of Cellular Physiology* 97, 267–284.

Anderson, N.L. and Anderson, N.G. (1998) Proteome and proteomics: New technologies, new concepts and new words. *Electrophoresis* 19, 1853–1861.

Arias, I.M., Doyle, D. and Schimke, R.T. (1969) Studies on the synthesis and degradation of proteins of the endoplasmic reticulum of rat liver. *Journal of Biological Chemistry* 244, 3303–3315.

Attaix, M., Manghebati, A., Grizard, J. and Arnal, M. (1986) Assessment of *in vivo* protein synthesis in lamb tissues with [^{3}H] valine flooding doses. *Biochimica et Biophysica Acta* 882, 389–397.

Bacchino, F.M., Tessitore, L., Cecchini, G., Messina, M., Zuretti, M.F., Bonelli, G., Gabriel, L. and Amenta, J.S. (1982) Control of cell catabolism in rat liver. *Biochemical Journal* 206, 395–405.

Ballard, F.J., Tomas, F.M. and Stern, L.M. (1979) Increased turnover of muscle contractile proteins in Duchenne muscular dystrophy as assessed by 3-methyl histidine and creatinine excretion. *Clinical Science* 56, 542–552.

Baracos, V.E. and Goldberg, A.L. (1986) Maintenance of normal length improves protein balance and energy status in isolated rat skeletal muscles. *American Journal of Physiology*, 251, C588–596.

Bates, P.C., Grimble, G.K., Sparrow, M.P. and Millward, D.J. (1983) Myofibrillar protein turnover: synthesis of protein-bound 3-methylhistidine, actin, myosin heavy chain and aldolase in rat skeletal muscle in the fed and starved states. *Biochemical Journal* 214, 593–605.

Bohé, J., Low, A., Wolfe, R.R. and Rennie, M.J. (2003) Human muscle protein synthesis is modulated by extracellular not intramuscular availability: a dose response study. *Journal of Physiology* 552, 315–324.

Boirie, Y., Gachon, P., Corny, S., Fauquent, J., Maubois, J.-L. and Beaufrère, B. (1996) Acute post-prandial change in leucine metabolism as assessed with an intrinsically labelled milk protein. *American Journal of Physiology* 271, E1083–1091.

Botbol, V. and Scornik, O.A. (1997) Measurement of protein degradation in live mice by accumulation of bestatin-induced peptides. *American Journal of Physiology* 273, 1149–1157.

Brachet, J. (1941) La détection histochimique et le microdosage des aciles pentosenucléiques. *Enzymologia* 10, 87–96.

Buchanan, D.L. (1961) Total carbon turnover measured by feeding a uniformly labelled diet. *Archives of Biochemistry and Biophysics* 94, 500–511.

Campbell, R.M. and Kosterlitz, H.W. (1947) Ribonucleic acid as a constituent of labile liver cytoplasm. *Journal of Physiology* 106, 12–13P.

Caspersson, T. (1941) Studien über den Eiweissumsatz der Zelle. *Naturwissenschaften* 29, 33–47.

Chinkes, D.L., Rosenblatt, J. and Wolfe, R.R. (1993) Assessment of the mathematical issues involved in measuring the FSR of protein using the flooding dose technique. *Clinical Science* 84, 177–183.

Conde, R.D. and Scornik, O.A. (1977) Faster synthesis and slower degradation of liver protein during developmental growth. *Biochemical Journal* 166, 115–121.

Davis, T.A., Fiorotto, M.L., Nguyen, H.V. and Burrin, O.G. (1999) Aminoacyl-tRNA and tissue free amino acid pools are equilibrated after a flooding dose of phenylalanine. *American Journal of Physiology* 277, E103–109.

Dudley, M.A., Burrin, D.G., Wykes, L.J., Toffolo, G., Cobelli, C., Nichols, B.L., Rosenberger, J., Jahoor, F. and Reeds, P.J. (1998) Protein kinetics determined *in vivo* with a multiple-tracer single-sample protocol: application to lactase synthesis. *American Journal of Physiology* 274, G591–598.

Essén, P., McNurlan, M.A., Wernerman, J., Milne, E., Vinnars, E. and Garlick, P.J. (1992) Short-term starvation decreases skeletal muscle protein synthesis in man. *Clinical Physiology* 12, 287–299.

Garlick, P.J., Millward, D.J. and James, W.P.T. (1973) The diurnal response of muscle and liver protein synthesis *in vivo* in meal-fed rats. *Biochemical Journal* 136, 935–945.

Garlick, P.J., Waterlow, J.C. and Swick, R.W. (1976) Measurement of protein turnover in rat liver: analysis of the complex curve for decay of label in a mixture of proteins. *Biochemical Journal* 156, 657–663.

Garlick, P.J., McNurlan, M.A. and Preedy, V.R. (1980) A rapid and convenient technique for increasing the rate of protein synthesis in tissues by injection of [^{3}H] phenylalanine. *Biochemical Journal* 192, 719–723.

Garlick, P.J., Wernerman, J., McNurlan, M.A., Essén, P., Lobley, G.E., Milne, G.E., Calder, A.G. and Vinnars, E. (1989) Measurement of rate of protein synthesis in muscle of postabsorptive young men by injection of a 'flooding dose' of (1-^{13}C) leucine. *Clinical Science* 77, 329–336.

Garlick, P.J., McNurlan, M.A., Essen, P. and Wernerman, J. (1994) Measurement of tissue protein synthesis rates *in vivo*: a critical analysis of contrasting methods. *American Journal of Physiology* 266, E287–297.

Glass, R.D. and Doyle, R. (1972) On the measurement of protein turnover in animal cells. *Journal of Biological Chemistry* 247, 5234–5252.

Halliday, D. and McKeran, R.O. (1975) Measurement of muscle protein synthetic rate from serial muscle biopsies and total body protein turnover in man by continuous intravenous infusion of L-[α-^{15}N] lysine. *Clinical Science and Molecular Medicine* 49, 581–590.

Halliday, D., Pacy, P.J., Cheng, K.N., Dworzak, F., Gibson, K.N. and Rennie, M.J. (1988) Rate of protein synthesis in normal man and patients with muscular dystrophy. *Clinical Science* 74, 237–245.

Harris, C.I. (1981) Reappraisal of the quantitative importance of non-skeletal muscle source of N^τ-methylhistidine in urine. *Biochemical Journal* 194, 1011–1014.

Haverberg, L.N., Omsted, P.T., Munro, H.N. and Young, V.R. (1975) N^τ-methylhistidine content of mixed proteins in various rat tissues. *Biochimica et Biophysica Acta* 405, 67–71.

Henshaw, E.C., Hirsch, C.A., Morton, B.E. and Hiatt, H.H. (1971) Control of protein synthesis in mammalian tissues through changes in ribosome activity. *Journal of Biological Chemistry* 246, 436–446.

James, W.P.T., Garlick, P.J., Sender, P.M. and Waterlow, J.C. (1976) Studies of amino acid and protein metabolism in normal man with L-[U-^{14}C] tyrosine. *Clinical Science and Molecular Medicine* 50, 525–532.

Kean, E.A. (1959) Effect of a deficient diet on the rate of liver-protein labelling in rats. *Quarterly Journal of Experimental Physiology* 44, 351–356.

Kimball, S.R., Farrell, P.A. and Jefferson, L.S. (2002) Exercise effects on muscle insulin signaling and action: role of insulin in translational control of protein synthesis in skeletal muscle by amino acids or exercise. *Journal of Applied Physiology* 93, 1168–1180.

Koch, A.L. (1962) The evaluation of the rates of biological processes from tracer kinetic data. *Journal of Theoretical Biology* 3, 283–303.

Lobley, G.E., Harris, P.M., Skene, P.A., Brown, D., Milne, E., Calder, A.G., Anderson, S.E., Garlick, P.J., Nevison, I. and Connell, A. (1992) Responses in tissue protein synthesis to sub- and supra-maintenance intake in young growing sheep: comparison of large-dose and continuous infusion techniques. *British Journal of Nutrition* 68, 373–388.

Loftfield, R.B. and Eigner, E.A. (1958) The time required for the synthesis of a ferritin molecule in rat liver. *Journal of Biological Chemistry* 231, 925–994.

Loftfield, R.B. and Harris, A. (1956) Participation of free amino acids in protein synthesis. *Journal of Biological Chemistry* 219, 151–159.

McNurlan, M.A., Essén, P., Hays, S.D., Buchan, V., Garlick, P.J. and Wernerman, J. (1991) Measurement of protein synthesis in human skeletal muscle: further investigation of the 'flooding' technique. *Clinical Science* 81, 557–564.

McNurlan, M.A., Essén, P., Thorell, A., Calder, A.G., Anderson, S.E., Ljungqvist, O., Sandgren, A., Grant, I., Tjäder, J., Wernerman, J. and Garlick, P.J. (1994) Response of protein synthesis in human skeletal muscle to insulin: an investigation with L-[$^2H_5$] phenylalanine. *American Journal of Physiology* 267, E102–108.

Millward, D.J. (1970) Protein turnover in skeletal muscle. I. The measurement of rates of synthesis and catabolism of skeletal muscle protein using [^{14}C] Na_2CO_3 to label protein. *Clinical Science* 39, 577–590.

Millward, D.J. and Bates, P.C. (1983) 3-Methylhistidine turnover in the whole body, and the contribution of skeletal muscle and intestine to urinary 3-methylhistidine excretion in the adult rat. *Biochemical Journal* 214, 607–615.

Millward, D.J., Garlick, P.J., James, W.P.T., Nnanyelugo, D.O. and Ryatt, J.S. (1973) Relationship between protein synthesis and RNA content in skeletal muscle. *Nature* 241, 204–205.

Millward, D.J., Garlick, P.J., Stewart, R.J.C., Nnanyelugo, D.O. and Waterlow, J.C. (1975) Skeletal muscle growth and protein turnover. *Biochemical Journal* 150, 235–243.

Millward, D.J., Garlick, P.J., Nnanyelugo, D.O. and Waterlow, J.C. (1976) The relative importance of muscle protein synthesis and breakdown in the regulation of muscle mass. *Biochemical Journal* 156, 185–188.

Millward, D.J., Bates, P.C., Grimble, G.K., Brown, J.G., Nathan, M. and Rennie, M.J. (1980) Quantitative importance of non-skeletal muscle sources of N^{τ}-methylhistidine in urine. *Biochemical Journal* 190, 225–228.

Mortimore, G.E. and Pösö, A.R. (1984) Lysosomal pathways in hepatic degradation; regulatory role of amino acids. *Federation Proceedings* 43, 1289–1294.

Munro, H.N. (1969) Evolution of protein metabolism in mammals. In: Munro, H.N. (ed.) *Mammalian Protein Metabolism* Vol. III. Academic Press, New York, pp. 133−182.

Neuberger, A., Perrone, J.C. and Slack, H.G.B. (1951) The relative metabolic inertia of tendon collagen in the rat. *Biochemical Journal* 49, 199–204.

Obled, C., Barre, F., Millward, D.J. and Arnal, M. (1989) Whole body protein synthesis: studies with different amino acids in the rat. *American Journal of Physiology* 257, E639–646.

Peavy, D.E. and Hansen, R.J. (1976) Lack of effect of amino acid concentration on protein synthesis in the perfused rat liver. *Biochemical Journal* 160, 797–801.

Peragón, J., Barroso, J.B., de la Higuera, M. and Lupiañez, J.A. (1998) Relationship between growth and protein turnover rates and nucleic acids in the liver of rainbow trout (*Oncorphyncus mykiss*) during development. *Canadian Journal of Fish and Aquatic Science* 55, 649–652.

Peragón, J., Barroso, J.B., Garúa-Salguero, L., de la Higuera, M. and Lupiáñez, J.A. (2001) Growth, protein-turnover rates and nucleic acid concentrations in the white muscle of rainbow trout during development. *The International Journal of Biochemistry and Cell Biology* 33, 1227–1238.

Preedy, V.R., Smith, D.M. and Sugden, P.H. (1986) A comparison of rates of protein turnover in rat diaphragm *in vivo* and *in vitro*. *Biochemical Journal* 233, 279–282.

Ratner, S., Rittenberg, D., Keston, S. and Schoenheimer, R. (1940) Studies in protein metabolism XIV: the chemical interactions of dietary glycine and body proteins in the rat. *Journal of Biological Chemistry* 134, 665–676.

Rennie, M.J. and Millward, D. (1983) 3-methylhistidine excretion and the urinary 3-methylhistidine/creatinine ratio are poor indicators of skeletal muscle protein breakdown. *Clinical Science* 65, 217–225.

Rennie, M.J., Smith, K. and Watt, P.W. (1994) Measurement of human tissue protein synthesis: an optimal approach. *American Journal of Physiology* 266, E298–307.

Schaeffer, A.L., Davis, S.R. and Hughson, G.A. (1986) Estimation of tissue protein synthesis in sheep during sustained elevation of plasma leucine concentration by intravenous infusion. *British Journal of Nutrition* 56, 281–288.

Schimke, R.T., Sweeney, E.W. and Berlin, C.M. (1965) The role of synthesis and degradation in the control of arginase in rat liver. *Journal of Biological Chemistry* 240, 322–331.

Schöch, G., Topp, H., Held, A., Heller-Schöch, G., Ballauf, A., Manz, F. and Sander G. (1990) Interrelation between whole body turnover rates of RNA and protein. *European Journal of Clinical Nutrition* 44, 647–658.

Schoenheimer, R., Rather, S. and Rittenberg, D. (1939) Studies in protein metabolism. X. The metabolic activity of body proteins investigated with L-leucine containing two isotopes. *Journal of Biological Chemistry* 130, 703–732.

Scornik, O.A. (1974) *In vivo* rate of translation by ribosomes of normal and regenerating liver. *Journal of Biological Chemistry* 249, 3876–3883.

Shemin, D. and Rittenberg, D. (1944) Some interrelationships in general nitrogen metabolism. *Journal of Biological Chemistry* 153, 401–421.

Smith, K., Barua, J.M., Watt, P.W., Scrimgeour, C.M. and Rennie, M.J. (1992) Flooding with L-[1-^{13}C] leucine stimulates human muscle incorporation of continuously infused L-[1-^{13}C] valine. *American Journal of Physiology*, 262, E372–376.

Smith, K., Reynolds, N., Downie, S., Patel, A. and Rennie, M.J. (1998) Effects of flooding amino acids on incorporation of labelled amino acids into human muscle protein. *American Journal of Physiology* 275, E73–78.

Southorn, B.G., Kelly, J.M. and McBride, B.S. (1992) Phenylalanine flooding dose procedure is effective in measuring intestinal and liver protein synthesis in sheep. *Journal of Nutrition* 122, 2398–2407.

Stephen, J.M.L. and Waterlow, J.C. (1966) Use of carbon-14-labelled arginine to measure the catabolic rate of serum and liver proteins and the extent of amino acid recycling. *Nature (London)* 211, 978–980.

Swick, R.W. (1958) Measurement of protein turnover in rat liver. *Journal of Biological Chemistry* 231, 751–763.

Swick, R.W. and Handa, D.T. (1956) The distribution of fixed carbon in the amino acids. *Journal of Biological Chemistry* 218, 577–585.

Thompson, M.G., Palmer, R.M., Thom, A., Garden, K., Lobley, G.E. and Calder, G. (1996) N^{τ}-methylhistidine turnover in skeletal muscle cells measured by GC-MS. *American Journal of Physiology* 270, L1875–1879.

Thompson, R.C. and Ballou, J.E. (1956) V. The predominantly non-dynamic state of body constituents in the rat. *Journal of Biological Chemistry* 223, 795–800.

Tjader, I., Essen, P., Wernerman, J., McNurlan, M.A., Garlick, P.J., Smith, H.K. and Rennie, M.J. (1996) Comparison of constant infusion and flooding methods for measuring muscle protein synthesis: effect of conventional and total parenteral nutrition. *Proceedings of the Nutrition Society* 56, 151A.

Waterlow, J.C. and Stephen, J.M.L. (1968) The effect of low protein diets on the turnover rates of serum, liver and muscle protein in the rat, measured by continuing infusion of L-[^{14}C] lysine. *Clinical Science* 35, 287–305.

Young, V.R. and Munro, H.N. (1978) N^{τ}-methylhistidine (3-methylhistidine) and muscle protein turnover: an overview. *Federation Proceedings* 37, 2291–2300.

Young, V.R., Haverberg, L.N., Bilmazes, C. and Munro, H.N. (1973) Potential use of 3-methylhistidine excretion as an index of progressive reduction in muscle protein catabolism during starvation. *Metabolism* 22, 1429–1435.

Zak, R.E., Martin, A., Prior, G. and Rabinowitz, M. (1977) Comparison of turnover of several myofibrillar proteins and critical evaluation of double isotope method. *Journal of Biological Chemistry* 252, 3430–3435.

Zhang, X.J., Chinkes, D.L., Sakurai, Y. and Wolfe, R.W. (1996) An isotopic method for measurement of muscle protein fractional breakdown rate *in vivo*. *American Journal of Physiology* 270, E759–767.

15

Protein Turnover in Tissues: Effects of Food and Hormones

This chapter is concerned with comparisons between different animals and different tissues. Allometric relationships between species were discussed previously. There are also genetic differences in synthesis rates between strains within species, as shown by the work of Bates and Millward (1981) in rats, of Oddy (1993) in sheep, of Sève and Pointer (1997) in pigs and of Lobley *et al.* (2000) in cattle. Age is also important, as discussed earlier. Finally, there are differences in nutritional state and methods of measurement. It is not surprising therefore that for almost every number in the tables there is a twofold range of variation. All that can be attempted here is to produce a broad conspectus.

Up to 1980 most measurements were made by constant infusion (e.g. Waterlow and Stephen, 1968; Garlick *et al.*, 1975). After Garlick introduced the flooding dose method (q.v.) it was used almost exclusively. Most of the work has been done on rats, and after that on sheep. By far the most widely studied tissue is muscle, particularly in relation to the effects of diet and hormones. It cannot be claimed that the following tables cover the whole literature, but we hope that they give a reasonably balanced picture of what has been found.

15.1 Synthesis in the Normal State

15.1.1 Rats

Fractional synthesis rates and RNA activities for a number of tissues are shown in Table 15.1. Because synthesis rates vary with age (see below) data are given only for rats weighing between 100 and 200 g. To reduce variability all the data were obtained by the flooding dose method; even so for many tissues there is a 1.5–2-fold variation in the values of k_{S}. A factor that is not taken into account in the table is the feeding state of the animal. In most cases it was post-absorptive but sometimes it was not stated. The effect of feeding on turnover rates is considered in the next section. There is a tenfold difference in k_S between the fastest tissue, gut mucosa, and the slowest tissue, muscle, but a much smaller range in k_{RNA}. This relative constancy between tissues suggests that the synthesis rate in the normal animal is mainly determined by the concentration of RNA.

To come to individual tissues: the highest rate of synthesis, not shown in Table 15.1, is that of the pancreas, where Burrin *et al.* (1992) recorded a rate of 370% day^{-1} in suckling; and a very high value has also been found in the newborn lamb. In early postmortem studies on severely malnourished children (Waterlow, 1948) we observed extreme shrinkage of the pancreatic acinar cells. Measurements of synthesis were not available at the time, but we suggested that tissues with the most rapid turnover might be the most vulnerable to protein malnutrition.

Marway *et al.* (1993) studied synthesis rates in both mucosa and serosa of the gut from oesophagus to rectum. Rates progressively decreased from mouth to anus. Those of the serosa were about $^{2}/_{3}$ of the mucosal rates. In the gut mucosa synthesis has two components: the first is synthesis of the constitutive protein of cells, formed in the crypt and moving towards the tip of the villus, from which they are discharged into the lumen. This process presumably follows lifetime kinetics. The second component is the synthesis of enzymes

Table 15.1. Protein synthesis (k_S% d^{-1}) and RNA activity (k_{RNA} mg synthesized mg^{-1}) in tissues of rats weighing ~100–200 g.

	k_S mean	Range	k_{RNA} mean	Range	References
Liver	93	78–106	23.5	16–33	1, 3, 5–8
Small gut mucosa	133	109–164	16	11–20	4–9
Kidney	57	48,66	27	–	3, 6, 15–18
Spleen	76	–	–	–	6
Lung	34	25–43	10.5	9,12	3, 6, 15, 20
Brain	17	17,17	15	–	6, 15
Bone	73	54–90	14	13–15	11, 15, 19
Skin	63	62–64	–	–	2, 10, 12, 19
Heart	18	15–23	13.5	11,16	6, 15, 21
Muscle (gastrocnemius)	11.4	5–18.5	13.3	7–19	1, 6, 9–11, 14, 15, 17, 18 20, 22

k_{RNA} = RNA activity = mg protein synthesized per mg RNA.

References: 1. Ballmer *et al.*, 1993; 2. Breuillé *et al.*, 1998; 3. Hashiguchi *et al.*, 1996; 4. Marway *et al.*, 1993; 5. McNurlan *et al.*,1979; 6. McNurlan and Garlick, 1980; 7. McNurlan *et al.*, 1980; 8. McNurlan and Garlick, 1981; 9. McNurlan *et al.*, 1982; 10. Obled and Arnal, 1992; 11. Odedra *et al.*, 1983; 12. Preedy and Garlick, 1981; 13. Preedy and Waterlow, 1981; 14. Preedy *et al.*, 1983; 15. Preedy *et al.*, 1985a; 16. Preedy *et al.*, 1985b; 17. Preedy and Garlick, 1988; 18. Preedy *et al.*, 1988; 19. Preedy *et al.*, 1990; 20. Preedy and Garlick, 1993; 21. Preedy and Garlick, 1995; 22.Yahya *et al.*, 1994.

and mucoproteins which are 'exported'. The results in Table 15.1 were obtained by the flooding dose method, which, because it only takes about 30 minutes, includes synthesis of the secreted proteins. Van der Schoor *et al.* (2002) found in young piglets that one half of the dietary protein intake was taken up by the gut and a large proportion of this was used for glycoprotein synthesis and was recycled. A further problem is to define the precursor when amino acids are entering the intracellular pool from two sources, arterial blood and gut lumen. The synthesis rates recorded in Table 15.1 have all been based on the enrichment of the tracer amino acid in homogenates of the mucosa; results of Stoll *et al.* (1997, 2000) suggested that extracellular amino acids are channelled to protein synthesis without mixing freely in the intracellular pool. In view of these uncertainties the results in the table can only be regarded as approximations.

For liver a distinction has been made between measurements by constant infusion (CI) and flooding dose (FD). This is because of the difference in time-scale. A constant infusion lasts 4–6 h, by which time a considerable amount of protein, e.g. albumin and fibrinogen, has been synthesized and secreted, and so lost to analysis. In the FD method measurements are made after some 30 minutes only, so that most of the proteins due to be secreted are still preserved in the liver. It is generally estimated that 'export' protein accounts for 25–30% of total protein synthesis in the liver.

The other tissue that requires comment is skeletal muscle. There is, of course, a whole range of muscles with different functions and different fibre types. Table 15.1 takes no account of these differences. The results, for the sake of uniformity, relate mostly to the gastrocnemius, which has been a favourite for many rat studies. In many of the papers on which the table is based the muscle is not named. In four studies which allow a direct comparison, the ratio of k_S in soleus, a slow-twitch oxidative muscle, to that in gastrocnemius, a fast-twitch mixed glycolytic and oxidative muscle, was 2.2. The higher synthesis rates in soleus were achieved by a combination of greater RNA concentrations and somewhat higher efficiency. There is an even greater difference between synthesis in fast and slow muscles in fish. The ratio of fractional synthesis rates (FSR) in red (slow):white (fast) muscle in carp was shown to be between 5 and 6 over temperatures changing from 8 to 34°C (Goldspink *et al.*, 1984). A similar difference was found by Laurent *et al.* (1978) between two wing muscles of the fowl

(Table 15.2). The large difference in k_S is entirely due to the difference in [RNA], with no difference in k_{RNA} or in the relative amounts of DNA.

Thus for the synthesis rates of muscle protein the prevailing fibre type, slow or fast, is clearly very important, but apparently it is not the whole story. Garlick *et al.* (1989) measured k_S, [RNA] and k_{RNA} in nine different muscles of the rat and concluded that '... only part of the variation in protein turnover rates among different muscles can be attributed to their fibre-type composition ... A component of the variability is muscle-specific and relates to the function and environment of that particular muscle.'

15.1.2 Synthesis in tissues of sheep, pigs and cattle

The results (Tables 15.3 and 15.4) vary a great deal with age, which makes it difficult to compare different species. Comparisons should be made at the same stage of maturity: the rat weaned at 3 weeks might be compared with the newly ruminant sheep at 8 weeks. This comparison is shown in Table 15.5, from Lobley (1993). The main difference is in the k_S of muscle.

There is little information about k_{RNA}. Garlick *et al.* (1976) found in pigs a linear correlation between k_S and RNA concentration in different tissues, which implies that k_{RNA} was constant; only the liver was odd man out, because the results were obtained by constant infusion, so no account was taken of the synthesis of exported proteins. Table 15.5 also shows the data assembled by Lobley (1993) on k_{RNA}, which appears to be very similar in both species. Peragón and his colleagues have published many papers giving figures for k_{RNA} in white muscle of rainbow trout. A typical value, measured at 15°C, is 1.8 (Peragón *et al.*, 2001). Goldspink *et al.* (1984) showed that in the white muscle of carp k_S at 34°C was about 8 × that at 14°C. A similar figure for the temperature sensitivity of protein synthesis in Antarctic fish was obtained by Mathews and Haschemeyer (1978). If this correction is applied to Peragón's figure the k_{RNA} in the carp muscle at 37°C would come to 14.4.

Table 15.2. Synthesis rate and RNA concentration and activity in the anterior and posterior latissimus dorsi muscle of the fowl.

	k_S	[RNA]	k_{RNA}
Anterior	17.0	7.6	22.4
Posterior	6.9	3.4	20.4

k_S, g protein synthesized g^{-1} protein = fractional synthesis rate (FSR) – often expressed as a percentage, g per 100 g protein; [RNA], mg RNA g^{-1} protein; k_{RNA}, mg protein synthesized mg^{-1} RNA. Data of Laurent *et al.*, 1978.

Table 15.3. Protein synthesis rates in tissues of lambs and sheep.

		k_S, % d^{-1}			
		A	B	C	D
	Age	Fetal	Newborn	Early growth	Adult
	Weight, kg	~2	~5	16	35+
Pancreas		–	184	–	–
Gastrointestinal mucosa		71	90	–	68 (2)
Liver		78	110 (3)	54	29 (5)
Skin		–	30 (2)	35	12 (2)
Skeletal muscle		26	22 (3)	4.5	2.1 (3)

Values are means with number of measurements in brackets. If no brackets, one measurement only. References: A, Schaeffer *et al.*, 1984; B, Attaix *et al.*, 1986, 1988, 1989; Soltesz *et al.*, 1973; C, Davis *et al.*, 1981; D, Buttery *et al.*, 1975; Connell *et al.*, 1997; Lobley *et al.*, 1994; Nash *et al.*, 1994; Rocha *et al.*, 1993; Southorn *et al.*, 1992.

Table 15.4. Synthesis rates in tissues of pigs and cattle, % d^{-1}.

	Pigs			Cattle	
Weight, kg	30[a]	42[b]	75[c]	250[d]	630[d]
Small gut	51	43	–	52	29
Large gut	34	31	–	–	–
Pancreas	141	200	–	–	–
Liver	103	21	23	32	15
Kidney	37	25	20	–	–
Lung	–	–	18	–	–
Spleen	–	–	30	–	–
Brain	–	–	8.5	–	–
Heart	9	6	–	–	–
Whole skin	13	9	–	6	4
Skeletal muscle	78	4	–	2	0.9

Sources: [a]Simon *et al.*, 1978. [b]Simon *et al.*, 1982. [c]Garlick *et al.*, 1976. [d]Lobley *et al.*, 1980.

Table 15.5. Comparison of k_S and k_{RNA} in rat and lamb at comparable stages of maturity.

	k_S		k_{RNA}	
	Rat	Sheep	Rat	Sheep
Whole body	31	12	15	18
Liver	53–100	48	15–28	19
Gastrointestinal tract	97	69	17–24	23
Skeletal muscle	12–24	7	12–13	13

Rats measured at weaning, 3 weeks. Lambs measured when beginning to ruminate, 8 weeks. k_{RNA} = mg protein synthesized per mg RNA. Reproduced from data of Lobley, 1993.

It is difficult to avoid the conclusion that the rate of polypeptide chain assembly, which is represented by k_{RNA}, is rather constant not only across tissues but also across species. It certainly varies much less than k_S.

15.1.3 Synthesis in human tissues

In humans (Table 15.6) investigations are limited to tissues that can be safely sampled by biopsy – muscle, liver, white blood cells and the mucosa of the small intestine. The rates in gut mucosa, liver and muscle are much the same as those in the larger animals. As in the rat, the FSR of duodenal mucosa is more than twice that of the liver. The moderately high FSR of lymphocytes makes a significant contribution to whole body synthesis. Many years ago it was found that in the rat 16% of injected radioactivity remained in the carcass after muscle was accounted for by measurement of creatine. This was attributed to activity in lymphoid tissue, bone marrow, etc. (Waterlow and Stephen, 1966).

Table 15.7 compares the distribution of whole body synthesis between different tissues in man and rat. The agreement is remarkable.

15.1.4 The effects of age

There is quite a good database on the relationship of protein synthesis to age in the rat – see, for example, studies on many tissues by Goldspink and Kelly (1984) and Lewis *et al.* (1984) – but for other animals, except sheep, the information is scanty. Age may be divided into three phases: the phase of growth; adult age or a more-or-less steady state; and senescence or ageing. Figure 15.1 shows the contrasting behaviour of liver and muscle, particularly during the growth phase. In liver k_S and [RNA] change little up to 2 years, whereas in muscle there is an early and progressive fall in both variables. In the early phases there is little change in k_{RNA} in either tissue, since k_S and [RNA] change in parallel. Goldspink *et al.* (1984) and Lewis *et al.* (1984) compared rats at 3 and 44 weeks; the old:young ratio of FSR was 0.8

Table 15.6. Fractional rates of protein synthesis (FSR) in human tissues.

	FSR % d^{-1}	References
Liver	22 (4)	1, 7–9, 11
Gastrointestinal tract (mucosa)		
Stomach	50 (3)	2, 6
Duodenum + jejunum	50 (7)	2, 3, 5, 16–18
Ileum	22 (2)	19, 21
Colon	9.5 (4)	6, 9, 11, 18
Kidney	42 (1)	21
Lymphocytes	10 (3)	9, 15, 19
Muscle	1.86 (7)	4, 6, 7–9, 12–15

Number of studies in brackets.

References: 1. Barle *et al.*, 1999; 2. Bennet *et al.*, 1993; 3. Bouteloup-Demange *et al.*, 1998; 4. Case *et al.*, 2002; 5. Charlton *et al.*, 2000; 6. Essén *et al.*, 1992; 7. Garlick *et al.*, 1991; 8. Garlick *et al.*, 1989a; 9. Garlick *et al.*, 1995; 10. Janusiewicz *et al.*, 2001; 11. Heys *et al.*, 1992; 12. McNurlan *et al.*, 1991; 13. McNurlan *et al.*, 1994; 14. McNurlan *et al.*, 1996; 15. McNurlan *et al.*, 1997; 16.Nakshabendi *et al.*, 1995; 17. Nakshabendi *et al.*, 1996; 18. Nakshabendi *et al.*, 1999; 19. Park *et al.*, 1994; 20. Rittler *et al.*, 2000; 21.Tessari *et al.*, 1996.

Methods: Reference nos 2, 3, 5, 16–18: by constant infusion, 21 by cannulation of the renal vessels, others by flooding dose.

FSR, fractional synthesis rate.

Table 15.7. Contribution of different tissues to total protein synthesis in rat and human.

	Human[a] 70 kg	Rat[b] 200 g
Whole body synthesis, g d^{-1}	265	4.75
Per cent of whole body synthesis		
Liver	28	31
Gut	33	25
Muscle	31	23
Lymphocytes	5	–

[a]Rates of whole body synthesis from Chapter 6. Other data from Tables 15.1 and 15.6. Data for protein content of organs from Snyder, 1975.

[b]Data from Goldspink and Kelly, 1984 and Lewis *et al.*, 1984.

for liver and 0.5 for muscle. A rat at about 10 weeks is more or less mature and at 1 year it is senescent. During this interval there is a further, quite dramatic, fall in k_S in muscle, which is due mainly to a continuing loss of RNA, and partly, in some cases, to a decrease in k_{RNA} (Table 15.8). Perhaps lower RNA activity is an effect of ageing. It is interesting in this connection that Merry and Houlihan (1991) described a 27% increase in polypeptide chain assembly time in the livers of rats between the ages of 140 and 760 days. Another effect of age in the rat is that k_S decreases earlier and more rapidly in a fast muscle such as tibialis anticus than in a slow one such as soleus (Lewis *et al.*, 1984; El Haj *et al.*, 1986; Garlick *et al.*, 1989b). These findings complement those in human sarcopenia, in which fast fibres are more affected than slow ones (Chapters 11, 12). Unfortunately, studies of ageing in humans (Chapter 12) do not include any information about RNA activity.

15.2 The Effects of Food on Protein Turnover in Tissues

An excellent review of this subject is that of Grizard *et al.* (1995). The results in the previous

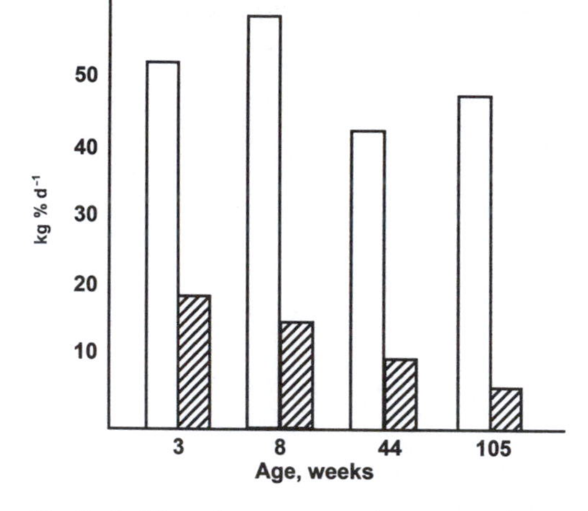

Fig. 15.1. Effect of age on rates of protein synthesis in liver and muscle of rats. Data of Goldspink and Kelly, 1984 and Lewis *et al.*, 1984.
□, liver; ▨, muscle.

section suggest that although in the normal state tissues differ in the synthesis rates of their mixed proteins, the efficiency of synthesis is remarkably similar. The picture changes when one looks at the effects of dietary and hormonal manipulations. It has been realized for a long time that there are metabolic exchanges between tissues: for example, that muscle releases alanine and glutamine for utilization by the splanchnic tissues (e.g. Felig, 1973, 1975). Reeds and co-workers in 1992 published a remarkable paper with the title 'Postnatal growth of gut and muscle – competitors or collaborators', in which they concluded: '... the skeletal muscle in effect functions as a protein store from which amino acids, such as glutamine and alanine, can be drawn in times of stress to preserve immune and perhaps gastrointestinal function. In other words, what could be seen as a "competition" is in fact a regulated "collaboration" in which the compromise of some functions ensures the maintenance of the organism as a whole.' As mentioned earlier, Waterlow and Stephen (1966) showed that on a low protein diet the initial loss of protein was from the liver, but after a time muscle and skin took over as the main source of loss. Whether this is simply a 'mechanical' result of the interplay between fast and slowly turning over systems (we attempted to model it, but without success), or whether there is some higher level of integration, we do not know. In the following sections we shall contrast changes in liver and muscle, because these are the tissues that have been most studied. Much important work was done on perfused livers (e.g. Flaim *et al.*, 1982) and on muscles *in vitro* (e.g. Tawa and Goldberg, 1992), but we concentrate here on studies *in vivo*.

15.2.1 Response to a meal

The changes in rat liver after a single high protein meal are illustrated in Fig. 15.2. Liver protein mass increased by more than 20% over the first 8 h and then began to fall. Net protein synthesis (deposition) rose sharply for about 4 h, and then fell, to become increasingly negative until the

Table 15.8. The effect of senility on muscle protein synthesis and RNA activity in the rat. Rates in old rats as percentage of rates in young adult rats.

Author	Muscle	Age, weeks old/young	k_S	[RNA]	k_{RNA}	Sex
Bates and Millward, 1981	quadriceps + gastrocnemius	46/14	87	92	94	?
Lewis *et al.*, 1984	soleus	44/8	65	70	91	M
	tibialis anticus	44/8	49	51	102	
El Haj *et al.*, 1986	soleus	52/7	41	54	92	M
	tibialis anticus	52/7	36	74	53	
Garlick *et al.*, 1989b	7 muscles	52/11	59	78	77	F
		Means:	56	70	85	

This table is an attempt to examine the changes that occur from adult age to senescence, but it is difficult to find, from the data, an age that represents early maturity.
[RNA] = RNA concentration, mg g^{-1} protein; k_{RNA} = RNA activity = mg protein synthesized mg^{-1} RNA; k_S = mg synthesized g^{-1} protein.

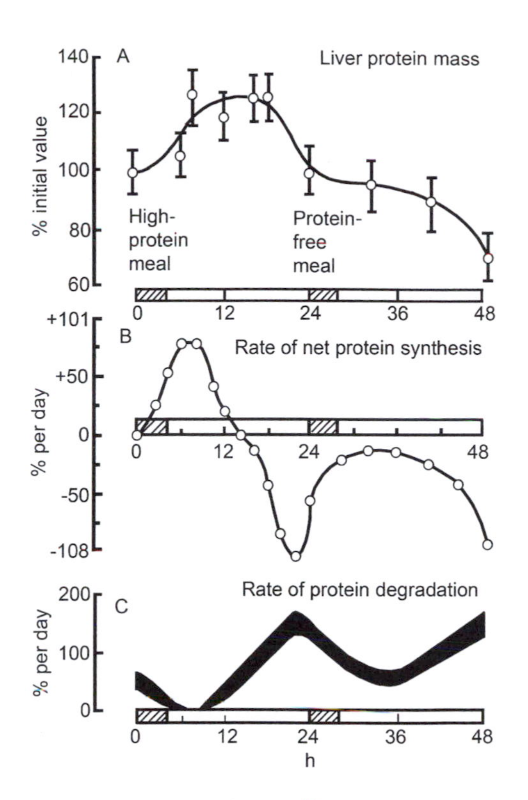

Fig. 15.2. Changes in the rat of liver protein mass and synthesis and degradation rates after a single high protein meal. The changes in total liver protein (A) were measured directly. Changes in net protein synthesis (B) were determined from the slope of curve A at appropriate time points. The range of rates of degradation was calculated as the difference between actual rates of synthesis measured by constant infusion of a labelled amino acid and net synthesis (curve B). Reproduced, by permission, from Millward, 1980, by courtesy of Elsevier-North Holland.

next meal. Degradation fell after the high protein meal and rose in the post-prandial phase. This diagram illustrates very clearly the interplay between synthesis and breakdown and the speed of their responses to food in the liver. Table 15.9 shows the changes in FSR in liver and muscle up to 52 h after a single meal. There was no fall in FSR in the liver over this whole time, which suggests that the changes in liver mass and protein deposition shown in Fig. 15.2 must have resulted from changes in degradation. This agrees with the finding of Condé and Scornik (1976), who found that on re-feeding protein-depleted rats the degradation rate in liver fell to zero. Muscle presents a rather different picture; for the first 24 h after the meal there was no immediate response of the FSR, but in the next 24 h, if fasting continued, the FSR fell to 60% of its initial rate.

15.2.2 Starvation

In rats starved for 1–2 d [RNA] and k_{RNA} in liver fell by only 10% and FSR by 20% (Fig. 15.3). In muscle the response was different: there was a large and immediate fall in FSR and RNA concentration, with k_{RNA} still relatively well preserved. A similar differential response in the two tissues was found in human patients with head injuries: the FSR of muscle protein was reduced by 50%, while that of liver-produced proteins – albumin and fibrinogen – increased (Mansoor *et al.*, 1997). With longer fasting the pattern changes. In a most instructive, and as far as we know unique, experiment, Cherel *et al.* (1991) fasted rats for 9 days. These rats were mature; if

Table 15.9. Rates of protein synthesis in liver and muscle of young rats at different times after a meal.

Time of infusion after last meal h	FSR, % d^{-1}	
	Liver	Muscle
0–6	44	6.03
6–12	42	7.06
12–18	47	7.36
18–24	64	6.86
24–30	46	4.78
48–54	47	3.31

Measurements by constant infusion of [^{14}C]-tyrosine with intracellular free tyrosine as precursor. From Garlick *et al.*, 1973.
FSR, fractional synthesis rate.

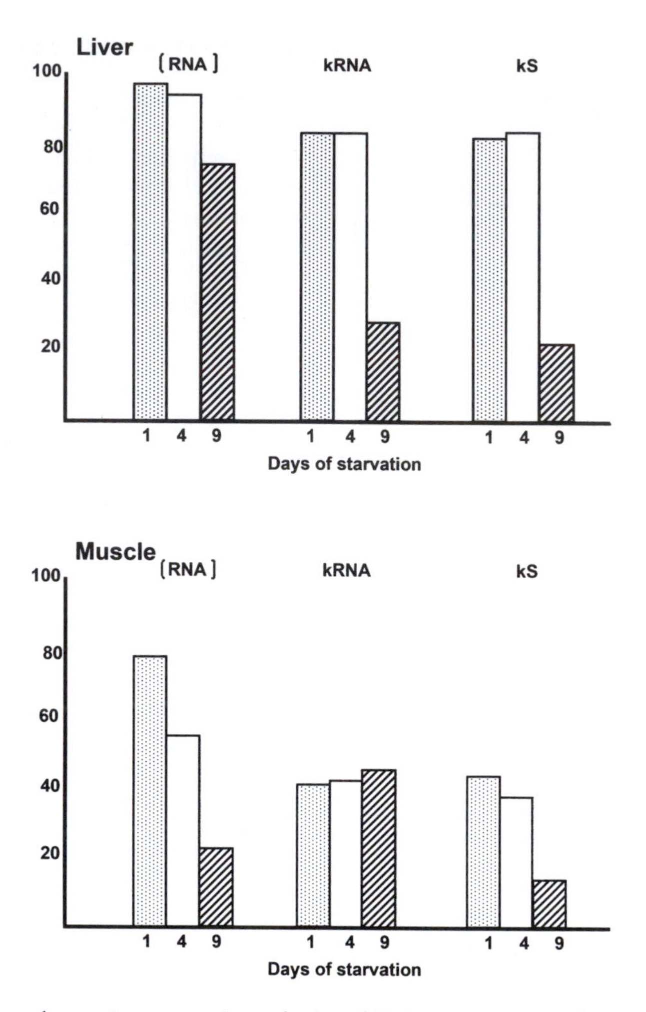

Fig. 15.3. The effects of starvation on protein synthesis and RNA concentration in liver and muscle of rats. The columns are per cent of initial before starvation. Compiled from data of Cherel, 1991; Garlick *et al.*, 1975; McNurlan *et al.*, 1979; Millward *et al.*, 1974, 1976; Preedy *et al.*, 1988. k_S = mg synthesized g^{-1} protein.

they had been young they would probably have died. In the liver there was a small fall in RNA concentration, but a profound reduction in efficiency, and hence of synthesis, to about $^1/_4$ of the normal level. In muscle there was a much greater fall in [RNA], with efficiency also low, so that synthesis was close to zero. In the early stages of starvation N excretion was low; the animal was clearly trying to adapt. At 9 days this adaptation had broken down and N excretion was greatly increased – what Whipple *et al.* (1947), working on dogs deprived of protein, called the 'pre-mortal increase'.

The response when food is restored to the starved animal is quite dramatic. The results in muscle after 4 days of starvation are shown in Table 15.10. Within 1 h k_S and k_{RNA} almost doubled, but there was as yet no change in RNA concentration. Insulin increased to a normal level and there was a huge decrease in corticosterone to well below the normal concentration. These findings suggest a strong control of the efficiency of translation, the mechanism of which, at least as regards insulin, has been elucidated by the work of Jefferson's group (see review by Kimball *et al.*, 2002).

Table 15.10. Synthesis rates, RNA activity and hormonal changes in skeletal muscle in rats refed after 4 days' starvation.

	k_S, % d^{-1}	[RNA] mg g^{-1}	k_{RNA} g synthesized g^{-1}	[Insulin] μ-units ml^{-1}	[Corticosterone] ng ml^{-1}
Fed	15.3	11.2	13.7	30.7	285
Starved 4 days	2.8	6.3	4.4	5.2	593
Re-fed 1 h	5.95	6.5	9.2	17.8	134
Re-fed 3 h	7.6	6.4	11.8	21.0	120

[RNA] = RNA concentration, mg g^{-1} protein; k_{RNA} = RNA activity = mg protein synthesized mg^{-1} RNA.
Data of Millward *et al.*, 1983.

15.2.3 Protein deprivation

The different responses of liver and muscle are seen particularly clearly with deprivation of protein. The results after 9 days on a protein-free diet are shown in Fig. 15.4. The reduction in k_S in liver resulted from a decrease in k_{RNA}, with no change in RNA concentration, whereas in muscle [RNA] was greatly reduced.

The effects of refeeding and the rapid response of k_{RNA} are shown in Fig. 15.5.

Another experiment, in which rats were fed a protein-free diet for 21 days, gave similar results in muscle, but rather different, though equally interesting, results in liver (Table 15.11). In muscle there was again a large fall in [RNA] which, together with a reduction in k_{RNA} led to a very low synthesis rate. In liver, in contrast, [RNA] was well maintained, and k_S and k_{RNA} increased to well above the initial level. Because degradation exceeded synthesis the liver lost 33% of its protein over the 21 days, but by increasing the fractional synthesis rate the absolute rate was almost maintained.

The rapid recovery of k_{RNA} and the slower recovery of [RNA] after a prolonged loss of protein are shown in Table 15.12.

Bone shows changes very like those in muscle. After 3 weeks on a diet containing almost no protein growth in tibial length came nearly to a halt and cartilage formation was greatly reduced; FSR and k_{RNA} fell to about 50% of control levels (Yahya *et al.*, 1994).

Zinc is an element whose effect on protein turnover might be important since it is a component of many metalloproteins, including DNA

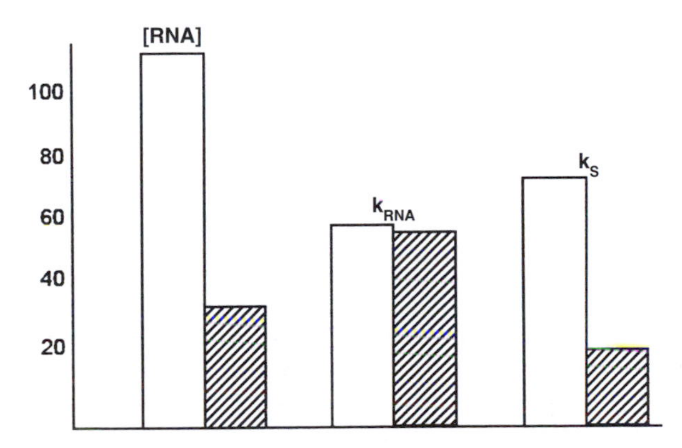

Fig. 15.4. The effect of 9 days on a protein-free diet on protein synthesis in liver and muscle of rats. The columns are per cent of controls. Data of Millward *et al.*, 1976.
□, liver; ▨, muscle.

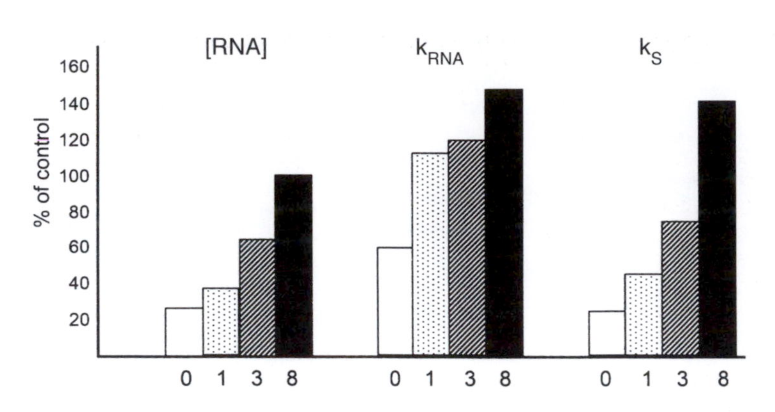

Fig. 15.5. The effect of refeeding on muscle protein synthesis in rats after 9 days on a protein-free diet. Data of Millward *et al.*, 1975.

and RNA polymerases. Zinc deficiency has been studied in rats by Giugliano and Millward (1984, 1987). They found that protein synthesis in muscle was depressed in rats after 10 days on a virtually zinc-free diet, with a small fall in [RNA] and a larger fall in k_{RNA}. There was much variation between individual zinc-deficient rats, but within this group there were correlations between depression in growth rate, k_S and k_{RNA}, with no differences in [RNA]. Degradation was also increased in the rats with the greatest depression in growth rate. The effect of micronutrients is an area that needs further exploration.

The conclusion is that in liver the ribosome content is little affected either by protein deprivation or by total starvation unless prolonged to a pre-mortal state. In contrast, in muscle ribosome content is reduced earlier and more severely in both conditions. Ribosomal activity, k_{RNA}, in muscle is greatly reduced in both states, but in liver only with prolonged starvation. Falls in protein synthesis reflect the interactions of these two components of the synthetic machinery.

From a teleological point of view it looks as if in muscle the whole process of turnover is shut down, degradation as well as synthesis, whereas in liver continuing synthesis is so important that it has to be maintained by amino acids supplied by breakdown of muscle protein.

15.2.4 Turnover rates of myofibrillar proteins

Any discussion of the effects of nutritional state on protein turnover in liver and muscle would be incomplete without reference to the obvious fact that in both tissues the total protein is a mixture

Table 15.11. Protein synthesis in liver and muscle of rats during 21 days on a protein-free diet. Numbers are per cent of initial value.

	k_S % d^{-1}		[RNA] mg g^{-1} protein		k_{RNA} synthesis, mg^{-1} RNA	
Day	Liver	Muscle	Liver	Muscle	Liver	Muscle
0	100	100	100	100	100	100
1	104	52	106	85	99	61
2	104	44	97	67	107	66
3	132	44	94	56	141	77
9	177	27	92	39	192	70
21	136	23	88	39	155	60

Measurements by constant infusion of $[^{14}C]$ tyrosine. Recalculated from Garlick *et al.*, 1975.
[RNA] = RNA concentration, mg g^{-1} protein; k_{RNA} = RNA activity = mg protein synthesized mg^{-1} RNA.

Table 15.12. Recovery of protein synthesis in muscle in rats after 30 days on a protein-free diet.

	FSR % d^{-1}	RNA/DNA mg mg^{-1}	RNA mg g^{-1} protein	Efficiency synthesis/RNA mg mg^{-1}
Initial (weight control)	13.8	2.28	10.2	13.8
30 days protein-free diet	2.7	0.67	3.2	8.0
Re-fed 1 day	5.9	0.80	3.9	15.3
3 days	10.4	1.48	6.3	16.6
8 days	19.8	2.72	10.1	19.2
14 days	17.1	2.67	7.9	20.3

FSR = fractional synthesis rate; [RNA] = RNA concentration, mg g^{-1} protein.
Data of Millward *et al.*, 1975.

of many different proteins. In the liver not much is known about the responses of individual proteins, except in the case of some enzymes, particularly those of the urea cycle (see Chapter 8). In muscle the proteins fall into two groups, sarcoplasmic and myofibrillar, and within the latter individual contractile proteins such as myosin, actin, etc. Pioneers in early studies on these proteins were Zak and Millward. More recently attention has been given to the mitochondrial proteins (Boirie *et al.*, 2001; Stump *et al.*, 2003).

Bates and Millward (1983) showed that the amounts of myofibrillar (M) and sarcoplasmic (S) protein in rat muscle are in the ratio of 2.4:1; in adult non-growing rats the fractional synthesis rate of M was about half that of S over a range of different muscles with different overall synthesis rates, e.g. 8% day^{-1} in the diaphragm, with slow oxidative fibres, and 4% day^{-1} in quadratus lumborum, mainly composed of fast glycolytic fibres. The relative rate, M/S, fell with age and with starvation (Table 15.13). This preferential loss of myofibrillar protein may explain in part the decline in muscle function with age (see Chapters 11 and 12.3).

A later study (Bates *et al.*, 1983) showed by a variety of methods that although in the fed state the synthesis rates of actin and myosin are the same, starvation causes a much greater reduction in the synthesis of actin (Table 15.14). In contrast, Zak *et al.* (1977) found by the double-isotope decay method (Chapter 14) that myosin heavy chains, α-actinin and tropomyosin turn over faster than actin or myosin light chains. Fiorotto *et al.* (2000) on the other hand found that undernutrition in suckling rats caused similar decreases, of about 30%, in the synthesis rates of seven different myofibrillar proteins. We think the evidence is strong, but not conclusive, that the myofibrillar proteins do turn over at different rates, or at least respond differently to dietary stimuli. Millward (1980) suggested that 'individual protein subunits might exchange between myofilaments throughout the myofibril and in the course of this exchange those in peripheral myofilaments interact with a degrading system located in the intermyofibrillar space'. This follows an idea originally proposed by Etlinger *et al.* (1975) and Zak *et al.* (1977). If component proteins are turning over at different rates it is

Table 15.13. Relative synthesis rates of myofibrillar and sarcoplasmic proteins in rat muscle at different stages of growth.

Growth rate % d^{-1}	Overall rate % d^{-1}	Relative synthesis rate myofibrillar:sarcoplasmic
6.3	29	0.73
2.3	14	0.72
0.7	5.2	0.65
0.3	4.9	0.56
0 (steady state)	4.5	0.50
−12.8 (4 days starved)	2.6	0.43

Data of Bates and Millward, 1978.

Table 15.14. Synthesis rates of a sarcoplasmic protein (aldolase) and of myofibrillar proteins in skeletal muscle of rats before and after 4 days' starvation.

	Synthesis, % d^{-1}	
Protein	Fed	Starved
Mixed	13.8	8.7
Sarcoplasmic – aldolase	17.3	9.3
Myofibrillar – myosin[a]	17.9	10.2
myosin[b]	14.9	6.7
actin[a]	16.8	5.7
actin[b]	15.4	2.7
actomyosin[b]	13.8	3.3

[a]Measurement by infusion of [^{14}C] tyrosine or [methyl-^{14}C] methionine, or both.
[b]Measurement by infusion of [methyl-^{14}C] methionine, with S-adenosyl methionine as precursor and protein-bound methylhistidine as product.
Data of Bates *et al.*, 1983.

strange how the myofibril maintains its contractile activity – perhaps because only a small percentage of myofibrils are affected at any one time.

15.3 The Effects of Hormones on Protein Turnover in Tissues

As mentioned earlier, the effects of hormones often overlap, inhibiting or supporting each other; they also interact with the effects of food, particularly in the case of insulin, because feeding stimulates insulin secretion. A good account of these interactions is given by Millward (1990). Useful general reviews are those of Sugden and Fuller (1991), Grizard *et al.* (1995) and Rooyackers and Nair (1997).

15.3.1 Insulin

As recently as 2001 a paper from Jefferson's laboratory began with the words: 'The feeding-induced stimulation of protein synthesis requires the hormone insulin and an adequate supply of amino acids. The relative contribution of each of these factors to the increase in protein synthesis continues to be a topic of investigation and controversy' (Anthony *et al.*, 2001). Investigation of the problem is complicated, because while amino acids stimulate the secretion of insulin, insulin tends to reduce plasma amino acid concentrations.

Most of the experimental work has been done on skeletal muscle, with synthesis measured either by constant infusion or by flooding dose; there is strong support for the role of insulin in promoting protein synthesis (k_S) in skeletal muscle of young rats, and increasing the efficiency of synthesis (k_{RNA}), provided that essential amino acids are also available. This has been shown in rats (Garlick and Grant, 1988; Jepson *et al.*, 1988) and in piglets (Wray-Cohen *et al.*, 1998; Davis *et al.*, 2000). Synthesis was reduced in animals made diabetic with streptozotocin (Millward *et al.*, 1976; Odedra *et al.*, 1982; Pain *et al.*, 1983; Ashford and Pain, 1986a) or by anti-insulin serum (Millward *et al.*, 1983), and was restored by insulin but not by infusion of amino acids (Odedra *et al.*, 1982). In addition, diabetes caused a reduction in RNA content in both liver and muscle, which was reversed by thyroid hormone (T_3) (Brown and Millward, 1983). Baillie and Garlick (1991b) found that in rat muscles insulin largely reversed the moderate depression of k_S produced by 12 h fasting, but had much less effect on the severe depression following a 36 h fast. With prolonged fasting the action of insulin is antagonized by corticosteroids, whose plasma levels are greatly increased (Table 15.10).

Papers by Garlick *et al.* (1988, 1992) go a long way towards resolving the problem of insulin versus amino acids as stimulators of synthesis. They found that in diabetic rats maximal increases in synthesis were only obtained with unphysiological levels of insulin, but when amino acids, particularly the BCAs, were infused the maximal response occurred earlier and with lower doses of insulin. Thus amino acids increase the sensitivity to insulin. This hypothesis is supported by later work at the molecular level, suggesting that amino acids have separate but complementary effects on the initiation of synthesis.

In both rat and piglet the stimulation of synthesis in skeletal muscle by insulin is highly dependent on age. In mature adult rats the response to insulin was less than half that found in young rats 1 month old (Baillie and Garlick, 1991a,b; Dardevet *et al.*, 1994); McNulty *et al.* (1993) found no response. In piglets the insulin response was greatly reduced even over the short interval between 7 and 26 days after birth. Studies on adult animals therefore tend not to be informative; in adult goats neither insulin nor amino acids had any effect on protein synthesis in skeletal muscle (Tauveron *et al.*, 1994). Likewise, in adult humans with type I diabetes restoration of insulin in the post-absorptive state had no effect on k_S in skeletal muscle (Pacy *et al.*, 1984). In diabetic animals the depression of synthesis rate results from a reduction in both RNA concentration and efficiency in fast-twitch muscles; in heart there is mainly a loss of RNA (Crozier *et al.*, 2003).

There is some information on the rate at which synthesis in muscle recovers when insulin deficiency is treated either directly or when stimulated by food. The results in Table 15.12 in the previous section are of interest: they show that after 4 days of starvation, when plasma insulin concentration as well as k_S and k_{RNA} were reduced to very low levels, some degree of recovery with a rise in insulin concentration had occurred after only 60 minutes of re-feeding. It is noteworthy also that starvation led to very high concentrations of corticosterone, and that administration of this glucocorticoid retarded recovery.

The response of skeletal muscle to insulin varies with the fibre type, being greater in fast-twitch glycolytic muscles than in slow-twitch oxidative muscles such as soleus (Odedra *et al.*, 1982; Baillie *et al.*, 1991b). As regards the different proteins in muscle, in newborn piglets insulin produced the same stimulation of synthesis in sarcoplasmic and myofibrillar fractions (Davis *et al.*, 2001). However, in mature miniature pigs insulin stimulated mitochondrial protein synthesis in muscle, with no effect on that of sarcoplasmic proteins or myosin heavy chain (Boirie *et al.*, 2001). This finding has been confirmed by Stump *et al.* (2003), who showed that in normal subjects infused with high doses of insulin there was increased synthesis of mitochondrial proteins, together with enhanced production of ATP and of the enzymes cytochrome oxidase and citrate synthase. This is an important demonstration of a link between mitochondrial protein synthesis and energy transduction. The effects were much smaller in patients with type 2 diabetes. Unlike Boirie, Stump *et al.* also found that insulin increased the expression of myosin heavy chain isoform I, the predominant isoform in oxidative muscle fibres.

There is general agreement that insulin has little or no effect on protein synthesis in the liver and other visceral organs, except perhaps on the production of albumin and other secreted proteins (Ashford and Pain, 1986a; Mosoni *et al.*, 1993a; Tauveron *et al.*, 1994; Boirie *et al.*, 2001; Davis *et al.*, 2001) and of ribosomal proteins (Anthony *et al.*, 2001). It was shown by Pain *et al.* (Pain and Garlick, 1974; Pain *et al.*, 1974a,b) that in the rat diabetes caused extensive disruption of the rough endoplasmic reticulum of the liver on which exported proteins are synthesized. This paper contains a useful analysis of the extensive work done on the problem up to that time, now mostly forgotten. In human diabetes biopsy of the small intestinal mucosa showed a small but significant decrease in k_S, which was restored to normal by treatment with insulin (Charlton *et al.*, 2000).

One would expect from the results in the whole body (Chapter 9) that the effect of insulin on protein breakdown would be more important than its effect on synthesis. Unfortunately there is very little information about breakdown in tissues, because of the difficulty of measuring it. In one study on diabetic rats, in which breakdown was measured as the difference between synthesis and growth, breakdown of muscle protein was decreased, but not as much as synthesis, so that the muscle remained in negative protein balance (Millward *et al.*, 1976). On the other hand Ashford and Pain (1986a,b) found in young rats made diabetic with streptozotocin a massive increase, within 2 days, in the degradation rate of ribosomal proteins, with a smaller effect on total muscle protein. The RNA concentration was reduced by 25–30% after 5 days of diabetes (Fig. 15.6). It has also been reported from measurements of methylhistidine and tyrosine release, that in the muscle of diabetic rats there is preferential degradation of myofibrillar proteins (Kadowaki *et al.*, 1989). More recently it has been shown that in rat muscle insulin deficiency increased protein degradation, with raised levels of mRNAs of the ubiquitin–proteosome pathway and increased transcription of the ubiquitin gene (Price *et al.*, 1996); the effect was potentiated by glucocorticoids (Rodriguez *et al.*, 1998; Mitch *et al.*, 1999). Current studies are elucidating the mechanism by which insulin acts on this degradative system (Rome *et al.*, 2004). To get a complete picture it is clearly perilous to ignore breakdown.

To summarize: insulin promotes, and lack of it depresses, protein synthesis in skeletal muscle of young animals; it has little effect in mature ones. Insulin deficiency also promotes protein breakdown, although whether there is an effect of age is not known. The action of insulin is tissue-specific: it has virtually no effect in liver and other visceral tissues. In liver synthesis responds to amino acids alone, whereas muscle, for maximal stimulation, needs insulin as well (Davis *et al.*, 2002). With insulin deficiency there is a reduction in the efficiency of synthesis in muscle, and there is some decrease in the RNA concentration as well. These are the observed effects: how they are brought about is another matter. Kimball *et al.* (1994), in a review of the subject, identified no less than 10 steps in the synthetic pathway that may be regulated by insulin. Now it is clear that breakdown is also involved. In the last 10 years much progress has been made at the molecular level – see, for example, articles by Anthony *et al.* (2001) and Crozier *et al.* (2003).

15.3.2 IGF-1

It has long been known that the polypeptide hormone IGF-1 is synthesized in the liver under the influence of growth hormone (GH), and that it circulates in the plasma mainly attached to a binding protein, IGF-BP3. This gave rise to the 'somatomedin hypothesis', that the effects of GH are all mediated by IGF-1. Secondly, it was found that IGF-1 is produced by most tissues, so that its action might be paracrine or autocrine, rather than

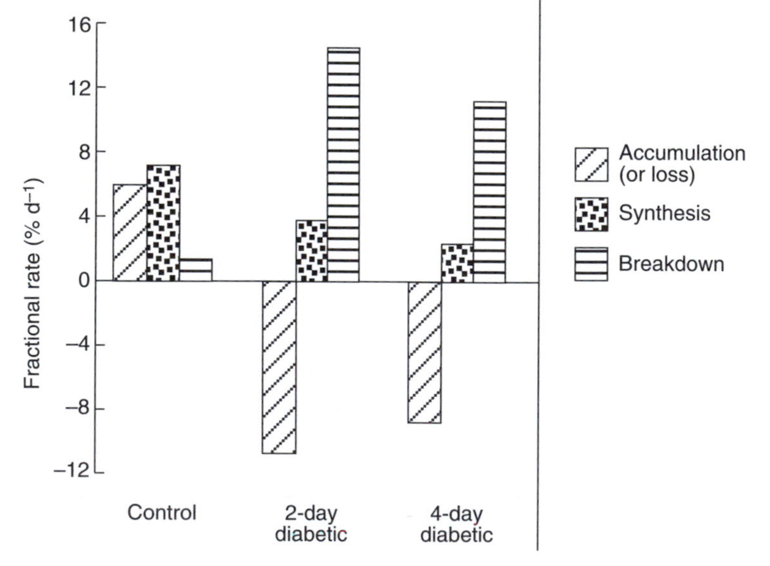

Fig. 15.6. Effect of insulin withdrawal on fractional rates of accumulation, synthesis and breakdown of ribosomal proteins in gastrocnemius muscle of rats. Reproduced, by permission, from Ashford and Pain, 1986a, by courtesy of the *Journal of Biological Chemistry*.

truly endocrine; this would follow if circulating IGF-1, complexed to a binding protein, was inactive, and only IGF-1 locally produced in the tissues was active. There are as yet no clear answers to these questions, but the observations of Fryburg (1994) mentioned in Chapter 9 tend to support the latter hypothesis. The relationship of IGF-1 to GH at the molecular level has been examined in some detail (for review see Thissen *et al.*, 1994).

IGF-1 reduces plasma concentrations of amino acids and insulin; in rodents it stimulates protein synthesis in muscle, particularly when amino acids are also supplied (Jacob *et al.*, 1996), with no effect on visceral tissues (Bark *et al.*, 1998). Similarly, in neonatal piglets synthesis was increased by IGF-1 in muscle but not in liver and other viscera except for the spleen (Davis, 2002). In cultured myocytes from young rats IGF-1 produced a large increase in synthesis with no effect on breakdown (Dardevet *et al.*, 1994).

Figure 15.7 shows an interesting relationship in muscle of pigs treated with growth hormone, between plasma IGF-1 and RNA activity.

Experimentally, protein restriction and fasting in humans both have a profound effect on circulating IGF-1 concentrations. A low protein diet decreased IGF-1 concentrations in tissues and transcription of the IGF-1 gene (Strauss and Takemoto, 1990; Grizard *et al.*, 1995). Fasting in rats reduced the plasma concentrations of IGF-1 and IGF-BP3 (Frystyk *et al.*, 1999) and total IGF mRNA levels in most tissues (Lowe *et al.*, 1989). These findings suggest the possibility that reductions in muscle protein synthesis, tibial growth and cartilage deposition produced by restriction in protein and energy intake (Yahya *et al.*, 1994) could be secondary to reduced IGF-1 activity. Against this is the finding by Thissen *et al.* (1991) that in rats when weight gain and growth in tail length and epiphysial cartilage were grossly restricted by a 5% protein diet, administration of IGF-1 produced no improvement.

15.3.3 Growth hormone

In rats hypophysectomy reduces growth partly because animals eat less. In hypophysectomized

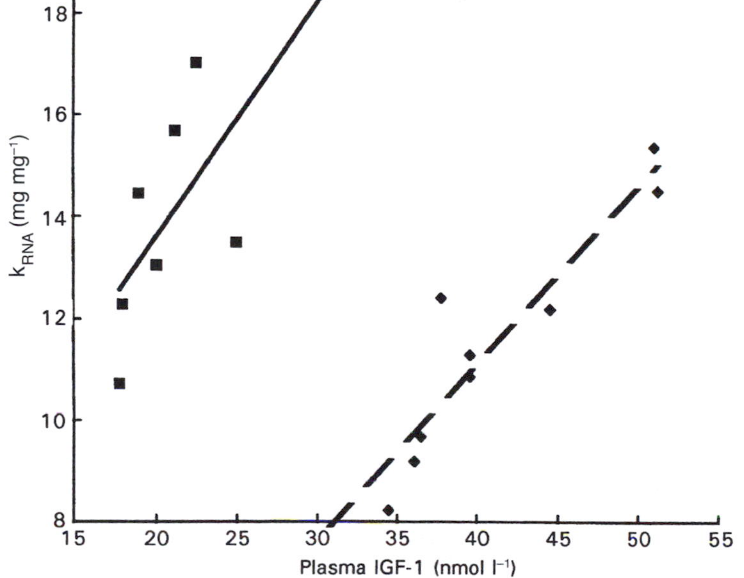

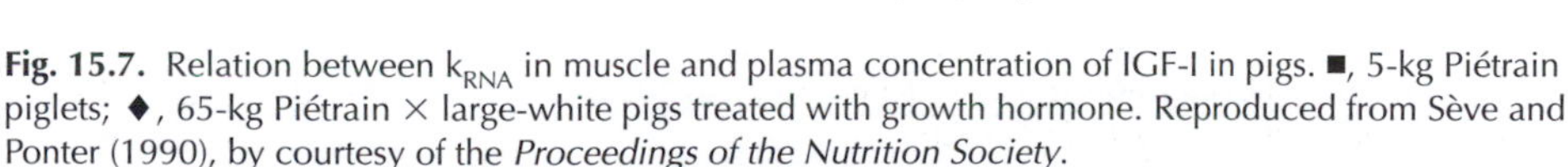
Fig. 15.7. Relation between k_{RNA} in muscle and plasma concentration of IGF-I in pigs. ■, 5-kg Piétrain piglets; ♦, 65-kg Piétrain × large-white pigs treated with growth hormone. Reproduced from Sève and Ponter (1990), by courtesy of the *Proceedings of the Nutrition Society*.

rats k_S and k_{RNA} were reduced in both muscle and liver, with no significant change in RNA concentration. It does not, of course, follow that providing GH to normal animals will increase protein synthesis above the normal level. Nevertheless, the use of GH to increase growth in lean body mass in farm animals suggested that the hormone might produce an increase in muscle protein synthesis. Pell and Bates (1987) showed in lambs that 100 $\mu g\ kg^{-1}\ day^{-1}$ of growth hormone produced a 30% increase in the FSR in a slow-twitch red muscle (biceps femoris) but no change in a fast-twitch white muscle (semitendinosus). This effect has not been found in man. Yarasheki *et al.* (1992, 1995) measured protein synthesis in muscle biopsies from both young and elderly men who were undergoing a course of resistance exercise training. The training led to an increase in muscle mass and strength, but the increase was no greater when growth hormone was administered as well, nor was there any measurable change in muscle protein synthesis. Reasons for the discrepancy with Pell's results may be that she used a dose of hormone that was $2^1/_2$ times that of Yarasheki, and her animals were growing, whereas Yarasheki's subjects were not. Similarly, Welle *et al.* (1996) tried to reverse depression of myofibrillar protein synthesis by administration of GH for 3 months. The treatment produced a substantial increase in muscle mass but no change in myofibrillar protein synthesis. However, this is not so paradoxical: an increase in fractional synthesis rate too small to be detected, if not counterbalanced by a change in breakdown, would by compound interest over a long period build up a substantial increase in muscle mass.

It should be added that none of this work resolves the question of whether GH acts via IGF-1, since it was clearly impossible to clamp IGF-1 over the long time period of these studies.

15.3.4 Thyroid hormones

In the 1980s there was much experimental work on the effects of thyroid hormones (TH) on tissue protein turnover, particularly skeletal muscle, the active hormone being triiodothyronine (T_3). A consensus emerged that in euthyroid animals, and even more in thyroidectomized ones, T_3 increases both synthesis and breakdown of muscle proteins, as measured by various methods (Goldberg *et al.*, 1980; Carter *et al.*, 1982, 1984; Hayashi *et al.*, 1986; Jepson *et al.*, 1988). In thyroidectomized rats the fall in muscle protein synthesis was caused by a reduced concentration of RNA, which was restored by treatment with T_3, k_{RNA} being only slightly depressed (Table 15.15) (Brown *et al.*, 1981; Brown and Millward, 1983). It is very difficult in these experiments to rule out the roles of food supply and other hormones, particularly insulin, growth hormone and insulin-like growth factor-1 (IGF-1). These relationships are reviewed by Dauncey and Gilmour (1996). At equal food intake hypothyroid pigs had higher IGF-1 levels than euthyroid controls (Morovat and Dauncey, 1998); in fetal sheep thyroid hormones had an important role in regulating IGF-1 gene expression. In piglets an increased energy supply or decreased demand for thermoregulation increased plasma T_3 concentrations and the number of T_3 binding sites in muscle (Dauncey, 1990) while protein had no effect. On the other hand, in rats fed diets of varying protein content, Jepson *et al.* (1988) showed that plasma levels of T_3 and of insulin were both correlated with protein intake; moreover, rates of muscle protein synthesis and breakdown were correlated with the concentrations of both hormones. Thus under physiological conditions the influence of protein and energy intake may be secondary, through their effects on T_3 concentration. However, in hypothyroidism hepatic 5′-monodeiodinase activity, which pro-

Table 15.15. Effect of thyroid hormone on protein turnover in rat skeletal muscle (gastrocnemius and quadriceps).

	k_S	[RNA]	k_{RNA}
Control	100	100	100
Thyroidectomized	60	68	87
Thyroidectomized + T_3	113	105	106

Numbers are per cent of control. Data of Brown *et al.*, 1981.

duces T_3 from thyroxin, was reduced – a response that would seem to be counter-regulatory (Harrison *et al.*, 1996).

Less has been done on liver than on muscle. A paper by Peavy *et al.* (1981) on perfused livers and isolated hepatocytes is very important. They showed that thyroidectomy decreased the synthesis of albumin and other secreted proteins by 50% so that the amount of albumin recovered from the liver was reduced, while the synthesis of non-secreted proteins fell by only 20%. The size distribution of polyribosomes was the same in thyroidectomized and normal rats and there was no difference in the proportion of polyribosomes that were synthesizing albumin. These results argued against a defect in initiation/elongation, in contrast to the finding by Mathews *et al.* (1973) that thyroidectomy doubled the time needed for assembly of polypeptide chains. The difference was attributed to technical factors. Peavy *et al.* concluded that TH deficiency caused a generalized incapacity to synthesize most proteins and that the extra deficit in the synthesis of secreted proteins might be due to co-translational breakdown before they are secreted. Broadly, this conclusion agrees with that derived from work on muscle (Table 15.15), that TH deficiency produces a fall in ribosomal concentration, and, in contrast to insulin, the effect is the same in muscle and liver.

The work of Dauncey and her colleagues has taken the action of thyroid hormones into the molecular age. They have been particularly interested in the early development of muscle and its myofibrils in the fetus and around the time of birth, showing that TH binds to a family of nuclear receptors that control the development of the myoblast into the myofibre (Dauncey and Gilmour, 1996). They have shown that in neonatal piglets TH promotes the conversion of slow to fast myosin heavy chain isoforms; it increases the abundance of mitochondria and mitochondrial enzymes and the concentrations of Na^+, K^+ and Ca^{2+} ATPases (Harrison *et al.*, 1996), confirming the earlier finding of Kjeldsen *et al.* (1986). Thyroidectomy or chemically induced hypothyroidism reverse these effects. In hypothyroid animals there is an increase in the nuclear T_3-receptor-binding capacity of muscle, which must have a compensatory effect (Harrison *et al.*, 1996). Thyroid hormones also modulate the gene expression of growth hormone receptors and IGF mRNAs in both liver and muscle, to an extent which varies somewhat with the stage of prenatal and postnatal development (Duchamp *et al.*, 1996; Forhead *et al.*, 2000, 2002). These findings reinforce the indications from the earlier work of the central role of TH in the hormonal orchestra. It remains to be seen how far the changes in myofibre type and composition found in the fetus persist into adult life.

15.3.5 Glucocorticoids

In an early study on rats Tomas *et al.* (1979) showed that high doses of corticosterone, producing plasma levels similar to those in stressed rats, led to an increase in muscle protein breakdown, determined by excretion of 3-methylhistidine. Muscles lost weight while liver gained weight – another example of the contrast between these two tissues. They observed also that the glucocorticoid treatment caused a substantial increase in plasma insulin concentrations. On the other hand, in starvation insulin falls and glucocorticoids rise (Table 15.10). Odedra *et al.* (1982), Millward *et al.* (1983), Pain *et al.* (1983), Ashford and Pain (1986a) and Southorn *et al.* (1990) suggested that corticosterone had two actions, one independent from and opposite to that of insulin and a second which caused resistance to insulin.

The effects of corticosterone on muscle in the rat were examined in detail by Odedra *et al.* (1983) and may be summarized as follows: (i) body weight and muscle weight were reduced while the relative liver weight increased; (ii) the k_S of muscle protein fell progressively to 60% of the control level after 12 days of treatment, the synthesis rate of actomyosin was reduced more than that of total muscle protein; (iii) RNA concentration fell *pari passu* with synthesis, but there was no change in efficiency; and (iv) the degradation rate of muscle protein doubled after 1 day of treatment, but after 4 days it began to fall to the control rate. Kayali *et al.* (1990) suggested that this adaptive fall in breakdown, measured as the difference between synthesis and growth, did not depend on insulin or thyroid hormones. The decrease in RNA concentration may result from increased degradation of ribosomal proteins, as was found by Ashford and Pain (1986b) in muscle of diabetic rats (*vide supra*). These effects of glucocorticoids are not confined to muscle; it is very interesting that in rats 8 days of treatment with corticosterone led to almost complete cessation of growth in the tibia, with massive reductions in k_S, k_{RNA} and the uptake of $^{35}SO_4$ into the proteoglycan of cartilage, but there was no change in RNA concentration (Yahya *et al.*, 1994).

In human tissues the effect of corticoids on synthesis has been examined more often than on breakdown, and then only in muscle. Wernerman *et al.* (1989) showed that infusions of cortisol, adrenalin or both together with glucagon – i.e. all the stress hormones – caused a reduction in the ribosome content of muscle and in the proportion of them that were in polyribosomes, implying a decreased capacity for synthesis as well as decreased activity. A similar finding was reported by Löfberg *et al.* (2002). On the other hand Gore *et al.* (1993) found that infusion of the three hormones together produced a 50% increase of protein synthesis in muscle biopsies; parallel studies by arteriovenous difference in the leg indicated that breakdown was increased more than synthesis. A unique study is that of McNurlan *et al.* (1996). They measured the synthesis rates of plasma albumin, muscle protein and lymphocytes in normal people after infusion for 6 h of a cocktail containing the three stress hormones. The results in Table 15.16 show in lymphocytes an immediate decrease in synthesis whereas decrease in muscle was delayed and albumin synthesis was increased. The timing of response is important when it comes to clinical situations. In a subsequent investigation these authors found that synthesis was reduced in both mononuclear cells and T lymphocytes, with a decrease in all the lymphocyte subpopulations except for the natural killer cells (Januszkiewicz *et al.*, 2001).

Interest has inevitably moved on to the molecular mechanism of glucocorticoid action and it was shown that initiation of synthesis was blocked by increased attachment of one of the initiation factors to its binding protein (Shah *et al.*, 2000). It is now possible to get indirect evidence of changes in breakdown by measuring the mRNAs that encode proteolytic enzymes although, as mentioned earlier, the validity of this approach has been questioned. Thus it has been shown that glucocorticoids activate the ubiquitin-proteasome system of protein degradation in muscle during fasting (Wing and Goldberg, 1993; Auclaire *et al.*, 1997). The activation seems to be less in old than in adult rats so it has been concluded that glucocorticoids do not play any role in the progressive muscular atrophy seen in old people (Dardevet *et al.*, 1995). This atrophy is a chronic process: activation of the ubiquitin-proteasome complex of skeletal muscle was found in patients with acute and severe head injuries who had increased whole body protein breakdown and 3-methylhistidine excretion (Mansoor *et al.*, 1996), but not in patients with Cushing's disease, in which there is muscle wasting with chronic excessive glucocorticoid production (Gore *et al.*, 1993).

Thus most of the work on the effect of glucocorticoid hormones has been done on muscle. In this tissue they most often decrease protein synthesis, and there is scope for re-examination of

Table 15.16. Rates of protein synthesis, % d^{-1}, during and after infusion of stress hormones (corticol, epinephrine, glucagon).

	Infusion		
	Saline	Hormones	
Experiment		During[a]	After[b]
1 Muscle	1.57	1.48	–
2 Muscle	1.46	1.60	–
3 Muscle	1.77	1.51	1.29
1 Albumin	6.68	6.90	–
3 Albumin	6.84	6.96	7.99
2 Lymphocytes	9.63	2.47	–
3 Lymphocytes	6.02	–	5.75

[a]Measured during last $1\frac{1}{2}$ h of 6 h infusion. [b]Measured 24 h after start of 6 h infusion. Infusion = 6.0 μg cortisol + 3 ng glucagon + 0.5 nmol epinephrine kg^{-1} min^{-1}. Data of McNurlan *et al.*, 1996.

the extent to which this results from decreased production of ribosomes, as indicated by several studies. There is no doubt that glucocorticoids increase muscle protein degradation, and recent work has concentrated on the enzymic machinery of proteolysis. It may be possible that these mechanisms are able to adapt to long-term stimulation by glucocorticoids.

15.4 References

Anthony, J.C., Lang, C.H., Crozier, S.J., Anthony, T.G., MacLean, D.A., Kimball, S.R. and Jefferson, L.S. (2001) Contribution to the translational control of protein synthesis in skeletal muscle by leucine. *American Journal of Physiology* 282, E1092–1101.

Ashford, A.J. and Pain, V.M. (1986a) Effect of diabetes on the rates of synthesis and degradation of ribosomes in rat muscle and liver *in vivo. Journal of Biological Chemistry* 261, 4059–4065.

Ashford, A.J. and Pain, V.M. (1986b) Insulin stimulation of growth in diabetic rats. *Journal of Biological Chemistry* 261, 4066–4070.

Attaix, D., Manghebati, A., Grizard, J. and Arnal, M. (1986) Assessment of *in vivo* protein synthesis in lamb tissues with [^{3}H] valine flooding doses. *Biochemica et Biophysica Acta* 882, 389–397.

Attaix, D.E., Aurousseau, E., Manghebati, A. and Arnal, M. (1988) Contribution of liver, skin and skeletal muscle to whole-body protein synthesis in the young lamb. *British Journal of Nutrition* 60, 77–84.

Attaix, D., Aurousseau, E., Bayle, G., Rosolowska-Huszkcz, D., Manghebati, A. and Arnal, M. (1989) Influences of age and weaning on *in vivo* pancreatic protein synthesis in the lamb. *Journal of Nutrition* 119, 463–470.

Auclaire, O., Garral, G.R., Zeronala, A.C. and Ferland, L.H. (1997) Activation of the ubiquitin pathway in rat skeletal muscle by catabolic doses of glucocorticoids. *American Journal of Physiology* 272, C1007–1016.

Baillie, A.G.S. and Garlick, P.J. (1991a) Response of protein synthesis in different skeletal muscles to fasting and insulin in rats. *American Journal of Physiology* 260, E891–896.

Baillie, A.G.S. and Garlick, P.J. (1991b) Attenuated responses of muscle protein synthesis to fasting and insulin in adult female rats. *American Journal of Physiology* 262, E1–5.

Ballmer, P.E., McNurlan, M.A., Grant, I. and Garlick, P.J. (1993) Responses of tissue protein synthesis to nutrient intake in rats exposed to interleukin $-1\ \beta$ or turpentine. *Clinical Science* 85, 337–342.

Bark, T.H., McNurlan, M.A., Lang, C.H. and Garlick, P.J. (1998) Increased protein synthesis after acute IGF-I or insulin infusion is localized to muscle in mice. *American Journal of Physiology* 275, E118–123.

Barle, H., Nyborg, B., Essen, P., Andersson, K., McNurlan, M., Wernerman, J. and Garlick, P.J. (1997) The synthesis rates of total liver protein and plasma albumin determined simultaneously *in vivo* in humans. *Hepatology* 25, 154–158.

Bates, P.C. and Millward, D.J. (1978) Changes in the relative rates of muscle synthesis and breakdown during muscle growth and atrophy. *Biochemical Society Transactions* 6, 612–614.

Bates, P.C. and Millward, D.J. (1981) Characteristics of skeletal muscle growth and protein turnover in a fast-growing rat strain. *British Journal of Nutrition* 46, 7–13.

Bates, P.C. and Millward, D.J. (1983) Synthesis rates of myofibrillar and sarcoplasmic protein fractions in different muscles and the changes observed during post-natal development and in response to feeding and fasting. *Biochemical Journal* 214, 587–592.

Bates, P.C., Grimble, G.K., Sparrow, M.P. and Millward, D.J. (1983) Synthesis of protein-bound 3-methylhistidine, actin, myosin heavy chain and aldolase in rat skeletal muscle in the fed and starved states. *Biochemical Journal* 214, 593–605.

Bennet, W.M., O'Keefe, S.J.D. and Haymond, M.W. (1993) Comparison of precursor pools with leucine, α-ketoisocaproate and phenylalanine tracers used to measure splanchnic protein synthesis in man. *Metabolism* 42, 691–695.

Boirie, Y., Short, K.R., Ahlman, B., Charlton, M. and Nair, K.S. (2001) Tissue specific regulation of mitochondrial and cytoplasmic protein synthesis by insulin. *Diabetes* 50, 2652–2658.

Bouteloup-Demange, C., Boirie, Y., Déchelotte, P., Gachon, P., and Beaufrère, B. (1998) Gut mucosal protein synthesis in fed and fasted subjects. *American Journal of Physiology* 274, E541–546.

Breuillé, D., Amal, M., Rambourdin, F., Bayle, G., Levieux, D. and Obled, C. (1998) Sustained modifications of protein metabolism in various tissues in a rat model of long-lasting sepsis. *Clinical Science* 94, 413–423.

Brown, J.G. and Millward, D.J. (1983) Dose response of protein turnover in rat skeletal muscle to triiodothyronine treatment. *Biophysica Acta* 757, 182–190.

Brown, J.G., Bates, P.C., Holliday, M.A. and Millward, D.J. (1981) Thyroid hormones and muscle protein turnover. *Biochemical Journal* 194, 771–782.

Burrin, D.G., Davis, T.A., Fiorotto, P.J. and Reeds, P.J. (1991) Stage of development and fasting affect protein synthesis activity in the gastro-intestinal tract of suckling rats. *Journal of Nutrition* 121, 1099–1108.

Burrin, D.G., Schulman, R.J., Reeds, P.J., Davis, T.A. and Gravitt, K.R. (1992) Porcine colostrum and milk stimulate visceral organ and skeletal muscle protein synthesis in neonatal pigs. *Journal of Nutrition* 122, 1205–1213.

Buttery, P.J., Beckerton, A. and Mitchell, R.M. (1975) The turnover rate of muscle and liver protein in the sheep. *Proceedings of the Nutrition Society* 34, 91A.

Carter, W.J., Benjamin, W.S. van de W. and Haas, F.H. (1982) Effect of experimental hyperthyroidism on protein turnover in skeletal and cardiac muscle as measured by [^{14}C] tyrosine infusion. *Biochemical Journal* 204, 69–74.

Carter, W.J., Benjamin, S.W. and Faas, F.H. (1984) Effect of a protein-free diet on muscle protein turnover and nitrogen conservation in euthyroid and hyperthyroid rats. *Biochemical Journal* 217, 471–476.

Case, G., Ford, G.C., Nair, K.S., Garlick, P.J. and McNurlan, M.A. (2002) Aminoacyl-tRNA enrichment after a flood of labeled phenylalanine: insulin effect on muscle protein synthesis. *American Journal of Physiology* 282, E1029–1038.

Charlton, M., Ahlman, B. and Nair, K.S. (2000) The effect of insulin on human small intestinal protein synthesis. *Gastroenterology* 118, 299–306.

Cherel, Y., Attaix, D., Rosolowska-Huszez, D., Belkhou, R., Robin, J.-P., Arnal, M. and Le Maho, Y. (1991) Whole body and tissue protein synthesis during brief and prolonged fasting in the rat. *Clinical Science* 81, 611–619.

Conde, R.D. and Scornik, O.A. (1976) Role of protein degradation in the growth of livers after a nutritional shift. *Biochemical Journal* 158, 385–390.

Connell, A., Calder, A.G., Anderson, S.E. and Lobley, G.E. (1997) Hepatic protein synthesis in the sheep: effect of intake as monitored by use of stable-isotope-labelled glycine, leucine and phenylalanine. *British Journal of Nutrition* 77, 255–271.

Crozier, S.J., Anthony, J.C., Schwover, C.M., Reitep, A.K., Anthony, T.G., Kimball, S.R. and Jefferson, L.S. (2003) Tissue-specific regulation of protein synthesis by insulin and free fatty acids. *American Journal of Physiology* 285, E754.

Cuthbertson, D., Smith, K., Babraj, J., Leese, G., Waddell, T., Atherton, P., Wackerhage, H., Taylor, P.M. and Rennie, M.J. (2005) Anabolic signalling deficits underlie amino acid resistance of wasting, aging muscle. *FASEB Journal* 19, 422–424.

Dardevet, D., Sornet, C., Attaix, D., Baracos, V.E. and Grizard, J. (1994) Insulin-like growth factor-1 and insulin resistance in skeletal muscles of adult and old rats. *Endocrinology* 134, 1475–1484.

Dardevet, D., Sornet, C., Taillandier, D., Savary, I., Attaix, D. and Grizard, J. (1995) Sensitivity and protein turnover response to glucocorticoids are different in skeletal muscle from adult and old rats. *Journal of Clinical Investigation* 96, 2113–2119.

Dauncey, M.J. (1990) Thyroid hormones and thermogenesis. *Proceedings of the Nutrition Society* 49, 203–215.

Dauncey, M.J. and Gilmour, R.S. (1996) Regulatory factors in the control of muscle development. *Proceedings of the Nutrition Society* 55, 543–559.

Davis, S.R., Barry, T.N. and Hughson, G.A. (1981) Protein synthesis in tissues of growing lambs. *British Journal of Nutrition* 46, 409–419.

Davis, T.A., Nguyen, H.H., Suryawan, A., Bush, J.A., Jefferson, L.J. and Kimball, S.R. (2000) Developmental changes in the feeding-induced stimulation of translation initiation in muscle of neonatal pigs. *American Journal of Physiology* 279, E1226–1234.

Davis, T.A., Fiorotto, M.L., Beckett, P.R., Burrin, D.G., Reeds, P.J., Wray-Cohen, D. and Nguyen, H.V. (2001) Differential effects of insulin on peripheral and visceral protein synthesis in neonatal pigs. *American Journal of Physiology* 281, E770–E779.

Davis, T.A., Fiorotto, M.L., Burrin, D.B., Reeds, P.J., Nguyen, H.V., Beckett, P.R., Vann, R.C. and O'Connor, P.M.J. (2002) Stimulation of protein synthesis by both insulin and amino acids is unique to skeletal muscle in neonatal pigs. *American Journal of Physiology* 282, E880–E890.

Duchamp, C., Burton, K.A., Herpin, P. and Dauncey, M.J. (1996) Perinatal ontogeny of porcine growth hormone receptor gene expression is modulated by thyroid status. *European Journal of Endocrinology* 134, 524–531.

Dudley, M.A., Wykes, L.J., Dudley, A.W., Burin, D.G., Nichols, B.L., Rosenberger, J., Jahoor, F., Heird, W.C. and Reeds, P.J. (1998) Parenteral nutrition selectively decreases protein synthesis in the small intestine. *American Journal of Physiology* 274, G131–137.

El Haj, A.J., Lewis, S.E.M., Goldspink, D.F., Merry, B.J. and Holehan, A.M. (1986) The effect of chronic and acute dietary restriction on the growth and protein turnover of fast and slow types of rat muscle. *Comparative Biochemistry and Physiology* 85A, 281–287.

Elia, M. (1991) The inter-organ flux of substrates in fed and fasted man, as indicated by arterio-venous balance studies. *Nutrition Research Reviews* 4, 3–31.

Essén, P., McNurlan, M.A., Wernerman, J., Milne, E., Vinnars, E. and Garlick, P.J. (1992) Short-term starvation decreases skeletal muscle protein synthesis in man. *Clinical Physiology* 12, 287–299.

Essén, P., McNurlan, M.A., Thorell, A., Tjader, I., Cass, G., Anderson, S.E., Wernerman, J. and Garlick, P.J. (1996) Determination of protein synthesis in lymphocytes after surgery. *Clinical Science* 91, 99–106.

Etlinger, J.D., Zak, R., Fishman, D.A. and Rabinowitz, R. (1975) Isolation of newly synthesized myofilaments from skeletal muscle homogenates and myofibrils. *Nature (London)* 255, 259–261.

Felig, P. (1973) The glucose-alanine cycle. *Metabolism* 22, 179–207.

Felig, P. (1975) Amino acid metabolism in man. *Annual Reviews of Biochemistry* 44, 933–955.

Fiorotto, M.L., Davis, T.A. and Reeds, P.J. (2000) Regulation of myofibrillar protein turnover during maturation in normal and undernourished rat pups. *American Journal of Physiology* 278, R845–854.

Flaim, K.E., Peavy, D.E., Everson, W.V. and Jefferson, J.S. (1982) The role of amino acids in the regulation of protein synthesis in the perfused rat liver. *Journal of Biological Chemistry* 257, 2932–2938.

Forhead, A.J., Li, J., Saunders, J.C., Dauncey, M.J., Gilmour, R.S. and Fowden, A. (2000) Control of ovine growth hormone receptor and insulin-like growth factor 1 by thyroid hormones *in utero*. *American Journal of Physiology* 278, E1160–1174.

Forhead, A.J., Li, J., Gilmour, R.S., Dauncey, M.J. and Fowden, A.L. (2002) Thyroid hormones and the mRNA of the GH receptor and IGFs in skeletal muscle of fetal sheep. *American Journal of Physiology* 282, E80–86.

Fryburg, D.A. (1994) Insulin-like growth factor 1 exerted growth hormone- and insulin-like actions on human muscle protein metabolism. *American Journal of Physiology* 267, E331–336.

Frystyk, J., Delhanty, P.J.D., Skjaerbaek, C. and Baxter, R.C. (1999) Changes in the circulating IGF system during short-term fasting and refeeding in rats. *American Journal of Physiology* 277, E245–252.

Garlick, P.J. and Grant, I. (1988) Amino acid infusion increases the sensitivity of muscle protein synthesis *in vivo* to insulin. Effect of branched-chain amino acids. *Biochemical Journal* 254, 579–584.

Garlick, P.J., Millward, D.J. and James, W.P.T. (1973) The diurnal response of muscle and liver protein synthesis *in vivo* in meal-fed rats. *Biochemical Journal* 136, 935–946.

Garlick, P.J., Millward, D.J., James, W.P.T. and Waterlow, J.C. (1975) The effect of protein deprivation and starvation on the rate of protein synthesis in tissues of the rat. *Biochimica et Biophysica Acta* 414, 71–84.

Garlick, P.J., Burke, T.L. and Swick, R.W. (1976) Protein synthesis and RNA in tissues of the pig. *American Journal of Physiology* 230, 1108–1112.

Garlick, P.J., Clugston, G.A., McNurlan, M.A., Preedy, V.R. and Fern, E.B. (1982) Nutrition and protein turnover. *Biochemical Society Transactions* 10, 290–291.

Garlick, P.J., Fern, M. and Preedy, V.R. (1983) The effect of insulin infusion and of food intake on muscle protein synthesis in postabsorptive rats. *Biochemical Journal* 210, 669–676.

Garlick, P.J., Wernerman, J., McNurlan, M.A., Essen, P., Lobley, G.E. Milne, E., Calder, G.A. and Vinnars, E. (1989a) Measurement of the rate of protein synthesis in muscle of postabsorptive young men by injection of a 'flooding dose' of [1-^{13}C] leucine. *Clinical Science* 86, 671–675.

Garlick, P.J., Maltin, C.A., Baillie, G.S., Delday, M.I. and Grubb, D.A. (1989b) Fiber-type composition of nine rat muscles. II. Relationship to protein turnover. *American Journal of Physiology* 257, E328–332.

Garlick, P.J., Wernerman, J., McNurlan, M.A. and Heys, S.D. (1991) Organ-specific measurements of protein turnover in man. *Proceedings of the Nutrition Society* 50, 217–225.

Garlick, P.J., McNurlan, M.A., Essen, P., Ballmer, P.E. and Hunter, K. (1995) Protein synthesis assessment in human tissues with stable isotopes. *Rivista Italiana di Nutritizione Parenterale ed Enterale* 13, 1–7.

Giordano, M., Castellino, P. and De Fronzo, R.A. (1996) Differential responsiveness of protein synthesis and degradation to amino acid availability in humans. *Diabetes* 45, 393–399.

Giugliano, R. and Millward, D.J. (1984) Zinc homeostasis in the severely zinc deficient rat. *British Journal of Nutrition* 53, 545–560.

Giugliano, R. and Millward, D.J. (1987) The effect of severe zinc deficiency on protein turnover in muscle and thymus. *British Journal of Nutrition* 57, 139–155.

Goldberg, A.L., Tischler, M., De Martino, G. and Griffin, G. (1980) Hormonal regulation of protein degradation and synthesis in skeletal muscle. *Federation Proceedings* 39, 31–36.

Goldspink, D.F. and Kelly, F.J. (1984) Protein turnover and growth in the whole body, liver and kidney of the rat from the foetus to senility. *Biochemical Journal* 217, 507–516.

Goldspink, G., Marshall, P.A. and Watt, P.W. (1984) Protein synthesis in red and white skeletal muscle of carp (*Cyprinus carpio*) measured *in vivo* and the effect of temperature. *Journal of Physiology* 361, 42P.

Gore, D.C., Jahoor, F., Wolfe, R.R. and Herndon, D.N. (1993) Acute response of human muscle protein to catabolic hormones. *Annals of Surgery* 218, 679–684.

Grizard, J., Dardevet, D., Papet, I.O., Mosoni, L., Mirand, P.P., Attaix, D., Tauveron, I., Bonin, D. and Arnal, M. (1995) Nutrient regulation of skeletal muscle protein in animals. The involvement of hormones and substrates. *Nutrition Research Reviews* 8, 67–92.

Halliday, D. and McKeran, R.O. (1975) Measurement of muscle protein synthesis rate from serial muscle biopsies and total body protein turnover in man by continuous intravenous infusion of L-(alpha ^{15}N) lysine. *Clinical Science and Molecular Medicine* 49, 581–590.

Halliday, D., Pacy, P.J., Cheng, K.N., Dworzak, F., Gibson, J.N.A. and Rennie, M.J. (1988) Rate of protein synthesis in skeletal muscle of normal man and patients with muscular dystrophy: a reassessment. *Clinical Science* 74, 237–240.

Harrison, A., Tivey, T.R., Clausen, T., Duchamp, C. and Dauncey, M.J. (1996) Role of thyroid hormones in early postnatal development of skeletal muscle and its implications for undernutrition. *British Journal of Nutrition* 87, 841–855.

Hashiguchi, Y., Molina, P.E., Preedy, V.R., Sugden, P.H., McNurlan, M.A., Garlick, P.J. and Atumrad, N.N. (1996). Central effects of morphine and morphine-6-glucuromide on tissue protein synthesis. *American Journal of Physiology* 271, R619–625.

Hayashi, K., Kayali, A.G. and Young, V.R. (1986) Synergism of triiodothyronine and corticosterone on muscle protein breakdown. *Biochimica et Biophysica Acta* 883, 106–111.

Heys, S.D., Park, K.G.M., McNurlan, M.A., Keenan, R.A., Miller, J.D.B., Evemin, O. and Garlick, P.J. (1992) Protein synthesis rates in colon and liver: stimulation by gastro-intestinal pathologies. *Gut* 33, 976–981.

Jacob, R., Hu, X., Niederstock, D., Hasan, S., McNulty, P.H., Sherwin, R.S. and Young, L.H. (1996) IGF-I stimulation of muscle protein synthesis in the awake rat: permissive role of insulin and amino acids. *American Journal of Physiology* 270, E60–66.

Januszkiewicz, A., Essén, P., McNurlan, M.A., Ringdén, O., Garlick, P.J. and Wernerman, J. (2001) A combined stress hormone decreases *in vivo* protein synthesis in human T lymphocytes in healthy volunteers. *Metabolism* 50, 1308–1314.

Jepson, M.M., Bates, P.C. and Millward, D.J. (1988) The role of insulin and thyroid hormones in the regulation of muscle growth and protein turnover in response to dietary protein in the rat. *British Journal of Nutrition* 59, 397–415.

Kadowaki, M., Harada, N., Takahashi, S., Nognuhi, T. and Naito, H. (1989) Differential regulation of the degradation of myofibrilar and total proteins in skeletal muscle of rats: effects of streptozotocin-induced diabetes, dietary protein and starvation. *Journal of Nutrition* 119, 471–477.

Kayali, A.G., Goodman, M.N., Lin, J. and Young, V.R. (1990) Insulin- and thyroid hormone turnover; independent adaptation of myofibrillar proteolysis to glucocorticoids. *American Journal of Physiology* 259, E699–705.

Kimball, S.R., Vary, T.C. and Jefferson, L.S. (1994) Regulation of protein synthesis by insulin. *Annual Reviews of Physiology* 56, 321–348.

Kimball, S.R., Farrell, P.A. and Jefferson, L.S. (2002) Role of insulin in translational control of protein synthesis in skeletal muscle by amino acids or exercise. *Journal of Applied Physiology* 93, 1168–1180.

Kjeldsen, K., Everts, M.E. and Clausen, T. (1986) The effects of thyroid hormones on ^{3}H-ouabain binding site concentration, Na,K-contents and ^{86}Rb-efflux in rat skeletal muscle. *European Journal of Physiology* 56, 529–535.

Laurent, G.J., Sparrow, M.P., Bates, P.C. and Millward, D.J. (1978) Turnover of muscle protein in the fowl (*Gallus domesticus*). *Biochemical Journal* 176, 343–405.

Lewis, S.M., Kelly, F.J. and Goldspink, D. (1984) Pre- and post-natal growth and protein turnover in smooth muscle, heart and slow- and fast-twitch skeletal muscles of the rat. *Biochemical Journal* 217, 517–526.

Lobley, G.E. (1993) Species comparisons of tissue protein metabolism: effects of age and of hormones. *Journal of Nutrition* 123, 337–343.

Lobley, G.E., Milne, V., Lovie, J.M., Reeds, P.J. and Pennie, K. (1980) Whole body and tissue protein synthesis in cattle. *British Journal of Nutrition* 43, 491–502.

Lobley, G.E., Connell, A., Milne, E., Newman, A.M. and Ewing, T.A. (1994) Protein synthesis in splanchnic tissues of sheep offered two levels of intake. *British Journal of Nutrition* 71, 3–12.

Lobley, G.E., Sinclair, K.D., Grant, C.M., Miller, L., Mantle, D., Calder, A.G., Warkup, C.C. and Maltin, C.A. (2000) The effects of breed and level of nutrition in whole-body and muscle protein metabolism in pure-bred Aberdeen Angus and Charolais beef steers. *British Journal of Nutrition* 84, 275–284.

Lòfberg, E., Gutierrez, A., Wernerman, J., Anderstam, B., Mitch, W.E., Price, S.R., Bergström, J. and Alvestrand, A. (2002) Effects of high doses of glucocorticoids on free amino acids, ribosomes, and protein turnover in human muscle. *European Journal of Clinical Investigation* 32, 345–353.

Lowe, W.L., Adamo, M.L., Werner, H., Roberts, C. and LeRoith, D.L. (1989) Regulation by fasting of rat insulin-like growth factor I and its receptor: effects on gene expression and binding. *Journal of Clinical Investigation* 84, 619–626.

Mansoor, O., Beaufrère, B., Roirie, Y., Ralliere, C., Taillandier, D., Aurousseau, E., Schoeffler, P., Arnal, M. and Attaix, D. (1996) Increased mRNA levels for components of the lysosomal Ca^{2+}-activated and ATP-ubiquitin-dependent proteolytic pathways in skeletal muscle from head trauma patients. *Proceedings of the National Academy of Sciences* 93, 2714–2718.

Mansoor, O., Cayol, M., Gachon, P., Boirie, Y., Schoeffler, P., Obled, C. and Beaufrère, B. (1997) Albumin and fibrinogen synthesis increase while muscle protein synthesis decreases in head-injured patients. *American Journal of Physiology* 273, E898–902.

Marway, J.S., Keating, J.W., Reeves, J., Salisbury, J.R.

and Preedy, V.R. (1993) Seromuscular and mucosal protein synthesis in various anatomical regions of the rat gastrointestinal tract and their response to acute ethanol toxicity. *European Journal of Gastroenterology and Hepatology* 5, 27–34.

Mathews, R.W. and Haschemeyer, A. (1978) Temperature dependency of protein synthesis in toadfish liver *in vivo. Comparative Biochemistry and Physiology* 61B, 479–484.

Mathews, R.W., Oronsky, A. and Haschemeyer, A.E.V. (1973) Effect of thyroid hormone on polypeptide chain assembly kinetics in liver protein synthesis *in vivo. Journal of Biological Chemistry* 248, 1329–1333.

McNulty, P.H., Young, L.E. and Barrett, E.J. (1993) Response of rat heart and skeletal muscle protein *in vivo* to insulin and amino acid infusion. *American Journal of Physiology* 264, E958–965.

McNurlan, M.A. and Garlick, P.J. (1980) Contribution of rat liver and gastrointestinal tract to whole body synthesis in the rat. *Biochemical Journal* 186, 381–383.

McNurlan, M.A. and Garlick, P.J. (1981) Protein synthesis in liver and small intestine in protein deprivation and diabetes. *American Journal of Physiology* 241, E238–245.

McNurlan, M.A., Tomkins, A.M. and Garlick, P.J. (1979) The effect of starvation on the rate of protein synthesis in rat liver and small intestine. *Biochemical Journal* 156, 185–188.

McNurlan, M.A., Pain, V.M. and Garlick, P.J. (1980) Conditions that alter rates of protein synthesis *in vivo. Biochemical Society Transactions* 8, 283–285.

McNurlan, M.A., Fern, E.B. and Garlick, P.J. (1982) Failure of leucine to stimulate protein synthesis *in vivo. Biochemical Journal* 204, 831–838.

McNurlan, M.A., Essen, P., Heys, S.P., Buchan, V., Garlick, P.J. and Wernerman, J. (1991) Measurement of protein synthesis in human skeletal muscle: further investigation of the flooding technique. *Clinical Science* 81, 557–564.

McNurlan, M.A., Essen, P., Milne, E., Vinnars, E. and Garlick, P.J. (1993) Temporal responses of protein synthesis in human skeletal muscle to feeding. *British Journal of Nutrition* 69, 117–126.

McNurlan, M.A., Essén, P., Thovell, A., Calder, A.G., Anderson, S.C., Ljungkvist, O., Sandgren, A., Grant, I., Tjäder, I., Ballmer, P.E., Wernerman, J. and Garlick, P.J. (1994) Response of protein synthesis in human skeletal muscle to insulin: an investigation with L-[$^{2}H_5$] phenylalanine. *American Journal of Physiology* 267, E102–108.

McNurlan, M.A., Sandgren, A., Hunter, K., Essen, P., Garlick, P.J. and Wernerman, J. (1996) Protein synthesis rates of skeletal muscle, lymphocytes and albumin with stress hormone infusion in healthy man. *Metabolism* 45, 1388–1394.

McNurlan, M.A., Garlick, P.J., Steigbigel, R.T., DeCristoforo, K.A., Frost, R.A., Lang, C.M., Johnson, R.W., Santasier, A.M., Cabahug, C.J., Fuhrer, J. and Gelato, M.C. (1997) Responsiveness of muscle protein synthesis to growth hormone administered in HIV-infected individuals declines with severity of disease. *Journal of Clinical Investigation* 100, 2125–2132.

Merry, B.J. and Houlihan, M. (1991) Effect of age and restricted feeding on polypeptide chain assembly kinetics in liver protein synthesis *in vivo*. *Mechanics of Ageing and Development* 58, 139–150.

Millward, D.J. (1980) Protein degradation in muscle and liver. In: Florkin, M., Neuberger, A. and Van Deenan, L.L.M. (eds) *Comprehensive Biochemistry* Vol 19B (i). Elsevier/North Holland, Amsterdam, pp. 153–252.

Millward, D.J. (1990) The hormonal control of protein turnover. *Clinical Nutrition* 9, 115–126.

Millward, D.J., Nnanyelugo, D.O., James, W.P. and Garlick, P.J. (1974) Protein metabolism in skeletal muscle: the effect of feeding and fasting on muscle RNA, free amino acids and plasma insulin concentrations. *British Journal of Nutrition* 32, 127–142.

Millward, D.J., Garlick, P.J., Stewart, R.J.C., Nnanyelugo, D. and Waterlow, J.C. (1975) Skeletal muscle growth and protein turnover. *Biochemical Journal* 150, 235–243.

Millward, D.J., Garlick, P.J., Nnanyelugo, D.O. and Waterlow, J.C. (1976) The relative importance of muscle protein synthesis and breakdown in the regulation of muscle mass. *Biochemical Journal* 156, 185–188.

Millward, D.J., Brown, J.G. and Odedra, B. (1981) Protein turnover in individual tissues with special emphasis on muscle. In: Waterlow, J.C. and Stephen, J.M.L. (eds) *Nitrogen Metabolism in Man.* Applied Science Publishers, London, pp. 475–494.

Millward, D.J., Odedra, B. and Bates, P.C. (1983) The role of insulin, corticosterone and other factors in the acute recovery of muscle protein synthesis on refeeding food-deprived rats. *Biochemical Journal* 216, 583–587.

Mitch, W.E., Bailey, J.L., Wang, X., Jurkovitz, C., Newby, D. and Price, S.R. (1999). Evolution of signals activating ubiquitin-proteasome proteolysis in a model of muscle wasting. *American Journal of Physiology* 276, C1132–1138.

Morovat, A. and Dauncey, M.J (1998) Effects of thyroid states on insulin-like growth factor-I, growth hormone and insulin are modified by food intake. *European Journal of Endocrinology* 138, 95–103.

Mosoni, L., Houlier, M.-L., Mirand, P.P., Bayle, G. and Grizard, J. (1993a) Effect of amino acids alone or with insulin on muscle and liver protein synthesis in adult and old rats. *American Journal of Physiology* 264, E614–620.

Mosoni, L., Patureau, M.P., Houlier, M.L. and Arnal, M. (1993b) Age-related changes in protein synthesis measured *in vivo* in rat liver and gastrocnemius muscle. *Mechanisms of Ageing and Development* 68, 209–220.

Munro, H.N. (1964) General aspects of the regulation of protein metabolism. In: Munro, H.N. and Allison, J.B. (eds) *Mammalian Protein Metabolism*. Academic Press, New York, pp. 381–481.

Nakshabendi, I.M., Obeidat, W., Russell, R.I., Downie, S., Smith, K. and Rennie, M.J. (1995) Gut mucosal protein synthesis measured using intravenous and intragastric delivery of stable tracer amino acids. *American Journal of Physiology* 269, E996–999.

Nakshabendi, I.M., Downie, S., Russell, R.I. and Rennie, M.J. (1996) Increased rates of duodenal mucosal protein synthesis *in vivo* in patients with untreated coeliac disease. *Gut* 39, 176–179.

Nakshabendi, I.M., McKee, R., Downie, S., Russell, R.I. and Rennie, M.J. (1999) Rate of small intestinal protein synthesis in human jejunum and ileum. *American Journal of Physiology* 277, E1028–1031.

Nash, J.E., Rocha, H.J.G., Buchan, V., Calder, G.A., Milne, E., Quirke, J.F. and Lobley, G.E. (1994) The effect of acute and chronic administration of the β-agonist, cimaterol, on protein synthesis in ovine skin and muscle. *British Journal of Nutrition* 71, 501–513.

Nicholas, G.A., Lobley, G.E. and Harris, C.I. (1977) Use of the constant infusion technique for measuring rates of protein synthesis in the New Zealand white rabbit. *British Journal of Nutrition* 38, 1–17.

O'Keefe, S.J.D., Haymond, M.W., Bennett, W.M. and Shorter, R.G. (1993) The measurement of gastrointestinal mucosal protein turnover in man. *European Journal of Gastroenterology and Hepatology* 5, S29–S32.

Obled, C. and Arnal, M. (1992) Contribution of skin to whole-body protein synthesis in rats at different stages of maturity. *Journal of Nutrition* 122, 2167–2173.

Oddy, V.H. (1993) Regulation of muscle protein metabolism in sheep and lambs: nutritional, endocrine and genetic aspects. *Australian Journal of Agricultural Research* 44, 901–913.

Odedra, B.R., Dalal, S.S. and Millward, D.J. (1982) Muscle protein synthesis in the streptozotocin-diabetic rat. *Biochemical Journal* 202, 363–368.

Odedra, B.R., Bates, P.C. and Millward, D.J. (1983) Time course of the effect of catabolic doses of corticosterone on protein turnover in rat skeletal muscle and liver. *Biochemical Journal* 214, 617–627.

Pacy, P.J., Nair, K.S., Ford, C. and Halliday, D. (1984) Failure of insulin infusion to stimulate fractional muscle protein synthesis in type 1 diabetic patients. *Diabetes* 38, 618–624.

Pain, V.M. and Garlick, P.J. (1974) Effect of streptozotocin diabetes and insulin treatment on the rate of protein synthesis in tissues of the rat *in vivo*. *Journal of Biological Chemistry* 249, 4510–4514.

Pain, V.M., Lanoix, J., Bergeron, J.J.M. and Clemens, M.J. (1974) Effect of diabetes on the ultrastructure of the hepatocyte and on the distribution and activity of ribosomes in the free and membrane-bound populations. *Biochemica et Biophysica Acta* 353, 487–498.

Pain, V.M., Albertse, S.E. and Garlick, P.J. (1983) Protein metabolism in skeletal muscle, diaphragm and heart of diabetic rats. *American Journal of Physiology* 245, E604–610.

Park, K.G.M., Heys, S.D., McNurlan, M.A., Garlick, P.J. and Eremin, O. (1994) Lymphocyte protein synthesis *in vivo*: a measure of activation. *Clinical Science* 86, 671–675.

Peavy, D.E., Taylor, J.M. and Jefferson, L.J. (1981) Alterations in albumin secretion and total protein synthesis in livers of thyroidectomized rats. *Biochemical Journal* 198, 289–299.

Pell, J.M. and Bates, P.C. (1987) Collagen and non-collagen protein turnover in skeletal muscle of growth-hormone treated lambs. *Journal of Endocrinology* 115, R1–4.

Peragón, J., Barroso, J.B., Garcia-Salguero, L., de la Higuera, M. and Lupianez, J.A. (2001) Growth, protein-turnover rates and nucleic-acid concentrations in the white muscle of rainbow trout during development. *The International Journal of Biochemistry and Cell Biology* 33, 1227–1238.

Preedy, V.R. and Garlick, P.J. (1981) Rates of protein synthesis in skin and bone and their importance in the assessment of protein degradation in the perfused rat hemicorpus. *Biochemical Journal* 194, 373–376.

Preedy, V.R. and Garlick, P.J. (1988) Inhibition of protein synthesis by glucagon in different rat muscles and protein fractions *in vivo* and in the perfused rat hemicorpus. *Biochemical Journal* 251, 727–732.

Preedy, V.R. and Garlick, P.J. (1993) Response of muscle protein synthesis to parenteral administration of amino acid mixtures in growing rats. *Journal of Parenteral and Enteral Nutrition* 17, 113–118.

Preedy, V.R. and Garlick, P.J. (1995) Ventricular muscle and lung protein synthesis *in vivo* in response to fasting, refeeding and nutrient supply by oral and parenteral routes. *Journal of Enteral and Parenteral Nutrition* 19, 107–113.

Preedy, V.R. and Peters, T.J. (1989) An investigation into the effects of chronic ethanol feeding on hepatic mixed protein synthesis in immature and mature rats. *Alcohol and Alcoholism* 24, 311–318.

Preedy, V.R. and Waterlow, J.C. (1981) Protein synthesis in the young rat – the contribution of skin and bones. *Journal of Physiology* 317, 45–46P.

Preedy, V.R., McNurlan, M.A. and Garlick, P.J. (1983)

Protein synthesis in skin and bone of the young rat. *British Journal of Nutrition* 49, 517–523.

Preedy, V.R., Smith, D.M., Kearney, N.F. and Sugden, P.H. (1984) Rates of protein turnover *in vivo* and *in vitro* in ventricular muscle of hearts from fed and starved rats. *Biochemical Journal* 222, 395–400.

Preedy, V.R., Smith, D.M., Kearney, N.F. and Sugden, P.H. (1985a) Regional variation and differential sensitivity of rat heart protein synthesis *in vivo* and *in vitro*. *Biochemical Journal* 225, 487–492.

Preedy, V.R., Smith, D.M. and Sugden, P.H. (1985b) The effect of 6 hours of hypoxia on protein synthesis in rat tissues *in vivo* and *in vitro*. *Biochemical Journal* 228, 179–185.

Preedy, V.R., Paske, L., Sugden, P.H., Schofield, P.S. and Sugden, M.C. (1988) The effects of surgical stress and short-term fasting on protein synthesis *in vivo* in diverse tissues of the mature rat. *Biochemical Journal* 250, 179–188.

Preedy, V.R., Marway, J.S., Salisbury, J.R. and Peters, T.J. (1990) Protein synthesis in bone and skin of the rat are inhibited by ethanol: implications for whole body metabolism. *Alcoholism: Clinical and Experimental Research* 14, 165–168.

Price, S.R., Bailey, J.L., Wang, X., Jurkovitz, C., England, B.K., Ding, X., Phillips, L.S. and Mitch, W.E. (1996) Muscle wasting in insulopenic rats results from activation of the ATP-dependent, ubiquitin-proteasome proteolytic pathway by a mechanism including gene transcription. *Journal of Clinical Investigation* 98, 1703–1708.

Rathmacher, J.A., Link, G. and Nissen, S. (1993) Measurement of 3-methlyhistidine production in lambs by using compartmental-kinetic analysis. *British Journal of Nutrition* 69, 743–755.

Reeds, P.J. and Harris, C.I. (1981) Protein turnover in animals: man in his context. In: Waterlow, J.C. and Stephen, J.M.L. (eds) *Nitrogen Metabolism in Man*. Applied Science Publishers, London, pp. 391–408.

Rittler, P., Demmelmair, H., Koletzko, B., Schildberg, F.W. and Hartl, W.H. (2000) Determination of protein synthesis in human ileum *in situ* by continuous [1-(13)C] leucine infusion. *American Journal of Physiology* 278, E634.

Rocha, H.J.G., Nash, J.E., Connell, A. and Lobley, G.E. (1993) Protein synthesis in ovine muscle and skin: sequential measurements with three different amino acids based on the large-dose procedure. *Comparative Biochemistry and Physiology* 105B, 301–307.

Rodriguez, T., Busquets, S., Alvarez, B., Carb, N., Agell, N., Lopez-Soriano and Argils, M. (1998) Protein turnover in skeletal muscle of the diabetic rat: activation of ubiquitin-dependent proteolysis. *International Journal of Molecular Medicine* 1, 971–977.

Rome, S., Margnier, E. and Vidal, H. (2004) The ubiquitin-proteasome pathway is a new partner for the control of insulin signalling. *Current Opinion in Nutrition and Metabolic Care* 7, 249–254.

Rooyackers, O.E. and Nair, K.S. (1997) Hormonal regulation of human muscle protein metabolism. *Annual Reviews of Nutrition* 17, 457–485.

Sampson, D.A., Hunsaker, H.A. and Jansen, G.R. (1986) Dietary protein quality, protein quantity and food intake: effects on lactation and protein synthesis and tissue composition in mammary tissue and liver in rats. *Journal of Nutrition* 116, 365–375.

Samuels, S.E., Taillandier, D., Auroussean, E., Cherel, Y., Le Maho, Y., Arnal, M. and Attaix, D. (1996) Gastrointestinal tract protein synthesis and mRNA levels for proteolytic systems in adult fasted rats. *American Journal of Physiology* 271, E232–238.

Schaeffer, A.L., Davis, S.R. and Hughson, G.A. (1986) Estimation of protein synthesis in sheep during sustained elevation of plasma leucine concentration by intravenous infusion. *British Journal of Nutrition* 56, 281–288.

Schaeffer, A.L. and Krishnamurti, C.R. (1984) Whole body and tissue fractional protein synthesis in the ovine fetus *in utero*. *British Journal of Nutrition* 52, 359–369.

Schimke, R.T. (1964) The importance of both synthesis and degradation in the control of arginase in rat liver. *Journal of Biological Chemistry* 239, 3808–3817.

Schöch, G. and Topp, H. (1994) Interrelations between the degradation rates of RNA and protein and the energy turnover rates. In: Raiha, N.C.R. (ed.) *Protein Metabolism During Infancy*. Nestlé Nutrition Workshop Series vol. 33, Nestlé, Vevey/Raven Press, New York, pp. 49–52.

Scornik, O.A. (1984) Role of protein degradation in the regulation of cellular protein content and amino acid pools. *Federation Proceedings* 43, 1283–1288.

Sève, B. and Ponter, A.A. (1997) Nutrient-hormone signals regulating muscle protein turnover in pigs. *Proceedings of the Nutrition Society* 56, 565–580.

Sève, B., Reeds, P.J., Fuller, M.F., Cadenhead, A. and May, S.M. (1986) Protein synthesis and retention in some tissues of the young pig as influenced by dietary protein intake after weaning. Possible connection to the energy metabolism. *Reproduction, Nutrition, Development* 26, 849–861.

Sève, B., Ballèvre, O., Ganier, P., Nublet, J., Pruguand, J. and Obled, C. (1993) Recombinant porcine somatotrophin and dietary protein enhance protein synthesis in growing pigs. *Journal of Nutrition* 123, 529–540.

Shah, O.J., Kimball, S.A. and Jefferson, L.S. (2000) Acute attenuation of translational initiation and protein synthesis by glucocorticoids in skeletal muscle. *American Journal of Physiology* 278, E76–82.

Simon, O., Münchmeyer, R., Bergner, H., Zebrowska,

T. and Buraczewska, L. (1978) Estimation of protein synthesis by constant infusion of labelled amino acids in pigs. *British Journal of Nutrition* 40, 243–252.

Simon, O., Bergner, H., Münchmeyer, R. and Zebrowska, T. (1982) Studies on the range of tissue protein synthesis in pigs: the effect of thyroid hormones. *British Journal of Nutrition* 48, 571–582.

Smith, K., Barna, J.M., Watt, P.W., Scrimgeour, C.M. and Rennie, M.J. (1992) Flooding with L-[1-^{13}C] leucine stimulates human muscle protein incorporation of continuously infused L-[1-^{13}C] valine. *American Journal of Physiology* 262, E372–378.

Snyder, W.S. (1975) Report of the Task Group on Reference Man. *International Commission on Radiological Protection*, no. 23. Pergamon Press, Oxford, UK.

Soltész, G., Joyce, J. and Young, M. (1973) Protein synthesis rate in the newborn lamb. *Biology of the Neonate* 23, 139–148.

Southorn, B.J., Palmer, R.M. and Garlick, P.J. (1990) Acute effects of corticosterone on tissue protein synthesis and insulin-sensitivity in rats *in vivo*. *Biochemical Journal* 272, 287–291.

Southorn, B.G., Kelly, J.M. and McBride, B.W. (1992) Phenylalanine flooding dose procedure is effective in measuring intestinal and liver protein synthesis in sheep. *Journal of Nutrition* 122, 2398–2407.

Stoll, B., Burrin, D.G., Henry, J., Jahoor, F. and Reeds, P.J. (1997) Phenylalanine utilization by the gut and liver measured with intravenous and intragastric tracers in pigs. *American Journal of Physiology* 273, G1208–1217.

Stoll, B., Chang, X., Fan, M.Z., Reeds, P.J. and Burrin, D.G. (2000) Enteral nutrient intake level determines intestinal protein synthesis and accretion rates in neonatal pigs. *American Journal of Physiology* 279, G288–294.

Strauss, D.S. and Takemoto, C.D. (1990) Effect of dietary protein deprivation on insulin-like growth factor (IGF)-I and -II, IGF binding protein-2 and serum albumin gene expression in the rat. *Endocrinology* 127, 1849–1860.

Stump, C.S., Short, K.R., Bigelow, M.L., Schimke, J.M. and Nair, K.S. (2003) Effect of insulin on human skeletal muscle mitochondrial ATP production, protein synthesis and mRNA transcripts. *Proceedings of the National Academy of Sciences, USA* 100, 7996–8001.

Sugden, P.H. and Fuller, S.J. (1991) Regulation of protein turnover in skeletal and cardiac muscle. *Biochemical Journal* 273, 21–37.

Tauveron, I., Larbaud, D., Champedron, C., Debras, E., Tesseraud, S., Bayle, G., Bonnet, Y., Thiéblot, P. and Grizard, J. (1994) Effect of hyperinsulinaemia and hyperaminoacidaemia on muscle and liver protein synthesis in lactating goats. *American Journal of Physiology* 267, E877–E885.

Tawa, N.E. and Goldberg, A.L. (1992) Suppression of muscle protein turnover and amino acid degradation by protein deficiency. *American Journal of Physiology* 263, E317–325.

Tessari, P., Garibotto, G., Inchiostro, S., Robando, C., Saffioti, S., Vettore, M., Zanetti, M., Russo, R. and Deferarri, G. (1996) Kidney, splanchnic and leg protein turnover in humans: insight from leucine and phenylalanine kinetics. *Journal of Clinical Investigation* 98, 1481–1492.

Thissen, J.P., Underwood, L.E., Maiter, D., Maes, M., Clemmons, D.R. and Keterslegers, J.-M. (1991) Failure of IGF-I infusion to promote growth in protein-restricted rats despite normalization of serum IGF-I concentrations. *Endocrinology* 128, 885–890.

Thissen, J.P., Ketelslegers, J.-M. and Underwood, L.E. (1994) Nutritional regulation of the insulin-like growth factors. *Endocrine Reviews* 15, 80–101.

Tomas, F.M., Munro, H.N. and Young, V.R. (1979) Effect of glucocorticoid administration on the rate of muscle protein breakdown *in vivo* in rats, as measured by urinary excretion of N^{τ}-methylhistidine. *Biochemical Journal* 178, 139–146.

Van der Schoor, S.R., Reeds, P.J., Stoll, B., Henry, J.F., Rosenberger, J.R., Burrin, D.G. and van Goudoever, J.B. (2002) The high metabolic cost of a functional gut. *Gastroenterology* 123, 1931–1940.

Volpi, E., Jeschke, M.G., Hemdon, D.N. and Wolfe, R.R. (2000) Measurement of skin protein breakdown in a rat model. *American Journal of Physiology* 279, E900–906.

Waterlow, J.C. (1948) Fatty liver disease in infants in the British West Indies. *Medical Research Council Special Reports Series* No. 263. HMSO, London.

Waterlow, J.C. and Stephen, J.M.L. (1966) Adaptation of the rat to a low protein diet: the effect of a reduced protein intake on the pattern of incorporation of L-^{14}C-lysine. *British Journal of Nutrition* 20, 461–484.

Waterlow, J.C. and Stephen, J.M.L. (1968) The effects of low protein diets on the turnover rates of serum liver and muscle proteins in the rat, measured by continuous infusion of L-[^{14}C] lysine. *Clinical Science* 35, 287–305.

Welle, S., Thorntone, C., Staff, M. and McHenry, B. (1996) Growth hormone increases muscle mass and strength but does not rejuvenate myofibrillar protein synthesis in healthy subjects over 50 years old. *Journal of Clinical Endocrinology and Metabolism* 81, 3239–3243.

Welle, S., Bhatt, K. and Thornton, C.A. (1999) Stimulation of myofibrillar synthesis is mediated by more efficient translation of mRNA. *Journal of Applied Physiology* 86, 1220–1225.

Wernerman, J., Botta, D., Hammerqvist, F., Thunell, S., von der Decken, A. and Vinnars, E. (1989) Stress hormones given to healthy volunteers alter the concentration and configuration of ribosomes in skeletal muscle, reflecting changes in protein synthesis. *Clinical Science* 77, 611–616.

Whipple, G.H., Miller, L.L. and Robschet-Robbins, F.S. (1947) Raiding of body tissue protein to form plasma protein and hemoglobin. *Journal of Experimental Medicine* 85, 277–286.

Wing, S. and Goldberg, A.L. (1993) Glucocorticoids activate the ATP-ubiquitin-dependent proteolytic system in skeletal muscle during fasting. *American Journal of Physiology* 264, E668–676.

Wray-Cohen, D., Nguyen, H.V., Burrin, D.G., Beckett, P., Fiorotto, M.L., Reeds, P.J., Wester, T.J. and Davis, T.A. (1998) Response of skeletal muscle protein synthesis to insulin in suckling pigs decreases with development. *American Journal of Physiology* 275, E602–E609.

Yahya, Z.A.H., Tirapegui, J.O., Bates, P.C. and Millward, D.J. (1994) Influence of dietary protein, energy and corticosteroids on protein turnover, proteoglycan sulphation on growth of long bone and skeletal muscle in the rat. *Clinical Science* 87, 607–618.

Yarasheki, K.E., Campbell, J.A., Smith, K., Rennie, M.J., Hólloszy, J.O. and Bier, D.M. (1992) Effect of growth hormone and resistance exercise on muscle growth in young men. *American Journal of Physiology* 262, E261–267.

Yarasheki, K.E., Zachwieja, J.J., Campbell, J.A. and Bier, D.M. (1995) Effect of growth hormone and resistance exercise on muscle growth and strength in older men. *American Journal of Physiology* 268, E268–276.

Zak, R.E., Martin, A., Prior, G. and Rabinowitz, M. (1977) Comparison of turnover of several myofibrillar proteins and critical evaluation of double isotope method. *Journal of Biological Chemistry* 252, 3430–3435.

16

Plasma Proteins

In the 1950s and '60s there was intense research on plasma protein metabolism. Scientists were stimulated by the theoretical and practical challenges presented by isotopes, and particularly by their application to man. Plasma proteins were a way in, because they could be sampled so easily. Many books were produced (e.g. Birke *et al.*, 1969; Rothschild and Waldmann, 1970; Wolstenholme and O'Connor, 1973; Allison, 1976) and many comprehensive reviews, e.g. Freeman (1968) and Rothschild *et al.* (1972).

The plasma proteins can be divided into three main groups. Table 16.1 lists some of those whose kinetics have been studied. There is, of course, a host of other circulating proteins and polypeptides, produced in various organs and tissues, about whose kinetics little or nothing is known: insulin, insulin-like growth factors and their binding proteins, growth hormone, parathyroid hormone, osteocalcin, melanin, caeruloplasmin, metallothionine, interleukins and many others.

Table 16.1. Some different plasma proteins.

1. Those synthesized and secreted by the liver:	
A.	Nutrient-transport proteins:
	albumin
	transferrin
	retinol-binding protein
	transthyretin
	apolipoproteins
B.	Acute-phase proteins:
	fibrinogen
	α_1-antitrypsin
	haptoglobin
	interleukin 6
2. Those produced by cells of the immune system:	
	IgG, IgA, IgM, IgE

16.1 Albumin

Albumin is the most important of the plasma proteins, being present in the largest amount – about 3 g kg^{-1} body weight, if intravascular (IV) and extravascular (EV) pools are taken together – or ~1.5% of total body protein. The IV pool contains about 40% of the albumin; according to the compartmental analysis of Matthews (1957) the EV pool may be divided into two parts, one turning over faster than the other (Fig. 16.1). The slower pool may perhaps occupy the hydrated space in the extracellular matrix of connective tissues. It is strange how little was and still is known about the functions of albumin in the blood, although it has long been recognized that through its oncotic pressure it controls the exchange of fluid between IV and EV spaces. An interesting correspondence in the *Lancet* (Phillips *et al.*, 1989; Soeters *et al.*, 1990 *et seq*) attempted to establish a correlation between hypoalbuminaemia and mortality from various diseases such as cancer and cardiac infarction. This was attributed by some of the authors to the capacity of albumin to act as a carrier for lipids, tryptophan and possibly other nutrients and antioxidants; yet people with the rare condition of congenital lack of albumin (analbuminaemia) apparently have virtually no handicaps (Bennhold and Kallee, 1959; Waldmann *et al.*, 1964; Freeman, 1968; Ott, 1968).

The stages of albumin synthesis have been worked out in some detail. Synthesized by ribosomes attached to the rough endoplasmic reticulum of the liver cell, it is produced initially as a pre-pro-albumin; this passes to the smooth reticulum, and thence to the Golgi apparatus, from

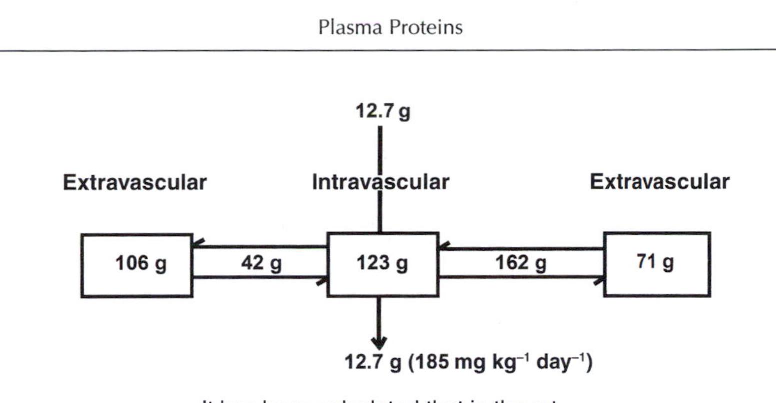

Fig. 16.1. Three-pool model of albumin distribution and exchange. Reproduced from Freeman, 1965, by courtesy of Pergamon Press, Oxford.

which it is secreted into the plasma (Geller *et al.*, 1972). During this passage polypeptide chains are detached by post-translational modification to produce the final molecule. Studies by the flooding dose method have shown that in man this process takes about 30 minutes and in one study it was found to be accelerated by infusion of nutrients (Ballmer *et al.*, 1995a).

There has been much controversy about the site of albumin breakdown, and the problem is still unresolved. Studies with perfused livers showed that only a small proportion is degraded in the liver (Cohen and Gordon, 1958). Other sites that have been proposed are the kidneys and the gastrointestinal tract. Wetterfors (1965) observed that over a range of five mammals from mouse to man the rate of albumin breakdown could be described by the allometric equation:

$$\text{breakdown} = 5.83 \times \text{wt}^{0.68}$$

Since this exponent is approximately that of body surface area, and since the absorptive area of the gut varies with body weight in the same way as surface area, they took this as an indication that the gut might be the site of breakdown. However, there has been no experimental support for this concept, and it seems likely that albumin is degraded throughout the body, perhaps in the endothelial cells of the blood vessels. When a large dose of albumin was given intravenously the transcapillary escape rate and the albumin breakdown rate were both increased in parallel (Rossing *et al.*, 1972).

16.1.1 Methods of measuring albumin turnover

The kinetics of albumin turnover have been studied *in vivo* in humans since the early 1950s. Initially the measurements were only of breakdown. The method was to inject a bolus of human albumin which had been labelled with ^{131}I by iodination *in vitro* of the tyrosine residues. Much attention had to be given to avoiding denaturation of the albumin during the process of iodination (McFarlane, 1964) and the International Atomic Energy Authority organized a multi-centre study to compare methods. The rate of albumin breakdown could be determined either from the decay curve of ^{131}I-albumin in plasma or from the urinary excretion of ^{131}I over periods of 10–28 days. A major complication is that the disappearance of label from the plasma results not only from its breakdown but also from exchange with the extravascular albumin pool, which is about 1.2–1.5 × the IV pool. As a consequence the decay curve had to be analysed by a multi-compartmental model. Nosslin (1973) described no less than 13 methods of calculating iodine turnover from ^{131}I data! A brilliant paper by Beeken *et al.* (1962) is a good example of how methods were compared. A computer analysis by Plantin (1966) compared four different models and found very few differences greater than 5% in the derived parameters. The calculation of fractional breakdown rate (FBR) from the urinary excretion of ^{131}I was complicated by much day-to-day variation in iodide excretion. This difficulty was overcome by whole-body counting of ^{131}I (Mouridsen, 1967; James and Hay, 1968); subtracting the intravascular ^{131}I-albumin from the total retained in the body gave the amount in the EV pool. From the amounts in the pools at different times it was possible to calculate separately rates of degradation, synthesis and net transfer between the pools. Shorter methods, depending on the urinary excretion of labelled iodide, were also developed, enabling albumin breakdown to be measured in about 1 day (Bianchi *et al.*, 1972b; Hoffenberg *et al.*, 1972).

A method for measuring synthesis of plasma proteins in the liver was developed independently by McFarlane (1963) and by Reeve *et al.* (1963). The principle, originally proposed by Swick (1958), was to give ^{14}C-carbonate, which labels the 6C atom of arginine and also the C atom of urea (see Chapter 5). A detailed account of the method is given by Jones *et al.* (1968) and by Tavill *et al.* (1968). The synthesis rate of albumin can then be derived from the precursor-product relationship:

$$\frac{\text{total incorporation of tracer in albumin}}{\text{mass of newly formed albumin}} = \frac{\text{total incorporation of tracer in urea}}{\text{mass of newly formed urea}}$$

Several authors used both methods in humans, and found agreement between the synthesis rate measured by ^{14}C-carbonate and the breakdown rate by ^{131}I (Tavill *et al.*, 1968; Wochner *et al.*, 1968; Bianchi *et al.*, 1972b). The new method had the great advantage that it required only 4 h instead of days or weeks, but it is not without problems, particularly because it cannot be assumed that urea is labelled instantaneously.

A variant of this method was to use [^{15}N] glycine as tracer, and measure ^{15}N in urea as a proxy for the precursor and the guanidino-N of arginine in albumin as product (Gersovitz *et al.*, 1980; Olufemi *et al.*, 1991).

Measurements of plasma protein turnover were greatly simplified by the introduction of the constant infusion and flooding dose methods. In normal adults, if a steady state can be assumed, there is satisfactory agreement with the results for breakdown obtained with ^{131}I. Because the timescales for measuring synthesis and breakdown are so different there are very few studies in which both have been measured in the same subjects at the same time (see Tavill *et al.*, 1968). A flooding dose of phenylalanine (Chapter 14) has evidently become the gold standard for this measurement, and the results from different laboratories are consistent with those by other methods.

Table 16.2 shows the results obtained by the different methods. With each method there is a nearly twofold range between the means of different studies, but the overall agreement is remarkable.

16.1.2 The regulation of albumin metabolism

It was suggested by Hoffenberg (1970) and Tavill *et al.* (1973) that a change in the size of the albumin pool, resulting from a change in synthesis rate, will be compensated if the fractional breakdown rate (FBR) is constant, since less will be removed if the pool is small and more if it is large. However, experiments in which the intravascular albumin pool (IVM) was decreased by plasmapheresis showed that this was not so; both fractional and absolute breakdown rates fell (Matthews *et al.*, 1961; Hoffenberg *et al.*, 1966). When in humans the IVM was increased by infusions of plasma the FBR increased (Andersen and Rossing, 1967; Tavill *et al.*, 1968). Clearly, the FBR is not constant. Rothschild proposed a mechanism by which albumin synthesis is controlled by the intravascular colloid osmotic pressure, an increase in pressure causing a decrease in synthesis; he showed that when this pressure was raised by infusions of albumin or dextran, the synthesis rate of albumin fell. A full account of this work is given in Oratz (1970) and in Rothschild's reviews of 1970 and 1972; they have given evidence of stimulation of albumin synthesis by thyroid hormones and cortisone. Other authors, however, found a positive correlation between colloid osmotic pressure and the rate of albumin synthesis (Rossing and Andersen, 1967), which is contrary to the theory.

Studies in rats have shown that infusions of albumin or of dextran decreased the transcription rates of the albumin gene and downregulated the level of albumin mRNA (Pietrangelo *et al.*, 1992). Similarly, but in the opposite direction, in rats in which a toxin, adriamycin, produced a model of the nephrotic syndrome, the levels of

Table 16.2. Albumin turnover in normal adults measured by different methods.

	n	N	FBR or FSR % d^{-1}	Range of FSR	References
A. ^{131}I-albumin	147	10	9.5	7.4–12.2	1–10
B. Glycine/carbonate-urea-arginine	28	6	8.1	4–12	10–15
C. ^{13}C leucine infusion VLDL-apo-B 100 as precursor		2	11.8	10.0–13.7	16, 17
D. Constant infusion Leucine/KIC	74	10	7.3	4.8–12.0	18–27
E. Flooding dose leucine or phenylalanine	88	13	7.9	5.8–10.4	28–37

n, number of subjects; N, number of studies; FBR and FSR, weighted means of each study. References: A: 1, Beeken *et al.*, 1962; 2, Cohen *et al.*, 1961; 3, Dykes, 1968; 4, Hoffenberg, 1966; 5, Jarnum and Lassen, 1961; 6, McFarlane, 1963; 7, Mouridsen, 1967; 8, Rossing and Andersen, 1967; 9, Wilkinson and Mendenhall, 1963; 10, Wochner *et al.*, 1968. B: 11, Gersovitz *et al.*, 1980; 12, McFarlane, 1963; 13, Olufemi *et al.*, 1991; 14, Tavill *et al.*, 1968; 15, Wochner *et al.*, 1968. C: 16, Cayol *et al.*, 1967, 17, Jackson *et al.*, 2001. D: 18, Ballmer *et al.*, 1995a; 19, Cayol *et al.*, 1995; 20, De Feo *et al.*, 1992; 21, de Sain-Van der Velden *et al.*, 1998b; 22, Fu and Nair, 1998; 23, Mansoor *et al.*, 1997; 24, Olufemi *et al.*, 1991; 25, Smith *et al.*, 1994; 26, Volpi *et al.*, 1966; 27, Zanetti *et al.*, 2001. E: 28, Ballmer *et al.*, 1990, 1992, 1993, 1995b; 29, Barber *et al.*, 2000; 30, Barle *et al.*, 1997; 31, Brutomesso *et al.*, 2001; 32, Caso *et al.*, 2000; 33, Fearon *et al.*, 1998; 34, Hunter *et al.*, 1995; 35, McMillan *et al.*, 1996; 36, McNurlan *et al.*, 1998; 37, Smith *et al.*, 1994.

mRNAs encoding albumin, transferrin and apo A1 lipoprotein were increased in a coordinated fashion (Kang *et al.*, 1999). These results fit with the demonstration by Häussinger *et al.* (1991) and Stoll *et al.* (1992) that hyperosmotic cell shrinkage stimulates protein breakdown and inhibits synthesis, while hypoosmotic swelling produces the opposite effects.

The findings in pathological states, in which the intravascular albumin mass (IVM) is reduced (Table 16.3), tend to support the theory of control by oncotic pressure. When breakdown is measured, as in Jensen *et al.*'s study (1967) of 30 patients with nephrosis, the *fractional* rates of urinary loss, breakdown and synthesis were all higher the smaller the IVM. For breakdown (Fig. 16.2A) this makes no physiological sense. It may be more illuminating to consider absolute rates. For breakdown the absolute rate (ABR) decreased a little, from 8.5 g day^{-1} in the highest quintile of IVM (mean 74 g) to 5.4 g day^{-1} in the lowest quintile (mean 22 g). This could be regarded as an adaptive response. The absolute synthesis rates (ASR) showed no consistent increase (Fig. 16.2B).

Jensen's is the only study in nephrotics in which breakdown was measured directly; the absolute breakdown rate was reduced. Synthesis, calculated on the assumption of a steady state, as the sum of loss by breakdown + loss in the urine, was slightly increased. In all the other studies on nephrosis and on those in protein-losing enteropathy and cancer cachexia, the absolute synthesis rate was increased, sometimes by a large amount (Table 16.3). In cirrhosis of the liver in two studies, breakdown, measured with [^{131}I] albumin, was decreased; in the one direct measurement synthesis was also reduced, presumably as a result of liver damage. Ballmer *et al.* (1993) showed a relation between reduction in the rate of albumin synthesis and the clinical assessment of the severity of liver damage. Thus the response to a low IVM could be regarded as 'purposeful' – to restore the IVM by a fall in breakdown and a rise in synthesis, and both could result from swelling of the liver cells from low oncotic pressure. Whether this is really the mechanism remains to be seen.

16.1.3 Effect of food on albumin synthesis

In 1954 Steinbock and Tarver injected into rats plasma from donor rats in which the proteins were

Table 16.3. Changes in intravascular albumin mass (IVM) and in absolute rates of synthesis (ASR) or breakdown (ABR) in various pathological states. Ratio patients:controls.

	Patients:controls		
	IVM	ABR	ASR
Nephrosis			
Jensen *et al*, 1967, [131] albumin	0.42	0.57	1.13
Bianchi, 1972a[a], [131] albumin	1.03	0.86	
Ballmer *et al.*, 1992, flood	0.58	–	1.47
de Sain-van der Velden *et al.*, 1998a, flood	0.63	–	2.25
de Sain-van der Velden *et al.*, 1998b, flood	0.53	–	2.07
Demant *et al.*, 1998, flood	0.60	–	1.36
Zanetti *et al.*, 2001, constant infusion	0.87	–	2.55
Giordano *et al.*, 2001b, constant infusion	0.78		1.70
Protein-losing enteropathy			
Schomerus and Mayer, 1975, [14C]-carbonate	0.76	–	3.12
Wochner *et al.*, 1968, [14C]-carbonate	0.48	–	1.24
Kerr, 1967, [131I] albumin	0.60	0.50	–
Cancer cachexia			
Fearon *et al.*, 1998, flood	0.78	–	1.28
Cirrhosis			
Wilkinson and Mendenhall, 1963, [131I]	0.67	0.72	–
Dykes, 1968, [131I]	0.70	0.32	–
Ballmer *et al.*, 1993, flood	0.89	–	0.81

The control values for this table are those given by each author.
[a]In this study although the IVM was normal the EVM was greatly reduced.

biologically labelled with [^{35}S] methionine prepared from yeast. The experimental rats were fed for 3 weeks on diets containing different amounts of protein. The half-lives on these diets were: 0% protein: 17 days; 25%: 5.1 days; 65%: 2.9 days.

Perfusion of the liver in rabbits showed that albumin synthesis was increased twofold when the donor rabbits were fed compared with fasted (Rothschild *et al.*, 1968). Albumin synthesis was almost doubled in perfused livers when the medium was supplemented with amino acids, insulin, growth hormone and cortisol (John and Miller, 1969). In rats starved for 2 days the absolute synthesis rates of both total liver protein and albumin fell in parallel with the decrease in amount of liver protein, so that there was no change in the fractional rates, nor in the proportion of total liver synthesis contributed by albumin – ~15% (Pain *et al.*, 1978). A different picture was produced by protein deficiency. Jeejeebhoy *et al.* found that in rats on a protein-free diet the FSR of albumin fell from 48 to 7% day^{-1}. In Pain's study (1978), albumin synthesis, as a fraction of total protein synthesis, was reduced by almost 50% after 9 days on a protein-free diet. There was a parallel fall in the concentration of translatable albumin mRNA. A study on rats by Kirsch *et al.* (1968) is of particular interest because synthesis was measured by the ^{14}C-carbonate method and breakdown with ^{131}I-albumin. The time-course of the changes was observed over 12 days on a protein-free diet, followed by another 12 days on a normal diet. During the first period synthesis fell steadily, but breakdown remained unchanged for 6 days and then fell abruptly, so that the final absolute rate was only 30% of normal. On refeeding synthesis increased rapidly and reached about twice the normal rate in 3–4 days, whereas breakdown rose to normal more slowly. Hoffenberg *et al.* (1966) made a similar study on human volunteers, who were switched from a high to a low protein diet. The low diet led to a small fall (13%) in IV albumin mass and serum concentration but a reduction of ~30% in rates of both breakdown and synthesis + transfer. The important points are that synthesis responds more rapidly than breakdown, and that these dynamic changes are more sensi-

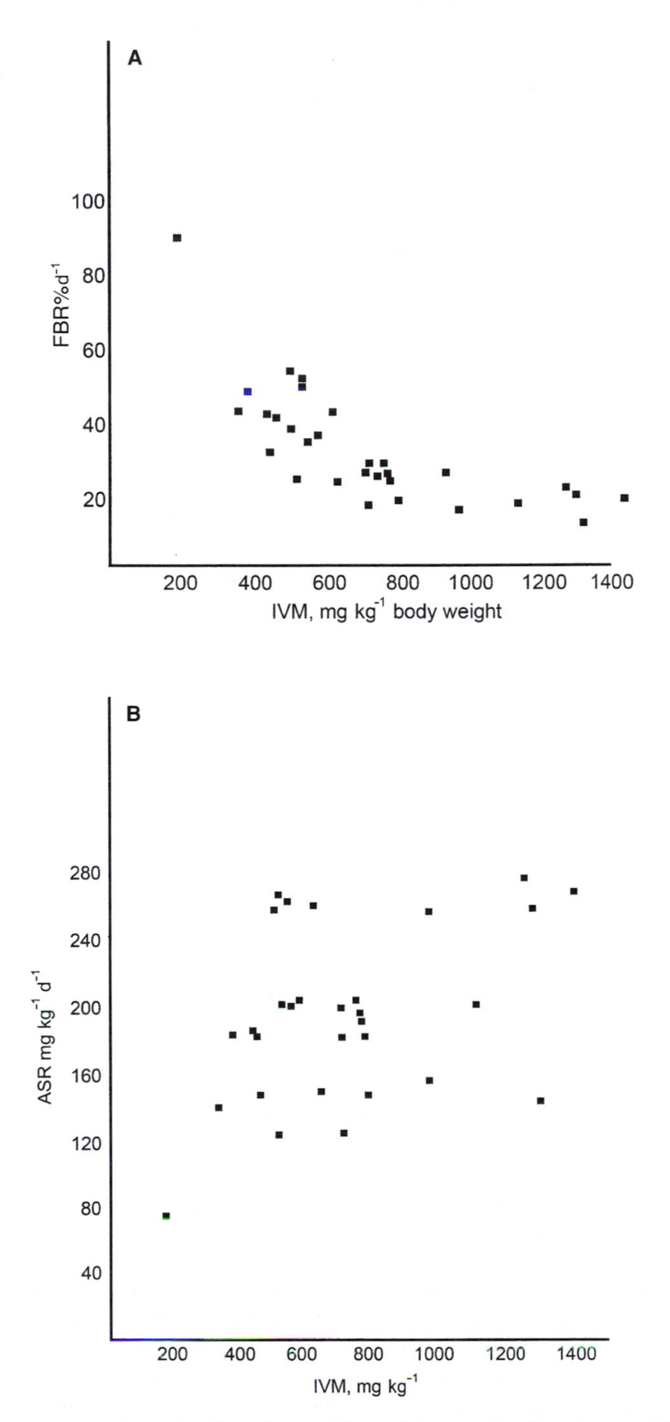

Fig. 16.2. Albumin turnover in nephrotic patients with low intravascular albumin mass (IVM). A, Fractional breakdown rate (FBR, % d^{-1}) of albumin *vs.* IVM; B, Absolute synthesis rate (ASR, mg kg^{-1} d^{-1}) *vs.* IVM. Drawn from data of Jensen *et al.*, 1967.

tive to protein deprivation than static changes in albumin concentration. These experimental observations are very relevant to the findings in malnourished children described below.

The literature on the effect of adding protein or a meal of amino acids on albumin synthesis is contradictory, ranging from no effect (Ballmer *et al.*, 1995b), a small effect (Hunter *et al.*, 1995; Barber *et al.*, 2000; de Meyer *et al.*, 2000) to a large increase (De Feo *et al.*, 1992; Cayol *et al.*, 1997; Brutomesso *et al.*, 2001). It is interesting that as long ago as 1978, Pain *et al.* showed that in rats on a protein-free diet there was a 50% reduction in the synthesis of albumin mRNA – an early example of the application of molecular biology to nutritional problems. These were all short-term studies; in the longer term the level and quality of dietary protein have also been found to influence the FSR of albumin. On a vegetable protein diet the FSR was 13% lower than on one containing animal protein (Caso *et al.*, 2000; Garlick and Contaldo, 2000). On a marginally adequate protein intake (0.6 g kg^{-1} day^{-1}) the FSR of albumin was only 60% of that on a generous intake (1.4 g kg^{-1} day^{-1}) (Jackson *et al.*, 2001).

As a footnote, it is interesting that the black bear during hibernation takes in no food or water and passes no urine, but recycles nitrogen with extraordinary efficiency. Albumin turnover increases fourfold; evidently all the amino acids needed for synthesis are derived from protein breakdown (Lundberg *et al.*, 1976) (Fig. 16.3).

16.1.4 Albumin turnover in severely malnourished children

Five studies have been reported on severely malnourished infants and young children, who were thought to be suffering from a diet deficient in both energy and protein: one from Mexico (Gitlin *et al.*, 1958); one from South Africa (Cohen *et al.*, 1962); and three from Jamaica (Picou and Waterlow, 1962; James and Hay, 1968; and Morlese *et al.*, 1996). The first four studies used

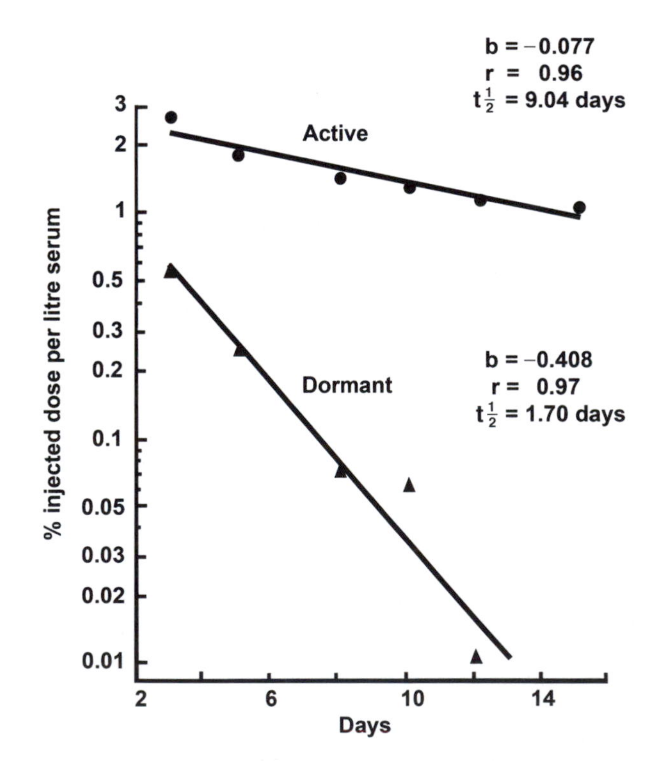

Fig. 16.3. Albumin turnover in the black bear before and during hibernation. Reproduced from Lundberg *et al.*, 1976, by courtesy of the *Mayo Clinic Proceedings.*

^{131}I-albumin; the fifth, some 30 years later, a constant infusion of ^{13}C-leucine. The two studies by Picou and Cohen had a similar protocol; children were studied about a week after admission to hospital, by which time they had lost oedema, and again a month or more later, when they were considered to be recovered. The only important differences between the two were that in South Africa about half the children had infections, whereas none were infected in Jamaica; and in South Africa the children were on a low protein diet for the first measurement and a high protein diet for the second, whereas in Jamaica they were on a high protein diet throughout. Table 16.4 shows remarkable agreement between the two studies. The outstanding finding is the great reduction in breakdown rate in the malnourished state – a compensatory change like that shown in Table 16.3. The study of James and Hay (1968) was rather complex. They used a small 4 pi whole body counter, which was more accurate than urine collection, to measure retained ^{131}I and hence by difference the size of the EV pool. They had two groups, malnourished and recovered, composed mainly of different children. Both groups were treated in the same way: albumin turnover was measured three times; after 10 days on a high protein intake, after a further week on low protein and a final week on either high or low protein. The 24-day study period allowed only partial recovery of the IV albumin mass in the malnourished group. The absolute synthesis and breakdown rates are shown in Fig. 16.4. Synthesis is much more sensitive than breakdown to changes in protein intake and the responses in the malnourished are greater than in the recovered children. Breakdown changes less and more slowly. During the low protein phase there was net transfer of albumin from the IV to the EV pool.

Morlese *et al.* (1996) used 8-h intragastric infusions of $^{2}H_3$-leucine, with the enrichment of apo-B 100 as precursor, to measure the synthesis rate of albumin in malnourished children. The measurements were made at 2 days after admission, at which time most of the children had an infection, at 8 days, when the infections had been treated, and at 60 days, when they had recovered. At the time of all three measurements the children received a low protein diet, although in between they had a good recovery diet. A comparison was made between children with and without oedema on admission. In both groups the albumin pool size, as judged by the plasma concentration, increased on treatment. Initially their absolute synthesis rates (ASRs) were the same; on recovery the ASR had increased by 40% in the children who had been originally oedematous, but was unchanged in those without oedema. It was concluded that restoration of the albumin pool in malnutrition is not due to an increase in albumin synthesis but possibly to a decrease in the rate of breakdown. Against this, however, are the findings of Picou and Cohen (Table 16.4) that breakdown rates were higher on recovery. An alternative explanation of Morlese's results might be that the low protein diet fed for 2 days before the final test reduced the rate of synthesis to a lower level than had prevailed during the period of recovery. The work of James and Hay (1968) referred to above suggests that this could happen, but does not tell us whether 2 days would be enough to produce the fall.

The general conclusion from these five studies is that in malnutrition in the long term the decrease in the albumin pool is compensated for by a fall in the rate of breakdown, which may be regarded as an adaptation; in the short term the level of protein intake has little effect on the breakdown rate but has an immediate effect on the synthesis rate.

Table 16.4. Albumin metabolism in children with protein-energy malnutrition.

		IVM, g	FBR, % d^{-1}	ABR, g d^{-1}
Cohen and Hansen, 1962:	malnourished	11.6	7.1	0.82
	recovered	18.1	15.6	2.82
Picou and Waterlow, 1962:	malnourished	10.7	8.9	0.95
	recovered	17.1	14.5	2.48

IVM, intravascular albumin mass; FBR, fractional breakdown rate; ABR, absolute breakdown rate.

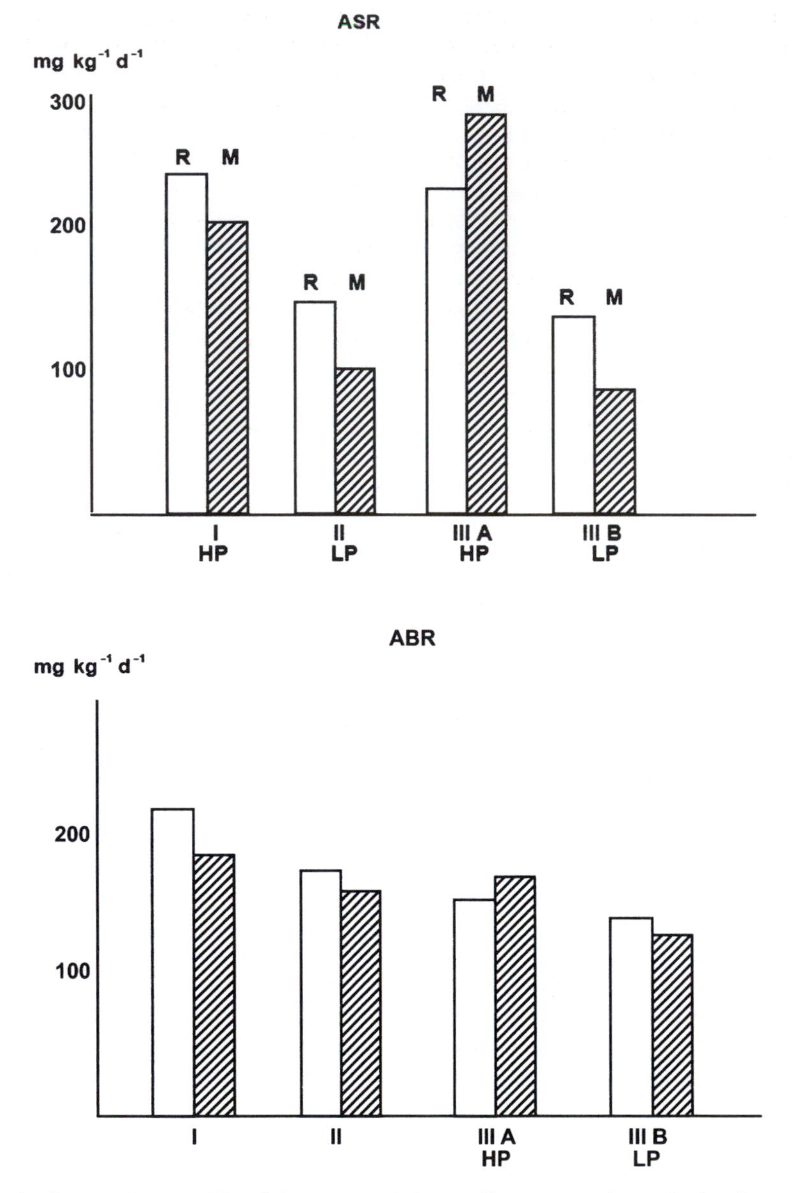

Fig. 16.4. Absolute synthesis and breakdown rates of plasma albumin in malnourished and recovered children fed different levels of protein for consecutive periods of 1 week.
I, 1st week – high protein; II, 2nd week – low protein; III A, 3rd week – high protein; III B, 3rd week – low protein.
▨, Malnourished □, recovered. Drawn from data of James and Hay, 1968.

16.2 Other Nutrient Transport Proteins

The other nutrient transport proteins (NTPs) synthesized by the liver include transferrin, retinol-binding protein (RBP), transthyretin (thyroxin-binding pre-albumin) and the apolipoproteins. These proteins have important physiological functions; Freeman (1968) gives an interesting discussion, which is still well worth reading even after more than 30 years. Many of them have been proposed as markers of the state of protein nutrition that might be more sensitive than albumin (Ingenbleek *et al.*, 1975; Reeds and Laditan, 1976; Shetty *et al.*, 1979), but none have been widely used for this purpose.

Some of the turnover rates of these proteins in adults and children are summarized in Table 16.5. The data give some support to the idea that the smaller the pool the faster it turns over, but there are many exceptions. The fractional turnover rate (FTR) of the IV transferrin pool is about twice that of albumin, and its plasma concentration is only $\frac{1}{10}-\frac{1}{20}$ that of albumin. The plasma concentration of VLDL-apo B 100 is 2–3 × that of RBP, but it turns over at more than twice the rate. More studies, with less variation between them, would be needed to establish a general relationship between pool size and turnover rate.

The turnover rates of the NTPs seem to be very sensitive to protein intake (e.g. Jeejeebhoy *et al.*, 1973). In subjects who changed from a high habitual protein intake to a lower one that still met the safe level (0.75 g kg^{-1} day^{-1}), within 10 days the absolute synthesis rates of transthyretin and retinol-binding protein fell to ~50% of their previous level, with no changes in plasma concentration (Afolabi *et al.*, 2004). These studies were made in the post-absorptive state after a 7-h infusion.

One may ask whether changes in the various NTPs are coordinated. Morlese *et al.* (1998a) found that in malnourished children, as the pool sizes increased on recovery, the ASRs of 3 NTPs – thyroxin-binding prealbumin, retinol-binding protein and HDL-apo A1 – all increased together, suggesting that they are coordinated. This is borne out by two experimental studies. In cultured hepatoma cells decreased oncotic pressure caused increases in the mRNA levels of both albumin and apo B100 (Yamanchi *et al.*, 1992). In rats made analbuminaemic by various methods there were coordinated rises in the mRNA levels of albumin, transferrin and apo A1 as well as of fibrinogen (Kang *et al.*, 1999). They claimed that analbuminaemia is the only known condition in which transcription of both NTPs and acute-phase proteins is increased simultaneously.

16.3 The Acute-phase Proteins

The acute-phase proteins (APP) are produced in the liver in response to infection and trauma. They include fibrinogen, C-reactive protein (CRP), α-1-acid glycoprotein (α1AG), haptoglobin and caeruloplasmin. Their structure and function are well summarized by Fleck *et al.* (1985) and Fleck (1989). Some estimates of their turnover rates in normal subjects are shown in Table 16.6.

The acute-phase response varies not only with the severity of the injury but with the time that has elapsed (Fig. 16.5). The response of fibrinogen is relatively small, whereas that of C-reactive protein, which is barely detectable in normal subjects, increases 100-fold or more. Davies *et al.* (1966) showed that in burned patients the synthesis rate of fibrinogen (ASR) after a lag of about 2 days increased 2–10 fold, and then declined again, so with acute injury the timing is critical.

In traumas or infections, except for fibrinogen, there is little information; in one study on symptom-free subjects with AIDS the ASRs of fibrinogen and α_1-antitrypsin were the same as that of controls, but that of haptoglobin was five times greater (Jahoor *et al.*, 1999). The responses of albumin and fibrinogen in patients in a few clinical conditions are shown in Table 16.7.

In many patients plasma albumin concentrations are low and it has been suggested that the APPs compete with albumin for the amino acids required for synthesis (e.g. Feldmann, 1986). The table shows that this is not always so: in the presence of an APP response, as shown by the increases in both concentration and ASR of fibrinogen, the ASR of albumin in most cases also increased to about twice the control level. An exception is a study of Moshage *et al.* (1987), in which the ASR of albumin was reduced in four patients with inflammatory disease. The low albumin concentrations may be due to the increased capillary leakage that is found with sepsis and trauma (Fleck *et al.*, 1985), albumin having a much lower molecular weight than fibrinogen (~80,000 *vs.* ~340,000). Kushner (1982) describes increased synthesis of a large number of liver enzymes during an acute phase response.

The effects of nutritional state on the APP response are also of interest. Dogherty *et al.* (1993) reported that the responses of C-reactive protein and serum amyloid A to triple vaccine were reduced in severely malnourished children. Two studies have compared the kinetics of APPs in children when they were admitted to hospital, severely malnourished with infections, and when

Table 16.5. Turnover rates of nutrient transport proteins in humans.

		Plasma concentration, g l^{-1}	Fractional turnover, % of IV pool day^{-1}	Absolute turnover, mg kg d^{-1}
Transferrin				
Jarnum and Lassen, 1961	normal	2.50	18.4	19
	infected	1.90	23.0	25
Katz, 1961	normal	2.30	9.2	25
Awai and Brown,1963	normal	1.90	7.9, 8.7	–
Wochner *et al.*, 1968	normal	0.70	9.6	6
Freeman, 1968	normal	–	12–19	–
Morlese *et al.*, 1997, children,				
	malnourished	1.2	23	15
	recovered	2.7	15	20
Transthyretin				
Oppenheimer *et al.*, 1965		0.29	36	10
Jackson *et al.*, 2001	high protein diet	0.25	50	
	maintenance protein	0.23	54	
Morlese *et al.*, 1998a, children,				
	malnourished	0.05	65	2.2
	recovered	0.14	60	3.7
Afolabi *et al.*, 2004, high protein		0.22	77	7.5
Retinol binding protein				
Morlese *et al.*, 1998, children,				
	malnourished	0.02	210	1.8
	recovered	0.03	200	3
Afolabi *et al.*, 2004, high protein		0.22	80	0.9
Lactoferrin				
Bennett and Kokocinski, 1979		0.0009	570	0.24[b]
VLDL-apo B-100				
Eisenberg and Levi, 1975	normal	0.075	40–60	–
Cryer *et al.*, 1986	normal		920	
Lichtenstein *et al.*, 1990	normal	0.018	315	11
de Sain-van der Velden *et al.*, 1998a	normal	0.095	636	13
Demant *et al.*, 1998[a]	normal	–	1380	17
Zanetti *et al.*, 2001	normal	0.075	1400	50
Jackson *et al.*, 2001 high protein		0.175	550	40
maintenance		0.14	550	32
Cummings *et al.*, 1995	normal	–	1470	9
	obese		710	20
HDL-apo A-1				
Jackson *et al.*, 2001, high protein		1.70	48	40
maintenance		1.55	41	33
Morlese *et al.*, 1998a, children,				
	malnourished	0.90	105	37
	recovered	1.25	80	45

All studies on normal adults unless otherwise stated. Studies up to and including 1975 used ^{131}I-labelled proteins. That of Cryer *et al.* (1986) used ^{15}N-glycine-hippurate. Later studies were done by constant infusion or flooding dose of labelled amino acids.

[a]The value given for this study represents catabolism of apo B 100 and does not include its transfer from VLDL-1 to VLDL-2. [b]Assumes body weight 70 kg.

Table 16.6. Some estimates of fractional turnover rates and plasma or serum concentrations of acute-phase proteins in normal adults.

	N	FBR or FSR % day^{-1}	Plasma concentration g l^{-1}	References
A. Fibrinogen				
1. With ^{131}I	3	27 (21–31)	2.4	1–3
2. With ^{14}C-carbonate	1	25	5.1	4
3. By constant infusion or flooding dose	11	22 (14–34)	2.4 (1.3–3.5)	5–15
B. α_1-anti-trypsin	2	53	1.0	10, 11
C. Haptoglobin	2	33–63	0.7	10–11
D. Caeruloplasmin	1	13	0.22	16
E. Fibronectin	2	35, 44	0.35	17, 18

N, number of studies. Values are the means and range of study means, not of individuals. References: 1, McFarlane *et al.*, 1964; 2, Davies *et al.*, 1966; 3, Andersen and Rossing, 1967; 4, Wochner *et al.*, 1968; 5, Balagopal *et al.*, 2001; 6, de Sain-van der Velden *et al.*, 1998b; 7, Fu and Nair, 1998; 8, Giordano *et al.*, 2001a; 9, Hunter *et al.*, 2001; 10, Jackson *et al.*, 2001; 11, Jahoor *et al.*, 1999; 12, McMillan *et al.*, 1996; 13, Mansoor *et al.*, 1997; 14, Preston *et al.*, 1998; 15, Zanetti *et al.*, 2001; 16, Sternlieb *et al.*, 1961; 17, Thompson *et al.*, 1989; 18, Carraro *et al.*, 1991.

they had recovered (Morlese *et al.*, 1998b; Reid *et al.*, 2002). The protocol of this work was the same as that described above for albumin. Plasma concentration and ASR of both α_1-antitrypsin and haptoglobin were initially about twice those found on recovery. There were no significant changes in fibrinogen. Thus the children, in spite of being very malnourished, were able to mount a modest and partial APP response. It is interesting that the response was attenuated in those children who were oedematous (Reid *et al.*, 2002). If the cause of nutritional oedema is a diet specifically deficient in protein, that could be limiting the APP response. However, the paper of Jackson *et al.* (2001) does not support this view. As discussed above, they measured the responses in normal adults to two levels of dietary protein, generous and maintenance. On the lower intake, although there were no differences in plasma concentration, the ASRs of fibrinogen and hapto-

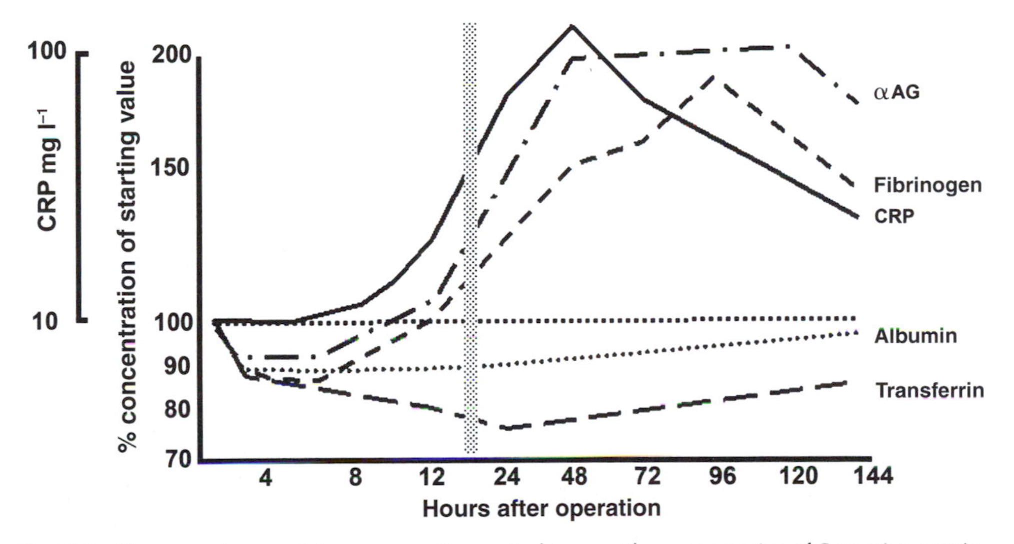

Fig. 16.5. The acute phase protein response after surgical trauma. The concentration of C-reactive protein (CRP) is shown as ml l^{-1}; for other proteins the percentage of the starting value is shown as a semilog plot. Reproduced from Fleck *et al.* (1989), by courtesy of the *British Medical Bulletin* (British Council).

Table 16.7. Ratios of plasma concentration and absolute synthesis rates of albumin and fibrinogen in patients compared with controls.

	Ratio patients to controls			
	Albumin		Fibrinogen	
	concentration	ASR	concentration	ASR
Acute				
Mansoor *et al.*, 1997, head injury	0.79	1.46	3.7	8.3
Chronic				
de Sain-van der Velden *et al.*, 1998b, nephrosis	0.59	2.25	2.1	1.48
Zanetti *et al.*, 2001 nephrosis	0.72	2.45	1.85	2.39
Giordano *et al.*, 2001 nephrosis (haemodialysis patients)	0.94	1.33[a]	1.35	1.68[a]
Preston *et al.*, 1998, cancer cachexia	0.74	–	1.33	3.23
Jahoor *et al.*, 1999, HIV	1.00	1.09	1.35	2.05
Cayol *et al.*, 1995, vaccination	1.0	1.0	1.24	1.54
Wochner *et al.*, 1968, gastrointestinal protein loss	0.5	1.24	0.87	1.68

[a]Based on assumed body weights of 70 kg.

globin were almost twice as high as on the generous intake. There was also a significant increase in the circulatory level of interleukin-6, one of the mediators of the APP response. It is suggested that the increased synthesis of fibrinogen and haptoglobin and the reduced synthesis of albumin and the other nutrient-transport proteins represent a low-grade inflammatory response with an increased tendency to atherogenesis – a result that recalls the correspondence in the *Lancet* referred to at the beginning of this chapter (Phillips *et al.*, 1989).

An interesting hypothesis was proposed by Reeds *et al.* (1994) to relate the APP response to the muscle wasting that occurs in so many clinical conditions. At its peak total synthesis of APPs after a stress, from the data of Fleck *et al.* (1985), could amount to as much as 0.85 g of protein per kg per day. The APPs contain larger amounts of the aromatic amino acids than muscle, and Reeds calculated that it would require the breakdown of 2 g of muscle protein to provide amino acids for the synthesis of 850 mg of APPs. The other amino acids that are produced in excess from muscle in order to meet the need for the aromatics will be oxidized and the nitrogen excreted in the urine. This would account for a large part of the 'catabolic' loss after trauma or in septic states. From experiments on rats (Chapter 15) starvation produces more severe breakdown of muscle protein, at least initially, than a protein-free diet. Therefore one would expect a specific deficiency of dietary protein to reduce the APP response, since less amino acids would be available from the breakdown of muscle protein. The evidence summarized above is conflicting – in favour of this hypothesis are some of the results in children, against it the findings in adults.

Again there is the question of whether the responses of the different APPs are coordinated. In the studies discussed above one protein out of three failed to respond, but the data are too scanty to allow any general conclusions. Better evidence comes from animal experiments. In a study on rats in which mRNAs were measured after an inflammatory stimulus (Milland *et al.*, 1990), all the APP mRNAs behaved in the same way, peaking at about 20–30 h and then decreasing rapidly. The only difference was in the height of the peak, which ranged from 2 × the control level for α_1-antitrypsin, 5 × for haptoglobin and 10 × for fibrinogen. These differences do not correspond with those mentioned earlier (page 259). The mRNAs of the nutrient transport proteins, albumin, RBP and transthyretin, behaved in the opposite way, falling after 20 h to levels of 10–20% of control. Only the mRNA of transferrin behaved differently, showing no change for 40 h, and then increasing slightly. These results suggest that within each group of proteins synthesized by the liver, the NTPs and the ATPs, the responses are coordinated, but in opposite directions, at least in an inflammatory state. There is a clear conflict between these findings and the data of Table 16.7.

16.4 The Immunoglobulins

There was great interest in the turnover rates of the various classes of immunoglobulins in the 1960s and '70s, but this seems to have disappeared with the rise of molecular immunology. A search of 23 volumes of *Annual Reviews of Immunology* produced no information, so we have to rely on the older work.

Table 16.8, from Waldmann *et al.* (1970), shows values for plasma concentrations and breakdown rates (FBR) in normal subjects. In pathological states the concentrations may increase 30–40-fold. According to Waldmann there are three patterns of relationship between plasma concentration and FBR: with IgM and IgA the FBR is independent of the concentration; for haptoglobin, transferrin and to a lesser degree IgD and IgE there is an inverse relationship between concentration and FBR; and for IgG the FBR increases with the concentration. Some of these relationships are summarized visually in Fig. 16.6, from Freeman (1965). This figure does not do justice to the behaviour of IgG, because, as Fig. 16.7 shows, the FBR rises very steeply in patients with hypogammaglobulinaemia (Waldmann and Jones,

Table 16.8. Turnover rates of immunoglobulins in healthy adults.

	FBR, % of IV pool day^{-1}		Serum concentration, g l^{-1}
	A	B	A
IgG	6	4.7	12
IgM	15	10	0.85
IgA	25	13	2.5
IgD	37	–	0.023
IgE	89	–	0.0005

A: from Waldmann *et al.*, 1970; B: from Birke, 1966. A paper by Birge-Jensen in *Protein Metabolism* (CIBA Foundation Symposium, New Series No. 9, pp. 249–266) gives information about the turnover rates of IgM and other immunoglobulins in a number of pathological conditions of the GI tract.

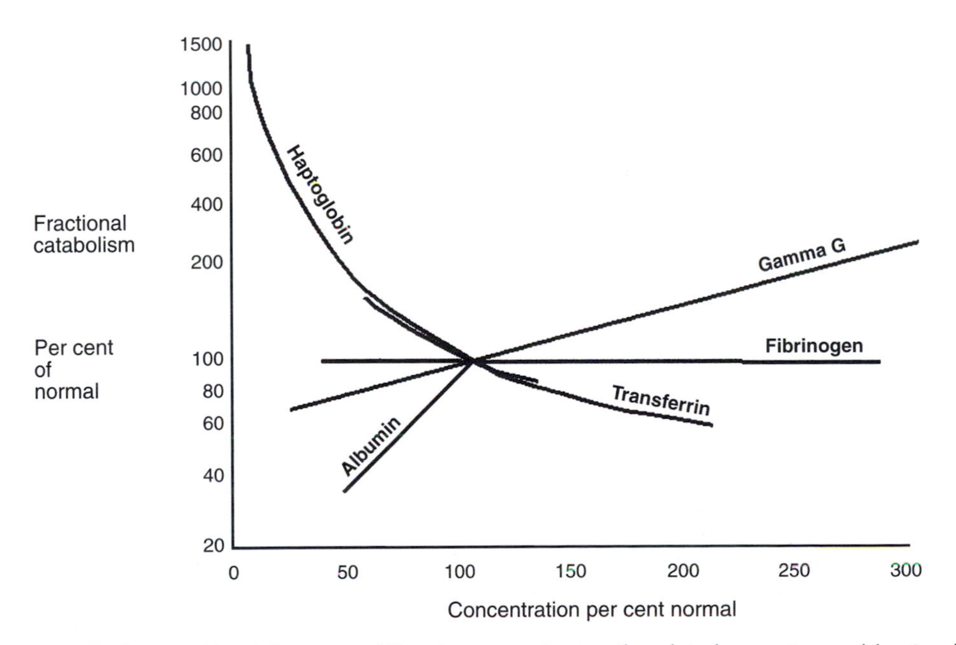

Fig. 16.6. The fractional breakdown rate of five plasma proteins: semilog plot of per cent normal fractional breakdown rate *vs.* per cent normal plasma concentration. Reproduced from Freeman, 1965, by courtesy of the *Series Haematologica*.

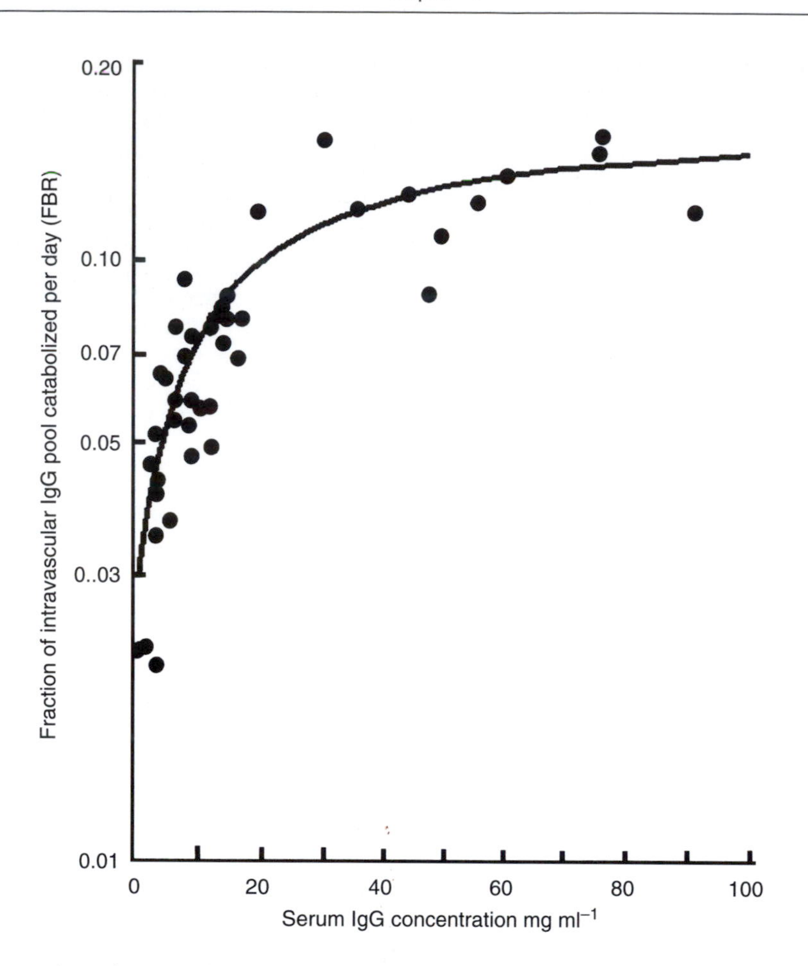

Fig. 16.7. Relation between the fractional breakdown rate of intravascular IgG and the serum IgG concentration in various pathological states. Reproduced from Waldman and Jones, 1973, by courtesy of CIBA Foundation, Novartis.

1973). They concluded that on the microvillous surfaces of enterocytes there were specific receptors which protected a fixed number of circulating IgG molecules from breakdown, so that when IgG concentrations were low, a higher proportion were protected and the FBR correspondingly decreased. This is a good example from the early work of the attempt to find a clue to the mechanism of breakdown from the FBR-concentration relationship.

16.5 References

Afolabi, P.R., Jahoor, F., Gibson, N.R. and Jackson, A.A. (2004) Response of hepatic proteins to the lowering of habitual dietary protein to the recommended safe level of intake. *American Journal of Physiology* 287, E327–330.

Allison, A.C. (1976) *The Plasma Proteins.* Plenum Press, New York.

Andersen, S.B. and Rossing, N. (1967) Metabolism of albumin and γ-globulin during albumin infusions and during plasmapheresis. *Scandinavian Journal of Laboratory and Clinical Investigation* 20, 183–184.

Awai, M. and Brown, E.G. (1963) Studies of the metabolism of I^{131}-labeled human transferrin. *Journal of Laboratory and Clinical Medicine* 61, 363–396.

Balagopal, P., Sweeten, S. and Mauras, N. (2002) Increased synthesis rate of fibrinogen as a basis for its elevated plasma levels in obese female adolescents. *American Journal of Physiology* 282, E899–904.

Ballmer, P.E., McNurlan, M.A., Milne, E., Heys, S.D., Buchan, V., Calder, A.G. and Garlick, P.J. (1990) Measurement of albumin synthesis in humans: a new approach including stable isotopes. *American Journal of Physiology* 259, E797–803.

Ballmer, P.E., Weber, B.K., Roy-Chaudhuri, P., McNurlan, M.A., Watson, H., Power, D.A. and Garlick, P.J. (1992) Elevation of albumin synthesis rates in nephrotic patients measured with [1-^{13}C] leucine. *Kidney International* 41, 132–138.

Ballmer, P.E., Walshe, D., McNurlan, M.A., Watson, H., Brunt, P.W. and Garlick, P.J. (1993) Albumin synthesis rates in cirrhosis: correlation with Child-Turcotte classification. *Hepatology* 18, 292–297.

Ballmer, P.E., McNurlan, M.A., Essen, P., Anderson, S.E. and Garlick, P.J. (1995a) Albumin synthesis rates measured with [$^2H_5$ ring] phenylalanine are not responsive to short-term intravenous nutrients in healthy humans. *Journal of Nutrition* 125, 512–519.

Ballmer, P.E., McNurlan, M.A., Grant, I. and Garlick, P.J. (1995b) Down-regulation of albumin synthesis in the rat by human recombinant interleukin-1β or turpentine and the response to nutrients. *Journal of Parenteral and Enteral Nutrition* 19, 266–271.

Barber, M.D., Fearon, K.M.C., McMillan, D.C., Slater, C., Ross, J.A. and Preston, T. (2000) Liver export protein synthetic rates are increased by oral meal in weight-losing cancer patients. *American Journal of Physiology* 279, E707–714.

Barle, H., Nyberg, G., Essen, P., Andersson, K., McNurlan, M.A., Wemerman, J. and Garlick, P.J. (1997) The synthesis rates of total liver protein and plasma albumins determined simultaneously *in vivo* in humans. *Hepatology* 25, 154–158.

Beeken, W.L., Volwiler, W., Goldsworthy, P.D., Garby, L.E., Reynolds, W.E., Stogsdill, R. and Stemler, R.S. (1962) Studies of I^{131}-albumin catabolism and distribution in normal young male adults. *Journal of Clinical Investigation* 41, 1312–1333.

Bennett, R.M. and Kokocinski, T. (1979) Lactoferrin turnover in man. *Clinical Science* 57, 453–460.

Bennhold, H. and Kallee, E. (1959) Comparative studies on the half-life of I^{131}-labelled albumin and non-radioactive human serum albumin in a case of analbuminemia. *Journal of Clinical Investigation* 38, 863–872.

Bianchi, R., Mariani, G., Pilo, A. and Donato, G. (1972a) Albumin metabolism in patients with chronic renal failure on low protein balanced diet. *Protides of the Biological Fluids* 19, 483–487.

Bianchi, R., Mariani, G., Pilo, A., Toni, M.G. and Donato, G. (1972b) Short-term measurement of plasma protein turnover. In: Wolstenholme, G. (ed.) *Protein Turnover.* CIBA Foundation Symposium, New Series, no. 9. Associated Scientific Publishers, Amsterdam, pp. 47–65.

Birke, G. (1967) Factors regulating the plasma protein levels in normal and pathological states. *Protides of the Biological Fluids* 14, 253–266.

Birke, G., Norberg, R. and Plantin, I.O. (eds) (1969) *Physiology and Pathophysiology of Plasma Protein Metabolism.* Pergamon Press, Oxford, UK.

Brutomesso, D., Iovi, E., Kiwanuka, E., Zanetti, M., Pianta, A., Vettove, M., Tiengo, A. and Tessaril, P. (2001) Insulin infusion normalizes fasting and post-prandial albumin and fibrinogen synthesis in type 1 diabetes mellitus. *Diabetic Medicine* 18, 915–920.

Carraro, F., Rosenblatt, J. and Wolfe, R.B. (1991) Isotopic determination of fibronectin synthesis in humans. *Metabolism* 40, 553–561.

Caso, G., Scalfi, L., Marra, M., Covino, A., Muscaritoli, M., McNurlan, M.A., Garlick, P.J. and Contaldo, D. (2000) Albumin synthesis is diminished in men consuming a predominantly vegetarian diet. *Journal of Nutrition* 130, 528–533.

Cayol, M., Tauveron, J., Rambourdin, F., Prugnaud, J., Gachon, P., Thieblot, P., Grizard, J. and Obled, C. (1995) Whole-body protein turnover and hepatic protein synthesis are increased by vaccination in man. *Clinical Science* 89, 389–396.

Cayol, M., Boirie, Y., Rambourdin, R., Prugnaud, J., Gachon, P., Beaufrère, B. and Obled, C. (1997) Influence of protein intake on whole body and splanchnic leucine kinetics in humans. *American Journal of Physiology* 272, E584–591.

Cohen, S., and Gordon, A.H. (1958) Catabolism of plasma albumin by the perfused rat liver. *Biochemical Journal* 70, 544–551.

Cohen, S. and Hansen, J.D.L. (1962) Metabolism of albumin and γ-globulin in kwashiorkor. *Clinical Science* 23, 351–359.

Cohen, S., Freeman, T. and McFarlane, A.S. (1961) Metabolism of ^{131}I labelled human albumin. *Clinical Science* 20, 161–170.

Cryer, D.R., Matsushima, T., Marsh, J.B., Yuskoff, M., Coates, P.M. and Cortner, J.A. (1986) Direct measurement of apolipoprotein B synthesis in human very low density lipoprotein using stable isotopes and mass spectrometry. *Journal of Lipid Research* 27, 508–516.

Cummings, M.H., Watts, G.F., Pal, C., Umpleby, M., Henessy, R., Naoumova, R. and Sönksen, P.H. (1995) Increased hepatic secretion of very-low-density lipoprotein apoliprotein B-100 in obesity: a stable isotope study. *Clinical Science* 88, 225–233.

Davies, J.W.L., Ricketts, C.R. and Bull, J.P. (1966) Studies of plasma protein metabolism: III. Fibrinogen in burned patients. *Clinical Science* 30, 305–314.

De Feo, P., Horber, F.F. and Haymond, M.W. (1992) Meal stimulation of albumin synthesis: a significant contributor to whole body protein synthesis in humans. *American Journal of Physiology* 263, E794–799.

Demant, T., Mathes, C., Gütlick, K., Bedynek, A., Steinhauer, H.B., Bosch, T., Packard, C.J. and Warwick, G.L. (1998) A simultaneous study of the metabolism of apolipoprotein B and albumin in nephrotic patients. *Kidney International* 54, 2064–2080.

de Meer, K., Smolders, H.C., Meesterburrie, J., de Sain-van der Velden, M., Voorbij, H.A.M., Okken, A., Reijngoud, D.-J. and Kulk, M. (2000) A single food bolus stimulates albumin synthesis in growing piglets. *Journal of Pediatric Gastroenterology and Nutrition* 31, 251–257.

de Sain-van der Velden, M.G., Kaysen, G.A., Barrett, H.A., Stellaard, F., Gadellaa, M.M., Voorbij, H.A. and Reijngoud, D.-J. (1998a) Increased VLDL in nephrotic patients results from a decreased catabolism while increased LDL results from increased synthesis. *Kidney International* 53, 994–1001.

de Sain-van der Velden, M.G., Kaysen, G.A., de Meer, K., Stellaard, F., Voorbij, H.A., Reijngoud, D.-J., Rabelink, T.J. and Koomans, H.A. (1998b) Proportionate increase of fibrinogen and albumin synthesis in nephrotic patients: measurements with stable isotopes. *Kidney International* 53, 181–188.

Dogherty, J.F., Golden, M.H.N., Raynes, J.G., Griffin, J.E. and McAdams, K.P.W.J. (1993) Acute-phase protein response is impaired in severely malnourished children. *Clinical Science* 84, 165–175.

Dykes, P.W. (1968) The rates of distribution and catabolism of albumin in normal subjects and in patients with cirrhosis of the liver. *Clinical Science* 34, 161–183.

Eisenberg, S. and Levi, R.J. (1975) Lipoprotein metabolism. *Advances in Lipid Research*, 13, 1–89.

Fearon, K.C.H., Falconer, J.S., Slater, C., McMillan, D.C., Ross, J.A. and Preston, T. (1998) Albumin synthesis rates are not decreased in hypoalbuminaemic cachetic cancer patients with an ongoing acute phase protein response. *Annals of Surgery* 227, 249–254.

Feldmann, G. (1986) Influence of synthesis of plasma proteins – in particular acute phase proteins – by hepatocytes on the production of other hepatic proteins. *Protides of the Biological Fluids* 34, 227–229.

Fleck, A., Colley, C.M. and Myen, M.A. (1985) Liver export proteins and trauma. *British Medical Bulletin* 41, 265–273.

Freeman, T. (1965) Gamma globulin metabolism in normal humans and in patients. *Series Haematologica* 4, 76–86.

Freeman, T. (1968) The function of plasma proteins. *Protides of the Biological Fluids* 15, 1–14.

Freeman, T. (1972) Albuminaemia – a study of albumin and transferrin metabolism. *Protides of the Biological Fluids* 19, 75–85.

Fu, A. and Nair, N.K. (1998) Age effect on fibrinogen and albumin synthesis in humans. *American Journal of Physiology* 275, E1025–1030.

Garlick, P.J. and Contaldo, D. (2000) Albumin synthesis is diminished in men consuming a predominantly vegetarian diet. *Journal of Nutrition* 130, 528–533.

Geller, D.M., Judah, J.D. and Nicholls, M.R. (1972) Intracellular distribution of serum albumin and its possible precursors in rat liver. *Biochemical Journal* 127, 865–874.

Gersovitz, M., Munro, H.N., Udall, J. and Young, V.R. (1980) Albumin synthesis in young and elderly subjects using a new static isotope methodology: response to level of protein intake. *Metabolism* 29, 1075–1086.

Giordano, M., De Feo, P., Lucidi, P., de Pascale, E., Giordino, G., Infantone, L., Zuccolo, A.M. and Castellino, P. (2001a) Increased albumin and fibrinogen synthesis in haemodialysis patients with normal nutritional status. *Journal of the American Society of Nephrology* 13, 349–354.

Giordano, M., De Feo, P., Lucidi, P., de Pascale, E., Giordino, M., Cirillo, D., Dardo, G., Silvorelli, S.S. and Castellino, P. (2001b) Effects of dietary protein restriction on fibrinogen and albumin metabolism in nephrotic patients. *Kidney International* 60, 235–242.

Gitlin, D., Cravioto, J., Frenk, S., Montano, E.L. Ramos-Galvan, R., Gomez, F. and Janeway, C.A. (1958) Albumin metabolism in children with protein malnutrition. *Journal of Clinical Investigation* 37, 682–686.

Häussinger, D., Hallbrucker, C., vom Dahl, S., Decker, S., Schweitzer, U., Lang, F. and Gerok, W. (1991) Cell volume is a major determinant of proteolysis control in liver. *FEBS Letters* 238, 70–72.

Hoffenberg, R. (1970) Control of albumin degradation *in vivo* and in the perfused liver. In: Rothschild, M.A. and Waldmann, T.A. (eds) *Plasma Protein Metabolism*, Chapter 15. Academic Press, New York.

Hoffenberg, R., Black, E. and Brock, J.F. (1966) Albumin and γ-globulin tracer studies in protein depletion states. *Journal of Clinical Investigation* 45, 143–152.

Hoffenberg, R., Zalin, A., McFarlane, A.S., Black, E.G. and Carson, E. (1972) A consideration of short-term measurement of plasma protein catabolic rates. *Protides of the Biological Fluids* 19, 467–471.

Hunter, K.A., Ballmer, P.E., Anderson, S.E., Broom, J., Garlick, P.J. and McNurlan, M.A. (1995). Acute stimulation of albumin synthesis rate with oral meal feeding in healthy subjects measured with [ring-$^2H_5$] phenylalanine. *Clinical Science* 88, 235–242.

Hunter, K.A., Garlick, P.J., Broom, I., Anderson, S.E. and McNurlan, M.A. (2001) Effects of smoking and abstention from smoking on fibrinogen synthesis in humans. *Clinical Science* 100, 459–465.

Ingenbleek, Y., van den Schnick, H., De Nayer, P. and De Visscher, M. (1975) Albumin, transferrin and the thyroxine-binding prealbumin/retinol-binding protein (TBPA-RBP) complex in assessment of malnutrition. *Clinica Chimica Acta* 63, 61–67.

Jackson, A.A., Phillips, G., McClelland, I. and Jahoon, F. (2001) Synthesis of hepatic secretory proteins in normal adults consuming a diet marginally adequate in protein. *American Journal of Physiology,* G1179–1187.

Jahoor, F., Gazzard, B., Phillips, G., Shapstone, D., Delrosario, M., Frazer, M.E., Heird, W., Smith, R. and Jackson, A.A. (1999) The acute-phase protein response to human immunodeficiency virus infection in human subjects. *American Journal of Physiology* 276, E1092–1098.

James, W.P.T. and Hay, A.M. (1968) Albumin metabolism: effect of the nutritional state and the dietary protein intake. *Journal of Clinical Investigation* 47, 1958–1972.

Jarnum, S. and Lassen, L.A. (1961) Albumin and transferrin metabolism in infections and toxic diseases. *Scandinavian Journal of Clinical and Laboratory Investigation* 13, 357–368.

Jeejeebhoy, K.N., Bruce-Robertson, A., Ho, J. and Sudtke, U. (1973) The comparative effects of nutritional and hormonal factors on the synthesis of albumin, fibrinogen and transferrin. In: Wolstenholme, G. (ed.) *Protein Turnover. CIBA Foundation Symposium*, New Series, no. 9. Associated Scientific Publishers, Amsterdam, pp. 217–238.

Jensen, H., Rossing, N., Andersen, S.B. and Jarnum, S. (1967) Albumin metabolism in the nephrotic syndrome in adults. *Clinical Science* 33, 445–457.

John, D.W. and Miller, L.L. (1969). Regulation of net biosynthesis of serum albumin and acute phase plasma proteins. *Journal of Biological Chemistry* 244, 6134–6142.

Jones, E.A., Craigie, A., Tavill, A.S., Simon, W. and Rosenoer, V.M. (1968) Urea kinetics and the direct measurement of the synthetic rate of albumin utilizing [^{14}C] carbonate. *Clinical Science* 35, 553–564.

Kang, J., Holland, M., Jones, H. and Kaysen, G.R. (1999) Coordinate augmentation in expression of genes encoding transcription factors and liver secretory proteins in hypo-oncotic states. *Kidney International* 56, 452–460.

Katz, J.H. (1961) Iron and protein kinetics studied by means of doubly labeled human crystalline transferrin. *Journal of Clinical Investigation* 40, 2143–2152.

Kerr, R.M., DuBois, J.J. and Holt, P.R. (1969) Use of ^{125}I- and ^{51}Cr-labelled albumin for the measurement of gastrointestinal and total albumin catabolism. *Journal of Clinical Investigation* 46, 2064–2082.

Kirsch, R., Frith, L., Black, E. and Hoffenberg, R. (1968) Regulation of albumin synthesis and catabolism by alteration of dietary protein. *Nature (London)* 217, 578–579.

Kushner, I. (1982) The phenomenon of the acute phase response. *Annals of the New York Academy of Sciences* 389, 39–48.

Lichtenstein, A., Cohn, J.S., Hackey, D.L., Millar, J.S., Ordovas, J.M. and Schaefer, E.J. (1990) Comparison of deuterated leucine, valine and lysine in the measurement of human apolipoprotein A-1 and B-100 kinetics. *Journal of Lipid Research* 31, 1693–1701.

Lundberg, D.A., Nelson, R.A. and Jones, J.D. (1976) Protein metabolism in the black bear before and during hibernation. *Mayo Clinic Proceedings* 51, 716–722.

Mansoor, O., Cayol, M., Gachon, P., Boirie, Y., Schoeffler, P., Obled, O. and Beaufrère, B. (1997) Albumin and fibrinogen synthesis increase while muscle protein synthesis decreases in head-injured patients. *American Journal of Physiology* 273, E898–902.

Matthews, C.M.E. (1957) The theory of tracer experiments with ^{131}I labelled plasma proteins. *Physics in Biology and Medicine* 2, 36–53.

Matthews, C.M.E. (1961) Effects of plasmapheresis on albumin pools in rabbits. *Journal of Clinical Investigation* 40, 603–610.

McFarlane, A.S. (1963) Measurement of synthesis rates of liver-produced plasma proteins. *Biochemical Journal* 89, 277–290.

McFarlane, A.S. (1964) Appendix on preparation of labelled proteins. In: Munro, H.N. and Allison, J.B. (eds) *Mammalian Protein Metabolism*, Vol. I. Academic Press, New York, pp. 331–338.

McFarlane, A.S., Todd, D. and Cromwell, S. (1964) Fibrinogen catabolism in humans. *Clinical Science* 26, 416–420.

McMillan, D.C., Slater, C., Preston, T., Falconer, J.S. and Fearon, K.C. (1996) Simultaneous measurement of albumin and fibrinogen synthetic rates in normal fasted subjects. *Nutrition* 12, 602–607.

McNurlan, M.A., Garlick, P.J., Frost, R.A., Decristofaro, K., Lang, C.H., Steigbigel, R.T., Fuhrer, J. and Gelato, M. (1998) Albumin synthesis and bone collagen formation in human immunodeficiency virus-positive subjects: differential effects of growth hormone administration. *Journal of Clinical Endocrinology and Metabolism* 83, 3050–3055.

Milland, J., Taylor, A., Thomas, T., Aldred, A.R., Cole, T. and Schreber, G. (1990) Gene expression in regenerating and acute-phase rat liver. *American Journal of Physiology* 259, G340–347.

Morlese, J.F., Forrester, T., Badeloo, A., Del Rosario, M., Frazer, M. and Jahoor, F. (1996) Albumin kinetics in edematous and nonedematous protein-energy malnourished children. *American Journal of Clinical Nutrition* 64, 952–959.

Morlese, J.F., Forrester, T., Del Rosario, M., Frazer, M. and Jahoor, F. (1997) Transferrin kinetics are altered in children with severe protein-energy malnutrition. *Journal of Nutrition* 127, 1469–1474.

Morlese, J.F., Forrester, T., Del Rosario, M., Frazer, M.

and Jahoor, F. (1998a) Repletion of the plasma pool of nutrient transport proteins occurs at different rates during the nutritional rehabilitation of severely malnourished children. *Journal of Nutrition* 128, 214–219.

Morlese, J.F., Forrester, T. and Jahoor, F. (1998b) Acute-phase protein response to infection in severe malnutrition. *American Journal of Physiology* 275, E112–117.

Moshage, H.J., Janssen, J.A.M., Franssen, J.H., Hafkenscheid, J.C.M. and Yap, S.H. (1987) Study of the molecular mechanism of decreased liver synthesis of albumin in inflammation. *Journal of Clinical Investigation* 79, 1635–1641.

Mouridsen, H.T. (1967) The turnover of human serum albumin before and after operations. *Clinical Science* 33, 345–354.

Nosslin, B. (1973) Analysis of disappearance time-curves after single injection of labelled proteins. In: *Protein Turnover*, CIBA Foundation Symposium, New Series, no. 9. Associated Scientific Publishers, Amsterdam, pp. 113–128.

Olufemi, O.S., Whittaker, P.G., Halliday, D. and Lind, T. (1991) Albumin metabolism in fasted subjects during late pregnancy. *Clinical Science* 81, 161–168.

Oppenheimer, J.H., Surks, M.I., Bernstein, G. and Smith, J.C. (1965) Metabolism of iodine-^{131}-labeled thyroxine-binding prealbumin in man. *Science* 149, 748–750.

Oratz, M. (1970) Oncotic pressure and albumin synthesis. In: Rothschild, M.A. and Waldmann, T. (eds) *Plasma Protein Metabolism*, Chapter 14, Academic Press, New York and London.

Ott, H. (1968) Congenital defects of protein synthesis and homoeostasis in analbuminaemia. *Protides of the Biological Fluids* 14, 307–310.

Pain, V.M., Clemens, M.J. and Garlick, P.J. (1978) The effect of dietary protein deficiency on albumin synthesis and on the concentration of active albumin messenger ribonucleic acid in rat liver. *Biochemical Journal* 172, 129–135.

Phillips, A., Shaper, A.G. and Whincup, P.H. (1989) Association between serum albumin and mortality from cardiovascular disease, cancer and other causes. *Lancet* ii, 1434–1436.

Picou, D. and Waterlow, J.C. (1962) The effect of malnutrition on the metabolism of plasma albumin. *Clinical Science* 22, 459–468.

Pietrangelo, A., Panduro, A., Chowdhury, J.R. and Shefritz, D.A. (1992) Albumin gene expression is down-regulated by albumin or macromolecular infusion in the rat. *Journal of Clinical Investigation* 89, 1755–1760.

Plantin, L.O. (1966) Aspects on the interpretation of protein turnover studies and the application of computer techniques. *Protides of the Biological Fluids* 14, 231–239.

Preston, T., Slater, C., McMillan, D.C., Falconer, J.S., Shenkin, A. and Fearon, K.C.H. (1998) Fibrinogen synthesis is elevated in fasting cancer patients with an acute phase response. *Journal of Nutrition* 128, 1355–1360.

Reeds, P. and Laditan, O. (1976) Serum albumin and transferrin in protein-energy malnutrition. Their use in the assessment of marginal undernutrition and the prognosis of severe undernutrition. *British Journal of Nutrition* 36, 255–63.

Reeds, P.J., Fjeld, C.R. and Jahoor, F. (1994) Do the differences between the amino acid compositions of acute-phase and muscle proteins have a bearing on nitrogen loss in traumatic states? *Journal of Nutrition* 124, 906–910.

Reeve, E.B., Pearson, J.R. and Martz, D.C. (1963) Plasma protein synthesis in the liver; method for measurement of albumin formation *in vivo*. *Science (Wash. DC)* 139, 914–916.

Reid, M., Badaloo, A., Forrester, T., Morlese, J.F., Heird, W.C. and Jahoor, F. (2002) The acute-phase protein response to infection in edematous and nonedematous protein-energy malnutrition. *American Journal of Clinical Nutrition* 76, 1409–1415.

Rossing, N. and Andersen, S.B. (1967) Colloid osmotic pressure and albumin metabolism. *Protides of the Biological Fluids* 14, 319–322.

Rossing, N., Krasilnikoff, P.A. and Christiansen, C. (1972) Protein metabolism in the capillary wall? *Protides of the Biological Fluids* 19, 437–441.

Rothschild, M.A. and Waldmann, T. (eds) (1970) *Plasma Protein Metabolism: Regulation of Synthesis, Distribution and Degradation*. Academic Press, New York and London.

Rothschild, M.A., Oratz, M., Wimer, E. and Schreiber, S.S. (1961) Studies on albumin synthesis: the effects of dextran and cortisone on albumin metabolism in rabbits studied with albumin I^{131}. *Journal of Clinical Investigation* 40, 545–554.

Rothschild, M.A., Oratz, M., Mongelli, J. and Schreiber, S.S. (1968) Effects of a short-term fast on albumin synthesis studied *in vivo* in the perfused liver and on amino acid incorporation by hepatic microsomes. *Journal of Clinical Investigation* 47, 2591–2599.

Rothschild, M.A., Oratz, M. and Schreiber, S.S. (1972) Albumin synthesis, Parts I and II. *New England Journal of Medicine* 286, 748–757 and 816–821.

Schomerus, H., and Mayer, G. (1975) Synthesis rates of albumin and fibrinogen in patients with protein-losing enteropathy and in a patient recovering from protein malnutrition. *Digestion* 13, 201–208.

Shetty, P.S., Watrasiewicz, K.E., Jung, R.T. and James, W.P.T. (1979) Rapid turnover transport proteins: an index of subclinical protein-energy malnutrition. *Lancet* 2, 230–232.

Smith, K., Downie, S., Barmu, J.M., Watt, P.W., Scrimgeour, C.M. and Rennie, M.J. (1994) Effect of a flooding dose of leucine in stimulating incorporation of constantly infused valine into albumin. *American Journal of Physiology* 266, E640–644.

Soeters, P.B., von Meyenfeldt, M.F., Meijerink, W.J.H.J., Fredrix, E.W.H.M., Wouters, E.M.F. and Schols, A.M.W.J. (1990) Serum albumin and mortality and subsequent correspondence. *Lancet* i, 348–351.

Steinbock, H.L. and Tarver, H. (1954) Plasma protein. V. The effect of the protein content of the diet on turnover. *Journal of Biological Chemistry* 209, 127–132.

Sternlieb, I., Morell, M.G., Tucker, W.D., Greene, M. and Scheinberg, I. (1961) The incorporation of copper into caeruloplasm *in vivo*: studies with copper^{64} and copper^{67}. *Journal of Clinical Investigation* 40, 1834–1840.

Stoll, B., Gerok, W., Lang, F. and Häussinger, D. (1992) Liver cell volume and protein synthesis. *Biochemical Journal* 287, 217–222.

Swick, R.S. (1958) Measurement of protein turnover in rat liver. *Journal of Biological Chemistry* 231, 751–764.

Tavill, A.S., Craigie, A. and Rosenoer, V.M. (1968) The measurement of the synthetic rate of albumin in man. *Clinical Science* 34, 1–28.

Tavill, A.S., East, A.G., Black, E.G., Nadkarni, D. and Hoffenberg, R. (1973) Regulatory factors in the synthesis of plasma proteins by the isolated perfused rat liver. In: *Protein Turnover*, CIBA Foundation Symposium no. 9. Elsevier, Amsterdam, pp. 155–171.

Thompson, C., Blumenstock, F.A., Saba, T.M., Fenstel, P.J., Kaplan, J.E., Fortune, J.B., Hough, L. and Gray, V. (1989) Plasma fibronectin synthesis in normal and injured humans as determined by stable isotope incorporation. *Journal of Clinical Investigation* 84, 1226–1235.

Volpi, E., Lucidi, P., Conciani, G., Monacchia, F., Rebold, G., Brunetti, P., Bolli, G.P. and De Feo, P. (1996) Contribution of amino acids and insulin to protein anabolism during meal absorption. *Diabetes* 45, 1245–1252.

Waldmann, T.A., Gordon, R.S. and Rosse, W. (1964) Studies on the metabolism of the serum proteins and lipids in a patient with analbuminemia. *American Journal of Medical Science* 37, 960–968.

Waldmann, T.A., Blaese, R.M. and Strober, W. (1970) In: Rothschild, M.A. and Waldmann, T. (eds) *Plasma Protein Metabolism*, Chapter 17. Academic Press, New York and London.

Waldmann, T.A. and Jones, E.A. (1973) The role of cell-surface receptors in the transport and catabolism of immunoglobulins. In: *Protein Turnover*, CIBA Foundation Symposium no. 9. Elsevier, Amsterdam, pp. 5–18.

Wetterfors, J. (1965) General aspects on the metabolism of serum albumin. In: Koblet, H., Vestia, P., Diggelmann, H. and Barandum, S. (eds) *Physiology and Pathophysiology of Plasma Protein Metabolism*. Huber, Berne, Switzerland, pp. 83–89.

Wilkinson, P. and Mendenhall, C.L. (1963) Serum albumin turnover in normal subjects and patients with cirrhosis measured by ^{131}I-labelled human albumin. *Clinical Science* 25, 281–292.

Wochner, R.D., Weissman, S.M., Waldmann, T.A., Houston, D. and Berlin, N.J. (1968) Direct measurement of the rates of synthesis of plasma proteins in control subjects and patients with gastrointestinal protein loss. *Journal of Clinical Investigation* 47, 971–982.

Wolstenholme, G.E.W. and O'Connor, M. (eds) (1973) *Protein Turnover*, Ciba Foundation Symposium no. 9. Elsevier, Amsterdam.

Yamanchi, A., Fukuhara, Y., Yamamoto, S., Yana, F., Takencker, M., Imai, E., Nogucti, T., Tanaka, T., Kamada, T. and Veda, N. (1992) Oncotic pressure regulates gene transcriptions of albumin and lipoprotein β in cultured rat hepatic cells. *American Journal of Physiology* 263, C307–404.

Zanetti, M., Barazzoni, R., Garibotto, G., Davanzo, G., Gabelli, C., Kiwanuka, E., Piccoli, A., Tosolini, M. and Tessari, T. (2001) Plasma protein synthesis in patients with low-grade nephrotic proteinuria. *American Journal of Physiology* 280, E591–597.

17
Collagen Turnover

'Animal connective tissues are systems of insoluble fibrils and soluble polymers evolved to take the stresses of movement and the maintenance of shape' (Scott, 1975). The soluble polymers are amino-sugars attached to proteins as proteoglycans: the fibrils are elastin and collagen.

There are about 20 different types of collagen, the commonest being types I and III. Prokop and Kirivikko (1995) have reviewed their structure and biosynthesis and the various diseases caused by mutations of the collagen genes. Collagens are defined as molecules with large domains of Glycine-X-Y sequences, coiled in a left-handed helix; three such chains are wrapped round each other to form a right-handed superhelix. Fibril-forming collagens are synthesized as larger precursors, procollagens, in which a peptide sequence is attached at both C- and N-terminal ends of the molecule. Assembly of the procollagens involves post-translational hydroxylation of Y-position proline and lysine and addition of galactose and/or glucose to some of the hydroxylysine residues. These processes take place within the fibroblast; the completed collagen is secreted into the extracellullar phase, where the N- and C-procollagen sequences are removed by specific proteinases. The collagen chains then self-assemble into fibrils, and finally lysine oxidase, an enzyme that depends on ascorbic acid, converts some lysine and OH-lysine residues to derivatives that form a complete series of cross-links. Degradation and remodelling of mature collagen involve a variety of proteinases in which the metalloproteinases play a pivotal role.

Collagen amounts to about 25% of body protein. Table 17.1 shows its concentration in the body and various tissues of the pig. In the whole body of the malnourished child the collagen concentration increases, showing that the soft tissues have shrunk round the skeleton and supporting structures (Picou *et al.*, 1966). The collagen of different tissues appears to respond to undernutrition in different ways: Dickerson and McCance (1964) found that it increases in muscle as a proportion of fat-free tissue, but decreases in skin.

17.1 Collagen Turnover

An early measurement of collagen turnover was that of Neuberger *et al.* (1951), who studied the decay rate of collagen from bone, skin, tendon and liver of rats after labelling with [^{15}N]-glycine. In old rats they found little loss of label after 7 weeks and concluded that mature collagen is virtually inert. Thompson and Ballou (1956) exposed rats to tritiated water for 6 months from conception and then killed them at intervals. They give a figure of 1000 days for the half-life of collagen. In both these studies the apparent rate of decay would be greatly slowed by recycling of labelled amino acids from the break-

Table 17.1. Collagen content of tissues in control and undernourished pigs.

	Collagen N as per cent of total N	
	Control	Undernourished
Whole body	27	32
Muscle	11	22
Skin	50	38
Skeleton	64	63

Data of Dickerson and McCance, 1964.

down of other proteins, so these must be regarded as underestimates of the real rate of degradation.

These results were obtained by measurement of decay after administration of a tracer. Nowadays the fractional synthesis rate (FSR) of collagen is determined by constant infusion or flooding dose of labelled proline from the ratio: [activity of hydroxyproline of total collagen (soluble + insoluble)]/[activity of proline precursor in plasma or tissue fluid]. With this method it is unnecessary to separate collagen from other tissue proteins, because hydroxyproline is found only in collagen, and the only precursor of hydroxyproline is proline. Proline is hydroxylated post-translationally and hydroxyproline is not re-utilized. Several authors have shown that in the rat in all the tissues studied the FSR of collagen decreases rapidly with age (e.g. Mays *et al.*, 1991). Some results are shown in Table 17.2.

In recent years workers have extracted tissues with various solvents – hot water, dilute saline with or without detergents, 0.5M acetic acid or 67% ethyl alcohol – to separate 'soluble' from 'insoluble' collagens. In rat skin the pool of soluble collagen, whether extracted with buffered saline or 0.5M acetic acid, expressed as a per cent of total collagen, reached a peak of about 80% while the rat was growing rapidly and then fell to a steady state of 20–30% as the rat and its collagen matured (Wirtschafter and Bentley, 1962; Molnar *et al.*, 1987). In protein-deficient rats soluble collagen fell earlier and to a lower level.

Babraj *et al.* (2005) made successive extractions of bone with buffered saline + detergent, 0.5M acetic acid, pepsin acetic acid and hot water. These fractions were shown to contain type I collagen and were described successively as procollagen, tropocollagen, highly cross-linked, and fully cross-linked or mature collagen. The amounts of these various fractions and their FSRs are set out in Table 17.3. The very small pro-collagen fraction has a high turnover rate, 10 × that of the mature collagen. The weighted FSR of all the fractions taken together is 0.06 h^{-1} in the post-absorptive state and nearly double – 0.11 h^{-1} – on feeding. It is interesting that collagen turnover in bone should be so sensitive to food. This is a very important paper from the point of view of methodology.

In man the FSR of collagen in muscle, bone, ligaments and tendon has been measured by Rennie and his colleagues in samples taken during surgery

Table 17.2. Changes with age in collagen synthesis and degradation in rat tissues.

Age, months	Fractional synthesis rate of tissue collagens % d^{-1}			Proportion of newly synthesized collagen degraded (%)		
	Muscle	Skin	Lung	Muscle	Skin	Lung
1	5	12	13	26	6	27.5
2	1.3	2.4	9	45	8	3.8
6	1.6	0.9	4.2	78	14	69
15	3.7	1.0	–	95	56	–
24	0.5	0.4	–	91	29	–

Tissues extracted with 67% ethyl alcohol. From Mays *et al.*, 1986, 1991.

Table 17.3. Synthesis rates of bone collagen fractions extracted successively with different solvents.

Fraction:	1	2	3	4
Solvent:	NaCl + detergent	0.5M acetic acid	Pepsin 0.5M acetic acid	Hot water (90°C)
Amount, μg mg^{-1} bone	~1	4	13	128
FSR, post-absorptive, h^{-1}	0.58	0.23	0.07	0.06
FSR, fed, h^{-1}	–	0.44	0.11	0.11

FSR, fractional synthesis rate. The overall weighted FSR was 0.069 post-absorptive, 0.12 fed. Data of Babraj *et al.*, 2005.

or by biopsy (Table 17.4) (Babraj *et al.*, 2002; Miller *et al.*, 2005). The rates in bone, ligament and tendon are remarkably high, compared with muscle collagen, reflecting the continual re-modelling that must be going on, even in elderly subjects.

The extent to which newly synthesized collagen is broken down was first shown by Kivirikko (1970) in intact animals and by Bienkowski *et al.* (1978) in cultured fibroblasts. They showed that 30–40% of newly synthesized collagen was degraded within minutes of its synthesis; its extent is measured as: [labelled OHPr, free or in low molecular weight fraction]/[labelled OHPr in high + low molecular weight fractions] (McNulty and Laurent, 1986). Some of these results are shown in Table 17.2. It is interesting that 6 days after one-sided pneumonectomy the FSR of collagen was almost 3 × that in unoperated control lungs, while 40% of the newly synthesized collagen was degraded (McNulty *et al.*, 1988). This shows that both synthesis and degradation are involved in remodelling. Laurent (1986) has a good review of the degradative process.

Procollagens are included in any so-called soluble collagen fraction, since they are 1000 × as soluble as mature collagen (Robins, 1979) and turn over very rapidly. Robins found the turnover rate of type I procollagen in rabbit skin to be about ~1.5 h^{-1} so that in labelling experiments it equilibrated very rapidly and closely with precursor proline in the tissue free pool (Robins, 1979). It would be convenient if procollagen could be taken as a proxy for the precursor of collagen synthesis, as VLDL apolipoprotein-B100 has been used as a precursor for the synthesis of proteins in the liver (Chapter 5); but the procollagen pool is very small – Robins (1979) found only 0.5 μg g^{-1} of tissue, and it is not as easily isolated as the apolipoprotein.

17.2 Markers of Synthesis and Breakdown

What is known about the biosynthesis and degradation of collagen has led to the development of markers that are increasingly used in clinical studies of bone diseases (for reviews see Eriksen *et al.*, 1995; Robins, 2003). An index of the formation of type I collagen, which accounts for 97% of the collagen in bone, is the concentration in plasma of type I procollagen propeptide (PICP) measured by radioimmunoassay. There are several indices of degradation, the classical one being the urinary excretion of hydroxyproline, which is produced as both free and in small peptides by collagen degradation. The free form is re-absorbed and rapidly degraded, but the peptides are excreted in the urine. This measure cannot be regarded as quantitative, but it was sufficient to show, many years ago, the extent of remodelling during recovery from malnutrition (e.g. Whitehead, 1965).

Other useful markers are the cross-links of pyrinolidine and deoxypyrinolidine, which normally hold the collagen fibrils together but are excreted in the urine during collagen breakdown (Robins *et al.*, 1991). Finally, a measure of degradation analogous to PICP is the carboxy-terminal telopeptide of type I collagen (ICTP) which, like PICP, appears in the plasma.

Table 17.4. Fractional synthesis rates of pro-collagen in human tissues.

	FSR, h^{-1}	Reference
Muscle collagen		
fed state, rest	0.018	Miller *et al.*, 2005
fed state, 24 h post exercise	0.053	
Skin	0.076	El-Harake *et al.*, 1998
Tendon		
rest	0.045	Miller *et al.*, 2005
24 h post exercise	0.077	
Bone		
fasting	0.06	Babraj *et al.*, 2002, 2005
fed	0.11	Babraj *et al.*, 2002, 2005

All extractions with 0.15M NaCl + detergent, except that of skin, which was with detergent only.

Thorsen *et al.* (1996) introduced a method for measuring by dialysis the release of prostaglandin E_2 in bone; this has been adapted by Langberg *et al.* (1999) to determine PICP and ICTP as well as prostaglandin E2 in the tissue fluid of the space round the Achilles tendon in human subjects before and after exercise. By these elegant experiments they showed that PICP, and hence collagen synthesis in the tendon, was increased threefold 72 h after a 3 h run, while ICTP showed a small but temporary fall. In the experiments of Miller *et al.* (2005) the FSRs of collagen in both muscle and tendon, measured with labelled proline, were substantially increased 24 h after unilateral leg exercise (Table 17.4); in this study there was no change in the serum concentration of the procollagen marker.

The work summarized here shows that collagen makes a far from negligible contribution to whole body protein turnover; in the near future we shall probably learn much more about how the structural components of the body – bone, tendons and ligaments – respond to mechanical and other stresses; see Scott (2003).

There is no real conflict between the old work establishing the very slow turnover rate of mature collagen and the most recent studies showing the relatively fast turnover rates of soluble or immature collagen. The recent literature is almost completely silent about the *amount* of soluble collagen in the fraction being studied. However, the paper of Wirtschafter and Bentley (1962) tells us that in rat skin the amount of soluble collagen parallels the growth rate, falling to about 0.5 mg g^{-1} skin at maturity. If skin collagen in the rat is about 120 mg g^{-1} skin (Waterlow and Stephen, 1966), the soluble fraction with its relatively high turnover rate is diluted 240 $\times$ by mature collagen. Moreover, in the adult rat some 50% of newly synthesized collagen is degraded (Table 17.2). Therefore it is not surprising that mature collagen appears to be almost completely inert.

17.3 References

Babraj, J.A., Cuthbertson, D.J., Rickhuss, P., Meier-Augenstein, W., Smith, K., Bohé, J., Wolfe, R.R., Gibson, J.N.A., Adams, C. and Rennie, M.J. (2002) Sequential extracts of human bone show differing collagen synthesis rates. *Biochemical Society Transactions* 30, 61–65.

Babraj, J.A., Smith, K., Cuthbertson, D.J.R., Rickhuss, P., Dorling, J.S. and Rennie, M.J. (2005) Human bone collagen is a rapid nutritionally modulated process. *Journal of Bone and Mineral Research* 20, 930–937.

Bienkoswski, R.S., Baum, B.J. and Crystal, R.G. (1978) Fibroblasts degrade newly synthesized collagen within the cell before secretion. *Nature* 276, 413–416.

Dickerson, J.W.T. and McCance, R.A. (1964) The early effects of rehabilitation on the chemical structure of the organs and whole bodies of undernourished pigs and cockerels. *Clinical Science* 27, 123–132.

El-Harake, W.A., Furman, M.A., Cook, B., Nair, K.S., Kukowski, J. and Brodsky, I.G. (1998) Measurement of dermal collagen synthesis rate *in vivo* in humans. *American Journal of Physiology* 274, E586–591.

Eriksen, E.F., Brixen, K. and Peder, C. (1995) New markers of bone metabolism: clinical use in metabolic bone disease. *European Journal of Endocrinology* 132, 251–263.

Kivirikko, K. (1970) Urinary excretion of hydroxproline in health and disease. *International Review of Connective Tissue Research* 5, 93–163.

Langberg, H., Skorgaard, D., Petersen, L.J., Bülow, J. and Kjaer, M. (1999) Type I collagen synthesis and degradation in peritendinous tissue after exercise determined by microdialysis in humans. *Journal of Physiology* 521, 299–306.

Laurent, G.J. (1986) Lung collagen: more than scaffolding. *Thorax* 41, 418–428.

Mays, P.K., McNulty, R.J. and Laurent, G.G. (1986) Age-related changes in lung collagen metabolism. *Biochemical Society Transactions* 14, 1084.

Mays, P.R., McNulty, R.J., Campa, J.S. and Laurent, G.J. (1991) Age-related changes in collagen synthesis and degradation in rat tissues: importance of degradation of newly synthesized collagen in regulating collagen production. *Biochemical Journal* 276, 307–313.

McNulty, R.J. and Laurent, G.J. (1986) Evidence *in vivo* for rapid and extensive degradation of newly synthesized collagen in tissues of adult rats. *Biochemical Society Transactions* 14, 776–777.

McNulty, R.J., Staple, L.H., Guerreiro, D. and Laurent, G.J. (1988) Extensive changes in collagen synthesis and degradation during compensatory lung growth. *American Journal of Physiology* 255, C754–759.

Miller, B.F., Olesen, J.L., Hansen, M., Døssing, S., Cramen, R.M., Welling, R.J., Langberg, H., Flyvbjerg, A., Kjaer, M., Babraj, J.A., Smith, K. and Rennie, M.J. (2005) Co-ordinated collagen and muscle protein synthesis in human patella tendon and quadriceps muscle after exercise. *Journal of Physiology* 567, 1021–1033.

Molnar, J.A., Alpert, N.M., Burke, J.F. and Young, V.R.

(1987) Relative and absolute changes in soluble and insoluble collagen pool size in skin during normal growth and with dietary protein restricted in rats. *Growth* 51, 132–145.

Neuberger, A., Perrone, J.C. and Slack, H.G.B. (1951) The relative metabolic inertia of tendon collagen in the rat. *Biochemical Journal* 49, 199–204.

Picou, D., Halliday, D. and Garrow, J.S. (1966) Total body protein collagen and non-collagen protein in infantile protein malnutrition. *Clinical Science* 30, 345–351.

Prokop, D.J. and Kivirikko, K.I. (1995) Collagens: molecular biology, diseases and potentials for therapy. *Annual Reviews of Biochemistry* 64, 403–434.

Robins, S.P. (1979) Metabolism of rabbit skin collagen. *Biochemical Journal* 181, 75–82.

Robins, S.P. (2003) Collagen turnover in bone diseases. *Current Opinion in Clinical Nutrition and Metabolic Care* 6, 65–71.

Robins, S.P., Black, D., Paterson, C.R., Reid, D.M., Duncan, A. and Seibel, M.J. (1991) Evaluation of urinary hydroxypyridinium cross link measurements as resorption markers in metabolic bone disease. *European Journal of Clinical Investigation* 21, 310–315.

Scott, J.E. (1975) Composition and structure of the pericellular environment: physiological function and chemical composition of pericellular proteoglycan (an evolutionary view). *Philosophical Transactions of the Royal Society, London*, series B, 271, 235–242.

Scott, J.E. (2003) Elasticity in extracellular matrix 'shape models' of tendon, cartilage, etc. A sliding proteoglycan-filament model. *Journal of Physiology* 553, 335–343.

Thompson, R.C. and Ballou, J.E. (1956) Studies of metabolic turnover with tritium as a tracer. *Journal of Biological Chemistry* 223, 795–809.

Thorsen, K., Kristoffersson, A.O., Lerner, U.H. and Lorentzon, R.P. (1996) *In situ* microdialysis in bone tissue. *Journal of Clinical Investigation* 98, 2446–2449.

Waterlow, J.C. and Stephen, J.M.L. (1966) Adaptation of the rat to a low-protein diet: the effect of a reduced protein intake on the pattern of incorporation of L-[^{14}C] lysine. *British Journal of Nutrition* 20, 461–484.

Whitehead, R.G. (1965) Hydroxyproline creatinine ratio as an index of nutritional status and rate of growth. *Lancet* ii, 567–570.

Wirtschafter, Z.O. and Bentley, A.S. (1962) The influence of age and growth rate on the extractable collagen of skin of normal rats. *Laboratory Investigation* 11, 316–320.

18

The Coordination of Synthesis and Breakdown

The object of this chapter is to examine how synthesis and breakdown might be coordinated, because if they were not coordinated the whole system would fall apart. It has been said that 'The emphasis during the last 50 years has been reductionist, explaining biological phenomena at the atomic and molecular level. The challenge for the next 50 years is to integrate these insights into an understanding of higher levels of temporal and spatial organization ...' (Henderson, 2003). There has indeed been an enormous literature on reductionism; however, it is necessary to move from the general to the specific. It would be impertinent for someone who has not worked in the field to go into detail, but it may be useful, as a beginning, to present an onlooker's view of the two processes that have to be coordinated.

18.1 Synthesis

Synthesis of a protein begins, of course, with the gene – with a specific set of bases along the gene. A transcription factor, which is a protein, binds to a segment of the gene and stimulates a polymerase, also a protein, to produce an RNA transcript of the gene – the precursor of messenger RNA. The operation of the transcription factor is regulated or modified by other protein factors – promoters, enhancers and repressors – which are attached to the gene, sometimes at a distance from the site of the transcription factor. Gill (2001) writes of 'the stunning complexity in the eukaryotic transcription machinery, with over 50 proteins assembled at the core promoter ... '

Transcription is also modified by the histones, proteins that coat the genes, so that access of the transcriptor depends on the extent to which they are methylated or otherwise modified by acetylation or ubiquitination. These modifications depend again on the action of enzymes, which may be stimulated by nutrients. This whole subject, termed epigenetics, has been reviewed for the non-expert by Mathers (2003).

When all these factors have combined successfully, the precursor mRNA that is produced is modified by 'splicing', which again requires enzymes. The ribosomes that put together the polypeptide chains, using the messenger RNA as a template, are an assemblage of three types of RNA together with some 80 different proteins. The structure of the ribosome, as depicted by Alberts *et al.* (2002: 344) is extraordinarily complex. The final stage, that of translation of the messenger, depends on an array of initiation and elongation factors, all proteins, of which about eight have been identified. The activity of one or more of these factors is modified by signals arising from specific receptors of hormones or nutrients situated on the cell surface or in some cases within the nucleus. The receptor and its site of action is linked by a pathway of intermediate proteins in which the signal is often phosphorylation or dephosphorylation, activated by kinases and phosphatases. Again, the receptors and the components of the signalling pathways are all proteins.

One must suppose that the activity of the initiation complex affects the translation of all the mRNA that are present at any moment, thus controlling the overall rate of protein synthesis. There is no specificity here, unless it is determined by differential rates of decay of mRNA, which we know are in general very fast – minutes rather than days. In many situations the amounts of specific mRNAs are increased or reduced: this is probably the result of transcriptional control, to which more and more attention is now being directed. Control of transcription has been likened to a key turning a lock, which is either open or closed. This is the type of control that seems to be particularly important in the development of the organism, determining whether a gene is active in one tissue and silent, perhaps for ever, in another.

Lastly there is the ribosome. The information presented in earlier chapters suggests that in some conditions, such as starvation, the amount of ribosomal RNA can vary, but that these changes occur more slowly than changes in efficiency of translation, i.e. initiation. All the components of the ribosomal sub-unit are apparently assembled together (Stryer, 1988) but whether they are synthesized together is another matter.

It seems that synthesis of the proteins that make up the bulk of our tissues, and the regulation of that synthesis involve a vast array of proteins that themselves undergo turnover, which is presumably regulated by other proteins and so on *ad infinitum*. This is not a very satisfactory conclusion.

18.2 Breakdown

Many authors have considered the 'purpose' of protein breakdown. Although the teleological approach is attractive to a physiologist, it is more useful to ask how breakdown contributes to the functioning and homeostasis of the body. The usual answers are that it enables flexibility in the responses of enzymes to changing conditions; it removes proteins when they are no longer of use, such as those that control the cell cycle; it removes miscoded and even mutated proteins and proteins that have suffered oxidative damage; and it keeps the free amino acid pools filled up. A further point might be that without degradation there would be no check on synthesis – a situation that would not be sustainable.

18.2.1 Mechanisms of breakdown

Whereas synthesis is like a large mountain peak, complicated in detail but the same for all proteins, for breakdown there are at least three peaks, to some extent clouded in mist, and perhaps more will be discovered. The three systems are the calcium-activated proteinases (caspases or calpains), the lysosome with the cathepsins that it contains and the ATP-dependent ubiquitin-proteasome system (Ub-P). Often, in situations where degradation is increased, all three systems are involved. For example, in a study on wasting of the soleus muscle produced by weightlessness (Taillandier *et al*., 1996) the concentration of each degradative system was measured with selective inhibitors: about $^{2}/_{3}$ of the increase in degradation could be attributed to the Ub-P system, the other $^{1}/_{3}$ being divided equally between lysosome cathepsins and the calpains. It may be that the cathepsins and calpains are responsible for an initial stage of processing which is then carried forward by the proteasome. The rôle played by each system also depends on the tissue. In liver the lysosomal system predominates; in skeletal muscle lysosomes are few and degrade only the sarcoplasmic proteins, myofibrillar proteins being broken down by the Ub-system. Thus degradation provides another example of the contrast between liver and muscle.

Calcium-activated proteinases (calpains or caspases)

The calpains are cystein-activated proteinases described by Attaix and Taillandier (1998) as 'ubiquitous'. According to Salvesen (2002) they are involved in programmed cell death (apoptosis) and also in the activation of cytokines. In muscle they play only a minor part in the degradation that produces muscle wasting but may be qualitatively important for the degradation of specific but minor protein components of the myofibril, such as troponin and tropomyosin (Attaix and Taillandier, 1998).

Lysosomal degradation

Lysosomes, discovered by De Duve in 1956, are cytoplasmic vesicles with an acid internal pH containing a collection of some 50 proteinases, including a variety of endopeptidases. The natural history, as it may be called, of lysosomal proteolysis has been elucidated by electron microscopy, territory that is unfamiliar to the biochemist. Early studies by perfusion of the liver showed that a region of cytoplasm is sequestered in a vacuole surrounded by a double membrane derived from the rough endoplasmic reticulum (ER), without attached ribosomes. This vacuole, called an autophagic vacuole or phagosome, then fuses with the lysosome, where the contents are exposed to the proteolytic enzymes. The process is referred to as 'macroautophagy' because it appears that cytoplasm is being digested in bulk; the autophagic vacuole contains mitochondria, fragments of ER, glycogen etc., in the same proportions as they occur in the cytoplasm. There is a parallel pathway, termed microautophagy, in which cytoplasmic proteins are engulfed directly by invaginations of the lysosomal membrane (endocytosis). Other variants of the lysosomal pathway cover the breakdown of secreted proteins (crinophagy) and of extracellular proteins (heterophagy). For more detailed information the reader is referred to reviews by Seglen and Bohley (1992), Blommaart *et al.* (1997) and Knecht *et al.* (1998). Figure 18.1 illustrates the complexity of the lysosomal pathway and its components. Mortimore and Pösö (1984) suggested that microautophagy might be responsible for basal degradation in the steady state and macroautophagy for the great increase, about threefold, in degradation that occurred in the

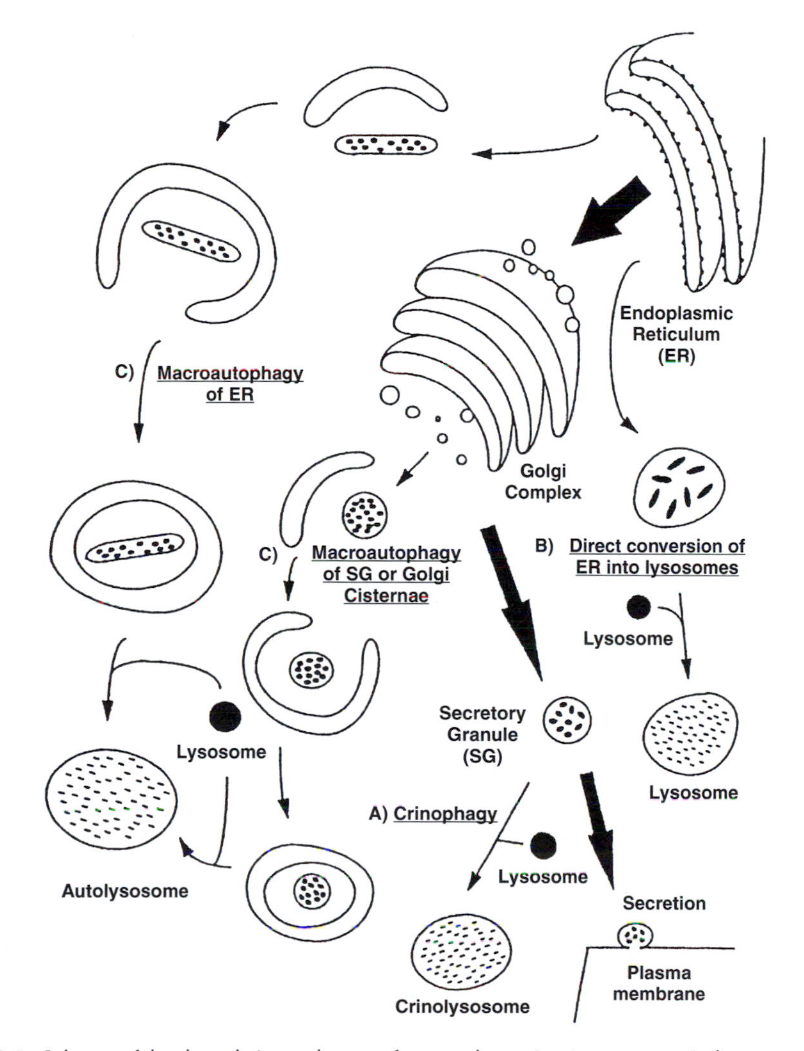

Fig. 18.1. Scheme of the degradative pathways of secreted proteins. Large arrows indicate protein traffic through the main secretory pathway. Reproduced from Knecht *et al.* (1998), by courtesy of *Advances in Molecular and Cellular Biology*.

starved liver. The key step in this increase is the amount of cytoplasm internalized in the autophagic vacuole; once inside, regardless of the amount, it is degraded at the very rapid rate of about 9% min^{-1}.

The overall fractional rate of protein breakdown in the liver at its maximum in the starved state is about 0.075% min^{-1} (Mortimore and Pösö, 1984). Compared with this, the intra-lysosomal degradation is virtually instantaneous. Therefore the limiting or rate-controlling step must be the rate of uptake into the vacuole. Another function of the lysosomal system is degradation of a large fraction of secretory proteins after they have been translated and before they are secreted (Bienkowski, 1983).

It seems, therefore, that at least in the liver, lysosomal protein degradation can be divided into two components, fixed and variable; the former is not acutely regulated, whereas the latter is subject to moment-to-moment control by amino acids, particularly leucine and glutamine, insulin, glucagon, glucocorticoids, prostaglandins and probably other effectors (Mortimore *et al.*, 1992). It is noteworthy that while glucagon stimulates degradation in liver it has the opposite effect in muscle.

Blommaart *et al.* (1997) point out the relevance of the studies of Häussinger and co-workers, showing that cell swelling inhibits degradation (see the superb review by Häussinger and Lang, 1991). It is suggested that swelling might be produced by influx of amino acids into the cell by sodium-coupled transport systems. There is evidence emerging that the protein kinase cascade may be involved in the signalling pathway. Phosphorylation of the ribosomal protein S6 increases ribosomal binding to the ER and thus inhibits macroautophagic sequestration (Blommaart *et al.*, 1997).

The ubiquitin-proteasome pathway

In 1971 Herschko discovered an ATP-dependent proteolytic system in reticulocytes (Herschko and Tomkins, 1971). Resolution of the components of this system showed that a small heat-stable protein was needed for its activity. This was identified as ubiquitin, a 76-residue protein present in all eukaryotic cells and with a sequence highly conserved in evolution. The first step in the proteolytic pathway is the activation of ubiquitin, which is then transferred to a carrier-protein. The final transfer of activated ubiquitin to the protein substrate requires a ubiquitin-protein ligase. Many of these ubiquitin-protein conjugates contain poly-ubiquitin chains; the proteins are then degraded to small peptides by a complex of proteases in a body called the proteasome, and the ubiquitins released and re-used. There are several species of transfer enzymes, each of which can act with one or more of a growing family of ligases. These play a major rôle in recognition and selection of proteins for degradation by the proteasome (Ciechanover and Schwartz, 1998) (see below).

The 20S proteasome, so-called from its sedimentation characteristics, has multiple sites of catalysis by subunits arranged in the form of a hollow cylinder. It is associated with 19S regulatory complexes which bind at each end of the cylinder to form the 26S proteasome. The complexes contain approximately 20 proteins, six of which are ATPases. They are presumably involved in the recognition and unfolding of proteins for degradation in the catalytic core; thus they stimulate activity of the 20S protein subunits, but do not themselves possess proteolytic activity. More detailed accounts of the structure and functioning of the proteasome are given in reviews by Coux *et al.* (1996), Ciechanover and Schwartz (1998), Attaix and Taillandier (1998) and Attaix *et al.* (2003).

The Ub-P system is responsible for the degradation of a wide range of proteins in a variety of tissues (Herschko and Ciechanover, 1992). These include, as well as myofibrillar proteins, regulatory proteins of the cell-cycle (cyclins), surface receptors, ion channels, antigenic proteins and abnormal or mutated proteins. They are also involved in transcriptional control, signal transduction and downregulation of receptors. Of particular interest from the physiological point of view, as mentioned above, is the increased degradation of the myofibrillar proteins of skeletal muscle that is found in various conditions of muscle wasting. This problem has been studied by the groups of Goldberg and of Attaix, who have found that the greater part of the increased degradation can be attributed to the Ub-P system. Mitch and Goldberg (1996) is a valuable review. Some of these studies are summarized in Table 18.1. Similar tables have been presented by Attaix and Taillandier (1998) for the changes in

Table 18.1. Increases in total or myofibrillar protein breakdown by proteasomes of skeletal muscle in experimental muscular wasting.

Wasting condition	Method of study	Reference
Fasting	A	Wing and Goldberg, 1993
	B	Medina *et al.*, 1991
	C	Medina *et al.*, 1995
Acidosis	A,C	Mitch *et al.*, 1994
	B	Bailey *et al.*, 1996
Sepsis	A	Tiao *et al.*, 1994
	B	Tawa *et al.*, 1997; Hobler *et al.*, 1998
	C	Voisin *et al.*, 1996
Cancer	A,C	Temparis *et al.*, 1994
Denervation	A,B	Medina *et al.*, 1995
Diabetes	B,C	Price *et al.*, 1996
Glucocorticoid treatment	C	Price *et al.*, 1994; Dardevet *et al.*, 1995
Thyroid hormone	B	Tawa *et al.*, 1997
Weightlessness	C	Taillandier *et al.*, 1996
Head trauma	C	Mansoor *et al.*, 1996

A, blocking of increased total or myofibrillar proteolysis by ATP depletion; B, blocking of increased total or myofibrillar proteolysis by proteosome inhibitors; C, increased expression of 20S proteosome subunits. Based on table in Attaix and Taillandier, 1998.

lysosomal proteinases and calpains that are found in muscle wasting. It is interesting that in chronic wasting, as in Cushing's disease or Duchenne muscular dystrophy, these increases were not found (Rallière *et al.*, 1997; Combaret *et al.*, 1996).

One aspect of myofibrillar degradation that received much attention in earlier years (e.g. Millward, 1980) but since then has been somewhat neglected is the mechanism of breakdown and reassembly of a complex and organized structure such as the myofibril, which contains a number of proteins turning over at different rates. Millward suggested that there might be an exchange of proteins between peripheral and central filaments of the myofibril. At the periphery of the myofibril individual proteins might be disassembled at varying rates, producing a pool of proteins that are either degraded or taken up into new fibrils which then migrate into the core. This is a subject that needs further study.

Targeting of proteins for degradation

The central problem here is that different proteins are broken down at different rates. For breakdown there is no equivalent of the specific messenger RNA that 'targets' the synthesis of proteins. Schimke (1970), whose chapter on protein degradation in Munro's book is still a landmark, lists 26 enzymes in rat liver with half-lives ranging from about 1 h to 16 days. He also stated that all these enzymes decayed by first order kinetics, which perhaps can be reconciled with the complexities described above of the degradation systems on the basis of statistical randomness, however many steps may be involved. Schimke suggested that degradation is controlled by two factors: the first is the activity of the degrading system, which Millward called the 'degradative environment'. A good example is the demonstration by Fritz *et al.* (1969) that the lactic dehydrogenase isozyme LDH-5 had a half-life of 16 days in liver and 31 days in muscle. Another example in the same category might be the stabilization of an enzyme by its substrate, as in the case of tryptophan pyrolase.

Schimke's second factor is the properties of the protein being degraded: 'A given protein molecule exists at a thermodynamic equilibrium in which only in certain states or confirmations will it be subject to degradation' (Schimke *et al.*, 1968). Wheatley (Wheatley *et al.*, 1982; Wheatley, 1984) developed a hypothesis based on the observation that the decay of isotopically labelled proteins always followed a 2-exponential curve. His idea was that there is always a redundancy of protein synthesis: some part of this excess cannot be 'assimilated' and is rapidly broken down as nascent proteins. Those proteins that are assimilated are broken down later, degradation at both stages being random. He concluded that 'the stability of a protein is dependent on conditions which will produce different thermodynamic equilibria of the various conformers it can assume'. Grisolia and Wheatley (1984) appeared to accept this idea when they wrote: 'The pervasive proteolytic system of the cell continually clears the decks of all proteins which do not enter some haven in which they can be usefully employed', and 'proteins do not have rigidly defined forms, only a statistical probability of existing in one of many forms.' Bienkowski (1983) also emphasized that newly synthesized secretory proteins had a priority for degradation, citing as an example that 10–40% of newly synthesized procollagen was degraded rather than forming mature collagen.

There followed a succession of studies defining more precisely structural features of a protein that might determine the extent to which it is liable to degradation. Bachmair *et al.* (1986) produced mutations of β-galactosidase in yeast, and showed that the degradation rate depended on the nature of the N-terminal amino acid (Table 18.2) (see Varsharsky, 1996). However, opinion in recent years seems to be that the scope of the rule is rather limited, and that although a few proteins may be recognized by a destabilizing N-terminal residue, the vast majority are targeted through different signals (Ciechanover and Schwartz, 1998; Kwon *et al.*, 1998).

Similar proposals were put forward for the targeting of proteins for degradation by the lysosomal proteinases and the calpains. Dice (1987, 1990) suggested that rapidly degraded normal proteins contained peptide regions rich in proline, glutamate, serine and threonine – the PEST hypothesis – and that this sequence targeted them for breakdown by the calpains. In the same way lysosomal degradation was activated by proteins containing peptide regions rich in lysine, phenylalanine, glutamate, arginine and glutamine (KFREQ). This hypothesis was elaborated in great detail, and many modifications of the KFREQ sequence have been described that stimulate degradation.

These attempts to define specific amino acid sequences that target proteins for degradation seem, on the whole, to have run into the ground in recent years, with increasing recognition of the importance of folding and of the thermodynamic state of protein molecules. The result is a move back towards the rather nebulous ideas of Wheatley and Grisolia discussed above. A paper from Varsharsky's group concluded that a protein *in vivo* is not a fixed entity and that the kinetics of degradation are not first order: 'the probability of degradation of a nascent, partially unfolded chaperonin-associated protein should be in general different from the probability of degradation of a folded counterpart of this protein at a later time in the same cell' (Lévy *et al.*, 1996). A related possibility was suggested by an observa-

Table 18.2. Dependence of half-lives of protein on their N-terminal amino acid.

	$T^1/_2$	
Stabilizing	> 20 h	Methionine Glycine Serine Alanine Threonine Valine
Destabilizing	10–30 min	Isoleucine Tyrosine Proline Glutamic acid Glutamine
Highly destabilizing	2–3 min	Leucine Lysine Arginine Phenylalanine Aspartic acid

From Bachmair *et al.*, 1986.

tion by M.F. Perutz (personal communication) that a protein molecule may oscillate between different energy levels. Suppose that each oscillation imposes a strain, so that eventually the molecule begins to break up at a weak point. This would lead to the pattern of breakdown that Garlick (1978) called 'multiple event' kinetics (see Chapter 2).

Many nutritional and hormonal effectors influence the rate of degradation (Table 18.1) but there are still gaps in our understanding of the signalling pathway(s) by which the various effectors produce their effects. It would greatly simplify matters if, at the physiological level, the conditions in which degradation is stimulated or inhibited could be narrowed down to only one or two effectors, such as insulin and glucocorticoids. Liu and Barrett (2002) point out that 'proteolysis is more sensitive than synthesis to small changes in plasma insulin within its physiological range'. They go on to say that it is 'unknown where in the insulin-signalling system the pathway for regulation of synthesis and breakdown diverges'. But do they need to diverge, since the same signalling pathway may activate different genes (Hill and Treisman, 1995)? Moreover, as Salghetti *et al.* (2000) maintain, in many transcription factor proteins there is overlap between regions signalling proteolysis and regions activating transcription.

18.3 Coordination

The point of the previous two sections was to emphasize that the machinery of both synthesis and breakdown involves a large number of proteins. More emphasis has been given to breakdown because the situation there seems rather less clear.

The most important question in protein kinetics is how synthesis and breakdown are coordinated so that they are in balance, in a steady state. Admittedly it has been said, 'Nature abhors a steady state … what we actually see is a kind of oscillation. The only time that the difference between anabolism and catabolism is actually zero is when the oscillation is going through a maximum or minimum' (Miller, 1973). This observation is undoubtedly correct; in Chapter 9 we examined oscillations in whole body synthesis and breakdown; at the tissue level there were many papers in the 1960s describing diurnal oscillations in liver enzymes in the rat, with some uncertainty on whether they were controlled by protein intake or by hormones, particularly corticosteroids (e.g. Wurtman, 1970; review by Kenney, 1970).[1] Nevertheless, it cannot be denied that over a longer period such as 24 h a steady state is achieved, with synthesis and breakdown cancelling out.

I looked at this question in an earlier review (Waterlow, 1999), principally from the point of view of nitrogen balance. On the coordination of synthesis and breakdown the difficulty is that as far as is known the ribosome does not speak to the proteasome, but it is logically necessary that if two opposing processes are to be coordinated, there must be some functional link between them. This link is the amino acids, the substrate for one process and the product of the other. If, hypothetically, the sensitivities of synthesis and breakdown to amino acids and their concentrations are plotted together, the lines must cross at some point, which is that of balance (Fig. 18.2). This formulation is only building on an idea put forward by Scornik (1984) and by Toates (1975) in relation to the opposing processes in lipid metabolism of lipolysis and fatty acid synthetase. The lines in Fig. 18.2 represent sensitivities; if the sensitivities are altered, perhaps by insulin, the position and steepness of the slopes may alter, but the lines will still cross.

Insulin clearly plays a key role. Its part, together with amino acids, in the regulation of muscle protein synthesis has been very fully described in numerous reviews (Kimball *et al.*, 1994; Pain, 1996; Proud and Denton, 1997; Wang *et al.*, 1998; Campbell *et al.*, 1999; Kimball and Jefferson, 2000; Shah *et al.*, 2000; Kimball *et al.*, 2002; Boulter *et al.*, 2004). Amino acids and insulin collaborate in the translation not only of general mRNAs but, by different pathways, in the translation of specific messengers, including those that produce ribosomal proteins and transcriptional activators (Proud and Denton, 1997; Proud, 2001). These latter studies were made on hamster cells *in vitro*. Whether they can be extrapolated more generally is, of course, a question. Stimulation or repression of specific

[1] It would be interesting to repeat these experiments with measurements of synthesis and breakdown and of tissue amino acid concentrations.

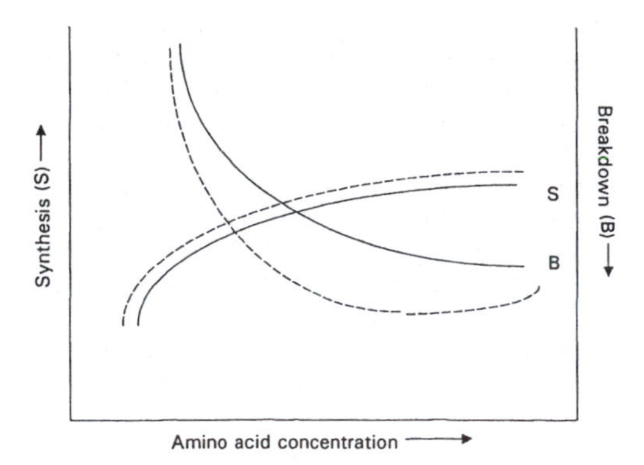

Fig. 18.2. Suggested model of the interaction between concentrations of amino acids and rates of protein synthesis (S) and breakdown (B). The dotted line shows the complementary effect of insulin. Reproduced from Waterlow, 1999, by courtesy of *Nutrition Research Reviews*.

mRNAs implies regulation of the transcription factors that produce them and will not be reflected by changes either in k_{RNA} or RNA concentration.

As for synthesis, so also for breakdown there is good evidence for control by amino acids and insulin. The classical study of Mortimore and Pösö (1984) showed very clearly that in the liver proteolysis by lysosomes was affected by the amino acid supply. In recent years the difficulties of measuring breakdown, discussed in Chapter 14, have been circumvented by determining increases in abundance of specific mRNAs of proteins involved in degradation, particularly of ubiquitin and the components of the proteasome, in response to starvation, insulin deficiency, sepsis, etc. (Table 18.1).

Conversely, amino acids and insulin reverse these changes (Larbaud *et al.*, 1996). Most of the work has been done on muscle, but the gut shows similar changes (Samuels *et al.*, 1996). There are differences of detail in the responses of the different elements of the Ub-P and of the other degradative systems and in the timing of these responses, which are well discussed by Attaix *et al.* (2003). Corticosteroids, of course, as discussed in Chapters 9 and 15, play an important role in the regulation of breakdown, but this rôle can be seen as additive or complementary to the underlying basic control by amino acids and insulin.

This picture of a system that is maintained, with fluctuations, in overall balance through amino acids/insulin as intermediates is simplistic and incomplete. There is still the conceptual problem that both synthesis and breakdown involve huge numbers of proteins that are turning over and so constantly having to be renewed; are these produced by other proteins, which require further proteins to produce them and so on *ad infinitum*? A solution to this problem would be if transcription factors could transcribe themselves. Secondly, the scheme tells us nothing about the quality of control. It seems probable that there is a quickly acting fine control, presumably acting on the initiation step of synthesis and manifesting itself by a change in k_{RNA}, and a coarse control, perhaps of the production and/or breakdown of ribosomes, manifested by changes in [RNA]. Whether there is a similar dichotomy of the control of breakdown is not yet clear. A further question is whether control is by on–off switching or graded. For a physiologist the control must be graded, but transcription is often described as being like a key in a lock, either on or off. However, apparently binding of proteins to DNA may be weak or strong (Ptashne and Gann, 2001). Moreover, most control pathways seem to involve phosphorylation, which is not an off–on process and can be a matter of degree – for example, the activity of the branched-chain keto acid dehydrogenase depends on the extent to which it is phosphorylated or dephosphorylated (Chapter 4).

The diagrams in books on molecular biology show the compartments of a system, the degree of their aggregation or of separation, the directions of their attachment, but nothing about their fluxes. They are like the old metabolic maps which showed the substrates and the enzymes, but nothing about the rates of flow. When I first read Haschemeyer's paper (Haschemeyer, 1969) telling us that the time needed at 21°C to synthesize a polypeptide chain of ~400 amino acids in the liver of an Antarctic fish was ~6.5 min, or ~1 second per peptide bond, the whole subject came alive.[1] More knowledge about the turnover rates of the most important proteins, such as transcription factors, ribosomes and proteins of the signalling pathways would surely increase our understanding of the mechanisms of control. Moreover, in molecular biology the proteins and other macromolecules are usually and inevitably studied in an artificial

[1] There were previous studies of assembly times of which I was not aware, such as those of Dintzis (1961) for haemoglobin chains and Loftfield and Eigner (1958) for ferritin.

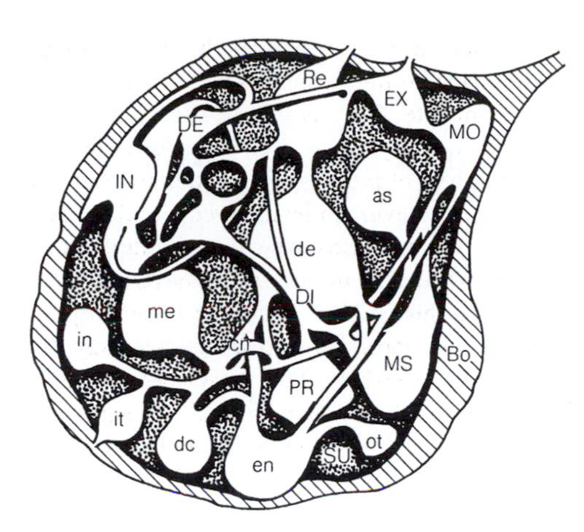

Fig. 18.3. Scheme of a generalized living system composed of 19 subsystems that process both matter/energy and information.
Bo, boundary; DE, decomposer; DI, distributor; EX, extruder; IN, ingestor; MO, motor; MS, matter/energy storage; PR, producer; SU, supporter; as, associator; cn, channel and net; de, decoder; en, encoder; in, internal transducer; it, input transducer; me, memory; ot, output transducer; Re, reproducer. Reproduced from Welch (1987), by courtesy of *Trends in Ecology and Evolution.*

milieu *in vitro*, whereas in reality, as is emphasized more and more, they are parts of a highly organized system. In the words of Hochachka (1999) '... recent developments in our understanding of intracellular myosin, kinesin and dynenin motors running on actin or tubulin tracks or cables supply a mechanistic basis for regulated intracellular circulating systems with cytoplasmic streaming rates varying over an approximately 80-fold range (1 to 780 μm sec^{-1})'.

This concept may provide a better explanation of many of the phenomena which in this book have been attributed to 'compartmentation'.

Figure 18.3 shows one concept of the internal organization of the cell. As I said in the Foreword, I hope that protein kinetics will contribute to a more physiological (I prefer this word to 'holistic') approach to studies of how the cell works.

18.4 References

Alberts, B., Johnson, A., Lewis, J., Raft, M., Roberts, K. and Walter, P. (2002) *Molecular Biology of the Cell*, 4th edn. Garland Science, New York.

Attaix, D. and Taillandier, D. (1998) The critical role of the ubiquitin-proteasome pathway in muscle wasting in comparison to lysosomal and Ca^{2+}-dependent systems. *Advances in Molecular and Cell Biology* 27, 235–266.

Attaix, D., Combaret, L., Kee, A.J. and Taillandier, D. (2003) Mechanisms of ubiquitination and proteasome-dependent proteolysis in skeletal muscle. In: Zempleni, J. and Daniel, H. (eds) *Molecular Nutrition*. CAB International, Wallingford, UK.

Bachmair, A., Finley, D. and Varshavsky, A. (1986) *In vivo* half-life of a protein is a function of its amino-terminal residue. *Science* 234, 179–186.

Bailey, J.L., Wang, X., England, B.E., Price, S.R., Ding, X. and Mitch, W.E. (1996) The acidosis of chronic renal failure activates muscle proteolysis in rats by augmenting transcription of genes encoding the ATP-dependent, ubiquitin-proteasome pathway. *Journal of Clinical Investigation* 97, 1447–1453.

Bienkowski, R.S. (1983) Intracellular degradation of newly synthesized secretory proteins. *Biochemical Journal* 214, 1–10.

Blommaart, A.F.C., Luiken, J.J.F.P. and Meijer, A.J. (1997) Autophagic proteolysis: control and specificity. *Histochemical Journal* 29, 365–385.

Boulter, D.R., Jefferson, L.S. and Kimbell, S.R. (2004) Regulation of protein synthesis associated with skeleletal muscle hypertrophy by insulin-, amino acid- and exercise-induced signalling. *Proceedings of the Nutrition Society* 63, 351–356.

Campbell, L.E., Wang, X. and Proud, C.G. (1999) Nutrients differentially regulate multiple translation factors and their control by insulin. *Biochemical Journal* 344, 423–441.

Ciechanover, A. and Schwartz, A.L. (1998) The ubiquitin proteasome pathway: the complexity and myriad functions of proteins death. *Proceedings of the National Academy of Sciences, USA* 93, 2727–2730.

Combaret, L., Taillandier, E., Voisin, L., Samuels, S.E., Bœspflug-Tanguy, O. and Attaix, D. (1996) No alteration in gene expression of components of the ubiquitin-proteasome proteolytic pathway in dystrophin-deficient muscles. *FEBS Letters* 393, 292–296.

Coux, O., Tanake, K. and Goldberg, A.L. (1996) Structure and functions of the 20S and 26S proteosomes. *Annual Reviews of Biochemistry* 65, 801–847.

Dardevet, D., Sornet, C., Taillandier, D., Savary, I., Attaix, D. and Grizard, D.J. (1995) Sensitivity and protein turnover response to glucocorticoids are different in skeletal muscle from adults and old rats: lack of regulation of the ubiquitin-proteasome pathway in aging. *Journal of Clinical Investigation* 96, 2113–2119.

Dice, J.F. (1987) Molecular determinants of protein

half-lives in enkaryotic cells. *FASEB Journal* 1, 349–357.

Dice, J.F. (1990) Peptide sequences that target cytosolic proteins for lysosomal proteolysis. *Trends in Biochemical Sciences* 15, 305–309.

Dintzis, H. (1961) Assembly of the peptide chains of hemoglobin. *Proceedings of the National Academy of Sciences* 47, 247–261.

Fritz, P.J., Versell, E.S., White, E.L. and Pruilt, K.M. (1969) The roles of synthesis and degradation in determining tissue concentrations of lactate dehydrogenase-5. *Proceedings of the National Academy of Sciences, USA* 62, 558–565.

Garlick, P.J. (1978) In: Waterlow, J.C., Garlick, P.J. and Millward, D.J. (eds) *Protein Turnover in Mammalian Tissues and the Whole Body.* North Holland/Elsevier, Amsterdam, p. 215.

Gill, G. (2001) Regulation of the initiation of eukaryotic transcription. *Essays in Biochemistry* 37, 33–44.

Grisolia, S. and Wheatley, D.N. (1984) Intracellular degradation of endogenous proteins and its regulation. *Life Chemistry Reports* 2, 257–297.

Haschemeyer, A.E. (1969) Rates of polypeptide chain assembly in liver *in vivo*: relation to the mechanism of temperature acclimation in *Opsanus tau. Proceedings of the National Academy of Sciences, USA* 62, 128–135.

Häussinger, D.T. and Lang, F. (1991) Cell volume in the regulation of hepatic function: a mechanism for metabolic control. *Biochimica et Biophysica Acta* 1071, 331–350.

Henderson, R. (2003) Half-time in a century of molecular biology. *Excellence in Science: News from the Royal Society.* April 2003, p. 10.

Herschko, A. (1991) The ubiquitin pathway of protein degradation and proteolysis of ubiquitin-protein conjugates. *Biochemical Society Transactions* 19, 726–729.

Herschko, A. and Ciechanover, A. (1992) The ubiquitin system for protein degradation. *Annual Reviews of Biochemistry* 61, 761–807.

Herschko, A. and Tomkins, G.M. (1971) Studies on the degradation of tyrosine aminotransferase in hapatoma cells in culture. *Journal of Biological Chemistry* 246, 710–714.

Hill, C.S. and Treisman, R. (1995) Transcriptional regulation by extracellular signals: mechanisms and specificity. *Cell* 80, 199–211.

Hobler, S.C., Tiao, G., Fischer, J.E., Monaco, J. and Hasselgren , P.O. (1998) Sepsis-induced increase in muscle proteolysis is blocked by specific proteasome inhibition. *American Journal of Physiology* 274, R30–37.

Hochachka, P.W. (1999) The metabolic implications of intracellular circulation. *Proceedings of the National Academy of Sciences, USA* 96, 12233–12239.

Kenney, F.T. (1970) Hormonal regulation of synthesis of liver enzymes. In: Munro, H.N. (ed.) *Mammalian Protein Metabolism* Vol. IV, Academic Press, New York and London, pp. 131–176.

Kimball, S.R. and Jefferson, L.S. (2000) Regulation of translation in mammalian cells by amino acids. In: Sonenburg, N., Hershey, J.W.B. and Mathews, M.B. (eds) *Translational Control of Gene Expression.* Cold Spring Harbour Laboratory Press, New York, pp. 561–580.

Kimball, S.R., Vary, T.C. and Jefferson, L.S. (1994) Regulation of protein synthesis by insulin. *Annual Reviews of Physiology* 56, 321–348.

Kimball, S.R., Farrell, P.A. and Jefferson, L.S. (2002) Role of insulin in translational control of protein synthesis in skeletal muscle by amino acids or exercise. *Journal of Applied Physiology* 93, 1168–1180.

Knecht, E., de Llano, J.J.M., Andreu, E.J. and Miralles, I.M. (1998) Pathways for the degradation of intracellular proteins within lysosomes in higher enkaryotes. *Advances in Molecular and Cell Biology* 27, 201–234.

Kwon, Y.T., Reiss, U., Fried, V.A., Hershko, A., Yoon, J.K., Gonda, D.K., Sangan, P., Copeland, N.G., Jenkins, A. and Varshavsky, A. (1998). The mouse and human genes encoding the recognition component of the N-end rule pathway. *Proceedings of the National Academy of Sciences, USA* 95, 7898–7903.

Larbaud, D., Debras, E., Taillandier, D., Samuels, S.E., Temperis, S., Champedron, C., Grizard, J. and Attaix, D. (1996) Euglycemic hyperinsulinaemia and hyperaminoacidemia decreased skeletal muscle ubiquitin in RNA in goats. *American Journal of Physiology* 271, E505–512.

Lévy, F., Johnsson, N., Rumenapf, T. and Varshavsky, A. (1996) Using ubiquitin to follow the metabolic fate of a protein. *Proceedings of the National Academy of Sciences, USA* 93, 4907–4912.

Liu, Z. and Barrett, E.J. (2002) Human protein metabolism: its measurement and regulation. *American Journal of Physiology* 283, E1105–1112.

Loftfield, R.B. and Eigner, E.A. (1958) The time required for the synthesis of a ferritin molecule in rat liver. *Journal of Biological Chemistry* 231, 925–943.

Mansoor, O., Beaufrère, B., Boirie, Y., Rellière, C., Taillandier, D., Aurousseau, E., Schoeffler, P., Arnal, M. and Attaix, D. (1996) Increased mRNA levels for components of the lysosomal, Ca^{2+}-activated, and ATP-ubiquitin-dependent proteolytic pathways in skeletal muscle from head trauma patients. *Proceedings of the National Academy of Sciences, USA* 93, 2714–2718.

Mathers, J.C. (2003) Nutrition and cancer prevention: diet-gene interactions. *Proceedings of the Nutrition Society* 62, 605–610.

Medina, R., Wing, S.S., Haas, A. and Goldberg, A.L.

(1991) Activation of the ubiquitin-APP-dependent proteolytic system in muscle during fasting and denervation atrophy. *Biochimica et Biophysica Acta* 50, 347–356.

Medina, R., Wing, S.S. and Goldberg, A.L. (1995) Increase in levels of polyubiquitin and proteasome mRNA in rat skeletal muscle and during denervation atrophy. *Biochemical Journal* 307, 631–637.

Miller, L.L. (1973) in discussion of a paper by Bianchi *et al.* In: Wolstenholme, G.E. (ed.) *Protein Metabolism.* CIBA Foundation Symposium, New Series, no. 9. Associated Scientific Publishers, Amsterdam, pp. 47–65.

Millward, D.J. (1980) Protein degradation in muscle and liver. In: Florkin, M., Neuberger, A. and Van Deenan, L.L.M. (eds) *Comprehensive Biochemistry* vol. 19B(1). Elsevier/North Holland, Amsterdam, pp. 153–232.

Mitch, W.E. and Goldberg, A.L. (1996) Mechanisms of muscle wasting: the role of the ubiquitin-proteasome pathway. *New England Journal of Medicine* 335, 1897–1905.

Mitch, W., Medina, R., Grieber, S., May, R.C., England, B.K., Price, S.R., Bailey, J.L. and Goldberg, A.L. (1994) Metabolic acidosis stimulates muscle protein degradation by activating the adenosine-triphosphate-dependent pathway involving ubiquitin and proteasomes. *Journal of Clinical Investigation* 93, 2127–2133.

Mortimore, G.E. and Pösö, A.R. (1984) Lysosomal pathways in hepatic protein degradation: regulatory role of amino acid. *Federation Proceedings* 43, 1289–1294.

Mortimore, G.E., Kadowski, M. and Haydrick, S.J. (1992) The autophagic pathway in liver: its regulation and role in macromolecular turnover. In: Nair, K.S. and Smith-Gordon, E. (eds) *Protein Metabolism in Diabetes Mellitus.* Nishimura, London, pp. 125–138.

Pain, V.M. (1996) Initiation of protein synthesis in mammalian cells. *European Journal of Biochemistry* 236, 747–771.

Price, S.R., England, B.K., Bailey, J.L., Van Vreede, K. and Mitch, W.E. (1994) Acidosis and glucocorticoids concomitantly increase ubiquitin and proteasome subunit RNA levels in rat muscles. *American Journal of Physiology* 267, C955–960.

Price, S.R., Bailey, J.L., Wang, X., Jurkovitz, C., England, B.K., Ding, X., Phillips, L.S. and Mitch, W. (1996) Muscle wasting in insulopenic rats results from activation of the ATP-dependent, ubiquitin-proteasome proteolytic pathway by a mechanism including gene transcription. *Journal of Clinical Investigation* 98, 1703–1708.

Proud, C.G. (2001) Regulation of mRNA translation. In: Chapman, K.E. and Higgins, S.J. (eds) *Essays in Biochemistry* 37, 97–108. Portland Press, Colchester, UK.

Proud, C.G. and Denton, R.M. (1997) Molecular mechanisms for the control of translation by insulin. *Biochemical Journal* 328, 329–341.

Ptashne, M. and Gann, A. (2001) Transcription initiation: imposing specificity by localization. *Essays in Biochemistry* 37, 1–16.

Rallière, C., Tauveron, I., Taillandier, D., Guy, L., Boiteaux, J.-P., Giraud, B., Attaix, D. and Thiéblot, F. (1997) Glucorticoids do not regulate the expression of proteolytic genes in skeletal muscle from Cushing's syndrome patients. *Journal of Clinical Endocrinology and Metabolism* 82, 3161–3164.

Salghetti, S.E., Muratani, M., Wijnen, H., Futcher, B. and Tanxy, W.P. (2000) Functional overlap of sequences that activate transcription and signal ubiquitin mediated proteolysis. *Proceedings of the National Academy of Sciences, USA* 97, 3118–3123.

Salvesen, G.S. (2002) Caspases and apoptosis. *Essays in Biochemistry* 38, 9–14.

Samuels, S.E., Taillandier, D., Aurrouseau, E., Cherel, Y., Le Maho, Y., Arnal, M. and Attaix, D. (1996) Gastrointestinal protein synthesis and mRNA levels for proteolytic systems in adult fasted rats. *American Journal of Physiology* 271, E232–238.

Schimke, R.T. (1970) Regulation of protein degradation in mammalian tissues. In: Munro, H.N. (ed.) *Mammalian Protein Metabolism* Vol. IV, Academic Press, New York, pp. 299–387.

Schimke, R.T., Ganschow, R., Doyle, D. and Arias, L.M. (1968) Regulation of protein turnover in mammalian tissues. *Federation Proceedings* 27, 1223–1230.

Scornik, O.A. (1984) Role of protein degradation in the regulation of cellular protein content and amino acid pools. *Federation Proceedings* 43, 1283–1288.

Seglen, P.O. and Bohley, P. (1992) Autophagy and other vacuolar protein degradation mechanisms. *Experientia* 49, 158–172.

Shah, O.J., Anthony, J.C., Kimbell, S.R. and Jefferson, L.S. (2000) 4E-BP1 and S6K1: translational integration sites for nutritional and hormonal information in muscle. *American Journal of Physiology* 279, E715–729.

Stryer, L. (1988) *Biochemistry*, 3rd edn. Freeman, New York.

Taillandier, D., Aurousseau, E., Meynial-Denis, D., Béchet, D., Ferrara, M., Cottin, P., Ducastaing, A., Bigera, Ax., Guezennec, C.Y., Schmidt, H.-P. and Attaix, D. (1996) Coordinated activation of lysosomal, Ca^{2+}-activated and ATP-ubiquitin-dependent proteinases in the unweighted rat soleus muscle. *Biochemical Journal* 316, 65–72.

Temparis, S., Asensi, M., Taillander, D., Aurousseau, E., Larbaud, D., Obled, A., Béchet, D., Ferrara, M., Estrela, J.M. and Attaix, D. (1994) Increased ATP-ubiquitin-dependent proteolysis in skeletal muscles of tumor-bearing rats. *Cancer Research* 54, 5568–5573.

Tawa, N.E., Odessey, R. and Goldberg, A.L. (1997) Inhibitors of the proteasome reduce the accelerated proteolysis in atrophying rat muscles. *Journal of Clinical Investigation* 100, 197–203.

Tiao, G., Fagan, J.M., Samuels, N., James, J.H., Hudson, K., Lieberman, M., Fischer, J.E. and Hasselgren, P.-O. (1994) Sepsis stimulates non-lysosomal, energy-dependent proteolysis and increases ubiquitin mRNA levels in rat skeletal muscle. *Journal of Clinical Investigation* 94, 2255–2264.

Toates, F.M. (1975) *Control Theory in Biology and Experimental Psychology.* Hutchinson, London.

Varsharsky, A. (1996) The N-end rule: functions, mysteries, uses. *Proceedings of the National Academy of Sciences, USA* 93, 12142–12149.

Voisin, L., Breuillé, D., Combaret, L., Pouyet, C., Taillandier, D., Aurousseau, E., Obled, C. and Attaix, D. (1996) Muscle wasting in a rat model of long-lasting sepsis results from the activation of lysosomal, Ca^{2+}-activated and ubiquitin- proteosome proteolytic pathways. *Journal of Clinical Investigation* 97, 1610–1617.

Wang, X., Campbell, L.E., Miller, C.M. and Proud, C.G. (1998) Amino acid regulates p70 S6 kinase and multiple translation factors. *Biochemical Journal* 334, 261–267.

Waterlow, J.C. (1999) The mysteries of nitrogen balance. *Nutrition Research Reviews* 12, 25–54.

Welch, G.R. (1987) The living cell as an ecosystem: hierarchical analogy and symmetry. *Trends in Ecology and Evolution* 2, 305–309.

Wheatley, D.N. (1984) Intracellular protein degradation: basis of a self-regulating mechanism for the proteolysis of endogenous proteins. *Journal of Theoretical Biology* 107, 127–149.

Wheatley, D.N., Grisolia, S. and Hernándes-Yago, J. (1982) Significance of rapid degradation of newly synthesized proteins in mammalian cells: a working hypothesis. *Journal of Theoretical Biology* 98, 283–30.

Wing, S.S. and Goldberg, A.L. (1993) Glucocorticoids activate the ATP-ubiquitin-dependent proteolytic system in skeletal muscle during fasting. *American Journal of Physiology* 264, E668–676.

Wurtman, R.J. (1970) Diurnal rhythms in mammalian protein metabolism. In: Munro, H.N. (ed.) *Mammalian Protein Metabolism*, Vol. IV, Academic Press, New York, pp. 445–479.

Index

It is expected that the Index will be used in conjunction with the table of contents. Some words, such as 'turnover' and 'synthesis' are not in the Index because they occur almost everywhere in the book.

Abbreviations: WBPT, whole body protein turnover; PS, protein synthesis; AA, amino acids; EAA and NEAA, essential and non-essential amino acids; FSR, fractional synthesis rate; 3-MeH, 3-methyl histidine.